AF798152

Applications of Human Performance Models to System Design

DEFENSE RESEARCH SERIES

Volume 1 Computer-Based Instruction in Military Environments
Edited by Robert J. Seidel and Peter D. Weddle

Volume 2 Applications of Human Performance Models to System Design
Edited by Grant R. McMillan, David Beevis, Eduardo Salas, Michael H. Strub, Robert Sutton, and Leo van Breda

Applications of Human Performance Models to System Design

Edited by

GRANT R. McMILLAN

Harry G. Armstrong Aerospace Medical Research Laboratory
Wright-Patterson Air Force Base, Ohio

DAVID BEEVIS

Defence and Civil Institute of Environmental Medicine
Downsview, Ontario, Canada

EDUARDO SALAS

Naval Training Systems Center
Orlando, Florida

MICHAEL H. STRUB

U.S. Army Research Institute for the Behavioral and Social Sciences
Fort Bliss, Texas

ROBERT SUTTON

Royal Naval Engineering College
Manadon, Plymouth, Devon, United Kingdom

and

LEO VAN BREDA

TNO Institute for Perception
Soesterberg, The Netherlands

SPRINGER SCIENCE+BUSINESS MEDIA, LLC

Library of Congress Cataloging in Publication Data

NATO Research Study Group 9 (DRG, Panel 8) Workshop on Applications of Human Performance Models to System Design (1988: Orlando, Fla.)

Applications of human performance models to system design / edited by Grant R. McMillan . . . [et al.].

p. cm. – (Defense research series; v. 2)

"Proceedings of a NATO Research Study Group 9 (DRG, Panel 8) Workshop on Applications of Human Performance Models to System Design, held May 9–13, 1988, in Orlando, Florida" – T.p. verso.

Includes bibliographies and index.

ISBN 978-1-4757-9246-1 ISBN 978-1-4757-9244-7 (eBook)

DOI 10.1007/978-1-4757-9244-7

1. Weapons systems – Design and construction – Congresses. 2. Human engineering-Congresses. 3. North Atlantic Treaty Organization – Armed Forces – Weapons systems – Design and construction – Congresses. I. McMillan, Grant R. II. NATO Research Study Group 9. III. NATO Defense Research Group. IV. Title. V. Series.

UF500.N38 1988 89-8783

355.8′2 – dc20 CIP

Proceedings of a NATO Research Study Group 9 (DRG, Panel 8) Workshop on Applications of Human Performance Models to System Design, held May 9–13, 1988, in Orlando, Florida

© 1989 Springer Science+Business Media New York

Originally published by Plenum Press, New York in 1989

Softcover reprint of the hardcover 1st edition 1989

All rights reserved

No part of this book may be reproduced, stored in a retrieval system, or transmitted in any form or by any means, electronic, mechanical, photocopying, microfilming, recording, or otherwise, without written permission from the Publisher

PREFACE

The human factors profession is currently attempting to take a more proactive role in the design of man-machine systems than has been characteristic of its past. Realizing that human engineering contributions are needed well before the experimental evaluation of prototypes or operational systems, there is a concerted effort to develop tools that predict how humans will interact with proposed designs. This volume provides an overview of one category of such tools: mathematical models of human performance. It represents a collection of invited papers from a 1988 NATO Workshop.

The Workshop was conceived and organized by NATO Research Study Group 9 (RSG.9) on "Modelling of Human Operator Behaviour in Weapon Systems". It represented the culmination of over five years of effort, and was attended by 139 persons from Europe, Canada, and the United States. RSG.9 was established in 1982 by Panel 8 of the Defence Research Group to accomplish the following objectives:

* Determine the utility and state of the art of human performance modelling.

* Encourage international research and the exchange of ideas.

* Foster the practical application of modelling research.

* Provide a bridge between the models and approaches adopted by engineers and behavioral scientists.

* Present the findings in an international symposium.

Both the Workshop and this volume were designed to acquaint potential users with a broad range of models that may be used to predict aspects of human performance during the system development process. This objective is addressed by overview papers, papers on specific model applications, and papers which provide details on modelling tools that are currently available. The members of RSG.9 sincerely hope that this volume will encourage interested users to try some of these tools in their own work. It is essential that these users give feedback to model developers on the strengths and weaknesses of the models that they try. Only by establishing such dialogue can the state of the art be truly advanced.

While the members of RSG.9 seek to foster the development and use of human performance models, we must state that inclusion of specific models or tools in this volume should not be construed as an endorsement by RSG.9. by NATO, or by any agency of its member nations.

As Chairman of RSG.9, I am indebted to many groups and individuals who played key roles in this endeavor. First, I must thank the members of RSG.9

who worked to develop the program, prepared papers, served as session chairman and co-editors, and have become good friends. The authors must be recognized for their efforts to produce a first-rate volume. I want to thank the members of NATO Panel 8 who patiently supported our efforts to accomplish our goals. The Workshop could not have been held without the funding provided by the U.S. Armed Services. I sincerely appreciate the support of the Aerospace Medical Panel of the Advisory Group for Aerospace Research and Development (AGARD), who made it possible for several European speakers to attend the Workshop. Ms. Rita Landis of Total Events and Meetings, Inc. is to be complimented on her excellent management of the administrative affairs of the Workshop. I especially want to thank Lt. Col. Allan Dickson and Mr. Charles Bates, Jr. for their consistent encouragement and the time to work on this project, and Ms. Laura Mulford who assisted in innumerable ways with the Workshop and the preparation of this volume.

Grant R. McMillan

Wright-Patterson Air Force Base, OH, USA
February 1989

CONTENTS

MODELS OF MULTI-TASK SITUATIONS

CREW PERFORMANCE MODELS

WORKSPACE DESIGN - ANTHROPOMETRICAL AND BIOMECHANICAL APPROACHES

MODELS OF TRAINING AND SKILL RETENTION

REVIEW AND CRITIQUE

KEYNOTE ADDRESS

INTRODUCTION

OBJECTIVES AND ORGANIZATION OF THE WORKSHOP

Grant R. McMillan

Armstrong Aerospace Medical Research Laboratory
Wright-Patterson Air Force Base, OH
USA

Human factors engineers are seeking the opportunity to have an early impact on the design of man-machine systems; to be involved from concept development onward. With this opportunity comes a responsibility. We must provide tools that allow the design team to predict human performance in the same manner that hardware and software performance is predicted. Mathematical models of human performance constitute one - perhaps the most intellectual - category of methods for addressing this requirement.

As the papers in this volume demonstrate, models have been developed for a broad range of human behaviors. Nevertheless, the available models have not fulfilled their potential as aids to system designers. In response to this shortcoming, Panel 8 of the NATO Defence Research Group established a Research Study Group (RSG.9) to investigate the issue and to foster the practical application of human performance models. While the definition of what constitutes a human performance model is subject to some controversy, RSG.9 focused on techniques that permit computer-based simulation of humans functioning in systems, as opposed to verbal-analytic or conceptual models.

In May of 1988, we organized a Technology Demonstration Workshop to provide potential model users with an overview and specific examples of tools that are available. This volume contains the technical papers that were presented there. The Workshop was attended by 139 individuals from Europe, Canada, and the United States (See Appendix A). The demographics of the attendees are discussed in a paper by Cody and Rouse in the Review and Critique section of this volume.

This volume is organized around six modelling areas (sections) which are described below. Each of these sections includes three types of presentations:

(1) Introductory papers by the Workshop session chairmen which summarize the technical papers and the discussion period that followed.

(2) Technical papers which present overviews or model applications.

(3) Technical papers on modelling software demonstrated at the Workshop.

The Workshop also had poster presentations, which could not be included in this volume. These presentations are listed in Appendix B. Because of the broad range of models addressed, the coverage in any one area is not exhaustive. Rather, there is an attempt to concentrate on the more developed, available, and promising technologies.

The first section, Task Allocation and Workload Analysis, focuses on techniques for estimating human workload when various tasks are assigned to the man or machine. This area has received much attention in recent years, and the activity is demonstrated by the ongoing developments discussed in the papers. The close tie between theoreticians developing multiple-resource theories of human task sharing and the developers of workload estimation software is both gratifying and promising.

The section on Models of Individual Tasks provides a sample of the techniques which represent the performance of a single operator performing a single task. This is perhaps the oldest area of human performance modelling and includes many truly "mathematical" models. This situation reflects the fact that we have sufficient understanding of human performance in some of these areas to attempt mathematical descriptions of the performance mechanisms.

Models of Multi-Task Situations primarily address a single operator performing multiple tasks. The techniques discussed here tend not to represent the mechanisms of human performance, but simulate instead the time, accuracy, and workload results using task network modelling tools such as SAINT (Systems Analysis of Integrated Networks of Tasks). This approach is required because the tasks make fewer constraints on human performance strategies, and because we do not have a good understanding of the performance mechanisms. Fortunately, many of the model applications in this area do not require a mechanistic level of analysis to provide the required answers.

This trend is seen even more strongly in the Crew Performance Models which represent multiple operators performing multiple tasks. These models tend to be an important tool for decision makers who are addressing issues of crew size and the effects of operational stressors such as fatigue and overload. The models have many characteristics in common with operations research simulations, but with much more emphasis on capturing the contribution of the human to system performance.

The section on Workspace Design - Anthropometrical and Biomechanical Approaches reviews models that address human performance at a rather basic level. They attempt to predict an operator's ability to see and reach controls and displays, to perform materials handling tasks without hazard, and to fit into and get out of workspaces. Although this is a very different level of analysis from the other models reviewed in this volume, one should not underestimate the importance of having reliable tools for studying such tasks.

Models of Training and Skill Retention represent techniques to aid in the design and utilization of training systems. The grain of analysis is relatively fine when attempting to model learning curves or to predict the acquisition and retention of specific skills. Other techniques provide a high level analysis of the expected benefit from specific training devices, or training device features.

The opening session of the Workshop set the stage for an evaluation process to be accomplished by the attendees. Criteria were proposed that research indicates designers use when selecting sources of information about human performance. The attendees were provided with a questionnaire, based upon these criteria, for evaluation of the modelling tools presented at the Workshop. This questionnaire was designed to generate feedback on the perceived utility of the models, on barriers to their use, and on the technologies judged to be most promising. The analysis results are presented by Cody and Rouse in the Review and Critique section.

The questionnaire was only one source of evaluation conducted at the Workshop. Current modelling technology suffers from many limitations when applied to real-world problems. Many of these shortcomings, as well as suggestions to remedy them, are also presented in the Review and Critique section.

The Workshop and this volume represent an initiative that we hope to continue in the future. If human performance models are ever to fulfill their potential as design aids, dialogue focused on applications must continue between model developers and model users. We trust that this endeavor is a step in that direction.

DESIGNERS' CRITERIA FOR CHOOSING HUMAN PERFORMANCE MODELS

William B. Rouse and William J. Cody

Search Technology, Inc.
4725 Peachtree Corners Circle
Norcross, Georgia 30092

INTRODUCTION

There are many reasons for developing human performance models. It is an intellectually satisfying and enriching experience to develop and experiment with models of human performance, chemical reactions, economic behavior, and so on. Such models help us to understand complex phenomena while also providing the means to communicate this understanding.

Although we have little doubt that the developers of models find joy and benefit from the process of modeling, we are less sanguine about the benefits that may accrue to potential users of the product of modeling -- the equations and/or routines typically embodied in a software package. This paper is concerned with human performance models as products, and with system designers as potential consumers of these products.

It is often argued that the availability and use of human performance models will enhance system design and result in more effective systems. This assertion is difficult to support in any conclusive manner. A more fundamental concern, however, is the apparent lack of use of many available human performance models. Why is it that modelers are so enthralled with models, while designers show little apparent interest? In this paper, we argue that this lack of success is due to modelers having ignored many of the criteria by which designers judge models in particular, and information sources in general. To support this premise, we first provide an overall perspective of system design.

INFORMATION SEEKING IN SYSTEM DESIGN

Our studies of designers in the aerospace industry have led us to conclude that: *Designers spend most of their time in the insular environment bounded by project co-workers and personal experience. Their information requirements span an enormous range, some of which can be satisfied only from project sources and others which can be satisfied in a variety of ways. Designers satisfy those requirements that can be met in more than one way by accessing sources with the least overhead in terms of their effort. These sources are social contact with co-workers and definitely not formally printed materials* (Rouse and Cody, 1988).

From this perspective, design can be viewed as an information transformation process. On one level, this process involves seeking, managing, and disseminating information. On another level, this process involves sub-processes for formulating, representing, associating, manipulating, and evaluating information as it is transformed and eventually manifested in a designed artifact (Rouse and Boff, 1987). Within this framework, it is useful to think of human performance models as information sources.

Figure 1 shows the scope of information needs for which human performance models might be useful. The rows of this figure are designated by the phases of the "standard" system acquisition process, while the columns denote the topics covered by the six technical sessions of this NATO Workshop on Applications of Human Performance Models to System Design. Clearly, there is a wide-spectrum of needs for human performance information that models <u>might</u> be able to help satisfy.

To motivate our later assertions regarding criteria for selecting information sources, it is useful to characterize the process of seeking information in design -- see Rouse (1986a, b) for reviews related to this topic. The process begins with the recognition of one or more information needs, potentially in terms of one or more elements of the stream of needs characterized in Figure 1. The next steps are identification of alternative information sources and selection among these sources. While this sounds very analytical, more often it is very intuitive with emphasis on "tried and true" sources.

Selection among information sources is also heavily influenced by "downstream" factors concerning how the information will be used -- selection, in itself, is only a minor component of design decision making, albeit the component which this paper addresses. Typical downstream concerns include whether or not use of the information will require learning new jargon, methods, etc.; the usability of the source in terms of ease of access and manipulation; and the interpretability of information from this source relative to the nature of the need motivating the search. Another concern is how information from this source can be used as a basis for advocating and defending design decisions.

USES OF INFORMATION

Beyond the above general concerns, identification of information sources and selection among these sources also depend on the specific uses intended for the information sought. Within the context of system design, it seems to us that there are five types of uses of information:

- o Specifying and clarifying design objectives and evaluation criteria (e.g., elaborating speed vs. accuracy tradeoffs).

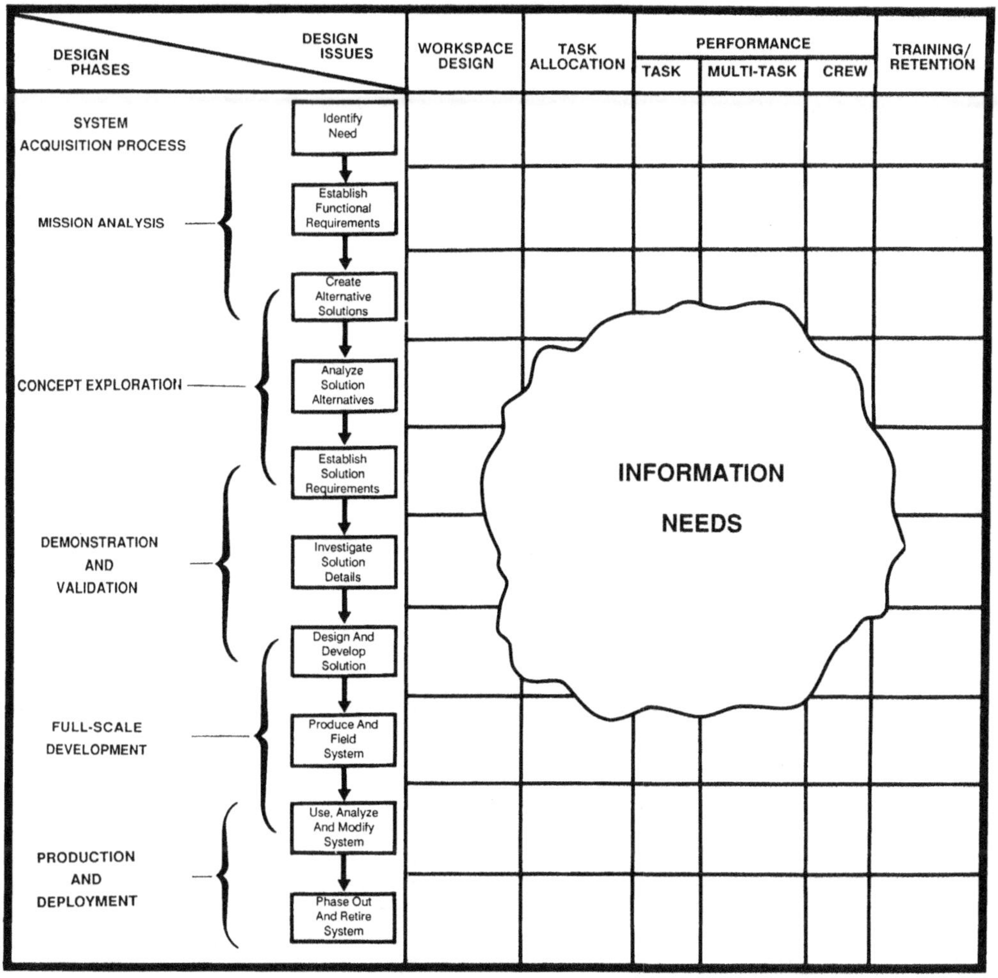

Fig. 1. Information Needs in System Design

- o Projecting outcomes or impacts of independent variables (e.g., analyzing sensitivity of performance to parameter variations).
- o Identifying and determining timing and location of events (e.g., analyzing mission to identify bottlenecks).
- o Identifying and determining attributes of things and events (e.g., choosing design parameters).
- o Understanding causes of outcomes and deficiencies (e.g., diagnosing performance shortfalls).

As shown in Figure 2, these five types of use are often integrated within an archetypical "what if" question concerned with achieving or not achieving design objectives as a function of internal and/or external choices. From this

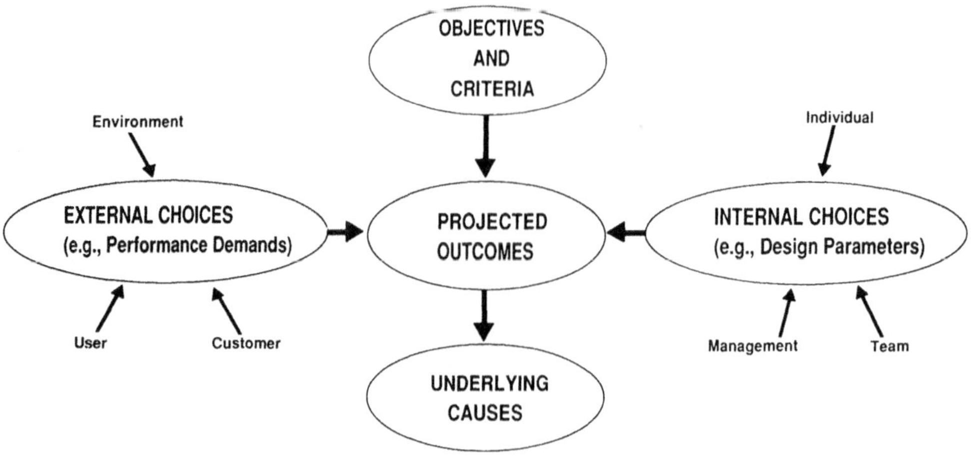

Fig. 2. Archetypical Question in System Design

perspective, the information seeker's questions may reflect a much richer context than the apparent simplicity of questions might lead one to perceive. The appropriateness of an information source may, therefore, depend on the relative importance of the relationships depicted in Figure 2.

The archetypical question depicted in Figure 2 occurs repeatedly as the elements of the matrix shown in Figure 1. External choices made by users, customers, or, in effect, the environment, as well as internal choices made by the individual designer, the design team, or management, motivate "what if" questions relative to outcomes satisfying criteria and achieving objectives. In addition, "why" questions are asked relative to the rationale and necessity of particular objectives and criteria, as well as explanations of outcome deficiencies.

IDENTIFICATION OF INFORMATION SOURCES

If one accepts the notion that design is predominantly an information transformation process, than it would seem that information seeking should be ubiquitous which, in turn, would suggest that identification of information sources and selection among these sources should be central. This set of related hypothesis has been explored by several investigators, e.g., Allen (1977), Cody (1988), and Rouse and Cody (1988), who have found a single class of information sources to be predominant. This class is <u>human judgement</u>.

Designers have many information needs and ask many questions. Not surprisingly, they answer many questions themselves. The next most likely source of information is colleagues. Other sources of human judgement include subject matter experts who may be incumbent system users (e.g., former aircraft pilots) or disciplinary experts.

Human judgement is the primary information source because the cost of access is low and the quality of the information received is perceived to be good, or at least adequate. An additional and important characteristic of this information source is that colleagues and subject matter experts often understand the context in which questions are asked -- that is, they have a contextually-based understanding of the relationships shown in Figure 2. In fact, humans often will not answer questions without first asking several contextually-clarifying questions themselves.

A second class of information sources is archives. Fact sheets, handbooks, textbooks, and journals are occasionally accessed to satisfy information needs -- "occasionally" is an important qualifier (Allen, 1977; Rouse and Cody, 1988). These sources of information often suffer from being difficult to identify, access, and interpret, with the possible exception of particular items with which the seeker is already familiar (e.g., the well-worn textbook or favorite journal). The possibility of pretty much ignoring the archives is an option that is much more tenable for designers than it is for researchers -- this never ceases to amaze researchers!

A third class of information sources is models. There are two subclasses: empirical and analytical. Empirical models include market surveys, user studies with mockups, experimental investigations with simulators, etc. These sources of information are models in the sense that the conditions and humans studied are, in effect, assumed to model the eventual actual conditions and real users of the system being designed. Analytical models primarily differ from empirical models in the sense that the human anthropometry, information processing, etc. of interest is modeled computationally rather than by other humans. This difference, of course, leads to analytical models being "packaged" rather differently than empirical models. The packaging of analytical models often presents difficulties for potential consumers of these models.

SELECTION AMONG INFORMATION SOURCES

The foregoing discussion of information seeking in system design and alternative information sources leads us to suggest a structured set of seven criteria that we believe strongly influence choices among information sources. This set of attributes is structured in the sense that it is weakly rank-ordered, e.g., failure to satisfy the first criterion results in rejection of an information source regardless of the potential of this source relative to subsequent criteria. The sequel to this paper (Cody and Rouse, 1989) will present the results of testing this model of "consumer behavior" in the context of selecting among human performance models.

The first criterion is applicability. Information seekers will tend to select sources that they perceive will provide information that is applicable to their question. More specifically, sources will be sought that will provide information relevant to one or more of the aspects of Figure 2 within the context of interest.

Information seekers are also concerned with credibility. It is of little value to base a line of reasoning on an information source that is not credible within the organization making the decisions of interest. For this reason, if analytical models, for example, are disdained organizationally they will seldom be a viable information source.

The next two criteria are availability and cost. These criteria are rather obvious. If the source is unavailable or the financial cost to acquire and use the model is prohibitive, the source is unlikely to be used. It is useful to note that acceptable levels of cost (i.e., financial expenditure) may interact with perceived credibility.

If an information source is feasible relative to the above criteria, three additional concerns are likely to be addressed. The first is interpretability. Does the nature of the answers provided by the information source directly satisfy the nature of the information need underlying the question? For example, if answers to a manual control question are needed in the frequency domain with 95% confidence intervals, does the source provide this type of answer or will some transformation/interpretation be necessary?

The next criterion is learnability. What will it take to understand the jargon, assumptions, processing, etc. associated with a particular source to the degree necessary to use the source effectively? This criterion presents little difficulty when utilizing human judgement, occasional difficulty for archival sources (assuming that they have satisfied the other selection criteria), and sometimes great difficulty when accessing models. It often can require a great deal of effort to truly understand an analytical model.

The seventh and final criterion for selecting among information sources is usability. This includes the ease with which inputs (questions) can be prepared, outputs (answers) accessed, and relevant operations performed. Obviously, human judgement typically scores fairly well relative to this criterion. Various data bases help to improve usability of archival sources, although not to the extent of sources of human judgement. Computer-based models have the potential to be very usable, but this potential is rarely realized.

ADDITIONAL ISSUES FOR ANALYTICAL MODELS

In light of the fact that the focus of this workshop is human performance models, the remainder of our comments will be premised on the assumption that one or more models have been chosen as an appropriate information source. In other words, we are assuming that one or more analytical models have passed the seven criteria hurdles outlined in the previous section.

Before discussing the use of analytical models in more detail, it is important to note that there are additional benefits to using models beyond the seven criteria elaborated earlier. One of these benefits is the natural tendency of a modeling effort to force complete and consistent formulations of the problem of concern. Assumptions and approximations must be explicitly chosen in order to develop a computational model. Other types of information source do not force this rigor.

The resulting model usually provides a framework for managing information. As knowledge is gained about the problem of interest, it often can, in effect, be encoded in the model or at least tabulated and classified in the context of the model. The model also can substantially influence what additional knowledge is sought or not sought -- in this way, empirical "fishing expeditions" can be avoided.

Models also provide rather unique ways of defending positions. In most design organizations, numbers and plots carry more weight than pure words. In the absence of empirical data (e.g., from market surveys or simulator studies), model-based analyses can provide powerful leverage. This possibility depends, of course, on the aforementioned credibility aspects of using models.

Criteria and benefits aside, we now want to consider a few detailed issues that may affect perceptions of human performance models in particular, if not information sources in general. We will pose these issues as a set of questions:

- Formulating/structuring problems: How easy is it to choose variables and structure relationships among variables? Is the range of available alternatives adequate?
- Estimating parameters: How easy is it to collect the input data necessary for the model? Are methods provided for choosing parameter values that provide appropriate fits to the data?
- Exercising model: How easy is it to perform the calculations/simulations necessary to producing the model's outputs? How does the computational time/cost increase as the size of the problem increases?
- Sensitivity analysis: How easy is it to test the effects of parametric and structural variations? Are methods provided for systematically assessing the effects of these variations?

These are not the types of question that one can usually answer just by hearing a presentation on how somebody else has used the model of interest. Answering these questions requires much more probing. Unfortunately, all too frequently one finds that developers of human performance models have not yet reached the point of providing these types of functionality to users. As a result, a model or modeling approach may be great in principle but awkward and cumbersome in practice.

CONCLUSIONS

This paper has described our "market analysis" of the criteria that system designers are likely to employ in choosing human performance models. In the sequel to this paper (Cody and Rouse, 1989), we discuss the results of testing our notions using the participants in this workshop.

REFERENCES

Allen, T.J., 1967, "Managing the Flow of Technology," MIT Press, Cambridge, MA.

Cody, W.J., 1988, "Recommendations for Supporting Helicopter Crew System Design," Search Technology, Inc., Norcross, GA.

Cody, W.J., and Rouse, W.B., 1989, A test of criteria used to select human performance models, this volume.

Rouse, W.B., 1986a, On the value of information in system design: A framework for understanding and aiding designers, Information Processing and Management, 22: 217-228.

Rouse, W.B., 1986b, A note on the nature of creativity in engineering: Implications for supporting system design, Information Processing and Management, 22: 279-285.

Rouse, W.B., and Boff, K.R., (Eds.), 1987, "System Design: Behavioral Perspectives on Designers, Tools, and Organizations," North Holland, New York.

Rouse, W.B., and Cody, W.J., 1988, On the design of man-machine systems: Principles, practices, and prospects, Automatica, 24.

TASK ALLOCATION AND WORKLOAD ANALYSIS MODELS

INTRODUCTION: TASK ALLOCATION AND WORKLOAD ANALYSIS MODELS

David Beevis

Defence and Civil Institute of Environmental Medicine
Downsview, Ontario
Canada

OVERVIEW OF THE PAPERS

Decisions made or assumed early in system development define the functions and tasks for which each operator will be responsible. Those decisions are generally referred to as "function allocation" or "allocation of functions" (North et al. 1982, Kantowitz and Sorkin 1987). Function allocation decisions define the task "load" imposed on the operator, and hence require verification in terms of the operator's ability to perform the assigned tasks. In this context the concept of "workload" is aimed at the question of whether or not the operator can sustain the "load" applied by the combination of task demands and the characteristics of the man:machine interface. This concept of workload is, in general, related to psychomotor and mental task demands, not to the physical task demands which are addressed by the anthropometrical and biomechanical models reviewed elsewhere in these proceedings.

As shown in the first paper by Linton et al., workload is a highly contentious topic. Whole symposia and publications have been devoted to it, (AGARD 1978, Fiedler 1987, Moray 1979, Roscoe 1987) but there is, as yet, no universally accepted definition. The Workshop organizers were aware of this problem, but considered that a Session should be devoted to models of workload. This reflected the importance of including human factors considerations in the early stages of system design, when decisions are being made about who or what hardware and software should perform specific system functions.

The ideal workload model, therefore, should permit task demands to be modelled early in the system development cycle before specific details of the man:machine interface have been determined. It should also permit iteration and refinement throughout the project development cycle, and be compatible with the workload metrics used during system test and evaluation. As noted by Linton et al., there are many potential approaches to such models several of which are covered elsewhere in these proceedings.

Due to the emphasis on application in the early design stages, the models reviewed in this Session fall into Linton et al's categories of "task analysis" and "computer simulation" techniques. One of the most well established techniques is to compare the time required by the operator to perform the assigned tasks with the time available. This approach has its roots in the principles of scientific management developed by F.W. Taylor, although the aim is somewhat different from the establishment of a "definite time and a definite manner" for each task. The time-based approach to human engineering operator tasks dates from at least 1959, as noted by Parks in a paper to an earlier NATO Advanced Study Institute (Parks 1979). The current paper by Parks and Boucek continues on from that review of the state of the art in time-line analysis of workload.

Given the trend towards more cognitive tasks, rather than behaviouristic or psychomotor tasks (Merriman 1984) it is not surprising that practitioners are seeking models that represent mental workload in other than time-based terms. Two workload metrics, the Subjective Workload Assessment Technique (SWAT) and the NASA-TLX, include both "time stress" and "mental effort load" as dimensions (see Fiedler 1987). Although SWAT has been used to predict workload during system development (Kuperman 1985) the two metrics have not been widely adopted to model workload in the early stages of design. They fall into Linton et al's category of empirical models, and appear more suited to the test and evaluation phases of a project.

One line of model development which does incorporate mental load in early design activities is based on the concept of mental resources (Wickens 1984). The third paper, by Aldrich et al. describes one such approach. It is a reflection of the need for such a technique that this model, which was originally developed to meet a pressing deadline on a major project, has been adopted by several other organizations and is in the process of refinement and development, as indicated in their paper.

A related approach, which deals somewhat differently with the summation of concurrent demands on mental resources, was demonstrated at the workshop by North and is reported as the fourth paper. The Workload Index (W/INDEX) treats shared mental resources as moderators of time-sharing behaviour. A demonstration of Sequiter's Workload Analysis System (SWAS) was also available to attendees.

Readers should also refer to the paper by Wickens in these proceedings which reports an attempt to correlate the results of these modelling approaches with operator workload ratings and performance.

DISCUSSION

Several issues arose from the discussions of the individual papers and demonstrations. Chief of these, and the one which provoked most debate, was the question of whether such normative models properly predict operator workload. Although this question applies to most of the models reviewed in the Workshop, it seemed to be particularly relevant to this class of models. This may have been due to the obvious link between predictions of workload and overall systems effectiveness, and the equally obvious link between workload and other performance factors such as motivation and stress.

Several human performance specialists pointed out that it is what operators actually do that determines workload, and what they actually do is moderated by factors such as boredom, emotional stress, and motivation as well as time-stress. Such factors are not well represented by the models reviewed here. Few workload models include a stress factor, and most represent a determined sequence of tasks, rather than reflecting the load-shedding and fluidity of task sequencing observed in practice.

The discussion highlighted some important differences between some of these factors. Motivation was seen to be a two-edged factor. It is desirable because it can improve performance quite dramatically, but it is undesirable because designers cannot confidently control it. They have to assume some baseline performance which might be improved by motivation. In contrast there was a sense that stress must be considered in systems design because it frequently worsens performance. In this context there was seen to be some merit in models which may underestimate an operator's abilities. The debate concluded with the argument that workload is already a difficult enough concept to manipulate without the addition of several other poorly defined and understood concepts such as motivation, stress and boredom.

The validity and reliability of the models is obviously dependent on both the data they use and on the task sequences which they represent. While the models have the merit of being logical progressions from the Function Flow Analyses and Task Analyses which are the stock in trade of human engineers, ideally they should reflect the fluidity of task sequencing mentioned above. To some extent such behaviour can be dealt with by successive iterations of normative models. For example, if the time-line approach used by Parks showed the operator to be occupied either significantly more or less than his criterion for acceptable load, then the task sequence or allocation of functions could be revised and the model run again. Aldrich et al. reported just such an approach. North refers to an interesting development of W/INDEX which will automatically change the task sequences as a result of the workload calculations. Linton et al. review other approaches under the heading of "simulation" models, one of which is SWAS.

The task sequences analysed obviously should also include reversionary mode operations, or failure mode operation if the operator will be required to perform in such situations. The whole question of manual and automatic operation must be looked at carefully, because, as North pointed out, the prudent operator will devote resources to monitoring the operation of supposedly automatic systems.

Linton's presentation on approaches to workload mentioned the topic of "Situational Awareness" and how such models might incorporate it; Parks and Boucek's presentation also referred to it in the context of the "explosion" of information management problems facing operators. Discussion on this ranged from the definition of Situational Awareness to whether or not it should be included in such models or treated separately. Some saw it as an integral factor in cognitive workload which should not be treated separately; others saw it as a distinct factor in an operator's ability to perform a task. Again the debate suffered from the lack of clearly agreed upon definitions.

A variety of other capabilities and limitations of workload models were discussed. Potential users were cautioned that the models, particularly the attentional demand type of model, can encourage users to make simplifying assumptions about the operability of the man:machine interface. They were also cautioned on the need for an in-house bank of task times (for time-line models) or attentional demand ratings, for compiling the models. The presentations also made it clear that the user's skill and experience are important factors in the successful use of a model. Experienced users not only have their own data banks, they know when and how to supplement the models with other techniques such as limited man-in-the-loop simulations.

On the positive side the models were seen as useful advances over previous approaches, particularly in their support of iterative development of the design concept. Although the models can be criticized on a number of grounds, they do have theoretical bases and do include empirical data. The speakers acknowledged the limitations which had been identified: nevertheless it was felt that available models can help the user a long way toward a design solution. They have the merit of being straightforward to use (even if they require a significant amount of task analysis), and they facilitate early identification of potential problems in the allocation of functions between human and machine. As with most other models they require some skill of the user, and they require further development and validation, and refinement of the underlying concepts on which they are based.

REFERENCES

AGARD, 1978, "Assessing Pilot Workload", NATO/AGARD, Neuilly sur Seine, AGARD-AG-233.

AGARD, 1978, "Methods to Assess Workload", NATO/AGARD, Neuilly sur Seine, AGARD-CP-216.

Fiedler, H.M., 1987, Proceedings of the DoD Workshop on Workload Assessment Techniques and Tools Held in Dayton Ohio, 27-28 September 1986, Naval Underwater Systems Center, Newport, NUSC-TD-6608.

Kantowitz, B.H., Sorkin, R.D., 1978, Allocation of functions, in: "Handbook of Human Factors", G. Salvendy, ed., John Wiley & Sons, New York.

Kuperman, G.G., 1985, Pro-SWAT applied to advanced helicopter crewstation concepts, in: Proceedings of the Human Factors Society 29th Annual Meeting, Baltimore, Human Factors Society, Santa Monica.

Merriman, S., (Ed), 1984, Workshop on Applications of Systems Ergonomics to Weapon System Development, Vols 1 & 2, NATO Defence Research Group, Brussels, DS/A/DR(84)408.

Moray, N., (ed), 1979, "Mental Workload: Its Theory and Measurement", Plenum Press, New York.

North, K., Stapleton, C., Vogt, C., 1982, "Ergonomics Glossary: Terms Commonly Used in Ergonomics", Bohn, Scheltema & Holkema, Utrecht/Antwerp.

Parks, D.L., 1979, Current workload methods and emerging challenges, in: "Mental Workload: Its Theory and Measurement", N. Moray, ed., Plenum Press, New York.

Roscoe, A.H., 1987, "The Practical Assessment of Pilot Workload", NATO/AGARD, Neuilly sur Seine, AGARD-AG-282.

Wickens, C.D., 1984, The multiple resources model of human performance: implications for display design, in: "Human Factors Considerations in High Performance Aircraft," NATO/ AGARD, Neuilly sur Seine, AGARD-CP-371.

OPERATOR WORKLOAD FOR MILITARY SYSTEM ACQUISITION

Paul M. Linton (1), Brian D. Plamondon (1), A.O. Dick (2), Alvah C. Bittner, Jr. (2), and Richard E. Christ (3)

(1) United Technologies Corporation
Sikorsky Aircraft Division
Stratford, Ct.

(2) Analytics Inc.
Willow Grove, Pa.

(3) U.S. Army Research Institute
Fort Bliss, Tx.

INTRODUCTION

The purpose of this paper is threefold. Initially, we take the liberty of introducing the subject of workload: what it means and why we as engineers and behavioral scientists are interested in it. We assume that practically all engineers involved in the design of new weapon systems are, by now, at least aware of the importance of operator workload (OWL). Indeed, the widely endorsed, yet poorly addressed, initiative to reduce operator workload is partly responsible for the seemingly mad rush to provide ever greater levels of automation in the cockpit, at the helm, or at the workstation. Those of us who practice the Human Factors Engineering profession realize the folly of providing automation simply because it is technologically feasible .. but that philosophical argument must remain the subject for another day. In any event, a brief discussion of workload is necessary to set the stage. The second purpose of this paper is to identify specific workload prediction and assessment models which we have reviewed in our research program, and present summary opinions as to the utility of these techniques. Lastly, we will suggest some top level questions, strategies, and issues which we all must confront when the time comes to actually select and apply a technique.

The research project on which we are reporting has involved a review and assessment of all common workload prediction and measurement techniques; subjective, analytic, and physiological. Insofar as the subject of this workshop is modeling, with the emphasis on practicality, we will generally limit our remarks to analytical techniques, specifically mathematical models, task analytic, and computer simulations. We cannot go into exhaustive detail on specific models as space does not permit. More to the point, many of the techniques we identify are ably represented at this workshop by their developers, who are far more qualified than ourselves to present them. Our intent here is to provide a broad context for review and evaluation.

BACKGROUND

Workload

The simple fact of the matter is that nobody seems to know what workload is. Numerous definitions have been proposed, and many of them seem complete and intuitively "right". Nevertheless, current definitions of workload all fail to stand the test of widespread acceptance or quantitative validation.

Discussions over possible underlying concepts of workload are fascinating, and provide an enjoyable intellectual exercise. Over the course of our program, we devoted a considerable amount of time and energy to just such philosophical considerations and discussions. In fact, approximately two dozen pages of our Task 3 report are devoted to presentation of what factors constitute workload, and the relationship between workload and performance. It would have been beneficial to extract the salient points from those discussions and reproduce them here, but time and space do not allow such a possibility. However, for the sake of completeness, and to assist in setting the stage for the remainder of this paper, we here present our four basic tenants of workload:

> I. Workload reflects relative, rather than absolute individual states. It depends on both the external demands and the internal capabilities of the individual. This relativity exists qualitatively as well as in dimensions of quantity and time.
>
> II. Workload is not the same as the individual's performance in the face of work or tasks, nor is it synonymous with our way of measuring performance.
>
> III. Workload involves the depletion of internal resources to accomplish the work. High workload depletes these resources faster than low workload.
>
> IV. Individuals differ qualitatively and quantitatively in their response to workload. There are several different kinds of task demands and corresponding internal capabilities and capacities to handle these demands. Persons differ in the amount of these capabilities which they possess, and their strategies for employing them.

Having at least acknowledged the hypothetical construct and multi-dimensional nature of workload, this paper will henceforth ignore the problem of definition. Our focus at this workshop is clearly on modeling; methodologies which can predict and measure workload are of interest here, and not a universally acceptable definition of their application domain. We will simply agree that workload is incompletely defined, certainly multi-faceted, and has a direct bearing upon an operator's ability to maintain or reach a desired performance level.

Operator Workload (OWL) Assessment Program for the Army

A strong argument can be made that the weakest link in the system effectiveness chain today is the operator's ability to understand and employ his system under battlefield conditions where stress, sheer number of required tasks, and enemy threats are at a maximum. The fact that workload can neither be completely defined nor adequately measured underscores the difficulty of developing valid predictive and measurement techniques, but the need for such techniques is so great that workload R&D in the past few years has actually been intensifying in the face of a seemingly intractable problem.

System integrators and design engineers have come to appreciate and advocate an intelligently designed operator interface as a keystone of overall system effectiveness. Such an interface is characterized by (1) a natural and consistent exchange of information between the operator and the system; (2) a workload which does not overburden the human with annoying, repetitive, or confusing tasks; (3) providing sufficient supervisory management, cognitive, and psychomotor tasks to maintain an active participation in the tactical situation. There is a growing awareness that high levels of hardware and software sophistication alone, and the automation which they promise, do not guarantee the technological edge to counter the enemy's vastly superior number of conventional forces and weapons. Technological capability will likely be under-utilized, and thus cost *ineffective* when not applied within an overall man-machine-mission context.

Against such a background, the Army Research Institute initiated a research project in October of 1986 to "develop and validate comprehensive methods for estimating and evaluating operator workload at different decision points in the systems acquisition process." Products of this program were envisioned to include a set of guidebooks which would provide Army personnel with selected methodologies for making decisions on operator workload during system acquisition.

There was no intent in this program to develop new or unique prediction or measurement methodologies. Rather, the program was directed at an examination of the state-of-the-art in workload techniques, a critical evaluation, and development of a formalized structure of applying appropriate techniques at appropriate key points within the material acquisition process. From the very beginning the emphasis was on practicality. Our concern was not whether any one of the dozens of existing techniques was "better" or "more valid" than the others. Our concern was suggesting the best way to utilize existing workload assessment technology to provide guidance on workload determination. Our programmatic goal was to aid Army managers tasked with developing and fielding the best system possible within defined time and budget constraints.

ANALYTIC WORKLOAD ASSESSMENT TECHNIQUES

Taxonomy

For the purposes of our work, we were comfortable in classifying workload techniques into analytical and empirical classes. Analytic techniques, the first of these, are those which can be used to estimate workload without a human operator in the loop. Empirical, the second class, include those techniques which gather data, either subjective opinion, physiological, or performance, from human operators. The intent of this arbitrary definition is only to differentiate between techniques which require that an operator interact with a system from those that do not. Obviously this workshop is principally concerned with the analytic class of techniques as this is the class which comprises various modeling methodologies.

We further subdivided the domain of analytic (no operator-in-the-loop) techniques into five descriptive categories. These categories include (1) Comparison, (2) Mathematical Models, (3) Expert Opinion, (4) Task Analysis, and (5) Computer Simulation. Of these five techniques, Comparison and Expert Opinion don't merit serious consideration within the context of this workshop, or at least how we perceive the purpose of the workshop. They can be valid methodologies and of definite value when properly applied, but they are outside the most generous definition of "model". Comparison and Expert Opinion are grounded in the elicitation of subjective opinions from operators and designers who have direct experience with either the equipment under investigation or equipment very similar. These techniques use "expert opinion" to project data from comparable systems, or estimate task difficulty based upon prior experience.

Benefit

Behavioral scientists and system designers should be concerned with utilizing analytic workload prediction techniques in the material acquisition process for a very simple reason. Military systems, along with their associated manpower, personnel, and mission requirements are defined very early .. in most cases before any hardware is available with which to apply empirical workload analysis. While analytic workload assessment can and should be exercised throughout a system's development cycle, it is especially important at early, pre-hardware stages. Good predictive techniques are hard to come by, and those that do exist have limitations. Nevertheless, the tremendous value of recognizing and diagnosing problems early on makes the use of these techniques imperative. Unfortunately, empirical evaluation of workload is frequently of value only to confirm a proper design, or to discover man-machine deficiencies. In either case, the cost-benefit tradeoff of empirical techniques is far less than that of a properly executed analytic technique applied in the concept formulation or exploratory development stage. Indeed, the ability to influence a design is a monotonically decreasing function with time, while the cost of a change is a monotonically increasing function with time. The analytic techniques, consequently, offer a potentially huge return on investment seldom, if ever, realized by the empirical techniques.

MATHEMATICAL MODELS

One of the more ambitious goals of early workload researchers was the development of a rigorous mathematical model which would predict operator and system performance. In principle, such a model would identify relevant variables and combine them appropriately so that workload-associated effects on performance could be accurately and reliably estimated. The major steps were as in all attempts to model human performance:

* Identify variables that influence workload either directly or indirectly;

* Determine the lawful relationships by which these variables interact;

* Establish how the resultant workload predictions drive predictions of performance.

To date, no fully comprehensive mathematical models have been developed, although several investigators have taken existing models from engineering application domains and extended them to some aspect(s) of workload-related operator performance. Of these models, the most prominent are manual control, information theory, and queuing theory techniques. Each model is proposed to contain some parameter or component that reflects the operator's load or effort under specified conditions. Some models contain specific parameters that are proposed to be an index of load; others presume loading by defining the environmental input characteristics that are assumed to affect workload and performance. The assumption in either case is that these mathematical models will predict workload-related drivers and resulting performance.

Many of the models described below are aimed at continuous control tasks or monitoring tasks which have information presented on separate displays. In part, this is because these tasks have been, and still are, important in complex system control. More importantly, the associated performance characteristics are definable and thus amenable to this realm of mathematical modeling. Today, with greater use of automated control systems and multifunction information displays, manual control task characteristics appear to be becoming relatively less important. The implication is that mathematical models need to be developed that reflect the current set of increasingly cognitive tasks.

Manual Control Models

The manual control models fall into two general categories; classical control theory and the more recently developed state-space estimation methods. Both were developed within the context of continuous manual control tasks, such as piloting a vehicle.

Classical Control Theory - Classical control theory utilizes closed-loop stability analysis to generate describing functions of the human operator engaged in a continuous control task. In essence, the human is considered to be a servomechanism attempting to eliminate feedback errors. Error, such as deviation from a flight path, is the input to the model, and operator response via some manipulator device is the output. Classical control theory provides a continuous prediction of operator output over time. In workload estimation applications, a baseline operator describing function is initially developed. To this, external perturbations (loading factors) are then applied which change the characteristics of the model in a manner believed to be indicative of workload. For example, system response lags to operator control inputs can be varied. Changes ascribed to increased loading may be used to predict OWL to the extent that the conditions under which the describing function was developed are generalizable.

Optimal Control Model - Modern control theory uses a system of differential equations containing state variables and control variables to describe the controlled system. The optimal control model (OCM), when given a process to control, does so by; (a) observing the state variables to the degree of accuracy possible, and (b) generating a control response to these variables while minimizing a scalar performance criterion or cost function. The criteria are usually defined as a function of error, control effort, or time. The OCM assumes that a well-trained human operator will behave as an optimal controller. This implies that the operator will be aware of his own and the system dynamics. That is, the operator has knowledge of human response capability, the disturbances affecting the system, and the criterion which defines optimal control. Variables such as observation noise and motor noise are used to introduce error and can be related to attentional scanning which is one variable considered to reflect difficulty, and hence workload. OCMs of the human operator have performed reasonably well in matching observed behavior, and are capable of handling complex multivariable systems (Baron, 1979). Within the appropriate context, the predictive validity of these models makes them very useful, although their mathematical complexity makes them inaccessible to many investigators. An excellent treatment of the applications of OCM to workload estimation may be found in Levison (1979). In this report, Levison traces the development of the model, defines the basic workload model, cites a number of validation studies, and suggests issues for further development. Additional examples of the model's application can be found in Rickard and Levison (1981) for the prediction of pilot ratings of different aircraft handling quality configurations, and in Wewerinke (1974) and Smit and Wewerinke (1978). The formulation predicts a workload index based on control effort which is developed in terms of OCM parameters. Levison (1970) defines an OCM model containing an attention parameter which influences the observation noise within the state variable estimator. This parameter can be used to determine the attention allocated to a display variable and hence its relative importance in a control task . The OCM model has also been used for display design evaluation (Baron & Levison, 1977; Gainer, 1979).

A recent development of the OCM approach is the Procedure-Oriented Crew (PROCRU) Model (Baron, Zacharias, Muralidharan, & Lancraft, 1980). PROCRU provides a framework for dealing with both discrete and continuous tasks. The OCM has considerable breadth and most of the studies have corresponding validation data. OCM is clearly a performance model with parameters which represent workload manipulations. These manipulations are of the form of amplitude, frequency, or phase lags in the equations. As a result, workload definitions are as varied as the manipulations employed.

Information Theory Models

Classic information theory is actually a mathematical formulation; it provides a metric of the transmission of information through an imperfect communications network. Information theory as applied to models of human activity achieved its height of popularity during the 1960's, and an excellent treatment of information theory can be found in Sheridan and Ferrell (1974).

Early applications of information theory in psychology can be found in Attneave (1959) and Garner (1962) while one of the first applications to the workload domain was that of Senders (1964). In this application, a model was used to describe the division of attention by an operator while monitoring information displays. It assumed that an operator, with a limited input channel capacity, sampled each information display at a frequency necessary to reconstruct the signal being presented on that display within specific error tolerances. The amount of time spent sampling each instrument is summed over all instruments to determine the fraction of the operator's time that must be spent observing. This time fraction is used as a measure of visual workload imposed by the information displays.

The use of information theory in the analysis and estimation of workload has been limited. Despite some efforts (e.g.,Crawford, 1979; Rault, 1976), applications in realistically complex environments are difficult to achieve due to the necessity to *a priori* establish all of the relevant simple and conditional stimulus and response probabilities. Because information theory provides output with respect to steady-state situations, it is not well suited for representing dynamic changes in workload. Nevertheless, the impact of information theory is strongly felt through the adoption of its concepts such as limited channel capacity, information transmission, and redundancy now contained in information processing approaches to behavior (Garner, 1974).

Queuing Theory Models

Queuing theory models of man-machine interaction characterize the operator as a single-channel processor sharing attentional resources serially among a variety of tasks. The human is conceptualized as a "server" processing multiple tasks and "server utilization" or "business" is used as a measure of workload. These models generally apply to situations in which performance times are critical. Within queing theory, performance times include both the time it takes to execute various tasks, as well as the time that tasks must wait before being performed. Rouse (1980) provides a good discussion of queuing theory and its application to man-machine modeling.

The emphasis in queuing models is more on when tasks are performed rather than how they are performed. As indicated by Rouse, these models are most appropriate in multitask situations in which the operator must cope with task priorities and performance requirements that vary among the tasks. Using Jahns' (1973) categorization of workload, these models are concerned primarily with the input load to the operator.

The queuing theory approach to workload estimation is generally associated with Senders' research on monitoring tasks (e.g., Senders, Elkind, Grignetti, & Smallwood, 1966; Senders & Posner, 1976). However, others such as Schmidt (1978), who analyzed the workload of air traffic controllers, also have applied queuing theory models. Walden and Rouse (1978), modeling pilot decision behavior, have also successfully applied this approach.

Summary

The application of manual control theory to workload estimation and prediction is generally restricted to environments involving continuous controlling tasks. During that period when workload was practically synonymous with vehicular control, manual control models were easily the most interesting and promising class of techniques providing predictions to system designers. In the present day, these models may be adapted to estimate measures generally associated with OWL, but the mathematical sophistication required to develop or even understand the models limits their applicability. Detailed system parameters must also be provided to exercise these models fully; these parameters are frequently not available during early concept development. Consequently, manual control models are generally not viable for most conceptual system evaluations.

The popularity of mathematical models seems to have waned. Information theory was most popular in the 1960's and manual control theory and queueing theory predominated during the 1970's. Although many of these models have experienced considerable success within the domain for which they were intended, they seem to have been supplanted in the 1980's by computerized task analysis and simulation models. A major problem with mathematical modeling is the absence of explicitly defined workload parameters. Thus while model outputs may identify and quantify particularly busy periods within a given time slice, or particularly high periods of information transfer, it is never quite clear how, or if, these phenomena relate to high workload. This observation, it should be pointed out, is not restricted to mathematical models alone and probably has relevance to most analytic techniques and methodologies.

There is always a place for a useful mathematical model, even if it is not as broad as one would like. An obvious and hopeful evolution would be that certain of these mathematical models, especially the optimal control model which can cover aspects of queuing theory, might be incorporated into the simulation models. It would certainly seem feasible to bring such models into simulations in a form which more people could use. The user will have to be careful, however, to define what is meant by workload as this determines the diagnosticity of the results.

TASK ANALYSIS METHODS

Task analysis techniques are the most commonly used of all analytic tools for estimating workload in the preliminary design process. Their frequent use probably springs from three reasons. In the first place they are relatively easy to understand and undertake. No extraordinary mathematical or simulation expertise is required to execute a task analysis. Relatively sophisticated task analyses can be completed with only an intimate knowledge of the hardware system being studied, a realistic and detailed mission scenario, and the willingness of the analyst to iterate and persevere. Secondly, military specification MIL-H-46855B requires a task analysis to be performed during all major system development efforts. It is a natural extension to move from this requirement to development of operator workload estimates based upon the results of the analysis. Lastly and simply, task analyses are useful for the analyst. Even if workload predictions are never derived, the wealth of knowledge gained from the task decomposition exercise educates the analyst and prepares him or her to provide future contributions.

Workload oriented task analytic methods generally examine operator performance requirements as a function of time within the context of a very specific mission scenario. The basic task analytic process begins with the definition of the mission scenario or profile. Next, general or top level mission requirements are systematically decomposed into operator tasks and these in turn are decomposed into more detailed sub-tasks or task elements. These task elements can then be translated into specific operator actions which are completely specified when placed within the context of the hardware system under consideration (i.e. cockpit, tactical operator station, work station etc.). Thus, the timing and sequencing of operator control activity will depend on the nature and layout of controls and displays. The result of the analysis is an operator activity profile as a function of mission time and phase, essentially a time-based analysis of required operator actions leading to successful mission completion.

Other approaches are more detailed in the analysis, and divide tasks into components based on sensory channel or body part (e.g., eyes, hand, foot, etc.). Recent methods have included a still more detailed analysis structure in an attempt to try to identify cognitive loads applied to the operator. However, these more detailed approaches typically contain time stress as a major contributor in the estimation of workload. Nevertheless, diagnosticity improves by virtue of identification of specific components that may be overloaded. There are many variations on the basic task analytic structure. The differences will be clarified in the discussions of each of the methods. The discussions presented here are intended to be illustrative of the types of information that can be integrated into the models and the nature of the results that can be obtained from them. A review of many task analytic techniques may be found in Meister (1985).

Timeline Task Analysis

The natural consequence of time-based task analysis is to define OWL as time stress. This is expressed as a ratio of Time required (Tr) to perform a task over the Time available (Ta), yielding Tr/Ta. In timeline task analysis, the situations of concern are those which cause the operator to approach the edges of the performance envelope, that is as Tr/Ta approaches 1.0. A technique incorporating such a time based definition is useful, but is probably best used as an initial coarse filter to identify gross design deficiencies and for cases in which the time required for a task is well defined. Diagnosticity, in the time-line approach, is limited to identifying general structural limitations where demands on the operator exceed his capacity to respond within some time frame.

A classic application of the timeline analysis technique employing the (Tr/Ta) metric is that described in Stone, Gulick and Gabriel (1987). They used this technique to identify workload encountered in overall aircraft operations, and with respect to specific body channels. The latter scheme was based on that developed in WAM (see below). Validation efforts are reported by the authors, with the results indicating that the procedure "...provides a reasonably accurate index for predicting the time required to complete observable tasks within the constraints of an actual mission."

Workload Assessment Model (WAM)

The Workload Assessment Model was first introduced in the 1970's as part of a more comprehensive man-machine system design aid: Computer Aided Function-Allocation Evaluation System (CAFES). While the overall success of CAFES was limited, its workload assessment module (WAM) was fully developed and performed admirably on several applications. In WAM, a mission timeline description is developed which indicates what tasks are performed during the mission and in what sequence they are performed. The individual sensory-motor channels (e.g., eyes, hands, feet, etc.) that are involved in the execution of each task are identified. WAM computes the channel utilization percentage, including the amount of time that each channel is occupied within a specific time segment. Percentages over a specified threshold level are considered excessive, and may be indicative of both function allocation and design inadequacies. A variant of WAM, the Statistical Workload Assessment Model (SWAM), allows time shifting of excessive workload tasks in an attempt to reduce the peak workload level. This, effectively, is a rescheduling of tasks to reduce time stress.

Linton, Jahns, and Chatelier (1977) report one application of SWAM. They examined a conceptual VF/VA-V/STOL aircraft in an attempt to determine whether a single pilot could manage the aircraft and systems in defined mission phases. The results indicated the potential single-pilot operability for the aircraft, but did not establish any validity measures for the assessment technique.

Time Based Analysis of Significant Coordinated Operations (TASCO)

TASCO analyzes cockpit workload during tactical missions using the standard time-based approach (Roberts & Crites, 1985; Ellison & Roberts, 1985). Two types of analysis are performed in TASCO. The first is crewstation task analysis, which is a design evaluation performed by a subject matter expert using a 5 point rating scale to judge design elements that are especially crucial to mission performance. The second is a Busy Rate Index analysis, which is essentially a Tr/Ta estimate over a set time interval.

The basic analytic component of the method is the Evaluation, Decision, Action, and Monitoring (EDAM) loop. Recognizing the cognitive aspects of today's cockpit design, the Evaluation part of EDAM accounts for the impact of information display design. The Decision part of the EDAM loop is made by a pilot based on training, experience, tactical doctrine and situational awareness. The decision results in an Action via the cockpit controls which is then Monitored to evaluate the outcome of the action. How the above mentioned EDAM loops are integrated into these analyses is unclear, as is the current state of development of the TASCO model. Early papers describing this technique appeared promising, however, and the technique probably merits further consideration.

Computerized Rapid Analysis of Workload (CRAWL)

CRAWL involves expert opinion superimposed upon a task analytic background with two basic sets of inputs (Bateman & Thompson, 1986; Thompson & Bateman, 1986). The first set of inputs includes task descriptions generated by subject matter experts (SMEs) on the proposed system under study, along with SME-generated workload ratings for four separate channels - visual, auditory, cognitive and psychomotor. Additionally, the average time for task completion is included. The second set of inputs contain timing information, including the starting time, for each occurrence of each task executed during the mission segment. Overall workload for each time segment is computed by summing the workload ratings for the four channels. In an effort to validate CRAWL, workload estimates, obtained while operators flew a single seat simulator, were compared to CRAWL predictions of workload for six combat mission scenarios. Overall, the authors report an average correlation of 0.74 between the predicted workload levels and subjective pilot workload ratings obtained during the simulation study.

Workload Index (W/INDEX)

W/INDEX combines mission, task, and timeline analyses with theories of attention and human performance to predict attentional demands in a crewstation (North, 1986). It differs from other task analytic techniques by providing estimates of the effect of time-sharing loads imposed by concurrent task demands. The model estimates workload demands for one-second segments based on individual task difficulty and time-sharing deficits. W/INDEX operates on the following data:

* Crewstation interface channels;
* Human activity list;
* Attention involvement levels;
* Interface conflict matrix; and
* Operator activity timelines

This technique has been applied to three different conceptual cockpit designs and was demonstrated to be sensitive to design changes, but apparently has not been validated against empirical studies.

The McCracken-Aldrich Approach

McCracken, Aldrich, and their associates have recently developed a task analysis approach that does not rely solely on the time-based definition of workload (McCracken & Aldrich, 1984; Aldrich, Craddock & McCracken, 1984; Aldrich & Szabo, 1986). These authors attempted to improve the diagnosticity of workload predictions by identifying behavioral dimensions which contribute to overall workload levels. They were also among the first to attempt to explicitly identify cognitive workload demands. This approach has impacted other task analytic and simulation methods, including CRAWL and Micro SAINT (Laughery, Drews, Archer, & Kramme, 1986 - see below).

The McCracken-Aldrich methodology involves performing mission and task analyses that generate a rough timeline of tasks which are divided into three categories: flight control, support, and mission. It is assumed that tasks within each category would be performed sequentially, but tasks across categories could be performed concurrently. It is also assumed that a flight control task would be performed at all times. These tasks are sub-divided into performance elements which, based on system characteristics, are used to generate workload estimates on five behavioral dimensions comprising cognitive, visual, auditory, kinesthetic, and psychomotor (Szabo et al. 1987).

Workload assessments are made by assigning numerical ratings for each of the five dimensions of the task. These ratings represent difficulty or effort. It is in the ratings that this technique differs most from other task analyses. The ratings are generated by comparing verbal descriptors of the task components with the verbal anchors identified with each scale value. The five workload dimensions are assigned scale values of one through seven. The scale and verbal anchors for the cognitive component are presented for illustrative purposes in Table 1. Similar tables exist for the other behavioral dimensions.

Table 1

Scale Value	Descriptors
1	Automatic, simple association
2	Sign/signal recognition
3	Alternative selection
4	Encoding/decoding, recall
5	Formulation of plans
6	Evaluation, judgement
7	Estimation, calculation, conversion

Cognitive workload component scale (McCracken & Aldrich, 1984)

Estimates of the duration of each task element are developed to the nearest one-half second after assigning numerical ratings. These durations are used to construct a strict task timeline using 10-second segments. Total workload is estimated by summing across concurrent entries for each workload dimension, visual, auditory, kinesthetic, cognitive, and psychomotor, during each 10-second interval. If this sum exceeds a threshold value, e.g., 7 on visual, then the operator is assumed to be overloaded. Computer automation allows workload estimates to be derived for each one-half-second interval. The frequency of overload segments can then be determined and the causative workload dimension identified.

Sikorsky Aircraft's Hamilton and Harper (1984) proposed a modification of the McCracken-Aldrich technique. Their variant replaces the summation method of workload estimation with an interference matrix approach for detailed workload analysis. This matrix defines acceptable, marginal, and unacceptable workload levels for each of the four dimension comparisons. A series of decision rules are then employed to define whether or not entire segments have acceptable, marginal, or unacceptable workload levels. This technique alleviates certain interpretive problems arising, for example, from having a total segment rating of 10 on visual tasks, with a scale range of only one to seven. Validation efforts with this technique indicated that it was sensitive to task differences and reflected pilot opinion ratings obtained in simulation studies. It was also found to predict slightly higher workload ratings than those obtained by the actual rating; this bias is certainly acceptable and even desirable for design purposes.

Cognitive Task Analysis

The idea that a more detailed task analysis structure can provide increased diagnosticity is an important one. This idea, combined with the fact of increased influence of cognitive tasking, leads to the approach of detailed decomposition of cognitive workload into component types. This approach has been developed and applied to several airborne military systems (Zachary, 1981). As in more traditional task analysis, operator tasks are decomposed and are grouped into four primary categories: cognitive, psychomotor, motor, and communicative/interactional. A mission scenario is independently developed with a variable timeline grain depending on mission phase (for example, an attack mission phase may be decomposed to second by second events whereas a return-to-base phase may be decomposed into five minute intervals). Operational personnel then work with cognitive scientists to map operator tasks onto the scenario timeline. Next, workload levels are assigned to each operator task as the scenario unfolds. Workload ratings for the same task may vary depending on the mission segment in which it is performed. Using this approach, the workload analysis is based on a set of workload rating scales that describe five distinct types of cognitive workload:

* planning difficulty,
* prediction difficulty,
* calculation difficulty,
* information processing complexity, and
* information absorption complexity

In addition, eight other workload scales are utilized in the categories of psychomotor (pointer movement and writing), motor (button-pushing frequency and keyset entry frequency), and interactional (interruption frequency, interruption magnitude, communication frequency, and communication complexity). Applications of this methodology for each time segment yield individual ratings on thirteen scales and averaged ratings for the four categories (cognitive, motor, psychomotor, and interactional), as well as an overall workload (average of 13 measures). This promising methodology has been applied to two Naval operators, the P-3C anti-submarine warfare tactical coordinator (Zaklad, Deimler, Iavecchia, & Stokes, 1982) and the F/A-18 single-seat attack pilot

(Zachary, Zaklad, & Davis, 1987). Little formal validation has as yet been accomplished, although the effort is still ongoing.

Task Analytical Model Summary

Task analysis has demonstrated high utility. The definition of workload within the various task analyses is not complete, but clearly the stress imposed by the requirement to complete tasks within an allotted time is an important part of OWL. Indeed, the criteria for most tactical missions contain a temporal component in the measures of effectiveness (MOE). It is also true if a task cannot be done within the time requirements, of what importance is accuracy? For those situations in which time required (Tr) is near or approaching the performance envelop boundaries (Ta), additional evaluations can and should be performed.

COMPUTER SIMULATION MODELS

The application of simulation models to the workload estimation problem is conceptually an extension of traditional operator-in-the-loop simulation procedures. The major difference, of course, is that the simulation effort is expanded to include a simulated operator. A simulated operator is most valuable when an unbiased comparison of candidate hardware systems is desired. In the best of all possible worlds, a valid and reliable simulated operator would eliminate contamination of workload data typically due to subjects' individual differences and motivations. Presumably, if a simulated operator were employed, differences in workload data should be entirely due to variations in system configuration. Naturally, no such human operator model presently exists, again due primarily to the lack of a workload definition and consequently its effect upon operator performance. Meister (1985) and Chubb, Laughery and Pritsker (1987) provide an expanded review of simulation models and their applications.

Accurate and detailed descriptions of the operator, system, and operational environment are prerequisites to a good simulation model. Given these inputs, the problem shifts to defining an appropriate workload index that can be used to compare differences across candidate system configurations or operational uses. In most instances, a task loading index such as time required/time available is used. Some simulation models can predict both operator workload and system performance for comparison with empirical measures of effectiveness (MOEs).

Simulation vs. Task Analysis

The distinction between the task analytic methods and the computer simulation methods is not always clear. Most computer simulation models employ a task analysis as part of the development effort, and most task analytic methods are now computerized. Simulation models may be characterized as elaborated task analytic methods which incorporate consideration of the statistical nature of the task elements. The basic distinction that is intended in our categorization is as follows:

Task analysis methods produce operator performance requirements as a function of fixed increments of time defined against a scenario background. Simulation models, in contrast, attempt to represent (simulate) operator behavior statistically for task and subtask execution within the system under study and produce measures of effectiveness for human-system performance.

In other words, running a computerized task analysis twice would yield identical answers. Running a simulation model twice would not necessarily yield the same results due to consequences of branching statements, statistical modification of task times and, where appropriate, performance accuracies.

Siegel-Wolf Network Models

The majority of today's simulation models are derivatives of the original network model developed by Siegel and Wolf in the mid-sixties. The basic utility of the Siegel-Wolf model is in providing system developers an indication of whether or not operators may be over-stressed or under-stressed by a proposed design. The model predicts task completion times and probabilities of successful task completion; it enters the realm of workload assessment by determining "stress" imposed upon the operator. Stress is caused by:

* Falling behind in time on task sequence performance;
* A realization that the operator's partner is not performing adequately; and
* The inability to successfully complete a task on the first attempt with the possible need for repeated attempts, or the need to wait for equipment reactions.

Input to the network model typically consists of 11 data items for each subtask and operator and are presented in Table 2. Although there may be several potential sources of the necessary data, including detailed task analysis, the major source is direct questioning of subject matter experts. Outputs from Siegel-Wolf models include a number of performance measures such as number of runs, average run time, number and percent of successful runs, average, peak, and final stress, and several others. The primary uses for these models are for the coarse prediction of system effectiveness and design analysis. Siegel-Wolf models are limited typically to discrete task modeling.

Table 2

1. Decision subtasks,
2. Non-essential subtasks,
3. Subtasks which must be completed before it can be attempted by another operator,
4. Time before which a subtask cannot be started,
5. The subtask that must be performed next,
6. Average task duration in seconds,
8. Average standard deviation of task duration,
9. Probability of being successful,
10. Time required for all remaining essential tasks, and
11. Time required for all remaining non-essential tasks.

The eleven data elements required for each subtask and operator for Siegel-Wolf Models (from Meister, 1985.)

SAINT/Micro SAINT

An important extension of the Siegel-Wolf model is called the System Analysis of Integrated Networks of Tasks (SAINT). SAINT, along with its microcomputer version Micro SAINT, is actually a task network simulation language. It contains a number of process branching rules, multiple distributions for modeling individual task operations, and a Monte Carlo sampling procedure for determining task execution. SAINT's underlying approach to estimating workload is the same as the Siegel-Wolf models; it defines stress as the ratio of time required to complete a task to the time available (Tr/Ta). However, unlike the original Siegel-Wolf, it can be used to model both discrete and continuous tasks. As a general purpose simulation language, it provides a comprehensive framework, but contains little implicit information toward a developed model. This means that the operator(s), system, and environmental characteristics must be entered by the modeler. Micro SAINT provides a simple menu-driven interface to facilitate this development effort.

Micro SAINT has been used in conjunction with other workload estimation methodologies. Laughery et al. (1986) used Micro SAINT to predict operator workload in four helicopter cockpit designs, utilizing a model which incorporated characteristics of the operator, the helicopter control and display layout, and the threat environment as task networks. Workload was assessed during the Micro SAINT simulation by adapting the McCracken-Aldrich (1984) technique. This task analytic methodology requires the assignment of workload demands for each of four operator dimensions, i.e. auditory, visual, cognitive, and psychomotor for each operator activity. Thus, each task is characterized by its requirements for each of the four dimensions. In the helicopter simulation, workload was assessed at 2-second intervals, tracking it through the simulated mission scenario. The results demonstrated that the methodology was sensitive to variations among helicopter cockpit designs, and that specific dimension overloads could be identified. The authors report that total development and execution time was on the order of 10 weeks, although subsequent development times may be substantially less. This integration of network simulation with more robust and diagnostic workload prediction methodologies is a promising development.

Simulation for Workload Assessment and Manning (SIMWAM)

Another simulation methodology is called the Simulation for Workload Assessment and Manning (SIMWAM) (Kirkpatrick, Malone & Andrews, 1984). While it is based on both SAINT and the Workload Assessment Model (WAM) (Edwards, Curnow, & Ostrand, 1977), it has been specifically developed to make it particularly suitable for examining manpower issues in complex multi-operator systems.

SIMWAM has been recently used to assess workload and manpower issues for an aircraft carrier's aircraft operations management system (Malone, Kirkpatrick & Kopp, 1986). The application focused on the effects of incorporating an automated status board (ASTAB) into the existing system. The simulation scenario involved 35 shipboard operators engaged in a launch/recovery cycle of 25 aircraft. Workload assessments were made on the existing baseline system and the improved system incorporating the ASTAB. Results of the analysis suggested that the introduction of ASTAB could allow a reduction in the number of required personnel by four operators. That conclusion was based on the workload having been reduced to near zero for these four individuals, where workload was defined by number of tasks performed and the amount of time that a particular operator was occupied with tasks. The number of operators who were heavily loaded (i.e., busy at least 75% of the time) was also reduced by one-half with the introduction of the ASTAB.

Sequiturs Workload Analysis System (SWAS)

Sequiturs Workload Analysis System is a hybrid model incorporating features of both network and production system models (Holley & Parks, 1987). In contrast to network models which are general simulation tools, this model has been developed specifically for workload analysis. The definition of workload is the familiar time required over time available (Tr/Ta); success is defined strictly in terms of the Tr/Ta ratio. SWAS contains a structured helicopter task database, organized according to task categories which in turn are broken into blocks containing sub-task elements. Each task element in the database has ten attributes including the mean time and standard deviation, and processing modes for discrete and continuous tasks. It also has built in assumptions about the organization and functioning of behavior, following the Wickens (1984) resource model. This model plays a major role in the organization, sequencing, and resource time-sharing for task elements as well as modification of performance times. (See Navon [1984] for a critical review of the resource model.) Additionally, SWAS contains a Methods Time Measurement (MTM) module which is used to assist the user in producing mean performance times. Finally, equations are built in to adjust for individual differences (on a scale from 1 = good to 9 = bad). Both means and standard deviations are adjusted in a multiplicative manner in the equations. The model has received several validation studies at Bell Helicopter comparing the simulation results with results from operator-in- the-loop studies in both simulation and actual flight for a single pilot helicopter. Authors of these studies report error rates predicted by SWAS differed from operator times by 1% to 8% (underestimate).

Human Operator Simulator (HOS)

The Human Operator Simulator (HOS) is a simulation model using a distinctly different approach than the Siegel-Wolf models (Wherry, 1969; Lane, Strieb, Glenn, & Wherry, 1981; Harris, Glenn, Iavecchia, & Zaklad, 1986). The HOS approach is based on four assumptions:

* Human behavior is predictable and goal oriented, especially for trained operators;

* Human behavior can be defined as a sequence of discrete micro-events, which can be aggregated to explain task performance;

* Humans are single channel processors, but can time-share (switch) among several concurrently executing tasks; and

* Fully trained operators rarely make errors or forget procedures

The implication of these assumptions is that HOS is deterministic: the outcomes of operator actions are derived from functional relationships formed as equations rather than by sampling from a probability distribution.

Although conceptually sound, early HOS models had limitations which restricted its usefulness to the research community at large. These limitations included both technical characteristics, (e.g. single operator only, deterministic vs. probabilistic) as well as transportability of the

HOS model to different host computers. HOS-IV, the most recent version, is implemented on a microcomputer, contains provisions for use of Monte Carlo simulations, and addresses several earlier criticisms (Harris, Iavecchia, Ross, & Shaffer, 1987). For example, the short and long-term memory elements are probabilistic and thus yield stochastic results that are expected of human operators. Provisions are currently being developed to incorporate other probabilistic factors into the operator model in order to deal with such aspects as incompletely trained or novice operators.

HOS-IV is a general purpose simulation facility. It allows whole system simulation, that is, simulation of the dynamic interactions of the environment, the hardware/software system as well as the operator. The HOS-IV user can build a model of the environment, hardware/software, or operator to any level of detail desired using a top-down approach. For example, task times can be crudely estimated and entered into the simulation if a very gross analysis is desired. Alternatively, tasks can be decomposed more finely such that subtasks are defined according to an appropriate set of basic human performance micro-events. For example, a visual detection could be modeled coarsely by merely specifying the overall task time, or it could be analyzed into micro-events such as an eye movement followed by a visual perception followed by a decision which results in a motor response.

HOS-IV contains a library of human performance models that can be used to simulate the timing and accuracy of particular human behaviors. The core set of micromodels, all of which are based on experimental literature, includes: models for eye movements; visual perception; decision time; short-term memory; listening and speaking; control manipulation; hand movement; and walking. The micromodels of the operator are available to the user so existing modules can be modified or replaced entirely with one of the user's creation. The microcomputer implementation of HOS contains an enhanced user interface to assist in defining, executing, and analyzing the simulation.

The result of the simulation is a detailed timeline of operator, hardware, and environmental events and actions which can be summarized and analyzed for a broad variety of purposes. Standard output analyses are available which provide statistics associated with performing tasks, subtasks, and basic behaviors. This includes the number of times a micromodel is executed, the mean and standard deviation of the time to complete a process, and the percent of simulation time spent on each process. Additionally, the user can access information on system measures of effectiveness.

Lane et al. (1981) identified a number of applications and validation efforts over a wide range of systems. HOS allows a very detailed model to be developed, providing a greater degree of diagnosticity than other simulation models. It is probably most useful as a follow-on analysis after less detailed analytic techniques have been used to refine the system design.

Model Human Processor (MHP)

Card, Moran and Newell (1983, 1986) have developed a potentially powerful collection of micromodels collectively called the Model Human Processor (MHP). Via MHP, they have established a framework for presenting data contained in the human performance literature in a manner which will make it more accessible to those involved in the engineering design process. They partition human behavior models according to their application to the perceptual, cognitive, or motor systems, and focus on simpler, more widely applicable models that capture the predominant characteristics of a problem. Models such as these can be used to define limits of operator-system effectiveness to practically any degree required. While the MHP micromodels are currently only described in the literature, some have been directly incorporated into the HOS library and are accessible to simulation modelers. Additionally, MHP has proven a fruitful model for analysis of computer interfaces, an area not covered by other models (Card, Moran & Newell, 1983). Further work in the development and application of human performance models is required.

Computer Simulation Model Summary

In recent years a number of new simulation tools have been developed which offer a unique opportunity to evaluate both time and accuracy of performance. Thus, combined with the task analysis which most simulations presume anyway, they are the best and most thorough of the analytic techniques. There is a cost, however, for gaining this improved capability, and that is the additional time and effort required for developing the simulation. In the long run this may be a small price to pay when contrasted against overall system development and life cycle costs. In many instances user friendly versions of older simulations have been developed in the last several years. As additional modules and computer tools are developed and databases are built, simulation techniques should move to the forefront of workload analytic techniques.

As with all the analytic workload tools surveyed, validation continues to a major issue facing the computer simulation methodologies.

GUIDANCE

There are several questions which must be considered when selecting an appropriate workload model for application. First and foremost, it is important to keep in mind both the nature of the workload issue being examined, and how the results are intended to be used. That is to say, the real needs of the user have to be clearly defined. It is costly and inefficient to implement workload analyses providing levels of detail which are neither appropriate nor wanted. It is a primary responsibility of the workload analyst to make these decisions in concert with the agency requesting the analysis.

In determining the needs of the user the following questions, at a minimum, will always be pertinent:

* What type of acquisition process describes the system under development? Is the program a formal full scale development effort, a product improvement program, non-developmental item, accelerated system acquisition program, etc.?

* Has a mission scenario been developed for the system in question? Does a clear understanding exist concerning how the system will be employed, what the nature and intensity of the projected threat will be, and where the operator(s) are expected to experience difficulty in employing the system?

* Has any operator workload analysis been done on predecessor or similar systems? There is no benefit in repeating past work. If subjective opinion of operators who have used similar systems is available, it may be the most valuable data possible.

* What is the stage of system development? Is the system closer to concept exploration and exist primarily on paper, or have prototypes been built and evaluated and the system is moving nearer to operational evaluation?

* Has any operator workload analysis been done on this system at an earlier stage of development?

The last question to be posed here is clearly the most important: although it is last on this list it must, in fact, be the first question asked by the workload researcher when considering the applicability of techniques to a particular situation. The question is, "*What are the real-world constraints on the analysis about to be undertaken?*" It is necessary to acknowledge that we don't operate in a perfect world. The type of information we can provide to system engineers and the design decisions we can influence are more often than not determined by such issues as:

* How much time is available to conduct our study?

* What are the levels of manpower and expertise available?

* What access will we have to field personnel?

* How much money has been allocated to our study?

* What computer facilities may be available for data collection and analysis?

This list of questions is by no means inclusive, but the point has been made. It is our opinion that any workload model, regardless of its reliability, sophistication, or validity, is utterly useless unless it can and will be used as a practical tool which provides a timely answer. Providing assistance to the system developer is the bottom line. Far more important than any model's attributes, and our critique of it, is the determination of whether it will be used to influence design. A usable model that influences design in a positive way can be considered successful.

We stated in the opening of this paper that one of our objectives was to provide guidance on which procedures are best suited to a given set of resources and measurement goals. Table 3 provides such guidance in an overview of the techniques and a consensual judgement of the authors about the data requirements, costs, diagnosticity, and subjectivity of each technique. The potential user may consult this table as a jumping off point; he or she is encouraged to investigate more fully those techniques which appear to be appropriate.

Resources

The questions posed above are certainly germane, but the issue of selecting a particular technique still remains even when all the answers are known. What would be most welcome would be some guidelines, or perhaps a computer assisted technique, to advance the beleaguered analyst from knowing "where he or she is" to knowing "where he or she has to go." To that end we will briefly mention three practical aids.

Table 3

Technique	Data Requirements	Cost/Effort* Requirements	Diagnosticity	Subjectivity
Comparison	System level	Low cost/ low effort	Low	High
Math Models	Detailed task	Low cost/ high effort	Low-Moderate	Low
Expert Opinion	Task level	Low cost/ low effort	Low-Moderate	High
Task Analysis				
Time Based	Task level	Low cost/ Moderate effort	Low-Moderate	Low
McCracken-Aldrich	Task level	Low cost/ Moderate effort	Low-Moderate	Moderate
Simulation				
Siegel-Wolf	Task level	Moderate cost/ High effort	Low	Moderate
SAINT	Task level	Moderate cost/ High effort	Low-Moderate	Moderate
Micro SAINT	Task level	Low cost/ Moderate effort	Low-Moderate	Moderate
SIMWAM	Task level	Moderate cost/ Moderate effort	Low	Moderate
SWAS	Task element level	High cost/ Moderate effort	Low	Moderate
HOS	Task element level	Low cost/ High Effort	Moderate-High	Low

* Cost refers to acquisition costs in dollars. Effort includes number of personnel and development time/effort.

The first of these is Workload Consultant for **Field** Evaluation (W.C. Fielde), an expert system developed by the Human Factors Division of the NASA Ames Research Center. (Casper, Shively, & Hart, 1987) W.C. Fielde is a microcomputer based decision support system which guides naive users to a selection of workload measures appropriate to his or her evaluation. The program is extremely simple to use and all rules involved in the decision process are made available to the user for review.. W.C. Fielde recommends several assessment methodologies in decreasing order of appropriateness, and provides additional information on each measure at the end of the program in the form of text files. We have been favorably impressed with the practicality of this program and highly recommend it. This aid is currently available from NASA.

The second and third aids are not yet available, but comprise part of our present research effort and will be available upon the completion of our work. One of these aids will be an extension of W.C. Fielde and is a similar rule based expert system to help the inexperienced analyst select the appropriate technique given the status of his program as described in the previous section. We call this expert system the OWL Matching Model. The last source of practical guidance will consist of a number of publications and handbooks which are deliverable under the provisions of our contract. The focus of the handbooks will be on the practical; methodology reviews and model critiques will be left in the scientific reports and made available to interested researchers. The guidebooks are intended for a different audience including both behavioral scientists who have little experience in workload methodology and prediction, and also military project managers who have little formal exposure to human factors engineering, but have identified their need for workload evaluations. It is assumed that these handbooks and pamphlets will be available to the general public once they have been approved for distribution by our Army sponsor.

SUMMARY

Few of the available analytical techniques may be considered to capture the full complexity of the workload issue. That is hardly startling given the inability of the scientific community to even define workload or its constituent components. Unfortunately, there is no "ideal" workload methodology, nor is one likely to be developed in the foreseeable future. What is in the foreseeable future, however, is a growing appreciation of the individual strengths of several workload prediction techniques and an intelligent integration and application of the proper techniques at the proper time. Also on the horizon are promising developments in computer simulation. Each of the analytic techniques can and does provide some useful information relating to workload which can be of value to the customer. In summary there are two axioms which are central to the utilization of the analytical techniques in particular, and workload techniques in general:

(1) A battery of techniques both analytical and, if possible, empirical is needed for each situation, and

(2) Different situations require a different mix of OWL assessment techniques.

Lastly we must mention that when all is said and done, success in modeling workload is primarily dependent upon a human who plans and conducts the work. Just as there is no magic technique which provides definitive and valid answers to workload questions, there is no magic technique which can perform independently of the creative, thoughtful and usually labor intensive participation of the analyst. To be blunt, all the analytic techniques we reviewed still require an analyst to do the dirty work. Considering the state-of-the-art in workload models, we should all be thankful that the well-informed and well-intentioned analyst is still the key factor in the decision making process.

"The views, opinions, and findings contained in this report are those of the author(s) and should not be construed as an official Department of the Army position, policy or decision unless so designated by other official documentation."

REFERENCES

Aldrich, T. B., Craddock, W., and McCracken, J. H., 1984, "A Computer Analysis to Predict Crew Workload During LHX Scout-Attack Missions: Volume 1," U.S. Army Research Institute, Ft. Rucker, Al.

Aldrich, T. B., and Szabo, S. M., 1986, A methodology for predicting crew workload in new weapon systems, in: "Proceedings of the Human Factors Society 30th Annual Meeting," Human Factors Society, Santa Monica, Ca.

Attneave, F., 1959, "Applications of Information Theory to Psychology," Holt, New York.

Baron, S., and Levison, W. H., 1977, Display analysis using the Optimal Control Model of the human operator, Human Factors, 19:437-457.

Baron, S., 1979, A brief overview of the theory and application of the optimal control model of the human operator, in: "Models of Human Operators in Vision Dependent Tasks," M. C. Waller, ed., NASA Conference Publication 2103, Washington, DC.

Baron, S., Zacharias, G., Muralidharan, R., and Lancraft, R., 1980, PROCRU: A model for analyzing flight crew procedures in approach to landing, in: "Proceedings of the Eighth IFAC World Congress," Tokyo.

Bateman, R. P., and Thompson, M. W., 1986, Correlation of predicted workload with actual workload measured using the subjective workload assessment technique. SAE AeroTech.

Card, S. K., Moran, T. P., and Newell, A., 1983, "The Psychology of Human-Computer Interaction," Lawrence Erlbaum Associates, Hillsdale, NJ.

Card, S. K., Moran, T. P., and Newell, A., 1986, The model human processor: An engineering model of human performance, in: "Handbook of Perception and Human Performance: Vol. 2. Cognitive Processes and Performance," K. R. Boff, L. Kaufman, and J.P. Thomas eds., Wiley, New York.

Casper, P. A., Shively, R. J., and Hart, S. G., 1987, Decision support for workload assessment: Introducing W C FIELDE, in: "Proceedings of the Human Factors Society 31st Annual Meeting," Human Factors Society, Santa Monica, Ca.

Chubb, G. P., Laughery, K. R., and Pritsker, A. B., 1987, Simulating manned systems, in: "Handbook of Human Factors," John Wiley and Sons, New York.

Crawford, B. M., 1979, Workload assessment methodology development, in: "Survey of Methods to Assess Workload," B.O. Hartman and R. E. McKenzie, eds., AGARD-AG-246, Neuilly Sur Seine, France.

Edwards, R., Curnow, R., and Ostrand, R., 1977, "Workload Assessment Model (WAM) User's Manual," Boeing Aerospace Co. Report D180-20247-3, Seattle, Wa.

Ellison, M. G., and Roberts, B. B., 1985, Timebased analysis of significant coordinated operations (TASCO): A cockpit workload analysis technique, in: "Proceedings of the Human Factors Society 29th Annual Meeting," Human Factors Society, Santa, Monica, Ca.

Gainer, P., 1979, Analysis of visual estimation of system state from arbitrary displays, in: "Models of Human Operators in Vision Dependent Tasks," M. C. Waller, ed., NASA Conference Publication 2103, Washington, DC.

Garner, W. R., 1962, "Uncertainty and Structure as Psychological Concepts," Wiley, New York.

Garner, W. R., 1974, "The Processing of Information and Structure," Lawrence Erlbaum Associates, Potomac, Md.

Hamilton, B. E., and Harper, H. P., 1984, Analytic methods for LHX mission and task analysis, in: "Proceedings of Advanced Cockpit Specialist Meeting," American Helicopter Society, Washington, DC.

Harris, R. M., Glenn, F., Iavecchia, H. P., and Zaklad, A., 1986, Human Operator Simulator, in: "Trends in Ergonomic/Human Factors III (Part A)," W. Karwoski, ed., North-Holland, Amsterdam.

Harris, R. M., Iavecchia, H. P., Ross, L. V., and Shaffer, S. C., 1987, Microcomputer Human Operator Simulator (HOS-IV), in: "Proceedings of the Human Factors Society 31st Annual Meeting," Human Factors Society, Santa Monica, Ca.

Holley, C. D., and Parks, R. E., 1987, Predicting man-machine system performance in predesign, in " American Helicopter Society National Specialist Meeting on Flight Controls and Avionics," American Helicopter Society, Washington, DC.

Jahns, D. W., 1973, Operator workload: What is it and how should it be measured?, in: "Crew System Design," K. D. Gross, and J. J. McGrath, eds., Anacapa Sciences, Inc, Santa Barbara, Ca.

Kirkpatrick, M., Malone, T. B., and Andrews, P. J., 1984, Development of an interactive microprocessor-based workload evaluation model (SIMWAM), in: "Proceedings of the Human Factors Society 28th Annual Meeting," Human Factors Society, Santa Monica, Ca.

Lane, N. E., Strieb, M. I., Glenn, F. A., and Wherry, R. J., 1981, The human operator simulator: An overview, in: "Manned Systems Design: Methods, Equipment, and Applications," J. Moraal and K. F. Kraiss, eds., Plenum Press, New York.

Laughery, K. R., Jr., Drews, C., Archer, R., and Kramme, K., 1986, A MicroSAINT simulation analyzing operator workload in a future attack helicopter, in: "Proceedings of the National Aerospace and Electronics Conference."

Levison, W. H., 1970, A model for task interference, in: "Proceedings of the 6th Annual Conference on Manual Control."

Levison, W. H., 1979, A model for mental workload in tasks requiring continuous information processing, in: "Mental Workload: Its Theory and Measurement," N. Moray ed., Plenum Press, New York.

Linton, P. M., Jahns, D. W., and Chatelier, P. R., 1977, Operator workload assessment model: An evaluation of a VF/VA-V/STOL system, in: "Proceedings of the AGARD Conference on Methods to Assess Workload," AGARD-CP-216, Neuilly Sur Seine, France.

Malone, T. B., Kirkpatrick, M., and Kopp, W. H., 1986, Human Factors Engineering impact of system workload and manning levels, in: "Proceedings of the Human Factors Society 30th Annual Meeting," Human Factors Society, Santa Monica, Ca.

McCracken, J. H., and Aldrich, T. B., 1984, "Analysis of Selected LHX Mission Functions: Implications for Operator Workload and System Automation Goals," Anacapa Sciences, Inc., (TNA ASI479-24-84), Fort Rucker, Al.

Meister, D., 1985, "Behavioral Analysis and Measurement Methods," John Wiley and Sons, New York.

North, R. A., 1986, A workload index for iterative crewstation evaluation, in: "Proceedings of the Eighth Annual Carmel Workshop: Workload and Training, an Examination of Their Interactions".

Rault, A., 1976, Pilot workload analysis, in "Monitoring Behavior and Supervisory Control," T. B. Sheridan and G. Johannsen eds., Plenum Press, New York.

Rickard, W. W., and Levison, W. H., 1981, Further tests of a model-based scheme for predicting pilot opinion ratings for large commercial transports, in: "Proceedings of the 17th Annual NASA-University Conference on Manual Control," University of California at Los Angeles.

Roberts, B. B., and Crites, C. D., 1985, Computer-aided cockpit workload analysis for all weather, multirole tactical aircraft, in: "Fourth Aerospace Behavioral Engineering Technology Conference Proceedings," Society of Automotive Engineers, Warrendale, Pa.

Rouse, W. B., 1980, "Systems Engineering Models of Human-Machine Interaction," Elsevier North Holland, New York.

Schmidt, D. K., 1978, A queing analysis of the air traffic controller's workload, IEEE Transactions on Systems, Man, and Cybernetics, SMC-8(6):492-493.

Senders, J. W., Elkind, J. I., Grignetti, M. C., and Smallwood, R., 1966, "An Investigation of the Visual Sampling Behavior of Human Observers," NASA CR-434, Washington, DC.

Senders, J. W., 1964, The human operator as a monitor and controller of multi-degree of freedom systems, IEEE Transactions on Human Factors and Electronics, HFE-5:2-5.

Senders, J. W., and Posner, M., 1976, A queueing model of monitoring and supervisory behavior, in: "Monitoring Behavior and Supervisory Control," T. B. Sheridan, and G. Johannsen, eds., Plenum Press, New York.

Sheridan, T. B., and Ferrell, W. R., 1974, "Man-Machine Systems: Information, Control, and Decision Models of Human Performance," MIT Press, Cambridge, Ma.

Smit, J., and Wewerinke, P. H., 1978, An analysis of helicopter pilot control behavior and workload during instrument flying tasks, in: "AGARD Aerospace Medical Panel Specialists Meeting on Operational Helicopter Aviation Medicine," AGARD, Neuilly Sur Seine, France.

Stone, G., Gulick, R. K., and Gabriel, R. F., 1987, Use of timeline analysis to assess crew workload, in: "The Practical Assessment of Pilot Workload," A. H. Roscoe, ed., AGARD-AG-282, Neuilly Sur Seine, France.

Thompson, M. W., and Bateman, R. P., 1986, A computer based workload prediction model, unpublished work.

Walden, R. S., and Rouse, W. B., 1978, A queing model of pilot decision making in a multi-task flight management situation, IEEE Transactions on Systems, Man, and Cybernetics, SMC-8(12):867-875.

Wherry, R. J., Jr., 1969, The development of sophisticated models of man-machine systems, in: "Proceedings of the Symposium on Applied Models of Man-Machine Systems Performance," North American Aviation, Columbus, Oh.

Wewerinke, P. H., 1974, Human operator workload for various control situations, in: "Tenth Annual Conference on Manual Control," Wright-Patterson Air Force Base, Oh.

Zachary, W., 1981, "Cost-benefit assessment of candidate decision aids for Naval Air ASW," Analytics Tech Report 1366-C, Willow Grove, Pa.

Zachary, W., Zaklad, A., and Davis, D. A., 1987, A cognitive approach to multisensor correlation in an advanced tactical environment, in: "Proceedings of the First Tri-Service Data Fusion Symposium," Johns Hopkins Univ., Columbia, Md.

Zaklad, A., Deimler, J., Iavecchia, H., and Stokes, J., 1982, "Multisensor Correlation and TACCO Workload in Representative ASW and ASUW Environments," Analytics Technical Report 1753A, Willow Grove, Pa.

WORKLOAD PREDICTION, DIAGNOSIS, AND CONTINUING CHALLENGES*

Donald L. Parks and
George P. Boucek, Jr.

The Boeing Company

INTRODUCTION

Many methods and models have evolved in recent years with the objective to improve our ability to predict and measure workload, based on the desire to assure operators can perform all tasks as required. With this evolution has come a better recognition of differences in need, language has become more precisely defined and there now is less confusion of concepts and purpose than 10 years ago. There is now a wide assortment of workload "tools" with widely varying degrees of complexity. Additionally, more attention is being given to whether the tools measure what they purport to measure, and whether the variety of tools now proposed actually do measure the same thing.

However, there is need for an increasing recognition of the designer's requirement to use workload evaluation results quite early for design diagnostics...to identify specific interfaces for attention in a given design approach, even in preliminary design. Even the earliest decisions are often too far reaching in impact to neglect an early appraisal...the simplest of time lines have been known to expose significant system problems. Furthermore, early and effective design integration of solutions is becoming more critical as systems become more and more complex, in order to avoid major change late in the design cycle for seemingly simple issues. Many of the tools that have emerged are unable to support this need.

Finally, a new issue is emerging as our ability to automate system operations improves. We need methods to appraise and control the consequences of underload as well as overload, and to produce accepted definitions of the upper and lower load limits - the "red lines" for marginal crew performance conditions.

---

* Originally released as Boeing Document D180-31116-1

Thus, although the reasons have changed, early and quantitatively defendable predictions of crew workload remains one of the most important issues for crew station design and development. In the past, the main interest was to avoid hardware changes late in the design-development cycle. Now, we keep hearing that it is "only" software. But consider the software nightmare that we face. There will be major issues to resolve in system and information integration, information management and display-control formatting, and in turn major workload questions as we attack the problem of comprehending and integrating: (a) an explosion of information associated with more sophisticated systems (b) a complex information network, and (c) complicated information integration display requirements. We need early simple-to-apply tools for this phase that set us up to transition to use of other tools as appropriate.

This paper reports the results of a continuing effort in the use of a timeline technique for workload prediction. It summarizes an evolution of our approach to timeline analysis and prediction (TLAP), and diagnostic applications. It presents methods to isolate the causes of high demand load conditions, including cause factors in peak load conditions and demand loads imposed by operation of each individual subsystem. It includes an approach to appraising cognitive workload as done in the past, and one using a new analytic technique. Efforts to demonstrate validity have been conducted and validation data are presented. Also, data is presented that demonstrates the degree to which results from use of the other tools correlate with this technique.

BACKGROUND AND PROBLEM SUMMARY

In cockpit design, we continue to emphasize the need for and use of an analytic predictive tool as a metric for workload evaluation. This is a critical aspect of our need to appraise workload before costly design, subsystem developments and cockpit integration have been committed. The cost of changes late in development is a major issue, but it is not until then that system development has progressed to the point where the context of the scenario can be realistically simulated for use of many of the other techniques now being proposed. Accordingly, we want to be able to forecast the effects of given design concepts and resolve related issues before spending extensive time and resources to design, develop and test such concepts.

In order to accomplish such objectives, typically, we want to compare new design concepts with existing designs in order to confirm that they are equal or better in terms of crew workload demands. This is especially important if major changes in design concept are being considered. Results of analyses at this level will provide an overall appraisal of relative workload for a new design concept.

However, the new design may feature areas of excessive workload. In turn, then, one of our strongest needs is for workload diagnostic techniques. In recent literature, there has been a lot of attention given to overall workload prediction and measurement, but little heed has been paid to techniques for resolving workload problems, once found. If there is a workload problem, how can it be resolved? The ability to identify heavy contributors to out-of-bounds conditions remains one of the major issues for the whole field of workload techniques. If a technique does not help to identify and focus on specific issues for resolving workload problems, it is not very useful to the designer. You've told him "There's a needle in the haystack somewhere. Go find it." So now, he has to redesign his concepts and needs to focus where the higher payoffs are indicated. He thus needs to follow some systematic procedure to identify, reduce or eliminate the problem. The more quantitative and defendable, the better.

If the problem is a workload peak, it should be made readily possible to identify the specific procedures, tasks or subsystems involved, and in turn possible solutions in the form of reallocation, redesign or both.

The point is, by the time a new airplane gets to the point where mission simulations and evaluations required for realistic use of many of the new tools can be made with a high level of confidence, a great number of design decisions, developments and subcontracts will have been made that go far beyond the cockpit in their impact. In fact, a great deal of interdisciplinary integration is involved in modern vehicles; a few of the areas impacted are sensors, avionics systems, details of subsystem design and levels of automation. As design and development progresses, the degrees of freedom for the cockpit designer become less and less until a late change of any consequence has a far reaching impact that is unacceptable. So we want to commence a new system with tools that can be used from the very early stages and proceed through design and development with increasing confidence that workload is manageable over the wide variety of circumstances that are of concern and will be demonstrated to be so in acceptance and certification testing.

We face, potentially, an explosion of information management problems, with greater attention to automation concepts. Such issues will become much more complex in the near future and workload management on our part will become more difficult. It is unlikely that we will change our requirements for pilot situation awareness and control, in order to assure he can monitor status and provide active and backup modes of operation as required. However, with new systems information management and flight management requirements, we can expect more extensive use of automation, expert system and artificial intelligence techniques as well as new approaches to information integration and display formatting in cockpit management. Adaptability and/or compatibility of our methodology with this part of the future should become part of our criteria for the workload tools of today.

More recently, another question is becoming relevant that was of little concern in the past. In the past, we have emphasized the need to minimize workload. With the old dials and gages, combined with limited technological ability to automate, we wanted to minimize workload. Now that we can better integrate and automate the display-control design, there is an increasing concern about the possibility of too little workload. Complacency and boredom could become increasingly important factors in cockpit operation. Accordingly, our metrics must feature the capability to establish upper and lower bounds for workload. Workload management to assure that pilot can readily accommodate surprises and that he will remain alert in a heavily automated system is a serious issue for the near future.

By now it should be clear that we feel a strong need for predictive, diagnostic and viable analytic tools. We also have a continuing interest in empirical performance measurement techniques. The reasons are two fold: (1) Our predictions must eventually be demonstrated in test. (2) There should be a correlation between empirical techniques and the predictive approach. If high correlations exist, predictive design and development tools can be used with greater confidence that they lead us down the right path.

TECHNICAL DISCUSSION

This section of this report presents an approach for timeline analysis and prediction (TLAP) of workload, and for diagnostic applications of the technique. The approach is based on time required to perform a task vs. time available within the task sequence. To clarify any misconceptions from past presentations, a step by step evolution of the techniques will be presented, with results as appropriate. Approaches for using this tool for workload diagnostics will be described. Techniques that have been applied to estimate the more elusive parameter---cognition---will also be discussed. Finally, data representing how robust this technique is for its purpose will be presented, including validation data and the degree to which other techniques have been found to correlate with it. Finally, discussion will address selected technological challenges that remain to be exploited.

Timeline Analysis, Prediction and Diagnosis

Timeline Analysis as an approach to workload evaluation started many years ago as a simple, manually developed layout of a task sequence on a timeline. This was then appraised for whether tasks could be performed within the time available, for stimulus-response compatibility and disparity, for any equipment performance characteristics that could impact completion of a given task series and for the ability to complete the task in time.

This approach to workload analysis has now progressed to a far more sophisticated level. While original features are still evident, data details and processing methodology has changed and application features have grown. Accordingly, and in order to minimize misconceptions about the context herein, an elaboration of the present approach and how it is used will be presented.

The first step in evolving the timeline based workload analysis is to develop a generic mission scenario for a typical flight, including major worst-case conditions. The scenario defines flight segments, altitudes, speeds, key events, typical operations requirements and both normal and degraded modes of operation. Constraints of operation, environmental variations, timing considerations and particularly critical situations are all outlined, and representative variations in the mission are also identified and injected into this generic scenario. The key objective is to develop a generic mission definition that encompasses all conditions and thus avoid oversights. The resulting scenario may be brief, serving as the baseline for an extensive task definition. Alternatively, it may be very extensive and detailed in which case it provides much of the structure for systems selection and detailing of the task definition.

In turn, the scenario is used to develop a detailed task definition and sequence. Given a scenario and a candidate cockpit configuration, operations procedures and specific display-control operation requirements are identified. Most typically, this is accomplished for both a baseline airplane with existing display-control concepts as a point of departure and for the new cockpit concept under development. This permits appraisal of workload on a relative rather than an absolute basis, for a higher level of confidence in the result.

Personal experience has demonstrated that the most convenient way to organize the task requirements is to construct a timeline that relates the display-control operations to the scenario and its key events as illustrated in figure 1. This can and has been done at varying levels of detailed definition depending on the level of cockpit definition. The main issue is to define when tasks have to occur, and when related sequencing must start in order to complete on time. This format is relatively easy for an analyst to develop and can become increasingly detailed as task analysis continues. It also provides a convenient outline for defining and inputting detailed data into the computer, as well as a checkpoint of the computer results against input data.

UNSHIFTED

MISSION TIMELINE
MISSION - FROM TETTLA F15E
TEST CASE
CONFIGURATION - CONFIG. A
FLIGHT PHASE - ENG START AND TAXI

CREWMEMBER - CAPTAIN

JUN 1, 1987

CODE	EVENT/PROCEDURE OR TASK DESCRIPTION	TASK DUR (SEC)
12A001	SET ENG MASTER SW	
560AI012	VIEW THROUGH CANOPY	2.00
800AC002	ENG MASTER SW	1.90
12A002	SET JFS HANDLE	
800AI003	JFS SW	1.40
800AC004	JFS HANDLE	1.80
12A003	SET R. THROTTLE FINGER LIFT	
800AI005	JFS READY LT	2.20
800AC006	R. THROTTLE FINGER LIFT	2.30
12A004	SET R. THROTTLE	
770AI007	R. RPM IND	2.30
730AC008	R. THROTTLE	2.80
12A005	SET R. GEN. SW	
770AI007	R. RPM IND	2.30
770AI009	R. EGT IND	2.00
770AI010	R. OIL PRESS IND	2.00
770BI011	R. FUEL FLOW IND	2.00
240FC014	R. GEN SW	2.10
12A006	SET L. THROTTLE FINGER LIFT	
240FI013	EMER BST LT	2.20
560AI012	VIEW THROUGH CANOPY	2.00
800AC015	L. THROTTLE FINGER LIFT	2.60
12A007	SET L. THROTTLE	
770AI016	L. RPM IND	2.30
730AC017	L. THROTTLE	2.90
12A008	SET L. GEN. SW	
770AI016	L. RPM IND	2.30
770AI018	L. EGT IND	2.20
770AI019	L. OIL PRESS IND	2.20
770BI020	L. FUEL FLOW IND	2.00
240FC021	L. GEN SW	2.30

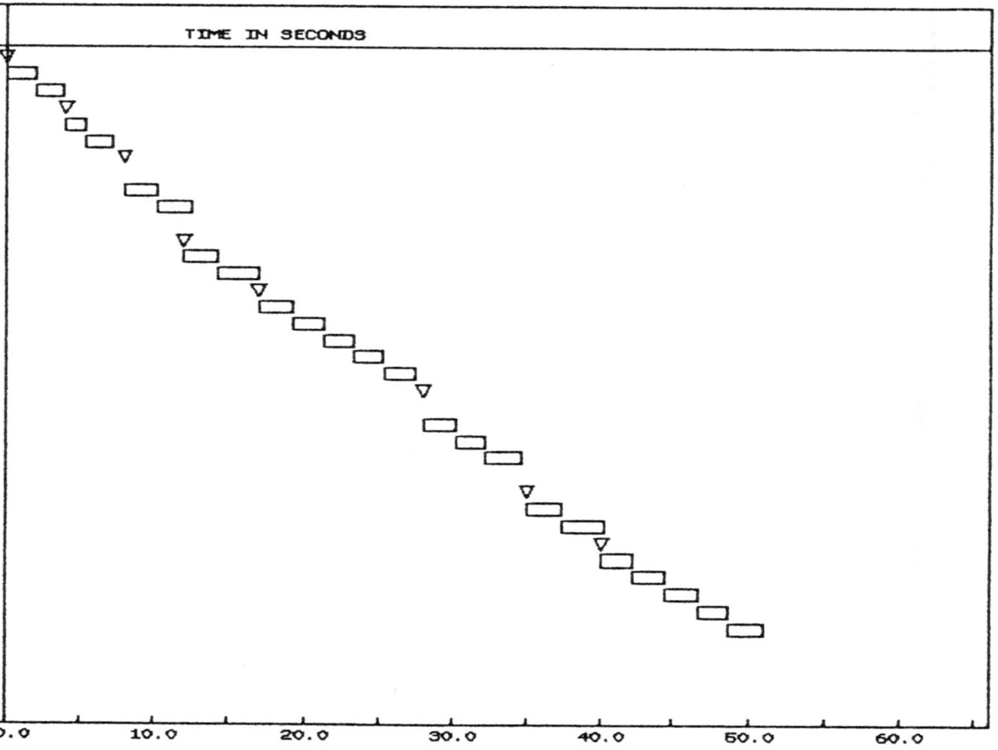

Figure 1. Timeline Definition Format

Other task characteristics that might be relevant can also be treated. These tasks that must be performed at a critical time can be identified. Those that can be slipped for earlier or later performance can also be defined. If tasks must be performed in a given sequence, this too can be specified. Other variations are feasible; many have been explored over the years, including task performance probabilities and task performance time variations (e.g., standard deviations).

The timeline is then further refined. Our development of timeline analysis recognizes that humans can do more than one thing at a time. For example, in driving a car, one processes visual and auditory information, makes decisions, steers, and operates the accelerator, brake (and clutch)---discrete, parallel, sequential and coordinated tasks are performed as an integrated network of responses. We use a human performance channel allocation scheme to handle such variations. This involves defining which aspects of a task are performed by vision (internal, external), hands (right, left), feet (right, left), audition (hearing, speech) and cognition. (This part can be automated if we have a data bank of task performance times and a definition of display-control task demands. For first use, derivation is manual. For initial application, the sources of data for estimating task performance times include information scattered through the literature, time and motion methods, appraisals in mockups, and simulation. Once in the data base, the task requirement is catalogued for future use.)

Most performance times are straight forward. However, cognitive workload has been another issue. Analyst estimates have been used by some analysts, but more quantitative approaches exist. One evolved from a flight test program some years ago wherein a consistent quirk was indicated in eye movement data. Pilots would look at a display, then there would be a pause before they took action. The action could be clearly related to the presented information, suggesting that processing was involved. This processing time came to be accepted as demonstrating cognitive operations. Furthermore, the pause time was essentially a constant percentage of task performance time. This relationship became the basis for a more quantitative estimate of cognitive workload in our approach. More recently, another estimating technique based on information theory has been developed. This latter technique will be discussed in a later portion of this paper.

At this point, data is ready for final computation and processing. By inspection, it can be determined if any task conflicts exist, in terms of conflicting or incompatible demands (like rubbing the head and patting the stomach at the same time). Processing through the computer produces a series of plots and reports for appraising workload demand and for detecting and diagnosing problem areas.

Workload estimates are produced by solving the equation:

$$W/L = \frac{T_R}{T_A} = \frac{\text{Time Required}}{\text{Time Available}} = xx\%$$

over small increments of time (in order to avoid a leveling effect). Processing produces a percent workload figure which can be plotted over time to produce a workload time history, as is illustrated in figure 2.

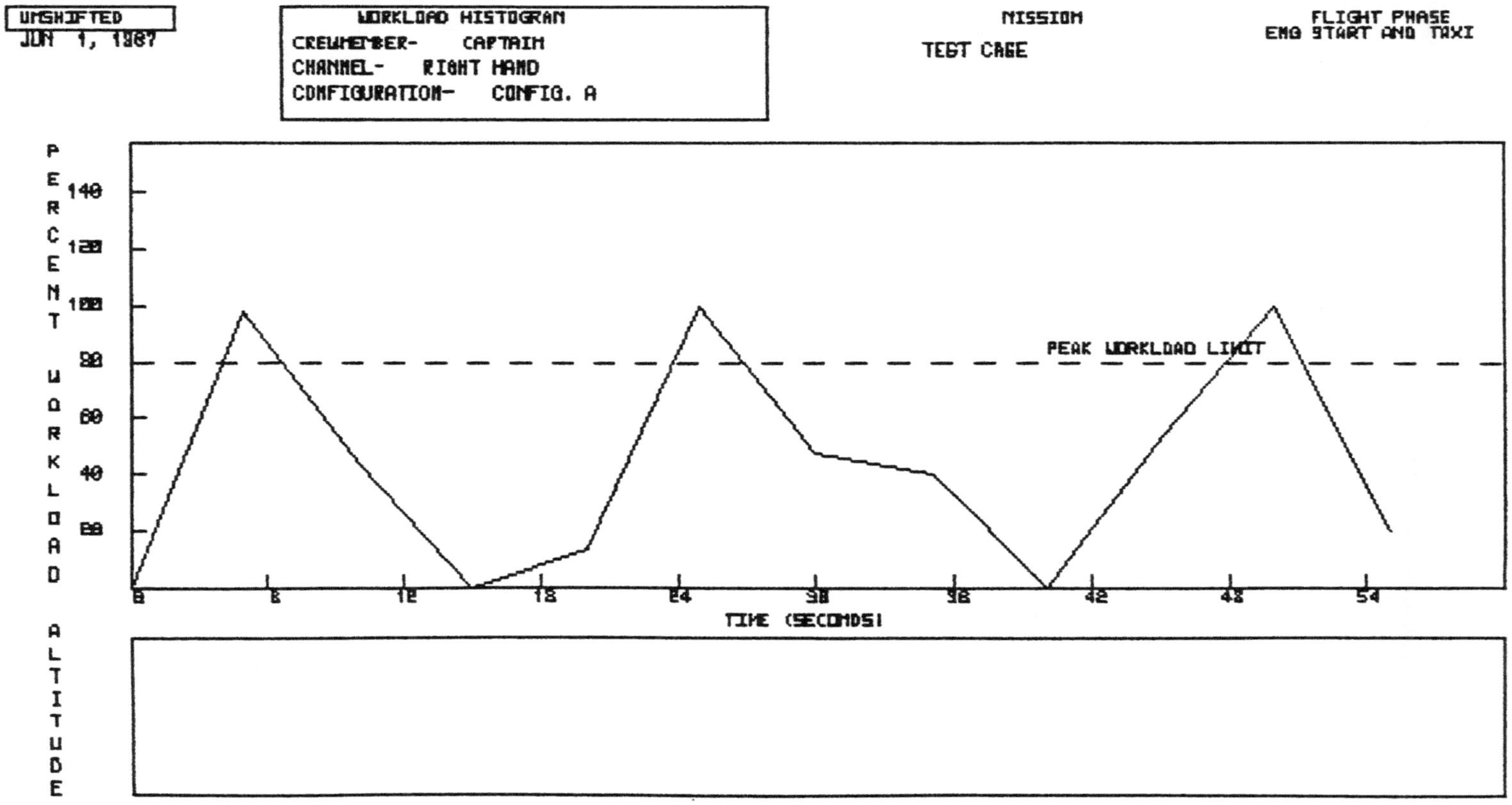

Figure 2. Example of Workload Time History Profile as Produced by TLAP

Referring to figure 2, an limit factor can be set in that is used as a cut-off for peak workload considerations. In between this case, the cut-off is 80%, which is that level of time demand where pilots have been observed to start dropping tasks in order to continue performing at a self imposed acceptable level, perhaps to allow for between-task transition time. For purposes of this paper, the 80% time might be considered to be the fully loaded condition.

As peaks exceeding the 80% time occur (or whatever limit might be chosen), diagnostics to determine the cause of the workload excesses are in order. There are several alternatives available to the analyst for this purpose.

- First, the task timeline (figure 1) for the particular time period can be inspected in detail for those tasks and events that cause the peaks. Analysis of attendant situations and conditions will provide necessary insights as to cause factors and alternative approaches.

- Second, a task sliding feature might be applied to determine if an artifact is present---if an analytic allocation requires performance at unduly stringent times and thus would be averted by a pilot in a real life condition. Some discretion may be desirable in interpretation here, to assume that the task sliding feature is reasonable.

- Third, a subsystem utilization feature can be used. Since the task time line involves given subsystems, the computer can accumulate the record of subsystem interface operations to produce a subsystems "time demand" summary, and can rank order the demand summary to facilitate diagnostics. This feature is illustrated in figure 3. The figure clearly illustrates where the heaviest demand loads exist and where the highest payoff could be attained in terms of design changes. Of course, it may be as convenient to automate some monotonous and less critical chore that applies throughout the mission (such as subsystem monitoring) to lower the overall workload level.

Other characteristics may also be of interest, such as the degree of variability in demands placed on a channel---it may be desirable to narrow the range of variability in demand for some channel in order to control a widely varying demand. Figure 4 illustrates a one sigma estimate of demand variability for each performance capability group for a mission segment. Casual inspection suggests there might be an advantage to reappraising the visual task allocations and making some adjustments in design. Admittedly, this is seldom feasible in practice, but the diagnostic implications are evident.

In overview, then, each of the diagnostic techniques offers a straight forward approach to supportable recommendations for changing procedures or design, or even for reallocating the tasks assigned to the crewmen.

<u>Cognition</u>

One of the most difficult components of workload to include in any analytic assessment technique is the cognitive component. Most early analytic techniques either ignored the cognitive factor and concentrated on the physical parameters or considered cognition as an all or none component. Newer techniques attempted to get at the variable nature of cognition.

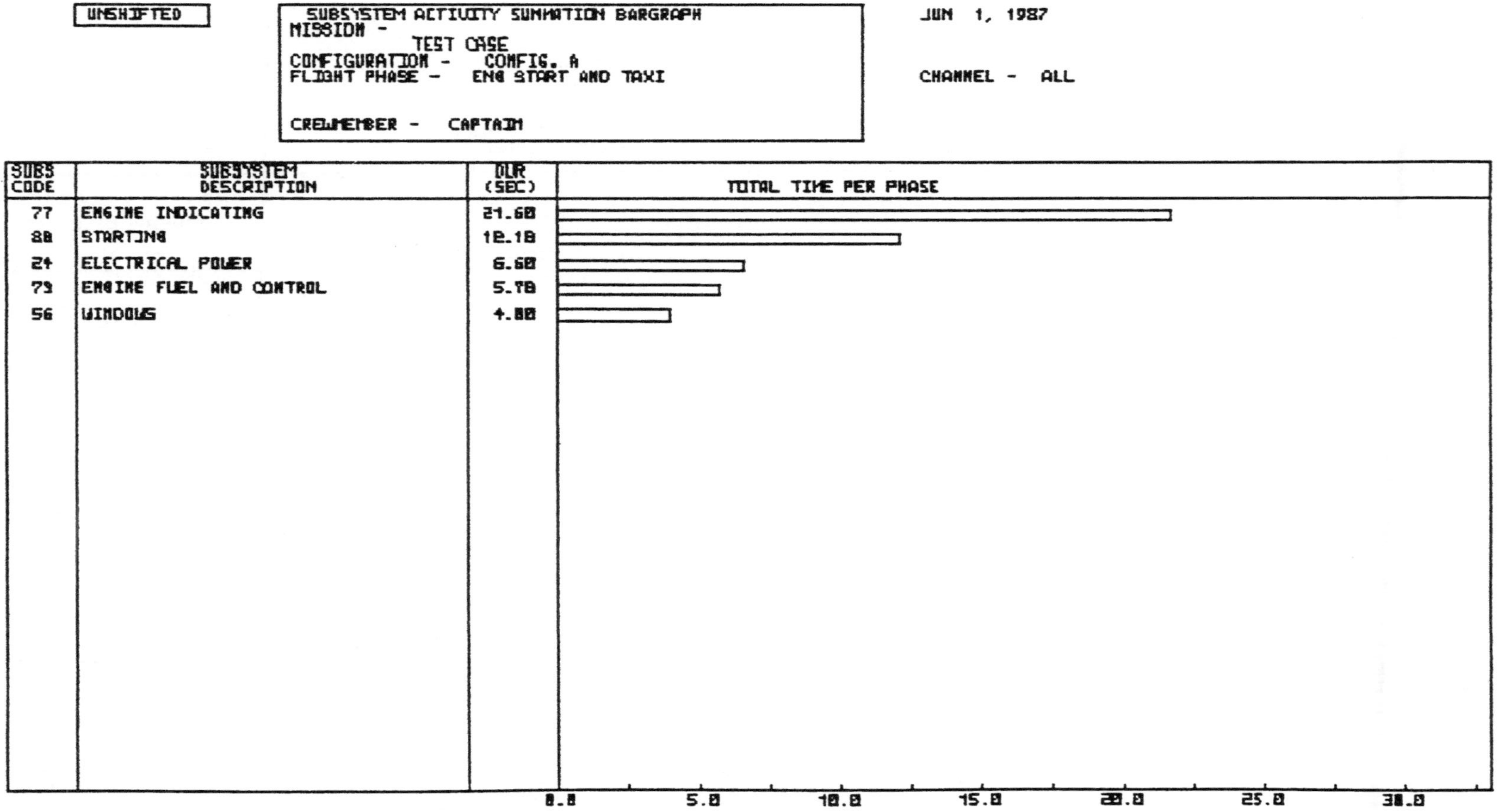

Figure 3. Subsystem Utilization Summary Produced by TLAP

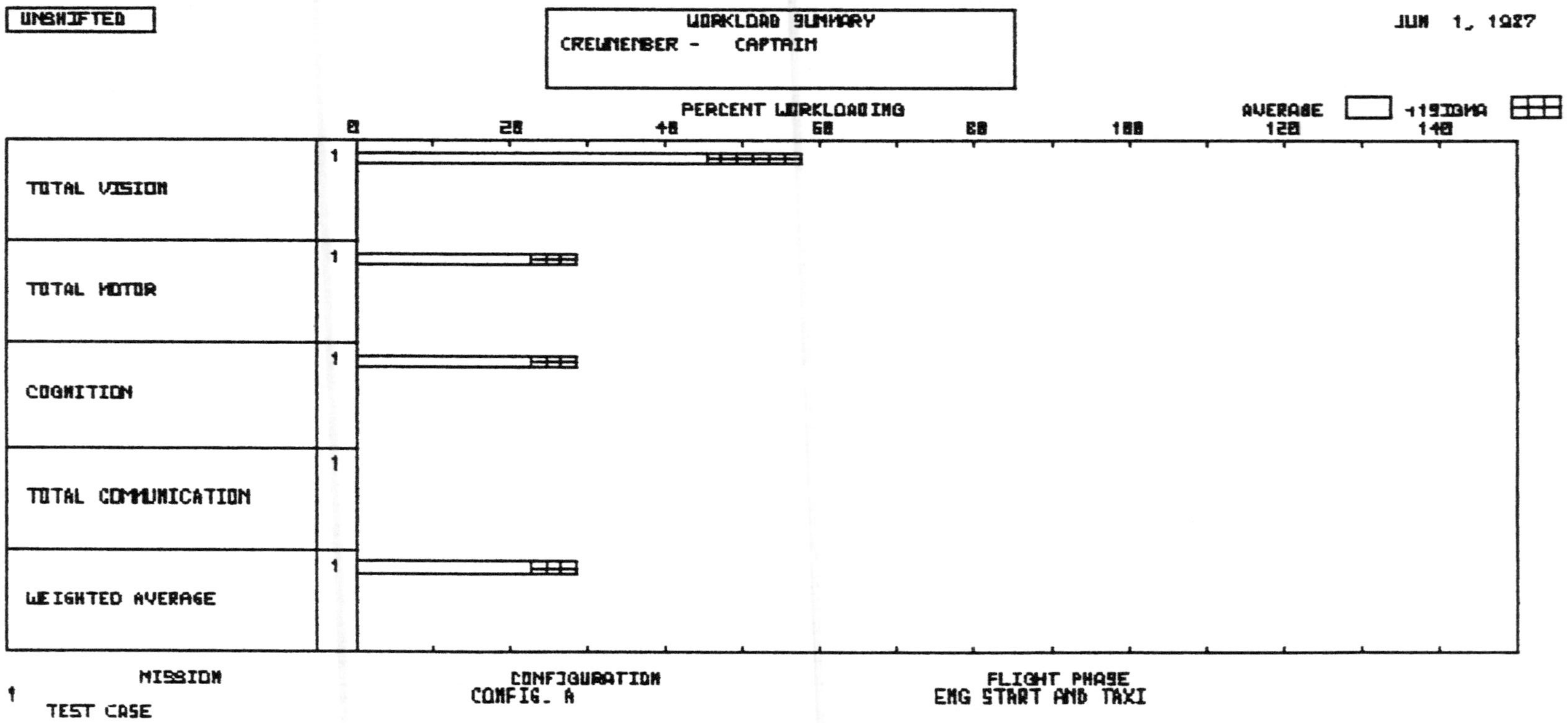

Figure 4. Summary of Channel Parameter Loads....Demand Average and Standard Deviation

One method used within Boeing and currently being used for a joint USAF/FAA program investigating workload assessment techniques uses a device complexity score as a basis for estimating cognitive workload. The device complexity score is based upon the information content of the possible states that the device presents; it is derived using the concepts of Information Theory. Information has been defined as the number of binary digits (BITS) into which an event can be encoded depending upon the number of alternatives it presents (figure 5). In this context, each display (i.e. gage, dial, flight display, etc.) is encoded based on the numerical units that have to be used. Each control is also encoded depending upon the number of choices (alternatives) that it enables. Procedure complexity is then equal to the sum of device complexity scores for all steps of a procedure. Summing device complexity scores could thus produce a figure representing total subsystem control complexity, i.e.,

$$\text{complexity} = \sum_i bi$$

where bi is the individual complexity score

$$\text{and cognition} = \sum_i Bi$$

where Bi is the corresponding cognitive workload score in seconds, from the equation

$$Bi = 0.27216 + 0.12456 \text{ bi if bi} \leq 8 \text{ bits, or}$$

$$Bi = 0.27216 + 0.22968 \text{ bi if bi} > 8 \text{ bits}$$

This technique includes information processing in the workload analysis as a major component of cognitive work. The advantages of this type of approach are that it: provides a common measure for relating the workload associated with different control and display devices; helps avoid the limitations associated with simply summing the numbers of devices; permits numerical evaluations to be made between different methods of interfacing a system or procedure with the crew. The cognition score for the timeline analysis is based upon the procedure complexity score and the communication channel time.

Timeline Robustnuess: Validation and Reliability Considerations

Although it is convenient to think of people as experiencing a similar level of workload in response to a set of fixed task demands (e.g. an average workload level), "constant" workload levels are not necessarily constant due to the individual nature of each persons actions. Certainly, many questions could be asked about the magnitude and influence of individual differences; these are typically resolved by allowing for variability in analyses and confirming the allowance in simulation. However, the current objective of the assessment task is to detect differences in workload levels, identify excessive peak loads and proceed into diagnosis and resolution. Manufacturers also need to assure that techniques to be used reliably discriminate between high and low levels of workload in order to make general conclusions. A brief discussion of status regarding some of the key questions is in order.

There are many types of validity, each affecting the ultimate usefulness and acceptability of the analytic technique. The questions to be answered are whether the timeline analysis technique really predicts what it says it predicts and how does its output relate to the actual workload being experienced in the procedures being followed by the crew? The most direct way to establish the answer to the first question is to perform the timeline analysis and then validate it by taking data in a flight test program. This was the method of choice for the Boeing 757 and 767 airplane programs. The results from these test programs indicate that the timeline analysis can accurately predict (with greater than 90 percent accuracy) the flight data for all phases of flight. Thus the technique is predicting what it says it is predicting. The second question is much more difficult to answer; discussion follows.

The Boeing Company teamed with Douglas Aircraft Company to work a program sponsored jointly by the U.S. Air Force and the Federal Aviation Administration to evaluate workload assessment techniques (Boucek, et al, 1987). One of the comparisons that was made in this study was the correlation between the predicted task loading (generated by timeline analysis) and the results of objective (physiological measures and secondary task performance) and subjective (NASA's TLX and USAF SWAT) measurement techniques. The purpose of this correlation analysis was to evaluate how much the workload scores overlap. If workload assessment techniques have been shown to be valid and reliable and they do not correlate with each other, then they are thought to be assessing different aspects of workload.

The results of this analysis show that the timeline data has not only consistent internal correlations but also interpretable correlations with the other assessment data (see figure 6). The results of the internal TLAP comparisons indicated that the visual channel data correlated well with everything (manual left .74, manual right .64, auditory .55, and cognitive .65) except the verbal channel. The manual channel data did not correlate with any other scale except the visual channel. Verbal and auditory channels were highly correlated (.78). Finally, the cognitive channel was highly correlated with the visual (.65), auditory (.85) and verbal (.65) channels.

The results of the comparison of the timeline data with the data from the other assessment techniques indicate that: the average inter-beat interval for the heart was correlated with the visual (-.55), manual left (-.64) and manual right (-.56) channels; the standard deviation for the interbeat interval correlated with the right manual (.54) channel. The Mulder Spectral Analysis Blood Pressure component correlates with the visual (.91) and the cognitive (.66) components. Wheel and column inputs correlate with the manual right (.55, .60) and the pedal inputs with the manual left (.59). Both of the subjective scales correlate with the verbal channel (SWAT .51, TLX .63). Finally, the accuracy of the response to the probe in the secondary task correlates with the manual right (.51),auditory (.51) and cognitive (.55).

SUMMARY AND CONCLUSIONS

This paper presents a progress report on use of Timeline Analysis and Prediction (TLAP) methods for workload analysis and diagnosis. It addresses questions that have been posed for the timeline approach to workload and demonstrates the credibility of the approach.

The report includes an outline of the approach and rationale to using timeline analysis as a cockpit development tool. The tool is adaptable in that it can be used in stages factored to the level of system definition. It can progress from a relatively simple level early in cockpit design when preliminary estimates might be all that are available, to very sophisticated levels with detailed design. A step by step procedure was described and methods of use were discussed. Approaches for incorporating cognition in a timeline framework were presented for further use and exploration. Validation experience was summarized, as was the extent to which other techniques correlated with applications of the tool.

<table>
<tr><th>Definition</th><th>Examples</th></tr>
<tr><td>Device complexity is the number of binary digits required to encode the possible number of alternatives associated with the device</td><td>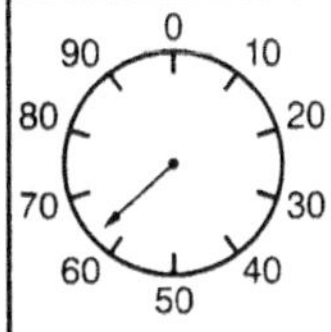

Instrument Reading
Number of Alternatives = $\frac{\text{Range}}{(0.5)(\text{Scale Unit})}$
Device Complexity = $\text{Log}_2 \frac{10}{(0.5)(1)}$ = 4.35 bits</td></tr>
<tr><td>Device Complexity =
Log_2(Number of Alternatives) bits</td><td>Rotary Selector Switch
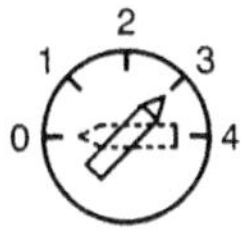

Discrete Control
Number of Alternatives = 5
Device Complexity = Log_2 (5) = 2.32 bits</td></tr>
</table>

Figure 5. Timeline Analysis Device Complexity Measure

a)	Visual	Manleft	Manright	Verbal	Auditory	Cognitive
VISUAL	1.00	**.74**	**.64**	.10	**.55**	**.65**
MANLEFT	-	1.00	.28	-.01	.25	.42
MANRIGHT	-	-	1.00	.21	.38	.49
VERBAL	-	-	-	1.00	**.78**	**.65**
AUDITORY	-	-	-	-	1.00	**.85**
COGNITION	-	-	-	-	-	1.00

b)	Visual	Manleft	Manright	Verbal	Auditory	Cognitive
HRM	**-.55**	**-.64**	**-.56**	.17	.14	-.02
HRSD	.14	.10	**.54**	-.13	-.13	-.31
APB	**-.91**	**-.87**	-.39	-.04	-.44	**-.66**
ARS	-.05	.12	.32	.11	-.08	-.26
EBK	-.11	-.03	.11	-.39	-.47	-.42
WHL	.34	.36	**.55**	.09	.15	-.01
STK	.38	.43	**.60**	-.13	-.07	-.12
PDL	**.54**	**.59**	.39	-.47	-.30	-.04
SWAT	.04	.05	.49	**.51**	.21	.20
TLX	-.02	.03	.40	**.63**	.28	.25
STRT	-.39	-.41	.14	.03	-.13	-.32
STRT%	.48	.19	**.51**	.31	**.51**	**.55**

Correlations greater ±.50 are in bold.

Legend	Task Channels	Legend	Task Channels
Visual	Eyes	Verbal	Spoken Communication
Manleft	Manual Left Hand Tasks	Auditory	Listening
Manright	Manual Right Hand Tasks	Cognitive	Cognitive Tasks

Legend	
HRM	Average Inter-beat Interval
HRSD	Standard Deviation for the Average Inter-beat Interval
APB	Mulder Spectral Analysis Blood Pressure Component from Heart Rate
ARS	Mulder Spectral Analysis Respiration Component from Heart Rate
EBK	Eye Blinks per Minute
WHL	Wheel Control Inputs per Minute
STK	Stick Control Inputs per Minute
PDL	Pedal Control Inputs per Minute
SWAT	Subjective Workload Assessment Technique
TLX	NASA Task Load Index
STRT	Secondary Task Reaction Time
STRT%	Probe Accuracy to Positive Probes for the Secondary Task

Figure 6. Correlation Matrices for the TLAP Variables: (a) Internally, and (b) with Other Methods

Both intra-tool correlations and correlations of results from applying other tools in the same context were encouraging. Magnitudes and trends were sufficiently consistent to warrant continued refinement and use of this tool. There was a very high degree of predictive accuracy compared to actual flight data and a respectable set of correlations between TLAP predictions and measurements using interbeat heartrate interval, the Mulder Blood Pressure Spectral Analysis, SWAT, and secondary task techniques.

In overview, then, continued research and development has led to beneficial refinements in the timeline approach to workload analysis, and demonstrated that the method is quite robust. It is a realistic, useful tool that has a large and growing experience base. It is useful early in design when data may be fragmentary and is easily adaptable as system changes occur. Continued evolution and refinement is warranted.

However, foreseeable changes in our work-a-day requirements will require that this and all other techniques be continually reexamined in the context of fast changing technology and issues. It will become more and more important for all techniques to be adaptable to this new environment in order to continue to produce diagnostic workload information. Experience to date with TLAP indicates that it has this capability.

Acknowledgement

This paper summarizes many years of progress on the timeline based approach to workload, as can easily be inferred from perusal of the bibliography. In addition to continuing company sponsorship, the most recent activity (Boucek, et al, 1987) was sponsored by USAF's Flight Dynamics Laboratory and the FAA, under the guidance of H. G. Britten-Austin, USAF's AFWAL/FIGR and P. K. Hwoschinsky, FAA's APS 450. Earlier efforts were sponsored by NASA, LRC (Miller, 1976), and USN, NADC (Parks and Springer, 1975).

REFERENCES

Auffret, R. (ed.), 1977, Studies on Pilot Workload, AGARD Conference Proceedings No. 217, Papers presented at the Aerospace Medical Panel Specialists' Meeting, Koln, West Germany, April.

Boucek, G. P., Jr., Sandry-Garza, D. L., and Logan, A. L., (Boeing) Biferno, M. A., Corwin, W. H. and Metalis, S., (Douglas), 1987, Proceedings of the Workshop on the Assessment of Crew Workload Measurement Methods, Techniques, and Procedures: Part Task Simulation Data Summary, AFWAL-TR-87-3103, Sept. 15-16.

Boyd, S. P., 1983, Assessing The Validity of SWAT as a Workload Measurement Instrument, United States Air Force Academy. Proceedings of the Human Factors Society, 27th Annual Meeting, 124-128.

Chiles, W. D., 1977, Objective Methods of Developing Indices of Pilot Workload, Civil Aeromedical Institute, Federal Aviation Administration, Oklahoma City, OK, July.

Cooper, G. E. and Harper, R. P., Jr., 1969, The Use of Pilot Ratings in the Evaluation of Aircraft Handling Qualities, Moffett Field, CA: NASA Ames Research Center, NASA TN D-5153.

Eggemeier, F. T., Crabtree, M. S., and Reid, G. B., 1982, Subjective Workload Assessment in a Memory Update Task, Proceedings of the Human Factors Society, 26th Annual Meeting.

Eggleston, R. G. and Kulwicki, P. V., 1984, A Technology Forecasting and Assessment Method for Evaluating System Utility and Operator Workload, Presented at the 1984 Annual Meeting of the Human Factors Society.

Fadden, D. M., 1982, Boeing Model 767 Flight Deck Workload Assessment Methodology, Paper presented at the SAE Guidance and Control System Meeting, Williamsburg, VA, November.

Gerathewohl, S. J., 1977, Inflight Measurement of Pilot Workload, Panelists: E. L. Brown, Douglas Aircraft Co., J. E. Burke, Vought Corporation, K. A. Kimball, USAMRL, S. P. Stackhouse, Honeywell, Inc. and W. Long, Bell Helicopter Co., Aerospace Medical Association Annual Scientific Meeting, Las Vegas.

Hart S. G. and Bortolussi, M. R., 1983, Pilot Errors as a Source of Workload, Paper presented at the Second Symposium on Aviation Psychology, Columbus, OH.

Hart S. G., Childress, M. E., and Hauser, J. R., 1982, Individual Definitions of the Term Workload, Paper presented at the 1982 Psychology in the DOD Symposium.

Hart S. G., and Staveland, L. E., In press, Development of a Multidimensional Workload Rating Scale: Results of Empirical and Theoretical Research, To appear in P. A.Hancock and N. Meshkati (Eds.), Human Mental Workload, Amsterdam: North Holland Press.

Hay, G. C. House, C. D., and Sulzer, R. L., 1978, Summary Report of 1977-1978 Task Force on Crew Workload. Report No. FAA-EM-78-15, U. S. Department of Transportation, Washington, D. C., December.

Linton, P. M., Jahns, D. W. and Chatelier, P. R., 1977, Operator Workload Assessment Model: An Evolution of a VF/VA-V/STOL System, AGARD-CPP-216, Aerospace Medical Panel Specialist Meeting, Koln, Germany, 18-22 April.

McLucas, J. L., Drinkwater, F. J. and Leaf, H. W., 1981, Report of the President's Task Force on Aircraft Crew Complement, Douglas Aircraft Company.

Miller, K. M., 1976, Timeline Analysis Program (TLA-1), Final Report, Boeing Document D6-42377-5, Prepared for National Aeronautics and Space Administration, Langley Research Center (NASA-CR-144942) April.

Moray, N. (Ed.), 1979, Mental Workload: Its Theory and Measurement, Plenum Press, New York.

Parks, D. L., 1978, Current Workload Methods and Emerging Challenges, in Mental Workload: Its Theory and Measurement, N. Moray (Ed.), New York: Plenum Press.

Parks, D. L. and Springer, W. E., 1975, Human Factors Engineering Analytic Process Definition and Criterion Development for CAFES, Boeing Document D180-18750-1, Prepared for Naval Air Development Center, Contract N62269-74-C-0693, June.

Parks, D. L. and Stern, P. H. and Niwa, J. S., 1965, Crew Number Study: Supporting Documentation for Advanced Manned Strategic Aircraft, Crew Factors Study, Volume III - Task Allocation Report; Boeing Document D6-16224-3, Prepared for U.S. Air Force Systems Command, Contract AF33(657)-15339, October.

Reid, G. B., Shingledecker, C. A., Nygren, T. E. and Eggemeier, F. T., 1981, Development of Multideminsional Subjective Measures of Workload, Proceedings of the 1981 IEFF International Conference of Cybernetics and Society, 403-406.

Roscoe, A. H., 1978, Assessing Pilot Workload, AGARD-AG 233, AD A051 587, Paris: NATO.

Roscoe, A. H., 1984, Assessing Pilot Workload in Flight, Paper reprinted from Conference Proceedings No. 373 Flight Test Techniques. AGARD, NATO, Neuilly-sur-Seine, France, April.

Ruggiero, F. T. and Fadden, D. M., 1987, Pilot Subjective Evaluation of Workload During a Flight Test Certification Program, Paper reprinted from Conference Proceedings No. 282, The Practical Assessment of Pilot Workload, AGARD, NATO.

Shingledecker, C. A. and Crabtree, M. S., 1982, Standardized Test for the Evaluation and Classification of Workload Metrics, Proceedings of the Human Factors Society, 26th Annual Meeting.

Sheridan, T. B. and Simpson, R. W., 1979, Toward the Definition and Measurement of the Mental Workload of Transport Pilots. Massachusetts Institute of Technology, January.

Siegel, A. I. and Wolf, J. J., 1961, Techniques for Evaluating Operator Loading in Man-Machine Systems, Applied Psychological Services.

Stone, G., Gulick, R. K. and Gabriel, R. F., 1985, Use of Task/Timeline Analysis to Assess Crew Workload, Douglas Paper 7592, Douglas Aircraft Company, Long Beach, CA.

Sulzer, R., Cox., W. J., and Mohler, S. R., 1981, Flight Crewmember Workload Evaluation, DOT/FAA/RD-82/83.

Wickens, C. D., 1980, The Structure of Attentional Resources, in Attention and Performance VIII, R. Nickerson and R. Pew (Eds.), Englewood Cliffs, New Jersey: Erlbaum.

Wierwille, W. W., 1979, Physiological Measures of Aircrew Mental Workload, Human Factors, 21, 575-593.

Wierwille, W. W., and Casali, J. G., 1983, A Validated Rating Scale for Global Mental Workload Measurement Applications. Proceedings of the Human Factors Society, 27th Annual Meeting.

Weiner, J. S., 1982, The Measurement of Human Workload, Ergonomics, 25, 953-965.

Whitley, L. C., and Vaughn, R. R., 1968, Man-Machine Stochastic Simulator TEN-708, MMSS Volume I, Boeing Document D6-29184-TN-1, April.

Whitley, L. C., and Vaughn, R. R., 1968, Man-Machine Stochastic Simulator TEN-708, MMSS Volume II, Boeing Document D6-29184-TN-II, April.

Whitmore, D. C. and Parks, D. L., 1974, Computer Aided Function-Allocation Evaluation System (CAFES), Phase IV, Final Report, Boeing Document D180-18433-1, Prepared for Naval Air Development Center, Johnsville, (Contract N62269-74-C-0274), December.

THE DEVELOPMENT AND APPLICATION OF MODELS TO PREDICT OPERATOR WORKLOAD DURING SYSTEM DESIGN*

Theodore B. Aldrich

Anacapa Sciences Inc.
Fort Rucker, Alabama

Sandra M. Szabo

Anacapa Sciences Inc.
St. Louis, Missouri

Carl R. Bierbaum

Anacapa Sciences Inc.
Fort Rucker, Alabama

INTRODUCTION

New weapons being developed for modern military forces feature advanced technology designed to extend and improve mission performance beyond the capability of existing systems. For example, aircraft systems are being developed with advanced technology designed to extend range, increase speed, provide for more precise navigation, avoid enemy threats, and acquire and engage enemy targets at night or in adverse weather.

In addition to improving mission capability, advanced technology also is designed to reduce crew workload. However, in some instances the tasks required to operate the technology may actually increase workload. The increased workload, in turn, may degrade human performance and, consequently, reduce rather than improve mission effectiveness.

Models that predict operator workload can be useful tools for human factors engineers who are attempting to address human capabilities and limitations as advanced technology is introduced into new weapon systems. In response to this requirement Anacapa Sciences, Inc. researchers, under contract to the U.S. Army Research Institute Aviation Research and Development Activity, have developed a series of models for predicting aviator workload. The work supports U.S. Army design studies for the following helicopter systems:

- a highly automated, multipurpose, lightweight helicopter, designated LHX (see Aldrich, Szabo, & Craddock,1986);

---

*Research reported in this chapter was performed in support of the Army Research Institute Aviation Research and Development Activity, Fort Rucker, Alabama, under Contract Numbers MDA903-81-C-0504 and MDA903-87-C-0523. Mr. Charles A. Gainer was the technical monitor. The views expressed are those of the authors and are not necessarily endorsed by the U.S. Army.

- the AH-64A, Apache (see Szabo & Bierbaum, 1986);
- the UH-60A, Blackhawk (see Bierbaum, Szabo, & Aldrich, 1987); and
- the CH-47D, Chinook.

The LHX models were used in advance of system design to predict single- and dual-crew workload under varying automation configurations. The AH-64A, UH-60A, and CH-47D models presently are being used for evaluating the impact of advanced technology modifications being proposed for each of these existing helicopters.

This chapter describes a four-phase research program aimed at the development and application of models to predict operator workload during system design. Phase 1 consists of the development of a mission/task/ workload analysis data base. Phase 2 consists of the development of computer models to predict operator workload. Phase 3 consists of applying the workload prediction models during system design studies. Phase 4 consists of research required to validate the workload predictions yielded by the models. Most of this chapter describes research performed in support of the LHX as reported by Aldrich, Szabo, and Craddock (1986). However, refinements in the methodology introduced by Szabo and Bierbaum (1986) in support of proposed AH-64A modifications, and by Bierbaum, Szabo, and Aldrich (1987) in support of proposed UH-60A modifications, also are included. Thus, this chapter presents the current state of the Anacapa Sciences, Inc. research directed at developing and validating operator workload prediction models.

THE MISSION/TASK/WORKLOAD ANALYSIS DATA BASE

The first phase of the workload prediction methodology requires the conduct of a comprehensive mission/task/workload analysis. In the case of the LHX, 24 proposed scout and attack mission profiles provided by the Directorate of Combat Developments (DCD) at the U.S Army Aviation Center (USAAVNC), Fort Rucker, Alabama were examined. Because of program schedule constraints only nine of the 24 mission profiles were selected for preliminary analysis. The nine mission profiles were subsequently divided into mission phases; the following three mission phases were selected for detailed analysis:

- Reconnaissance,
- Target Service (Air-To-Ground), and
- Target Service (Air-To-Air).

Each of the three mission phases listed above was further divided into segments; a limited sample of 29 mission segments was selected for the detailed task analysis. Each of the 29 mission segments, in turn, was divided into mission functions. Finally, each of the mission functions was divided into mission tasks. A total of 58 unique functions and 135 unique tasks were identified for the 29 mission segments that were analyzed.

The same general procedure was used to conduct the AH-64A, UH-60A, and CH-47D analyses. In the case of the AH-64A, Szabo and Bierbaum (1986) conducted a comprehensive analysis of an entire composite mission from preflight through postflight. They identified 52 unique segments, 159 unique functions, and 689 unique tasks. In the UH-60A analysis, Bierbaum, Szabo, and Aldrich (1987) identified 34 critical segments, which were subsequently divided into 48 unique functions and 138 unique tasks. In the CH-47D analysis, 37 critical mission segments were divided into 65 unique functions and 154 unique tasks.

For each unique task, the following additional data were derived:

- crewmember performing the task,
- subsystem equipment associated with the task,
- estimate of the time required to perform the task, and

- estimates of workload associated with the sensory (i.e., visual, auditory, kinesthetic) cognitive, and psychomotor components of the task.[1]

Procedures for deriving the additional task data are briefly described in the paragraphs that follow.

Crewmember Performing the Task

The first step in deriving the additional task data was to identify the crewmember performing each task. All flight control tasks were assigned to the pilot. Primary mission tasks (e.g., Align Sight, Activate Trigger) and most support tasks (e.g., Check Aircraft Systems, Transmit Message) were assigned to the second crewmember.

Subsystem Equipment Associated With the Task

The next step in the analyses was to identify the subsystem equipment associated with the performance of each task. In each analysis the identified subsystem equipments were categorized into major subsystem categories. The categories vary among the different systems depending upon (a) the mission assigned to the particular aircraft of interest and (b) the existing configuration of that aircraft.

Estimate of the Time Required to Perform the Task

The methods of estimating task times also varied somewhat for the different systems. Aldrich, Craddock, and McCracken (1984) describe the methods for estimating task times in the LHX analyses. In their analyses, each task was first categorized as discrete or continuous. Discrete tasks are characterized by actions having a definite, observable start and end point. Activation of switches, performance of procedures, and transmissions of radio messages are examples of discrete tasks. Existing helicopter task analyses for the OH-58D (Taylor & Poole, 1983) and for the AH-64 and the Advanced Helicopter Improvement Program (AHIP) (Siegal, Madden, & Pfeiffer, 1985) were used as references in deriving estimates of LHX discrete task times.

Continuous tasks do not have observable start and end points and cannot be reduced to procedures; mission requirements and conditions determine their duration. Examples of continuous tasks are flight control tasks and target tracking tasks. Aldrich, Craddock, and McCracken (1984) assigned times to continuous tasks so that each discrete task could be accomplished within the elapsed times assigned to concurrent continuous tasks. For example, the times assigned to the continuous tasks associated with the Hover Masked function, were long enough to allow the operator to complete all of the discrete tasks (e.g., Check Aircraft Systems, Transmit Message) performed concurrently with the continuous tasks in the Hover Masked function. All assigned times for discrete and continuous tasks were reviewed by subject matter experts (SMEs)

During the AH-64A analysis, Szabo and Bierbaum (1986) identified two types of discrete tasks. Specifically, they categorized discrete tasks as either "discrete fixed" or "discrete random". Discrete fixed tasks have definite start and end points within the function (e.g., Set SIGHT SEL Switch). Discrete random tasks are discrete tasks that occur intermittently and/or randomly during a portion of the function (e.g.,

---

[1]Estimates of the kinesthetic workload component of tasks were introduced during the AH-64A analysis by Szabo and Bierbaum (1986). The higher specificity of their task analysis, compared to the LHX analyses, required the kinesthetic estimates. The kinesthetic estimates were retained by Bierbaum, Szabo, and Aldrich (1987) in their UH-60A analysis and are currently being used in the CH-47D analysis.

Check Fuel Quantity Indicator). Szabo and Bierbaum derived most of their task times by timing the actual tasks as they were performed in the AH-64A Cockpit, Weapons, and Emergency Procedures Trainer. For tasks not trainable in the trainer they used estimates provided by AH-64A SME's. Bierbaum, Szabo, and Aldrich (1987) retained the refined categorization of discrete tasks for the UH-60A analysis. UH-60A task time estimates were obtained during interviews with UH-60A SMEs.

Estimates of Workload Associated With the Sensory, Cognitive, and Psychomotor Components of the Task

Workload, as the term is used in this research, is defined as the total attentional demand (i.e., mental workload) placed on the operator(s) as they perform the mission tasks. Consistent with Wickens theory that workload is a multidimensional construct, the research methodology addresses three different components of workload; sensory, cognitive, and psychomotor (Wickens, 1984). The sensory component refers to the complexity of the visual (V), auditory (A), or kinesthetic (K) stimuli to which an operator must attend; the cognitive (C) component refers to the level of information processing required from the operator; the psychomotor (P) component refers to the complexity of the operator's behavioral responses. The steps performed to determine the workload associated with each of these components for each of the mission tasks are described in the paragraphs that follow.

McCracken and Aldrich (1984) estimated LHX task workload by using 7-point ordinal scales for rating the visual, cognitive, and psychomotor workload components and a 4-point ordinal scale for rating the auditory workload components of each task. Szabo and Bierbaum (1986) added a kinesthetic sensory component to their analysis of workload, and developed an ordinal 7-point kinesthetic rating scale with verbal anchors similar to the visual, cognitive, and psychomotor rating scales. They also developed an ordinal 7-point auditory rating scale to replace the original 4-point auditory rating scale used by McCracken and Aldrich.

During the UH-60A analysis, Bierbaum, Szabo, and Aldrich (1987) added a second visual scale and converted the ordinal scale measures to interval scale measures. The second visual scale was added so that the attentional demand associated with the visual component of the mission tasks could be estimated under both naked eye (visual-unaided) and night vision goggle (visual-aided) conditions. Both visual scales retain the same verbal anchors used in the prediction of AH-64A crew workload.

The interval scales used in the UH-60A analysis were constructed by using a pair comparison survey methodology (Engen, 1971). The survey presented matched pairs of verbal anchors for the visual (both naked eye and night vision goggles), auditory, cognitive, and psychomotor workload component scales to 20 UH-60A instructor pilots (IPs) from the UH-60A Aviator Qualification Course (AQC) at the USAAVNC, Fort Rucker, Alabama. The frequency with which the IPs selected each verbal anchor was used to compute a value for each verbal anchor on an approximately equal-interval scale.

The matched pairs of verbal anchors for the kinesthetic workload component scale were similarly arranged in a questionnaire and administered by mail to a group of 22 human factors experts who have had extensive research experience in workload measurement. Pair comparison response frequencies were tabulated to develop interval scale values for the kinesthetic workload component scale. The six workload component interval scales used in the UH-60A analysis are presented in Table 1.

Once the workload component scales had been developed, a short verbal descriptor of each of the workload components was written for each task. The descriptors were then compared to the verbal anchors in the appropriate interval or rating scale. In each instance, a consensus was

Table. 1. Workload Component Scales for the UH-60A Mission/Task/Workload Analysis

Scale Value	Descriptors
	Visual-Unaided (Naked Eye)
1.0	Visually Register/Detect (Detect Occurrence of Image)
3.7	Visually Discriminate (Detect Visual Differences)
4.0	Visually Inspect/Check (Discrete Inspection/Static Condition)
5.0	Visually Locate/Align (Selective Orientation)
5.4	Visually Track/Follow (Maintain Orientation)
5.9	Visually Read (Symbol)
7.0	Visually Scan/Search/Monitor (Continuous/Serial Inspection, Multiple Conditions)
	Visual-Aided (Night Vision Goggles [NVG])
1.0	Visually Register/Detect (Detect Occurrence of Image) With NVG
4.8	Visually Inspect/Check (Discrete Inspection/Static Condition (With NVG)
5.0	Visually Discriminate (Detect Visual Differences) With NVG
5.6	Visually Locate/Align (Selective Orientation) With NVG
6.4	Visually Track/Follow (Maintain Orientation) With NVG
7.0	Visually Scan/Search/Monitor (Continuous/Serial Inspection, Multiple Conditions (With NVG)
	Auditory
1.0	Detect/Register Sound (Detect Occurrence of Sound)
2.0	Orient to Sound (General Orientation/Attention)
4.2	Orient to Sound (Selective Orientation/Attention)
4.3	Verify Auditory Feedback (Detect Occurrence of Anticipated Sound)
4.9	Interpret Semantic Content (Speech)
6.6	Discriminate Sound Characteristics (Detect Auditory Differences)
7.0	Interpret Sound Patterns (Pulse Rates, Etc.)
	Kinesthetic
1.0	Detect Discrete Activation of Switch (Toggle, Trigger, Button)
4.0	Detect Preset Position or Status of Object
4.8	Detect Discrete Adjustment of Switch (Discrete Rotary or Discrete Lever Position)
5.5	Detect Serial Movements (Keyboard Entries)
6.1	Detect Kinesthetic Cues Conflicting with Visual Cues
6.7	Detect Continuous Adjustment of Switches (Rotary Rheostat, Thumbwheel)
7.0	Detect Continuous Adjustment of Controls
	Cognitive
1.0	Automatic (Simple Association)
1.2	Alternative Selection
3.7	Sign/Signal Recognition
4.6	Evaluation/Judgment (Consider Single Aspect)
5.3	Encoding/Decoding, Recall
6.8	Evaluation/Judgment (Consider Several Aspects)
7.0	Estimation, Calculation, Conversion

(continued)

Table. 1. Workload Component Scales for the UH-60A Mission/Task/Workload Analysis (Continued)

Scale Value	Descriptors
	Psychomotor
1.0	Speech
2.2	Discrete Actuation (Button, Toggle, Trigger)
2.6	Continuous Adjustive (Flight Control, Sensor Control)
4.6	Manipulative
5.8	Discrete Adjustive (Rotary, Vertical Thumbwheel, Lever Position)
6.5	Symbolic Production (Writing)
7.0	Serial Discrete Manipulation (Keyboard Entries)

reached by the two analysts who initially had assigned the workload estimates independently. The consensual estimates were subsequently reviewed by SMEs for the selected system.

A complete summary of the data derived from the mission/task/workload analysis was entered on function analysis worksheets, such as the one selected from the AH-64A analysis (Szabo and Bierbaum, 1986) and depicted in Figure 1. A separate worksheet was prepared for each unique function identified in each analysis. The verb and object for each task within the function are presented in the first two columns, respectively. The crewmember performing each task is indicated by the letter (i.e. Pilot [P], Gunner [G] or Both [B]) in the third column. The subsystems associated with each task are shown in the fourth column. The verbal descriptors and the numerical estimates of workload for the sensory, cognitive, and psychomotor components (i.e., Visual-Unaided [V], Visual-Aided [G], Auditory, [A], Kinesthetic [K], Cognitive [C] and Psychomotor [P]) of each task are shown in the fifth, sixth, and seventh columns. For each task involving a specific switch, a switch description is presented in the eighth column. The estimated length of the discrete tasks is presented in the ninth column. The continuous tasks are identified in the tenth column with the letter "c." The function analysis worksheets thus provide a comprehensive summary of the information used to establish the data base for developing the workload prediction models in Phase 2 of the research.

FUNCTION 54 Designate Target (Autonomous) **TOTAL TIME (Approximate) 13.5 Seconds**

TASKS		ID #	SUBSYSTEM(S)	WORKLOAD COMPONENTS			SWITCH DESCRIPTION	DURATION (SECONDS) DISCRETE/ CONTINUOUS	
VERB	OBJECT			SENSORY	COGNITIVE	PSYCHOMOTOR			
Monitor	HAD Message (TOF)	G296	Fire Control Computer (AFC)	Read Symbolic Display V-7	Interpret Symbolic Readout and Make Judgment (Time to Lase) C-5			5	
Pull	Laser Trigger	G361	Laser (AL)	Feel Trigger Movement K-2	Verify Correct Position (Laser Activated) C-2	Lift Cover; Pull and Hold Trigger P-1	Springloaded Trigger (SPTR)	1	
Note	Weapon Impact	G639	Sensor Display (VSD)	Visually Detect Image V-1	Evaluate Sensory Feedback and Make Judgment (Target Destroyed) C-5			5	
Release	Laser Trigger	G362	Laser (AL)	Feel Trigger Movement K-2	S-R Association C-1	Release Trigger P-1	Springloaded Trigger (SPTR)	.5	

Figure 1. AH-64A function analysis worksheet.

DEVELOPMENT OF COMPUTER BASED WORKLOAD PREDICTION MODELS

Phase 2 of the methodology consists of developing computer models to predict total workload experienced in the performance of individual and concurrent tasks. Whereas the mission/task/workload analysis methodology follows a top-down approach, the computer models are developed using a bottom-up approach. The task data identified during Phase 1 constitute the basic elements of analysis. The steps required to develop the models follow:

- establish computer data files,
- write function and segment decision rules, and
- write computer programs.

Each of these steps is described briefly in the subsections below.

Establish Computer Data Files

The first step in developing each of the workload prediction models is to enter the mission/task/workload data derived during Phase 1 into computer files. Specifically, the information summarized on the function analysis worksheets is used to create the following data files:

- a list of segments,
- a list of functions,
- a list of tasks,
- a list of subsystem identifiers,
- workload ratings, and
- time estimates.

Develop Function and Segment Decision Rules

The next step in developing the workload prediction models is to write time-based decision rules for building the mission segments from the task data base. Function decision rules specify the sequence and time for the performance of each task within each function; segment decision rules specify the sequence and temporal relationships for combining the functions to form mission segments. For the LHX analyses, Aldrich, Craddock, and McCracken (1984) developed one set of segment decision rules for a one-crewmember configuration and a second set of decision rules for the two-crewmember LHX configuration. Szabo and Bierbaum (1986) developed a single set of segment decision rules for the AH-64A analysis, and Bierbaum, Szabo and Aldrich (1987) developed another set of segment decision rules for the UH-60A analysis.

Write Computer Programs

The time-based function and segment decision rules are the blueprints for placing the tasks performed by the operator(s) at the appropriate point on the mission timeline. Computer programs are developed to implement the function and segment decision rules. The timeline produced by programming the function and segment decision rules enables the identification of all tasks performed by the operator(s) at each half-second interval in the mission segment.

Computer programs also are developed for producing estimates of total workload associated with the performance of concurrent and sequential tasks. The total workload for concurrent tasks is computed by summing the workload component ratings (i.e., visual, auditory, kinesthetic, cognitive, and psychomotor) assigned during the task analyses. The specific half-second intervals when excessive workload occurs can be identified on the segment timeline by referring to the workload component sums. Four indices of overload producible by the model have been developed (Aldrich, Craddock, & McCracken 1984) and are listed and defined below:

- A component overload occurs whenever the sum of the ratings assigned to a given workload component (i.e., visual, auditory, kinesthetic, cognitive, or psychomotor) for concurrent tasks equals "8" or higher. Thus as many as five component overloads may occur for two or more concurrent tasks. A value of "8" was chosen as the criterion for an overload because it exceeds the maximum value on any of the workload component rating scales.

- An overload condition occurs whenever a component overload, as defined above, occurs in at least one component of the concurrent tasks. In theory as many as five component overloads (i.e., visual, auditory, kinesthetic, cognitive, and psychomotor) may occur within a single overload condition.

- Overload density is the percentage of time during a mission segment that a component overload occurs. It is calculated by dividing the number of timelines with component overloads by the total number of timelines in the segment.

- The term subsystem overload is used to describe the relationship between a component overload and a subsystem. It is computed by tallying the number of times each subsystem is associated with a component overload.

The component overload, overload condition, and subsystem overload indices provide diagnostic information about excessive workload for concurrent tasks. The overload density index provides a potential diagnostic measure of cumulative workload associated with sequences of concurrent tasks.

Following the steps described above, Aldrich, Craddock, and McCracken (1984) developed both one- and two-crewmember baseline workload prediction models for LHX analyses. Workload prediction models also have been developed for the AH-64A (Szabo & Bierbaum, 1986) and for the UH-60A (Bierbaum, Szabo, & Aldrich, 1987). These baseline workload prediction models provide benchmarks for comparisons to be made when the models are exercised to predict workload for alternative crew configurations or proposed automation options.

APPLYING THE WORKLOAD PREDICTION MODELS DURING SYSTEM DESIGN STUDIES

The third phase of the research consists of exercising the workload prediction models and applying the results to system design studies. This section describes the third phase of the research and presents some of the results produced from applying the LHX workload prediction models.

Workload Predictions: One- vs Two-Crewmember LHX Baseline Configurations

The one- and two-crewmember baseline LHX workload prediction models were developed using the data base compiled during the LHX mission/task/workload analysis. The tasks, subsystems, workload ratings, and time estimates are identical in both models and the function and segment decision rules were written so that both models have identical timelines. The only difference between the two models is the allocation of the functions between the crewmembers. Thus, workload predictions produced by the one-crewmember baseline model can be compared with workload predictions produced by the two-crewmember model to provide estimated differences in operator workload between the one- and two-crewmember LHX configurations.

Results summarized in Table 2 indicate that, for the 29 LHX segments, there were 263 overload conditions in the baseline one-crewmember configuration and 43 overload conditions in the baseline two-crewmember configuration. The 263 overload conditions in the

Table 2. Frequency of Overload Conditions and Component Overloads: One and Two-Crewmember LHX Baseline Configuration

	Number of Overload Conditions	Number of Component Overloads				
		V	A	C	P	Total
One-Crewmember	263	79	--	54	203	336
Two-Crewmember	43	21	--	17	15	53

one-crewmember configuration are composed of 79 visual component overloads, 54 cognitive component overloads, and 203 psychomotor component overloads, for a total of 336 component overloads. The 43 overload conditions in the two-crewmember configuration are composed of 21 visual component overloads, 17 cognitive component overloads, and 15 psychomotor component overloads, for a total of 53 component overloads. In the one-crewmember configuration overload conditions were predicted in each of the 29 segments that were analyzed. In the two-crewmember configuration, overload conditions were predicted in only 15 of the 29 segments; the pilot was overloaded in only three of these 15 segments.

Workload Predictions: One- vs Two-Crewmember Configurations With Proposed Automation Options

The next step in the LHX analyses was to exercise the one- and two-crewmember models to predict how much operator workload would be reduced by individual automation options and combinations of options being considered for the LHX design. The methodology consists of three tasks:

- selecting automation options to be exercised by the models,
- revising the estimates of workload for each task, and
- exercising the one- and two-crewmember computer models.

Selecting the automation options. As part of the Army's LHX trade-off studies, the DCD at the USAAVNC, Fort Rucker, Alabama, developed alternative mission equipment packages (MEP) and aircraft survivability equipment (ASE) packages for the LHX. The MEP and ASE consisted of advanced technology equipments designed to automate many of the crew functions. The MEP and ASE descriptions were reviewed by Anacapa analysts and human factors specialists assigned to the DCD. Twenty-six individual automation options of interest were selected for analysis.

Revising the workload estimates. The next step in applying the methodology was to determine how each of the automation options would affect operator workload. A review of the task descriptions and the generic subsystems reported on the function analysis worksheets provided clues about how the workload would be affected by each of the proposed automation options. Based on the review, new descriptors of the operator's activities were entered into the sensory, cognitive, and psychomotor columns of the worksheets. The revised descriptors were then used to assign new estimates of workload to each component of those tasks affected by the automation options. In cases where automation completely eliminated a task, zero ratings were assigned to the workload components. No time estimates were changed as a function of automation; therefore, the decision rules for building functions from tasks and for building segments from functions remained unchanged.

Exercising the models with the automation options. Following revision of the workload estimates, new computer files were built to reflect the impact of each of the 26 automation options. Subsequently, the one- and two-crewmember models were exercised using the new files to predict workload associated with each of the 26 individual automation options and 16 different combinations of the individual automation options.

Table 3 presents results from exercising the one-crewmember workload model with the five individual automation options that produced the greatest reductions in workload. The Hover Hold and Automatic Sight Alignment options ranked highest with a 41.8% and 33.5% reduction in overload conditions and a 41.7% and 30.1% reduction in component overloads, respectively.

Table 4 presents results from exercising the two-crewmember model with the five automation options that produced the greatest reductions in workload. The Automatic Sight Alignment and Automatic Target Tracking options ranked highest with a 37.2% and 32.6% reduction in overload conditions and a 39.6% and 35.8% reduction in component overloads, respectively. The highest ranking option in the one-crewmember analysis, Hover Hold, reduced no overload conditions in the two-crewmember analysis.

Table 5 presents results from exercising the one- and two-crewmember models with a combination of all 26 individual automation options. The combination of 26 automation options reduced overload conditions 96.2% and component overloads 97% in the one-crewmember analysis. Reductions in psychomotor component overloads contributed the most (62%) to the reduction in total component overloads. The combination of 26 automation options reduced all of the overload conditions and component overloads in the two-crewmember analysis[2]. Reductions in visual and cognitive component overloads contributed more (39.6% and 32.1%, respectively) to the reduction in total component overloads than the reductions in psychomotor component overloads (28.3%).

Table 3. Workload Reduction From Five Highest Ranking Automation Options, One-Crewmember LHX Configuration

Automation Configuration	% Reduction in Overload Conditions N = 263	% Reduction in Total Component Overloads N = 336
Hover Hold	41.8	41.7
Automatic Sight Alignment	33.5	30.1
Automatic Target Tracking	16.0	19.6
Voice Recorder for Message Entry During Low Workload Intervals	5.7	4.8
Automatic Updating of Position	5.3	5.7

Table 4. Workload Reduction From Five Highest Ranking Automation Options, Two-Crewmember LHX Configuration

Automation Configuration	% Reduction in Overload Conditions N = 43	% Reduction in Total Component Overloads N = 53
Automatic Sight Alignment	37.2	39.6
Automatic Target Tracking	32.6	35.8
Automatic Updating of Position	18.6	20.8
Automatic Maneuver NOE	16.3	13.2
Automatic Display of Location Relative to Selected Waypoints	11.6	15.1

[2]Results from another analysis indicated that all of the overload conditions in the two-crewmember baseline LHX model could be eliminated with a combination of only nine automation options (Aldrich, Szabo, & Craddock, 1986).

Table 5. Workload Reduction From a Combination of 26 Automation Options: One- and Two-Crewmember Analyses

	% Reduction in Overload Conditions	% Reduction in Component Overloads	Relative Contribution to Overload Reductions (%)		
			V	C	P
Automation Configuration	N = 263	N = 336	N=79	N=54	N=203
A Combination of 26 Automation Options--One-Crewmember Analysis	96.2	97.0	23.3	14.7	62.0
Automation Configuration	N = 43	N = 53	N = 21	N = 17	N = 15
A Combination of 26 Automation Options--Two-Crewmember Analysis	100.0	100.0	39.6	32.1	28.3

The results presented in Tables 2 through 5 demonstrate how the models can be used to conduct comparability analyses of operator workload for various crew and automation configurations. Similar analyses will be conducted for automation options being proposed for the AH-64A, UH-60A, and CH-47D aircraft.

RESEARCH REQUIRED TO VALIDATE THE WORKLOAD PREDICTION MODELS

Phase 4 consists of research required to validate the workload parameters used to develop the models and the workload predictions yielded by the models. Workload parameters that require validation include the:

- workload ratings assigned to each task,
- total workload estimates for concurrent tasks,
- estimated time required to perform each task,
- threshold for excessive workload,
- temporal relationships among tasks, and
- sequential relationships among tasks.

Specific predictions yielded by the models that require validation include the four indices of excessive workload described above.

A research plan (Aldrich & Szabo, 1986) describes the research required to validate the LHX workload prediction model. Although the research plan was developed specifically for the LHX, it can also guide research required to validate the AH-64A or UH-60A workload prediction models.

The validation research consists of three phases. During Phase 1, the reliability of the workload rating scales and the workload predictors are established. During Phase 2, validation data are collected through a series of studies employing part-mission and full-mission simulation. During Phase 3, the results from Phases 1 and 2 are used to refine the workload prediction model. Each of the three phases is described briefly below.

<u>Establish the Reliability of the Workload Rating Scales</u>

To accomplish this objective two surveys are required. The first survey presents pair comparisons of the verbal anchors for each workload rating scale to SMEs. The SMEs must choose the anchor in each pair that imposes more attentional demand. The survey results indicate the degree of agreement among the SMEs and also can be used to produce equal-interval scales (Engen, 1971) to replace the ordinal scales that were used in the original workload analysis.

The first survey has been conducted for the LHX, AH-64A and UH-60A workload prediction models. In the case of the LHX and AH-64A, a consensus set of verbal anchors was developed for each of the five workload component scales. A survey instrument, comprising all pair comparison combinations of the consensus verbal anchors from each workload rating scale, was produced. The pair comparison survey was mailed to 71 human factors researchers and practitioners who are SMEs in workload research. The data from 38 completed surveys were used to develop each rater's rank order judgments of the verbal anchors. The rank ordered judgments were analyzed using Kendall's Coefficient of Concordance (Siegal, 1956) to assess the degree of agreement among the SMEs. The Coefficients of Concordance for the five scales are as follow:

- Visual - .39,
- Auditory - .46,
- Kinesthetic - .38,
- Cognitive - .69, and
- Psychomotor - .47

All of the above Coefficients of Concordance are significant at the .001 level, indicating a degree of consensus among the SMEs.

Bierbaum, Szabo, and Aldrich (1987) performed a similar analysis for the UH-60A workload component rating scales. They developed a pair comparison survey and personally presented the matched pairs of verbal anchors for the visual (both naked eye and night vision goggles), auditory, cognitive, and psychomotor workload component scales[3] to UH-60A IPs from the UH-60A AQC at the USAAVNC, Fort Rucker, Alabama. The data were used to develop each rater's rank order judgments of the verbal anchors. The Coefficients of Concordance for the five scales are as follow:

- visual, no goggles - .25 (19 IPs),
- visual, with night vision goggles - .18 (19 IPs),
- auditory, - .32 (14 IPs),
- cognitive, - .45 (11 IPs), and
- psychomotor,- .46 (15 IPs).

Although these Coefficients of Concordance are smaller than those computed from the LHX and AH-64A data, they also are significant at the .001 level. Thus, the coefficients indicate some degree of agreement among the IPs who provided the ratings.

The second survey has not yet been developed. It will ask SMEs to use the verbal anchors in the workload scales to rate the short descriptors of visual, auditory, kinesthetic, cognitive, and psychomotor workload components for each task in the model. Correlational techniques will be used to evaluate the interrater reliability of the workload ratings.

<u>Employ Flight Simulation Research to Validate the Workload Prediction Model Parameters</u>

Part-mission and full-mission simulation experiments will be required to validate the workload estimates produced by the models. For the part-mission simulation, mini-scenarios will be generated by selecting concurrent and sequential tasks from the mission and task

[3]The survey did not include verbal anchors from the kinesthetic scale because the analysts doubted that IPs would be able to distinguish between levels of attentional demand for the kinesthetic verbal anchors.

analysis. For the full-mission simulation, a composite mission scenario will be developed by selecting segments from the mission and task analysis.

The part-mission simulation will be conducted using a repeated measures experimental design in which each subject will fly the mini-scenarios multiple times. Results will be analyzed to assess the correlation between the workload model predictions and measures of the operators' performance on the concurrent and sequential tasks. The correlation coefficients will serve as the primary measure of how accurately the workload predictions forecast excessive workload at the task level of specificity. To assess the validity of the time estimates used in the model, the actual amount of time required to perform the various tasks in the mini-scenarios will be compared with the times estimated during the task analysis. The sequential relationships among the tasks will be evaluated by noting the subjects' ability to progress through the mini-scenarios following the sequence of tasks specified by the model.

During the full-mission simulation experiments, each trial will start at the beginning of a composite scenario and continue without interruption to the end. Analysis of results will include all of the analyses performed during the part-mission simulation data analyses. In addition, an analysis will be performed to assess the effects of inserting secondary tasks into the composite mission scenario.

The planned experiments have not been conducted because a flight simulation facility capable of supporting the part-mission and full-mission simulation studies has not been available. However, the new Crew Station Research and Development Facility (CSRDF), located at the Army's Aeroflightdynamics Directorate, NASA Ames, California recently procured a high-technology generic flight simulator that is ideally configured for validating the LHX workload prediction model. A high fidelity AH-64 flight simulator at McDonnell Douglas Helicopter Company or an Army AH-64A Combat Mission Simulator may become available for performing research required to validate the AH-64A workload prediction model.

<u>Refine the Workload Prediction Models</u>

Refinement of the workload prediction models has been on-going since the original LHX workload prediction models were completed. Improvements introduced during the development of the AH-64A model include:

- a model of the entire AH-64A combat mission, from preflight through postflight,
- a more granular mission/task/workload analysis at the switch and display element level of specificity,
- development of a scale for rating the kinesthetic workload component of mission tasks,
- expansion of the existing 4-point scale to a 7-point scale for rating the auditory component of mission tasks,
- categorization of discrete tasks into discrete fixed and discrete random tasks,
- analysis of visual workload component specifiers, internal viewing vs. external viewing, for identifying possible visual workload clashes for concurrent operator tasks,
- analysis of psychomotor workload component specifiers, left hand vs right hand, for identifying possible psychomotor workload clashes for concurrent operator tasks, and

- a listing of the type of switch for each task that involves a switch operation.

Improvements introduced during the development of the UH-60A workload prediction model include:

- development of a visual-aided workload component scale for rating visual workload while using night vision goggles, and
- development of equal-interval rating scales to replace the ordinal scales in the LHX and AH-64A workload models.

During the validation research, additional refinements will occur. The data from the pair comparison survey will be used to produce equal-interval rating scales to replace the ordinal scale values in the LHX and AH-64A data bases. The models will be exercised to produce refined workload predictions based upon the new scale values.

As the part-mission and full-mission simulation results are analyzed, additional refinements will be made to the workload prediction models. The researchers will make necessary corrections to the workload estimates, time estimates, and decision rules. Refined workload predictions will be produced using the empirically derived workload estimates and time values.

DISCUSSION AND CONCLUSIONS

The workload prediction methodology described above provides a systematic means for predicting human operator workload in advance of system design or system modifications. This section of the chapter (a) discusses some of the weaknesses and strengths of the methodology so that the reader may better judge the value of the workload prediction models, and (b) offers some conclusions for the reader to consider.

Methodological Weaknesses

In all of the workload analyses described in this chapter, the workload estimates assigned during the mission/task/workload analysis phase are the basic units of analysis. The greatest weakness in the methodology stems from the subjective nature of these estimates.

As previously described, the workload estimates consist of numerical values assigned to the sensory, cognitive and psychomotor components of each task. The assigned estimates are derived by comparing verbal descriptors of the tasks with verbal anchors judged to represent increasing levels of attentional demand. Until the scales are demonstrated to be both reliable and valid, any results from exercising the models can be questioned.

Another methodological weakness exists in the procedure that sums the subjective values of the task workload components to derive total workload estimates for a given component of concurrent tasks. In the LHX and AH-64A analyses, the subjective values are clearly ordinal. Summing ordinal values to derive total estimates is a questionable procedure. The development of interval scales will eliminate this weakness.

A related methodological weakness stems from the treatment of each of the different types of workload components, (i.e., visual, auditory, kinesthetic, cognitive, and psychomotor) as separate and independent entities. It seems doubtful that, in reality, psychomotor workload can exist independently of concurrent cognitive and visual workload.

The analysts' decision to designate a total value of "8" as the threshold for identifying sensory, cognitive, and psychomotor overloads represents another subjective aspect of the research methodology. The selection of "8" is based solely upon the rationale that "7" is the upper limit of human capacity in the three workload modalities. Thus, it can be argued that the decision to use the value of "8" as the criterion for defining component overloads is an arbitrary one.

Methodological Strengths

The methodological weaknesses, considered by themselves, may lead the reader to question whether the methodology offers any advantages. However, certain strengths are believed to compensate for any impact that the weaknesses may have when applying the methodology to system design questions. The primary strength is that the methodology produces conservative estimates of workload.

First, whenever possible the decision rules are written to delay crew support functions on the timeline so that they will not conflict with high workload flight control and mission functions. In addition, the duration of flight control functions is extended so that all concurrent tasks can be presented on the timeline. To the extent that the function and task times are extended, the predictions of overload conditions and component overloads, produced by the stress of limited time are minimal.

Second, the criterion used to define excessive workload produces conservative estimates of component overloads. The methodology does not distinguish between varying degrees of overload when the sum of the ratings exceeds the threshold value of "8". The criterion value of "8" also precludes the recognition of instances in which a lower value may represent an overload condition. For example, a situation in which each of two or more workload components has a workload estimate of "6" may constitute a more critical overload condition than a situation in which only one component has a value of "8" or higher. In defining overload conditions, the methodology does not consider the total estimate for all three workload components.

A third way in which the predictions of excessive workload are conservative is that they predict workload under ideal operating conditions. The methodology does not consider increases in workload that will occur if mission performance is degraded due to visual obscuration, malfunctioning subsystems, or enemy activity. Obviously, such degradation would increase the workload beyond the level predicted by the present models.

A second major strength of the methodology is that it is designed to permit refinement during the analyses. Specifically, the methodology provides a means for refining the estimates of both workload and time as additional information becomes available. The workload estimates can be revised by assigning new verbal descriptors and numerical estimates to the workload components for each task; the timeline estimates can be revised by writing new decision rules.

Conclusions

The workload prediction methodology described above provides a systematic means of predicting human operator workload in advance of system design. The methodology predicts the attentional demand associated with the sensory, cognitive, and psychomotor components of individual and concurrent operator tasks. The workload predictions are computed and displayed on half-second timelines for both single- and dual-crew configurations. The workload estimates can be revised to predict the impact of (a) different crew configurations and (b) various automation options being considered during system design and system modifications.

In addition, the research methodology provides information for identifying emerging system personnel, manning, and training requirements. By assisting in the identification of these requirements, the methodology provides a means of developing early estimates of system personnel and training costs. The personnel and training cost estimates can then be factored into trade-off studies conducted during the early stages of system development.

REFERENCES

Aldrich, T. B., Craddock, W., & McCracken, J. H. (1984). A computer analysis to predict crew workload during LHX scout-attack missions (Draft Technical Report ASI479-054-84[B], Vol. I, II, III). Fort Rucker, AL: Anacapa Sciences, Inc.

Aldrich, T. B., & Szabo, S. M. (1986). Validation of the LHX one-crewmember workload prediction model (Draft Technical Memorandum ASI678-202-86[B]). Fort Rucker, AL: Anacapa Sciences, Inc.

Aldrich, T. B., Szabo, S. M., & Craddock, W. (1986). A computer analysis of LHX automation options and their effect on predicted workload (Draft Technical Report ASI479-063-85[B]). Fort Rucker, AL: Anacapa Sciences, Inc.

Bierbaum, C. R., Szabo, S. M., & Aldrich, T. B. (1987). A comprehensive task analysis of the UH-60 mission with crew workload estimates and preliminary decision rules for developing a UH-60 workload prediction model (Draft Technical Report ASI690-302-87[B], Vol. I, II, III, IV). Fort Rucker, AL: Anacapa Sciences, Inc.

Engen, T. (1971). Psychophysics II: Scaling methods. In J. W. Kling and L. A. Riggs (Eds.), Experimental psychology (3rd ed.), pp. 51-54. New York: Holt, Rinehart, and Winston.

McCracken, J. H., & Aldrich T. B. (1984). Analyses of selected LHX mission functions: Implications for operator workload and system automation goals (Technical Note ASI479-024-84). Fort Rucker, AL: Army Research Institute Aviation Research and Development Activity.

Siegal, A. I., Madden, E. G., & Pfeiffer, M. G. (1985). Task and training requirements for advanced Army helicopters (U.S. Army Research Institute Research Note 85-105). Fort Rucker, AL: Applied Psychological Services, Inc.

Siegal, S. (1956). Nonparametric statistics for the behavioral sciences. New York: McGraw-Hill.

Szabo, S. M., & Bierbaum, C. R. (1986). A comprehensive task analysis of the AH-64 mission with crew workload estimates and preliminary decision rules for developing an AH-64 workload prediction model (Draft Technical Report ASI678-204-86[B], Vol. I, II, III, IV). Fort Rucker, AL: Anacapa Sciences, Inc.

Taylor, R. R., & Poole, R. (1983). OH-58D MEP descriptions and workload analysis (Report No. 406-099-063). Fort Worth, TX: Bell Helicopter Division Textron.

Wickens, C. D. (1984). Engineering psychology and human performance. Columbus, OH: Merrill.

W/INDEX: A PREDICTIVE MODEL OF OPERATOR WORKLOAD

Robert A. North and Victor A. Riley

Honeywell Inc.
Systems and Research Center
Minneapolis, MN

INTRODUCTION

One of the major developments in weapon system design over the past decade is the emergence of technologies that enable single crewmembers to operate very complex systems in highly dynamic environments. However, the growing complexity of the tactical environment is keeping pace with the ability of new automation technologies to deal with it, producing a constant tension between the economic and practical forces that drive crew size reduction and the mission performance requirements that favor larger crew sizes.

The crucial factor in the middle of this tension is crew workload. When all of the decision-making, systems monitoring, planning, situation assessment, and control responsibilities have been allocated between human and machine, the question of whether the operator's workload capacity has been exceeded is likely to determine the feasibility of the design. In the most complex systems operating in the most complex environments, where a single automation failure may make the operator's job unmanageable, it is especially important to consider factors that will make the job unnecessarily difficult by contributing to the operator's workload.

System designers need a tool that allows them to derive the crewstation design and automation configuration that produces the most manageable workload levels. This tool should permit rapid consideration of a wide range of design options, and should be useful at any stage of the design process, from high level concept generation through advanced development. In order to be useful in analyzing a complex environment, where an operator's attention is likely to be shared between multiple tasks over much of the mission, the tool must employ a realistic model of attentional timesharing and impose appropriate levels of workload penalties to account for different levels of conflicts between multiple tasks.

Finally, the tool should facilitate detailed and systematic consideration of all the major task or design attributes that contribute to workload. As new automation technologies mature and the operator's task becomes more management and decision-making than control, this should include consideration of the operator's cognitive processing as well as manual and sensory demands.

WORKLOAD INDEX (W/INDEX)

Honeywell Systems and Research Center has developed a computer-based tool to predict operator workload produced by specific crewstation designs over the course of representative mission scenarios. The Workload Index (W/INDEX) software tool, which runs in both MS-DOS and VAX-VMS environments, allows system designers to consider the workload consequences of decisions involving the physical layout of the crewstation, the application of automation to specific crew tasks, the use of various human-machine interface technologies, and the sequence of crew task loading. It is intended to be used iteratively at any stage of the design process to identify the best crewstation configuration, that is, the best combination of geometry, automation, and crew interface technologies that produces the lowest predicted workload over a wide range of mission conditions.

To use W/INDEX, the analyst must supply information for three W/INDEX databases: a task timeline, an interface/activity matrix, and an interface conflict matrix. These are indicated in Figure 1, and will be described in the following paragraphs.

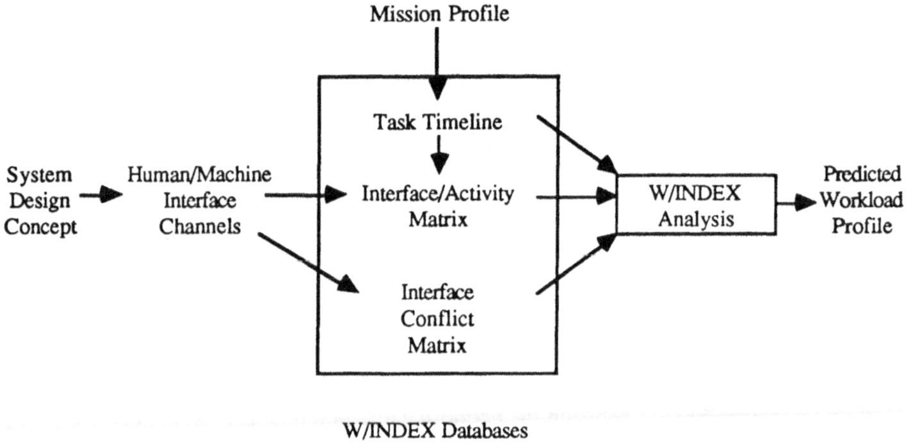

Figure 1. W/INDEX Data Flow

The task timeline is derived from a mission profile. It represents the specific tasks performed by the operator during mission performance and when they occur on the timeline. Each task may be assigned up to 20 start and stop times.

The interface/activity matrix is derived from the task timeline and from the human/machine interface channels, which represent the design concept. In the early stages of design, these interface channels may be very high level, such as "visual", "auditory", "manual", and "verbal". This permits the designer to predict potential overtaxing of one of these types of operator resource. During detailed design, specific controls and displays may be represented as channels. What the interface/activity matrix does is specify the amount of attention the operator must pay to each channel in the performance of each task. An example is shown in Figure 2. The numbers in the matrix are on a five point subjective scale, with 1 being very low attentional demand and 5 being very high. The two-letter identifier preceding the activity name referred to an aircraft system (e.g., FC = flight control, etc.)

The conflict matrix is derived solely from the design concept and represented in terms of interface channels. It specifies the degree of workload penalty that results when the operator must attend to multiple channels simultaneously, and gives the designer a means of considering the physical and cognitive capabilities of the operator in the context of the specific design being evaluated. The use of this matrix and the theory behind it will be discussed in detail in the next section.

Having provided each of these three types of data, the analyst simply commands W/INDEX to calculate a predicted workload level for each half second in the timeline. W/INDEX does this for each half second by summing the attentional demands in each

INTERFACE/ACTIVITY MATRIX

ACTIVITY NAME	CG	TH	TD	FC	FD	LT	LD	CD	RD	SP	HP
FC-AIR-TO-AIR	0	2	0	0	5	0	0	0	0	0	0
FC-HIGH-LEVEL	0	0	0	0	1	0	0	0	0	0	0
FC-INIT IFFC	0	0	2	0	0	0	0	0	0	0	0
NV-EVAL INTERCE	3	0	0	0	0	0	0	0	0	0	0
NV-FOLLOW STRG	2	0	0	0	0	0	0	0	0	0	0
NV-MONITOR INTE	1	0	0	0	0	0	0	0	0	0	0
NV-MONITOR PATH	2	0	0	0	0	0	0	2	0	0	0
NV-RECOG JM EFF	4	0	0	0	0	0	0	0	0	0	0
NV-RECOG TRACK	2	0	0	0	0	0	0	0	0	0	0
NV-VER FUEL	0	0	0	0	0	0	0	0	0	0	0
CO-ACC ASSIGN	1	0	0	0	0	0	0	2	0	0	0
CO-ACC MESSAGE	0	0	0	0	0	0	0	2	0	0	0
CO-AEW ALERT	3	0	0	0	0	0	0	0	0	0	0
CO-OB WING STAT	0	0	2	0	0	0	0	0	0	2	0
CO-REQ MESSAGE	0	0	0	0	0	0	0	3	0	0	0
CO-REV MESSAGE	5	0	0	0	0	0	0	0	0	0	0
CO-VER WING JAM	0	0	2	0	0	0	0	0	0	3	0
ASE-DETECT JAM	3	0	0	0	0	0	0	0	0	0	0
ASE-VER ECM	0	0	0	0	0	0	0	0	0	0	0
FRC-MAS ARM ON	0	0	0	0	0	0	0	2	0	0	0
FRC-VER WEAPON	0	0	0	0	0	0	0	0	0	2	0
MM-COMP FENCE	2	0	0	0	0	0	0	0	0	0	0
MM-EVAL SENS RN	2	0	0	0	0	0	0	0	2	0	0
MM-INIT PAS SRC	1	0	0	0	0	0	0	3	0	0	0
MM-MONITOR SYS	1	0	0	0	0	0	0	0	0	0	0
MM-PERFORM ID	2	0	0	0	0	0	0	2	0	0	0
MM-SELECT A/A	0	0	0	0	3	0	0	0	0	0	0
MM-VER DISPLAY	1	0	0	0	0	0	0	0	0	0	0

Figure 2. Example Interface/Activity Matrix.

interface channel across all the tasks that occur during that half second. These interface channel attentional demand totals are then summed to produce an additive workload estimate. Then, the additional costs of timesharing between multiple channels are determined by multiplying the sum of each pairwise combination of channels with the corresponding conflict matrix value. This is done both within channels and between channels. These timesharing costs are then added on to the current half second total to produce an instantaneous workload level for that time period.

The algorithm can be expressed as follows:

$$W_T = \sum_{i=1}^{l} \sum_{t=1}^{m} a_{t,i} + \sum_{i=1}^{l} \left[(n_{t,i} - 1)\, c_{ii} \sum_{t=1}^{m} a_{t,i} \right] + \sum_{i=1}^{l-1} \sum_{j=i+1}^{l} c_{ij} \sum_{t=1}^{m} (a_{t,i} + a_{t,j})$$

Where:

W_t = instantaneous workload at time T

i,j = 1..l are the interface channels

t = 1..m are the operator's tasks or activities

$n_{t,i}$ = number of tasks occurring at time t with nonzero attention to channel i

$a_{t,i}$ = attention to channel i to perform task t

$c_{i,j}$ = conflict between channels i and j

and where:

$a_{t,i}$ and $a_{t,j}$ are both nonzero.

The first term of this formula provides the purely additive workload level, while the second term provides the penalty due to demand conflicts within channels and the third term provides the penalty due to demand conflicts between channels.

Having calculated instantaneous workload for each half second in the mission timeline, W/INDEX then provides a five-second average. This average extends to two seconds before time t to two seconds after time T, and accounts for the operator's discretion to anticipate or delay tasks to smooth instantaneous workload changes. This provides a smooth workload profile over the mission timeline.

A sample of W/INDEX output is shown in Figure 3. This chart, (a Macintosh-drawn compilation of four W/INDEX runs), shows four different combinations of automation configurations and interface technology options. The minimum path represents the lowest workload crewstation design. In this case, it switches between design conditions, suggesting that the lowest workload can be achieved when several interface options are available to the operator and the operator can choose which one to use based on other concurrent activities.

In this way, W/INDEX considers the difficulties of the tasks, the geometry of the crewstation, the availability of interface technologies and automation options, and the physical and cognitive capabilities of the operator. This summarizes the capabilities of the current version of W/INDEX, as it has been used on a variety of programs. At the end of this paper, we will describe how W/INDEX is being modified to meet new sets of challenges while improving the usability, accuracy, and reliability of the tool. The next section describes the theory and use of the conflict matrix.

CONFLICT MATRIX

The interface conflict matrix is one of the major features that separates W/INDEX from other timeline workload analysis tools. In the matrix, each interface channel is paired off with itself and all the other channels, and for each combination, a conflict factor from 0 to 1 is assigned.

This approach recognizes that some types of attention combinations are more difficult to timeshare than others. For example, it is easy to speak and drive at the same time, but difficult to speak and comprehend speech at the same time. Similarly, it is easier to avoid traffic and monitor an oncoming stoplight simultaneously than to avoid traffic while adjusting the radio frequency. Both task combinations require simultaneous visual attention, but the latter one requires more divided attention.

To guide the assignment of these conflict values, W/INDEX uses a model of multiple attentional resources developed by Dr. Christopher Wickens and researchers at the University of Illinois. The basic idea of this model is illustrated in Figure 4. It represents a space in which task demands can be placed according to whether they are verbal or spatial, their input form visual or auditory, and their output form manual or vocal. The closer together two tasks or interface channels are in this space, the more they draw on the same attentional resources and, therefore, the more difficult they are to timeshare.

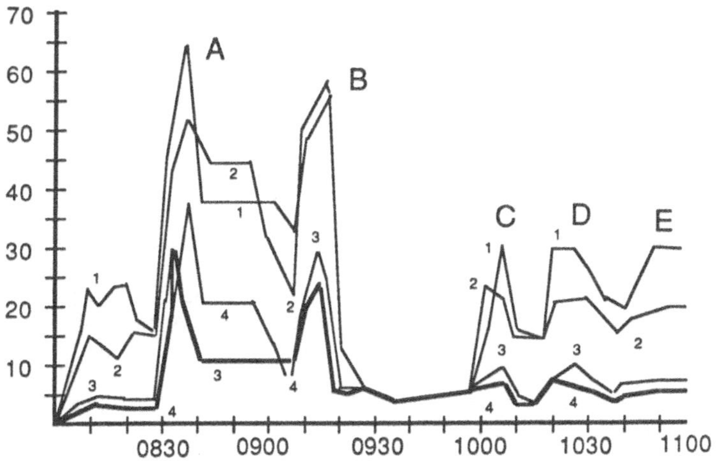

Figure 3. A Macintosh-Drawn Compilation of Four W/INDEX Workload Profiles. Each of the four profiles represents a different combination of automation and interface technology options.

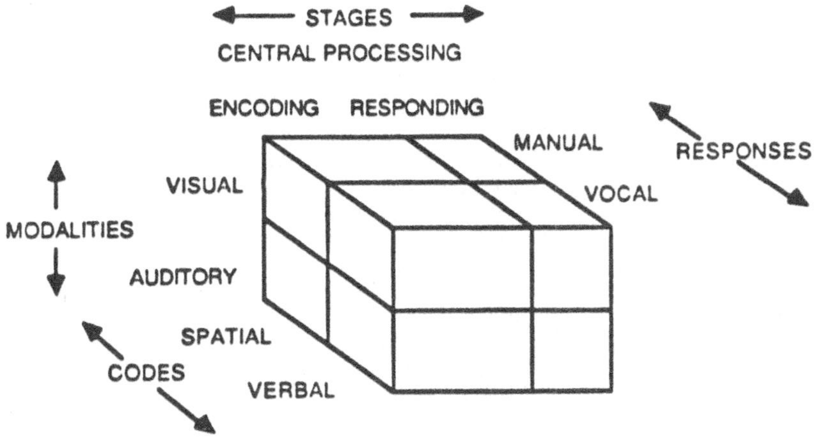

Figure 4. A Multiple Resources Model of Human Attentional Allocation

Figure 5 shows how this concept is used to assign conflict ranges to categories of interface channel combinations. Note that visual/visual conflicts are much higher than visual/speech conflicts. The analyst may then adjust the conflict value within the range given to account for physical separation or integration of controls and/or displays.

One of the primary strategies for reducing workload peaks, when they are produced, is to reduce the impacts of interchannel conflicts by substituting new interface channels for critical tasks. For example, a pilot required to operate a front panel switch while in air-to-air combat may experience an unmanageable demand conflict. If the switch task can be replaced by a speech command, the conflict is substantially reduced, and the result is lower workload.

APPLICATIONS

Since its inception in 1983, W/INDEX has been applied to a wide range of systems and questions. We used it to evaluate early concepts for the Army's LHX attack scout helicopter, then to evaluate degraded operations and pilot-vehicle interface alternatives later in LHX design. We also used W/INDEX to explore issues of one- versus two-man crews for the Apache advanced helicopter.

In 1986, we evaluated an Advanced Tactical Fighter (ATF) design against representative mission scenarios, and produced recommendations for changes in crewstation layout as well as automation and crew interface options.

Our latest effort looked at the feasibility of a crewstation design for the National Aerospace Plane (NASP). The purpose of this analysis was to compare workload predicted for the NASP cockpit with that for an existing transport, and thereby to estimate the feasibility of the proposed crewstation design and and task allocation. This analysis was able to address questions posed by the Air Force and NASA customers that were not approachable through any other means.

FREQUENTLY ASKED QUESTIONS ABOUT W/INDEX

One of the major issues accompanying any effort to predict workload is the degree to which the analysis can predict actual workload in the real world. This has proven to be an elusive goal, primarily because workload measurement itself is an inexact science even under the best experimental conditions. Several studies have been performed (Casali and Wierwille, 1983; Wierwille and Connor, 1983; and Wierwille, Rahimi, and Casali, 1985) to compare the accuracy of about fifteen different methods of measuring real workload during the performance of experimental tasks, including subjective ratings by the subjects, physiological measures, and dual task measures. Of these, only the subjective subject ratings have proven reliable in all conditions tested.

Furthermore, workload capacity can vary widely between individuals and even within individuals at different times. For all these reasons, it has not been possible to calibrate W/INDEX to a reliable measure of real workload, nor has it been possible to establish some predicted score as a reliable upper limit above which workload is unmanageable.

The use of subjective judgments of attentional demand raises questions about inter-rater reliability, since different analysts are likely to assign different ratings to these demand levels. Fatigue, which W/INDEX does not consider, is also an issue, as is the realism of conceptual mission timelines.

Honeywell is currently taking steps to address each of these issues. Table 1 shows how our current efforts relate to each concern raised above. Some of these approaches will be detailed in the next section. However, it is important to remember that the real goal of a W/INDEX analysis is to determine the best crewstation configuration, not to prove the ultimate feasibility of the design concept. As long as the analysis for a given design is internally consistent and appropriately representative and exhaustive, it should result in finding that best design solution, which is the tool's primary goal.

	Response	Task "B" Resources: Visual	Auditory	Manual	Verbal
Task "A" Resources	Visual	HIGH CONFLICT (.7-.9) Directly competing resources (e.g., two search tasks; less if tasks are adjacent or on same display areas.			
	Auditory	LOW CONFLICT (.2-.4) Noncompeting resources (e.g., search and listening).	HIGH CONFLICT (.7-.9) Highly competitive resources; some time-sharing if discriminability between inputs is high.		
	Manual	LOW CONFLICT (.1-.3) Noncompeting resources.	LOW CONFLICT (.1-.3) Noncompeting resources.	HIGH CONFLICT (.7-.9) Competing resources such as two tracking tasks or discrete choice tasks have shown high dual-task decrements.	
	Verbal	LOW CONFLICT (.1-.3) Noncompeting resources.	MEDIUM CONFLICT (.4-.6) More interfering if task requires voiced output.	LOW CONFLICT (.2-.4) Nonoverlapping resources showing little dual-task decrement in studies of tracking and voice input.	HIGH CONFLICT (1.0) Requires complete serial output; e.g., giving two messages or voice commands.

Figure 5. Prototype Conflict Matrix, used to Provide Conflict Levels Based on Resources Category.

W/INDEX PLANNED IMPROVEMENTS

Honeywell is currently working to improve W/INDEX in several areas. These efforts should provide better resolution to W/INDEX analyses while enabling subject matter experts to use W/INDEX without needing special knowledge of the tool or of the psychological theories behind it.

TABLE 1. RELATION OF W/INDEX IMPROVEMENTS TO WORKLOAD ANALYSIS ISSUES

Issue	Approach
Predictive Accuracy	Include consideration of multiple task dimensions that may affect workload
Cognitive Issues	Extend Multiple Resources Model to cognitive domain and verify inclusion of variable cognitive conflicts
Inter-rater Reliability	Automate database production by using question-answer dialogue
Timeline Realism	Connection with SAINT to generate and modify task schedules
Role in Design	Integrate W/INDEX with set of existing and emerging design tools to provide integrated analysis capability

W/INDEX's task difficulty rating method is being improved by incorporating a cognitive taxonomy into its structure that provides more detailed consideration of the variable conflict levels that may arise between tasks due to their cognitive differences or similarities. This extends the Multiple Resources Model into the strictly cognitive domain.

Furthermore, the next generation of W/INDEX will facilitate systematic consideration of the numerous task and interface channel characteristics that can contribute to workload. By taking more factors into account, W/INDEX should be able to represent specific design issues, such as display fidelity and fixation requirements, thereby improving its specificity.

Since the consideration of these factors will be systematic, the next generation of W/INDEX will be automated. Analysts will interact with the tool using a question-and-answer dialog, and W/INDEX will automatically construct the Interface/Activity and Conflict matrices based on the analyst's responses. Since the analyst will not have to directly insert values into these matrices, no special knowledge of the tool will be required to use it. Furthermore, systematic weighting of the various factors that contribute to workload will improve the tool's reliability between analysts and between analyses.

Finally, W/INDEX will be integrated into a systematic process Honeywell is developing to address a wide range of system design issues throughout the design process. This integrated approach will link tools, both Honeywell-developed tools and tools already available from other sources, so that the data flow from one tool to another follows the system development process. These tools will permit designers to elicit and prioritize function allocation issues and tradeoffs, predict the effects of automation on system performance, develop appropriate teaming strategies and information networks, develop mission profiles for individual operators, and determine individual operator workload levels.

The first step to integrating W/INDEX into this process has been to link the output from the Systems Analysis of Integrated Networks of Tasks (SAINT) tool developed by the Air Force to W/INDEX. SAINT considers the availabilities of operators in a team and schedules tasks for them so as to optimize the performance of the whole team network. Since each operator's task schedule constitutes a mission timeline for that operator, SAINT schedule output for a single operator can be fed to W/INDEX for workload estimation. Recently, we achieved this one-way linkage. Ultimately, W/INDEX and SAINT will interact, so that if undesirable workload levels result from one task schedule, the resulting peak periods can be sent back to SAINT for rescheduling.

CONCLUSION

We have described Honeywell's Workload Index (W/INDEX) crewstation design tool, both as it currently exists and as we foresee its future development. The currently available tool has been applied to a range of crewstation design problems with good results.

REFERENCES

Casali, J. G., and Wierwille, W. W. (1983) "A Comparison of Rating Scale, Secondary-Task, Physiological, and Primary-Task Workload Estimation Techniques in a Simulated Flight Task Emphasizing Communications Load." Human Factors, Vol. 25, No. 6,, pp. 623-642.

Wickens, C. D. (1984) "The Multiple Resources Model of Human Performance: Implications for Display Design," in the AGARD/NATO Proceedings, Williamsburg, VA.

Wierwille, W. W., and Connor, S. A. (1983) "Evaluation of 20 Workload Measures Using a Psychomotor Task in a Moving-Base Aircraft Simulator." Human Factors, Vol. 25, No. 1, pp. 1-16.

Wierwille, W. W., Rahimi, M., and Casali, J. G. (1985) "Evaluation of 16 Measures of Mental Workload Using a Simulated Flight Task Emphasizing Mediational Activity." Human Factors, Vol. 27, No. 5.

A MODEL FOR PERFORMING SYSTEM PERFORMANCE ANALYSIS IN PREDESIGN

Charles D. Holley

Sequitur Systems, Inc.
7527 Nine Mile Bridge Rd.
Fort Worth, TX 76135

INTRODUCTION

Two areas of digital simulation are particularly relevant to valid system performance during predesign: mission modelling and man-machine modelling. In this paper, the latter technique is emphasized. The architecture (including the artificial intelligence components) and validation data for a particular model are presented. The components of the predesign process are illustrated in Fig. 1. This process has evolved from, and been applied to a variety of weapons systems (cf. Refs. 1, 2 & 3) and documents (for example, Refs. 4, 5 & 6).

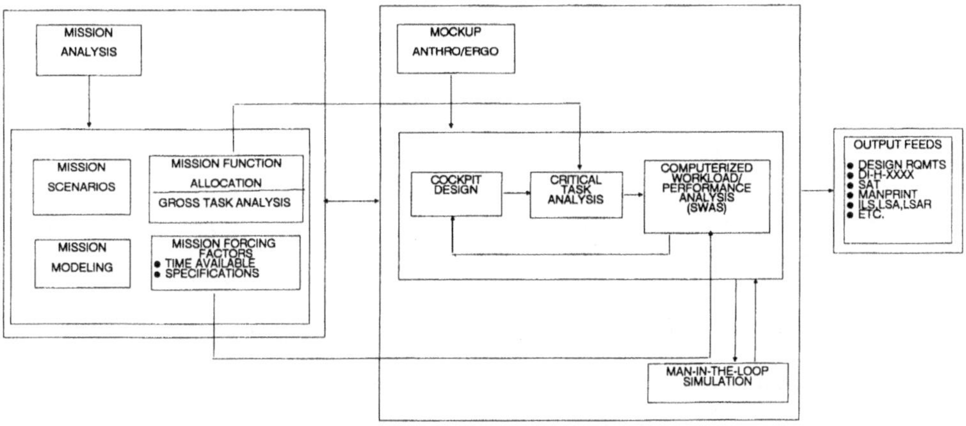

Fig. 1. System workload/performance analysis process: operator

A DIGITAL SIMULATION MODEL FOR PREDESIGN

Essentially, the pros and cons of digital simulation models can be evaluated along seven dimensions (cf. Ref. 7), which are not necessarily orthogonal. For example, a model may have good predictive validity for such a small range of tasks (i.e., Siegel-Wolf, HOS) that its application to system design is too restrictive to be of much value (i.e., it lacks generality). The seven dimensions and their definitions are as follows:

1) Predictive validity. This refers to the ability of the model to provide accurate predictions of system performance compared to subsequent, actual performance by human subjects - from the target population - using the equipment.

2) Generality. This refers to the range of tasks, systems, and populations to which the model can be applied.

3) Flexibility. This refers to the model's capability for representing complex task element interactions and is closely related to generality.

4) Transportability. This refers to the ease with which a model can be implemented on a variety of hardware or the wide-spread availability of hardware to support the model.

5) Theoretical underpinnings. This refers to the underlying theory, usually psychological, that the model purports to simulate. It also refers to theoretical propositions that were used in developing the model.

6) Generalizability. This refers to technical aspects of experimental design that are necessary for achieving predictive validity. It includes both internal and external control, although the latter concept is more relevant for digital simulation (compare Ref. 8).

7) Ease of use. This refers to the ease with which a model can be applied to a task by a user. In other words, the user interface with the model must be optimized to facilitate its use.

Sequitur's Workload Analysis System (SWAS) was developed from existing models, and rates well on all of the dimensions.

SWAS is a microcomputer-based, hybrid model (both a network model and production system), which was developed in the mid-eighties (Ref. 9). SWAS is a time-based model that provides system (man and machine) performance predictions, including probability of success, and also identifies "bottlenecks" to successful task completion. These bottlenecks include such entities as intra- and inter-operator processing difficulties and equipment delays. In addition, SWAS provides for assessing individual differences in operator performance; including the effects of wearing nuclear, biological, and chemical (NBC) protective clothing.

The user interface is menu-driven and has been designed to maximize ease of operation. For example, SWAS solicits user input only when the user's intentions cannot be inferred from information otherwise available; most instructions are accomplished by cursor selection or function keys; masks, prompts, and legal values are used for keyboard input; and local error correction is usually employed. In addition, online help is always available through a function key.

Software Architecture

As illustrated in Fig. 2, a modular architecture has been used in the development of SWAS. This approach allows existing modules to be revised or new modules to be added without creating major impacts on the system. Overall control is under the direction of a software executive, which also contains several subordinate executives; the most important of these is the simulation executive which controls all of the simulation and analysis activities of the model.

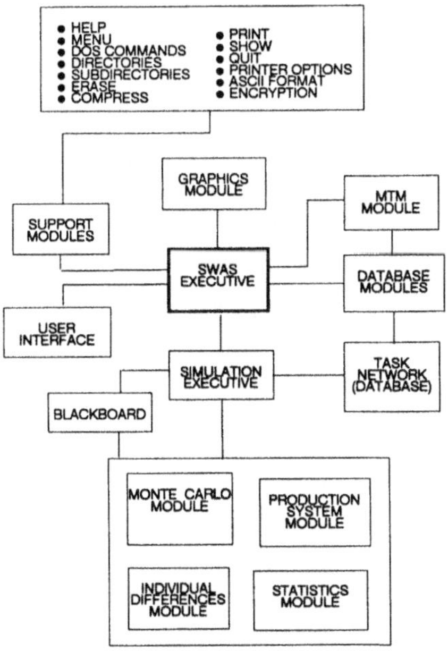

Fig. 2. SWAS software architecture

Database modules include such functions as creating and modifying task directories and task element data files, merging two or more data files, adding concurrent tasks, entering task element data, and editing data files. The MTM module is available on demand by the user. This function guides him through a task element description to produce a mean performance time using the Methods Time Measurement process (Refs. 10, 11).

Operational Description

From an operational perspective, SWAS may be viewed as a task database management system and a simulation and analysis system. The respective procedures are described in the following subsections.

The simulation executive controls four principal modules: Monte Carlo, production systems, individual differences, and statistics. The Monte Carlo module provides random draws from a truncated Gaussian distribution for each task element performance time during each iteration. The production system module provides adjustments to task element performance times, based on a human information processing model for concurrent task performance. The individual differences module makes adjustments to task element performance times, based on runtime variables that are set by the user. The statistics module collects data from each iteration and provides descriptive and inferential statistics regarding system performance/workload. From an operational perspective, SWAS may be viewed as a task database management system and a simulation and analysis system.

Task Database Management. SWAS maintains a task directory that is hierarchically organized into task categories (e.g., navigation, flight control, weapons delivery) and task blocks (for example, automatic bobup manoeuvre, laser firing). Task blocks represent subtasks and are comprised of the task elements that the system must perform to accomplish the subtask (see Table 1). Task blocks may be either independent or dependent. Independent task blocks contain task elements whose characteristics (e.g., execution time) remain the same regardless of the mission context. Conversely, dependent task blocks contain task elements whose attributes may vary with the mission context.

Table 1. Task element attributes used in SWAS

ATTRIBUTE	DESCRIPTION
Operator number	Models the interaction of 1 to 5 operators (for example, crew, maintainer, machine)
Task number	Sequential task numbers, by operator
Essentiality	Designates essential/nonessential task element
Processing modes (discrete task element)	Designates the human information processing required by the task
Processing modes (continuous tasks)	Designates the human information processing modes and difficulty level required for any concurrent, continuous task
Mean	Mean time to perform the task element
Standard deviation	Standard deviation for mean performance time
Subtask type	Human or machine task element
Task precedence	Task number designating a task element that must be completed prior to this task element
Task description	Verbal description of the task element

After the task element database has been developed, the user sequentially selects and merges task blocks to build a mission segment file. As part of this building process, codes for continuous concurrent tasks and codes for task precedences are appended to the task element data. Performance/workload simulation and analysis are then conducted on the segment file.

Performance/Workload Simulation and Analysis. As part of the performance workload simulation, the user specifies certain parameters that are to be considered by the model. These include three individual difference variables, NBC clothing (MOPP), and three general variables. The individual difference variables are labeled and set independently for each operator using a scale of 1 (good) to 9 (bad). These variables are assumed to interact linearly and result in task element performance times and standard deviations being adjusted by a multiplicative factor ranging from 0.7 to 1.3 (Ref. 12). NBC clothing effects are based on the selected MOPP level and result in task element performance times and standard deviations being adjusted by a multiplicative factor ranging from 1.0 to 2.0.

The general variables for the simulation are: (1) the number of iterations to be performed (reasonable stability appears to occur with at least 25 iterations, but a minimum of 100 is recommended), (2) the sample size that was used to determine the task element times and standard deviations (harmonic n), and (3) the time available (TA) for the system to perform all essential task elements in the segment. If the user does not have a firm time available (i.e., from the mission analyses), he may invoke the concept of nominal time available. In this case the highest value of time required, across all iterations, is used as time available; this results in a probability of success of 1.00 and the data are used for relative performance/workload comparisons.

SWAS initially adjusts the task element database to accommodate the individual differences and MOPP level selections. Then, for each iteration, the following sequential steps are performed:

1. Task element means and standard deviations are adjusted to accommodate timesharing with continuous concurrent tasks (multiplicative factor: 1.0 to 2.0).

2. Task element means and standard deviations are adjusted to accommodate timesharing between discrete concurrent tasks (multiplicative factor: 0.0 to 1.0).

3. Based on the task element means, standard deviations, and harmonic "n" random draws are made from a truncated Gaussian distribution and the task element means are replaced by the resulting values.

4. Timelines are calculated and include any wait times resulting from task precedences for other operators.

5. Summary data are temporarily stored, including the overall success or failure of the iteration (success means all essential task elements were completed within the time available).

After all iterations have been completed, SWAS provides a detailed summary output in tabular and graphical form to the analyst. Using the detailed output, the user can draw conclusions regarding such issues as the probability that the system can successfully complete the mission segment, relative proportions of time required by human operators vis-a-vis equipment operators, distributions of workload across the system, and bottlenecks resulting from task timesharing overloads or delays caused by waiting on other operators.

Human Information Processing

A fundamental capability of SWAS, which sets it apart from other models, is its inherent mechanisms for incorporating the notion of intra-operator timesharing into the performance/workload timelines, and to do so in a theoretically plausable manner. As previously stated, this capability is accomplished by the production system module, but the theoretical and empirical data underlying these rules form the essential elements of the implementation.

The human information processing (HIP) model used in SWAS is based on Wicken's Multimodal Theory (e.g., Refs. 13,14). According to Wicken's, information processing resources can be partitioned into separate and limited quantities or structures, and task performance is associated with resource demand (Ref. 14). Timesharing will improve or degrade to the extent that tasks compete for these resources. Wicken's proposed structures have been defined by stages of processing, modalities of input and output, and codes of central processing. A considerable amount of empirical work has been conducted that generally supports the premises of the theory (see, e.g., Ref. 15).

For simplicity, SWAS refers to the information processing resources as input, central, and output modes. Channels within these modes are:

Input: visual, auditory, other
Central: verbal, spatial, other
Output: vocal, manual, other

The output mode for discrete tasks may be further differentiated as left or right hand. The "other" code is used to indicate that the task element does not use the mode, thereby making it completely available for timesharing. In addition to the processing modes, a difficulty code is also used with continuous tasks.

Consistent with Wicken's theory, SWAS makes the assumption that performance time for a discrete task will increase or remain the same during concurrent performance with a continuous task, and vice versa during concurrent performance with another discrete task. The amount of adjustment to be made is based on the Euclidean distances between tasks, as defined by their processing mode requirements. These Euclidean distances were derived from a multidimensional scaling (MDS) study (Ref. 16) using the INDSCAL (Individual Scaling) method developed and validated by Carroll and Chang (Ref. 17). A similar application of MDS for validating some aspects of Wicken's Multimodal Theory was demonstrated by Derrick (Ref. 18). In SWAS, the production system module uses these distances, as well as other rules, to adjust task element performance.

Predictive Validation

As previously argued, a crucial criterion for digital simulation models is their predictive validity. In the case of time-based models such as SWAS, this means how well do the system performance predictions for time required compare to actual performance of the same tasks. Several predictive validation studies have been conducted for SWAS, three of which will be reported herein. The first two studies were conducted using high-fidelity, man-in-the-loop (MIL) simulation (see Fig. 3 and Table 2), and the third study was conducted during actual flight test (see Fig. 4). In all of the studies, experienced U.S. Army aviators were used as subjects, but different subjects were used for each study. In addition, each study employed a single pilot attack/ scout helicopter in an anti-armour role, although implementation of the scenario varied between studies. The SWAS predictions were derived independently from the MIL simulation and flight test data. Details of these studies have been reported elsewhere (Refs. 19, 20).

The data in Fig. 4 also illustrates an important adjunct capability

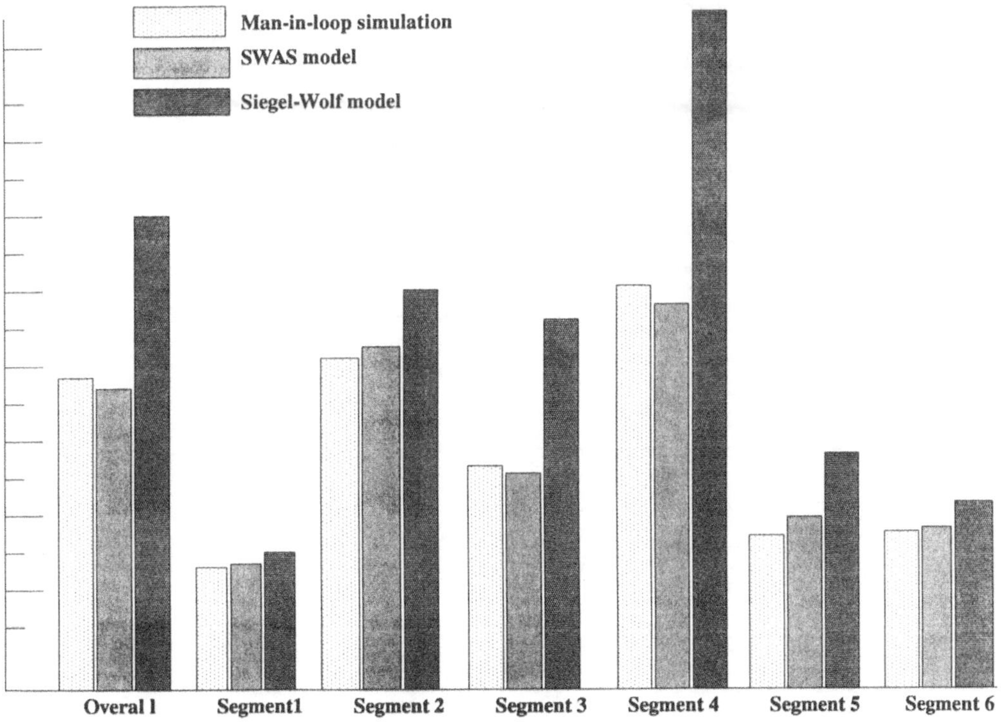

Fig. 3. Mean system performance/workload for three methods of evaluation

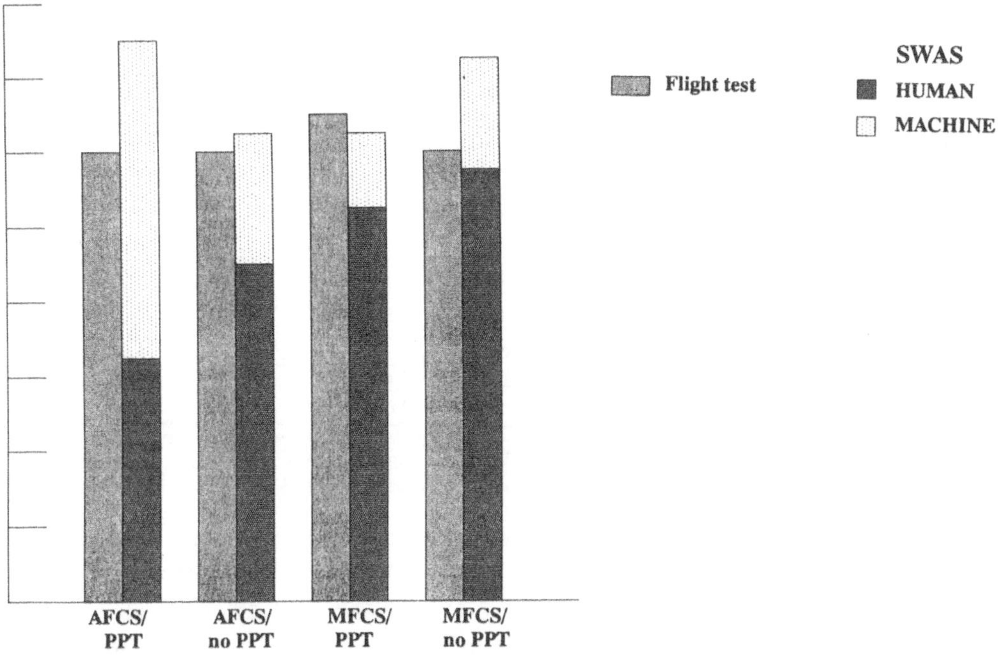

Fig. 4. Comparison of predicted and actual time required

of the SWAS model: partitioning system performance/workload into its man and machine components. Specifically, for both the SWAS predictions and the flight test results, no difference across experimental conditions was predicted or observed at the system performance level. However, the SWAS results illustrate how workload shifted from the man to the machine as the level of automation increased, essentially indicating a significant inverse linear relationship between levels of automation and crew workload.

Table 2. SWAS predictions compared to man-in-loop simulation results

		Mean Time (sec)	
Extracted Task	FCS Condition	Man-in-Loop Simulation	SWAS Prediction
Hover reconnaisance	Auto	7.80	8.02
	Manual	9.00	8.25
HELLFIRE engagement	Auto	12.95	13.36
	Manual	13.25	13.45
Off-axis gun engagement	Manual	6.61	6.82

CONCLUSIONS

Digital simulation can play a key role in the predesign process. With valid system performance predictions early in the design process, design changes can be implemented while they are relatively inexpensive. Additionally, these accurate, early-on predictions can ultimately lead to the fielding of a more effective system, potentially in a more rapid manner than occurs with the traditional life cycle model.

While digital simulation can play a crucial role in the system design process, its successful implementation depends on the validity of the model that is used. In this paper, a relatively new, hybrid model - SWAS - was reviewed. Seven evaluative dimensions were defined and the argument was advanced that the SWAS model performed favourably on all dimensions.

REFERENCES

1. Taylor, R.R. and Poole, E.R., A Mission Oriented Approach to Cockpit Design as Applied to Observation and Attack Helicopters, American Helicopter Society 40th Annual Forum, Arlington, VA, May 1984.

2. Holley, C.D. and Parks, R.E., Predicting Man-Machine System Performance in Predesign, Bell Helicopter, Textron, 1987.

3. Graf, V.A. and Holley, C.D., Human Factors Impact on the V-22 Osprey Cockpit Development: An Overview, American Helicopter Society, Washington, D.C., June 1988.

4. Human Engineering Requirements for Measurement of Operator Workload, ADS-30, U.S. Army Research and Technology Laboratories (AVSCOM), St. Louis, MO, November 1986.

5. O'Donnell, R.D. and Eggemeir, T., Workload Assessment Methodology, "Handbook of Perception and Human Performance," Vol. 2, Cognitive Processes in Performance, K.R. Boff, L. Kaufman and J.P. Thomas (eds), Wiley, New York, NY, 1986.

6. Human Engineering Design Criteria for Military Systems, Equipment, and Facilities, MIL-STD-1472C, Department of Defence, May 1981.

7. Sticha, P.J., Analytical Models of Performance of Procedure, IEEE, 0547-3578/84/0000-0841, 1984.

8. Campbell, D.T. and Stanley, J.C., "Experimental and Quasi-Experimental Designs for Research," McNally, Chicago, IL, 1966.

9. Sequitur's Workload Analysis System (SWAS): User's Manual, Sequitur Systems, Fort Worth, TX, 1986.

10. Karger, D.W. and Bayha, F.H., "Engineered Work Measurement," Industrial Press, New York, NY, 1966.

11. Generic Data System, Methods Time Measurement (MTM) Association, Fair Lawn, NJ, 1987.

12. Siegel, A.I. and Wolf, J.J., "Man-Machine Simulation Models," Wiley, New York NY, 1969.

13. Wickens, D.C., Processing Resources in Attention, "Varieties of Attention," R. Parasuraman and R. Davies (eds), Academic Press, New York, NY, 1984.

14. Wickens, D.C., The Structure of Attentional Resources, "Attention and Performance VIII," R. Nickerson and R. Pew (eds), Erlbaum, Hillsdale, NJ, 1980.

15. Wickens, D.C., Mountford, S.J. and Schreiner, W.S., Multiple Resources, Task-Hemispheric Integrity, and Individual Differences in Time-Sharing Efficiency, Human Factors, 23, 1981.

16. Multidimensional Scaling of Task Relationships, Sequitur Working Papers, Sequitur Systems, Fort Worth, TX, 1985.

17. Carroll, J. and Chang, J., Analysis of Individual Differences in Multidimensional Scaling via an N-way Generalization of Eckart-Young Decomposition, Psychometrika, 35, 1970.

18. Derrick, W.L., The Relationship Between Processing Resource and Subjective Dimensions of Operator Workload, "Human Factor Society 25th Annual Meeting, Proceedings," 1981.

19. Advanced Rotorcraft Technology Integration: Interim Report, BHTI Report 677-099-02, Bell Helicopter, Fort Worth, TX, May 1986.

20. Advanced Rotorcraft Technology Integration: Flight Test Report, BHTI Report 677-099-026, Bell Helicopter, Fort Worth, TX, August 1986.

MODELS OF INDIVIDUAL TASKS

INTRODUCTION: MODELS OF INDIVIDUAL TASKS

Robert Sutton

Control Engineering Department
Royal Naval Engineering College
Manadon, Plymouth, Devon, UK

OVERVIEW OF THE PAPERS

In a session which has such a general title as above, it is not possible to illustrate and discuss the broad spectrum of models that are available and fit into this category. However, the models considered herein will give an insight into the usefulness of such tools and their potential utility.

As an introduction to this field of study, Stassen reviews the modelling of manual control tasks. Initially, he explains the differences between manual and supervisory control tasks. Common to both forms of control, however, is the concept that the human operator possesses an internal representation of the system dynamics, the task and system disturbances. For the manual control problems, the operator's internal representation can be described by an internal model. The paper proceeds to discuss two types of manual control models, namely, the quasi-linear model and the optimal control model. The quasi-linear or describing function model is shown to be a useful tool mainly for frequency response analysis of single-input single-output manual control tasks. Using modern control theory, an optimal control model is also developed for use in multivariable systems. In addition, a brief outline is given of possible approaches which may be used in supervisory control models.

By using optimal control models to help describe the human monitoring and decision making process, Stein shows in his work the versatility of the control theoretic approach. The models are based on the information processing structure of an optimal model and extended by a decision sub-model, the decision process being formulated using classical decision theory. A variety of multivariable monitoring and decision making tasks are presented that show close agreement between experimental and theoretical results.

In considering the problem of target acquisition, Jagacinski suggests that speed-accuracy tradeoffs and stimulus-response compatibility are two important factors in determining the performance of discrete movements in such tasks. Traditionally, these factors have been applied to position control systems. However, the relative efficiency of various types of higher order dynamic systems for performing target acquisition tasks can also be understood in terms of these concepts. The importance of describing movement systems at multiple levels of abstraction is also discussed in this context.

The importance of auditory masking when listening for low level signals in the presence of high noise is the subject of the work presented by Rood, Patterson, and Lower. The paper describes the development of a model which can predict auditory thresholds over the frequency range 100 Hz to 4.5 kHz and its use in noise spectra with high low-frequency content. Validation of the model is based on the results in noise spectra experiments performed in a helicopter noise simulator. Applications of the model are discussed in relation to its current use for implementing auditory warnings in helicopters.

An important and integral part of the Workshop was to acquaint the participants with human performance models that are now being hosted on microcomputers and to demonstrate their usage. Two such demonstrations which relate to this session are detailed in the papers by Allen, McRuer, and Thompson; and Levison. Allen et al. give details of the application of two general purpose dynamic systems analysis programs that can be used to analyze human performance models. Finally, Levison in his paper describes the conceptual ideas relating to the development of an optimal control model and then proceeds to discuss its implementation in a specially designed computer program.

DISCUSSION

The discussion opened with a query from the session Chairman who asked to what extent the types of models that had been discussed are used in the design process, as there appeared to be a huge gap between the model developers and the users. It was considered by some psychologists that the manual control models, in particular, were not extensively used owing to their complexity. In order to use such models, it was felt that a background in electrical or control engineering was required to comprehend the underlying theory. Alternatively, there was a suggestion that models of this kind were necessary as many of the psychological tests lacked sophistication, and therefore were very restricted in their applications.

When using such a tool as the optimal control model, the main problem encountered was the number of parameters that are not uniquely defined. However, it was felt that even with limited knowledge of the parametric structure of the model, reasonable results can be obtained which give a good indication as to the way to proceed.

After further discussion, there was unanimous agreement that many of the problems highlighted could be overcome provided design teams adopted an interdisciplinary approach. Interdisciplinary work groups were considered to be the most versatile in solving man-machine design problems in an effective manner.

In addition, the general applicability of the control theory modelling techniques was questioned. One view expressed was that, while the techniques are useful when dealing with skill based behaviour with a continuous output as in a tracking task, they have limited use in rule based or knowledge based tasks which are discontinuous and do not have an easily quantifiable output. The reason for this limitation was considered to be due to the variability in humans operating at different behavioural levels. At the skill level, the task can be more or less well defined and is therefore more easily modelled. However, the higher level tasks as described by rule based or knowledge based behaviour allow the operator the opportunity to use one of several possible strategies. It was also stressed that the control models are based on linear theory and are used in situations where the standard deviation of the experimental error is expected to be small. Thus, the choice of modelling technique can also be related to the measurement procedure adopted.

Finally, there was general agreement that many of the software packages being demonstrated at the Workshop will help relieve many of the problems cited earlier in the discussion.

ON THE MODELING OF MANUAL CONTROL TASKS

Henk G. Stassen

Man-Machine Systems Group
Laboratory for Measurement and Control
Delft University of Technology, The Netherlands

INTRODUCTION

As long as complex technological systems are built, human beings are expected to control and supervise these systems. An important part of the design is the area where the human supervisor and the system come together, the so-called human-machine interface. In order to design the interface, the individual tasks to be executed by the supervisor and system have to be defined very precisely. However, this task allocation can only be achieved optimally when the capabilities and limitations of both the supervisor and the system to be supervised are known.

In system engineering and design, a well-proven approach to describing a system is to model the system's behavior. Such a description consisting of a model structure and parameters enables us to achieve the following important goals:

- Reduction of the data measured to a set of relevant parameters.
- The possibility to recognize analogies.
- The prediction of the dynamic behavior of a system under development.
- Quantitative knowledge required to optimize the system's performance.

Hence, models to describe the dynamic behavior of the human and machine would be very desirable.

In linear systems theory, it has been shown that the optimization of system performance can be achieved only if the dynamics of the system to be controlled are known (Conant and Ashby, 1970; Francis and Wonham, 1975). In optimal filter theory, one needs to be informed about the dynamics of the system and the statistics of the disturbances to be compensated (Kalman and Bucy, 1961). In analogy with these important statements, one may conclude that for correct supervision of a plant, the human supervisor has to be familiar with the plant, that is, he has to possess an internal representation of:

- The statistics and dynamics of the plant to be supervised.
- The tasks to be executed.
- The statistics of the disturbances to be compensated.

Without such an internal representation, one cannot count on a human supervisor to act in an optimal way. To represent the supervisor's internal representation, the researcher can construct an approximation to it for use in a human performance model. The modeler's approximation to the internal representation we call the internal model. Over the last decades, control, artificial intelligence, verbal and psychological performance models have been proposed, some quantitative, others qualitative. Interestingly enough, all concepts showed that the existence of an internal model is a necessary but non-sufficient condition for building human performance models (Stassen, Kok, van der Veldt and Heslinga, 1986; Borys, Johannsen, Hansel and Schmidt, 1987). That is, without a good internal model it is impossible to have a good human performance model, but a good internal model alone is not enough.

In optimizing a human-machine interface, one realizes that human performance is not all that matters in the design of an optimal system. It is just as important to consider what demands a certain task imposes on the operator's limited resources (Stassen, Johannsen, and Moray, 1988). This may or may not correspond with performance. In this review on modeling manual control tasks, we will restrict ourselves to human performance models.

MANUAL CONTROL

The history of human performance modeling is only four decades old, but a variety of models can be found. We prefer to base this review on the qualitative three-level model (Fig. 1), where Rasmussen distinguishes a Target Oriented Skill-Based Behavior (SBB), a Goal Oriented Rule-Based Behavior (RBB), and a Goal Controlled Knowledge-Based Behavior (KBB) (Rasmussen, 1983; 1986). The model can be used to classify human operator tasks. It is widely accepted that manual control tasks in stationary process conditions mainly lead to SBB, whereas procedurally oriented tasks which include monitoring, interpreting, and "teaching" a plant under supervision in stationary and non-stationary conditions are mainly RBB. Fault management and planning require not only knowledge of the tasks to be performed, but also require creativity and intelligence in the operator, hence they lead to KBB. In some cases, the situation is ambiguous. For example,

Table 1. The Relation Between Human Behavior and Human Operator Tasks. The Number of * Indicates the Significance of the Relation.

Human Operator Task / Human behavior	Intervening Manual Control	Supervisory Control			
		Interpreting Monitoring	Teaching	Fault Management	Planning
SBB	**	*	*	*	
RBB	*	**	**	*	*
KBB		*		**	**

"symptom based" emergency operating procedures are intended to be rule-based, even though the operator does not understand the true nature of the disturbance. This is an attempt to reduce KBB to RBB. Table 1 (Stassen, et al., 1986) shows that one type of behavior is more or less strongly related to a particular task. All three types of behavior may be relevant for the three phases (detection, diagnosis, and correction) in different fault management tasks (Johannsen, 1984). The importance of the relation between the contents of the internal representation, the task characteristics, and the level of behavior required is another reason for basing our review on Rasmussen's taxonomy.

At the SBB level, human performance models have been developed, mainly for manual control and detection tasks. Famous control models in the frequency domain and time domain, such as the Describing Function Model (McRuer and Jex, 1967) and the Optimal Control Model (Kleinman, Baron and Levison, 1971) have proven their value in the design of controls and displays (Johannsen and Govindaraj, 1980). Many non-linear models, restricted only to a particular configuration and hence not generalizable to new ones, are reported (Morris, 1982). Some manual tasks include a detection or decision aspect (Elkind and Miller, 1966; Gai and Curry, 1976), but in general all the tasks were well-defined and simple.

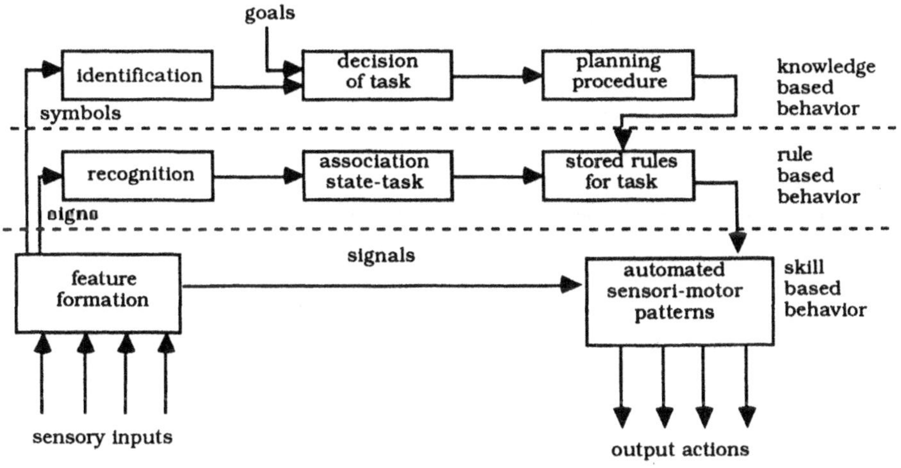

Figure 1. Schematic model of three different levels of human information processing (Rasmussen, 1983).

It is of interest to see that all those models have two very important properties in common.

- The tasks to be performed are well defined, and thus the operator is forced to act as a stationary controller; the operator has to possess a rather precise internal representation, and so in the development of human performance models, the use of accurate internal models could be assumed. In the most extreme form this is shown in the Optimal Control Model, where the Kalman-Bucy filter as well as the predictor and the optimal control law are based on a replica of the system to be controlled.

- Even in many nonlinear models, it is apparent from the literature that the internal model of the environment is an integral part of the human performance model. A good example is a model that describes the helmsman's behavior during the maneuvering of the very large crude oil carriers (Veldhuyzen and Stassen, 1977).

Because the Describing Function Model and the Optimal Control Model have proven to be powerful tools in the actual design of man-machine interfaces, both the models will be discussed in more detail in the next two sections.

DESCRIBING FUNCTION MODELS

The simplest kind of manual control is continuous one-dimensional tracking, where the operator's task is to make the output $y(t)$ of the controlled system correspond as closely as possible to the displayed input $u(t)$. In the case that the operator observes and acts upon the error, $e(t) = u(t) - y(t)$, and attempts to null this error, one speaks of compensatory tracking. The man acts as a servo-element (Fig. 2). His control behavior can be described by a linear model with a describing function $G(\omega)$ and a remnant $n(t)$. Such a model is called a quasi-linear model (McRuer and Jex, 1967). The important question now is whether it is meaningful to model the operator's behavior by a linear model $G(\omega)$ or, to say it in another way: Does the model output $x(t)$ count for the main part of the generated control signal $z(t)$ or not? Four major questions with reference to the applicability of linear models to human performance may be formulated (Sheridan and Ferrell, 1974):

- How repeatable is a given human operator's response?

- How linear is the human operator in a given task, i.e., in one with a definite type of input function and a definite controlled system? To what extent does the superposition theorem of linear systems hold in this case?

- Over how broad a set of tasks, i.e., input functions and/or controlled systems, can the same form of linear equation hold true?

- What is the variability among responses of different human operators for the same task?

Many researchers have asked these questions and the results have led to a rather broad area in which the tools can be applied (McRuer and Jex, 1967; McRuer and Krendel, 1974). It is interesting to review the arguments which are reported to prove that the variance of the remnant will not be equal to zero (McRuer and Jex, 1967; van Lunteren and Stassen, 1967; Stassen, Meijer and van Lunteren, 1970):

- Higher harmonics due to the non-linear behavior.

- Possible non-stationary operator behavior.

- Observation noise due to inaccurate observation.

- Motor noise due to inaccurate motor control.

- Generation of test signals by human operator in order to obtain more and/or better information about the system under control.

Man's performance as a controller depends upon a great number of variables which can be summarized as follows (McRuer and Jex, 1967):

- Environmental variables, such as additional tasks, vibration level, ambient illumination, and temperature.

- Procedural variables, such as instructions for the given task, order of presentation of trials, experimental design or measurement procedure, and control criteria specifying the payoff and effort, time or errors.

- Task variables, such as reference input signals and disturbances; the dynamics of the controlled system; what and how the information is displayed; control devices used; and the control configurations, such as steering, compensatory control, pursuit control, and preview control.

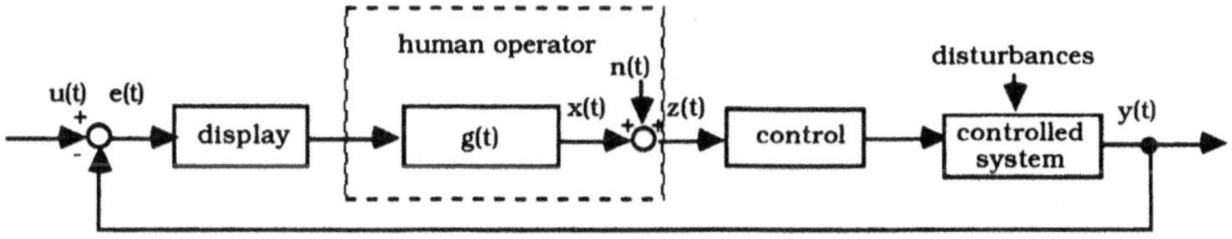

Figure 2. Compensatory Tracking

The most well-known Describing Function Model is the Cross-Over Model. It originates from the servo-control technique (McRuer and Krendel, 1959). The model is based on the stability criterion of a closed loop, hence an implicit internal model is used. McRuer concluded from many studies that the operator's behavior could be fitted best by the describing function $G(\omega)$ (McRuer and Krendel, 1959):

$$G(\omega) = K \frac{1+j\omega\tau_1}{1+j\omega\tau_2} \frac{1}{1+j\omega\tau_n} e^{-j\omega\tau_v} \tag{1}$$

as well as an additional remnant $n(t)$. In this formula, K is a gain factor, τ_1, τ_2 and τ_n are time constants, τ_v is a time delay and ω is the radial frequency. In later publications (McRuer, Graham, Krendel and Reisener, 1965; McRuer, Hofman, Jex, Moore, Phatak, Weir and Wolkovitch, 1968), it is shown that:

- The gain factor K is somewhere between 1 and 100.
- By means of the parameter values of τ_1 and τ_2 a phase margin of the closed loop system can be obtained between 60 and 100 degrees.
- The time constant τ_n is about 200 msec.
- The time delay τ_v is somewhere between 120 and 200 msec.

Furthermore, he argued that the parameters τ_n and τ_v are more or less physiologically determined, whereas K, τ_1 and τ_2 can be chosen in such a way that the best closed loop dynamics can be achieved; that is, the highest possible open loop gain with a phase lead less than 180 degrees. From servo control theory, it is known that for good control dynamics the open loop gain, $\mathring{H}(\omega) = G(\omega)H(\omega)$, should be like an integrator in the neighborhood where $|H(\omega)| = 1$. Thus, for manual control it follows that

$$\mathring{H}(\omega) = G(\omega)H(\omega) = \frac{\omega_c}{j\omega} e^{-j\omega\tau_v} \tag{2}$$

where $H(\omega)$ is the system transfer function and ω_c is called the cross-over frequency. In this case, the servo behavior for inputs with a bandwidth $\omega_u < \omega_c$ is good, since $|\mathring{H}(\omega)| >> 1$ for $\omega < \omega_c$, and also the suppression of disturbances is high, since $|\mathring{H}(\omega)| << 1$ for $\omega > \omega_c$. The model formulated by Eq. (2) is called the Cross-Over Model; the equation learns in what way starting with the describing function $G(\omega)$, Eq. (1), the values of the parameters τ_1, τ_2, and K will be adjusted by the human operator, so that the open loop gain $\mathring{H}(\omega) = G(\omega)H(\omega)$ harmonizes with Eq. (2).

Often the human operator's model is simplified by the approximation:

$$\frac{1}{1+j\omega\tau_n} \simeq e^{-j\omega\tau_n} \quad \text{for } \omega_u < \omega_c \tag{3}$$

so that Eq. (1) equals:

$$G(\omega) = K \frac{1+j\omega\tau_1}{1+j\omega\tau_2} e^{-j\omega\tau}, \text{with } \tau = \tau_n + \tau_v \qquad (4)$$

An overview of the results reported in the literature is given by Table 2 (McRuer, Graham, Krendel and Reisener, 1965). These results are based on the model given by Eq. (4). Note that the system with transfer function, $H(\omega) = [1/j\omega]^3$ cannot be controlled, since by adapting the values of τ_1, τ_2, and K no solution can be found which agrees with Eq. (1) and Eq. (4). Many applications of the Cross-Over Model are reported in the literature:

- Vehicle control, such as automobile driving (McRuer and Weir, 1969), bicycle riding (van Lunteren and Stassen, 1967), and space control and aviation (Summers and Ziedman, 1964; Costello and Higgins, 1966).

- The control of two- or more-dimensional input-output systems (Levison and Elkind, 1967; Weir, Heffley and Ringland, 1972; van Lunteren, 1979).

- Industrial management (Crossman, 1971).

- Biophysics and physiology (McRuer, Magdaleno and Moore, 1968; van Lunteren and Stassen, 1970; Repa, Albers, Rotvin, and Tourtelotte 1971; Allen, Jex, McRuer and Dimarco, 1974).

- Biomechanics (Jex, 1971).

- Secondary tasks as a tool for mental load studies (Jex, McDonnel and Phatak, 1966; Stassen, van Dieten and Soede, 1975; Jex, 1988).

Another interesting application of the Describing Function Model is reported in the field of manual control of suddenly changing system dynamics (Elkind and Miller, 1966, 1968; Kok, 1973; Stassen, 1976). The task of the human operator is to control a system, consisting of a number of linear constant subsystems, which at unpredictable moments switches from one subsystem to another. As long as a certain subsystem is under control, the Cross-Over Model provides a good fit to the human operator's behavior. At the moment, however, that a transition occurs from one subsystem to another, this model cannot be applied. In that case, the model should be extended with a decision element, since the task of the human operator is augmented with:

- The detection of a change in the dynamics of the system.

- The identification of the dynamics of the new subsystem.

- The selection of the optimal strategy for the subsystem under control according to the Cross-Over Model.

Table 2. Operator's Model Adaptation to the Dynamics of the Controlled System.

Controlled System	Human Operator Model	Parameters
1	$\frac{\omega_c}{j\omega} e^{-j\omega\tau}$	$\tau_1 = 0$ $K = \omega_c \cdot \tau_2$ $\tau_2 >> 1$ $\tau = \tau_v$
$\frac{1}{j\omega}$	$\omega_c \, e^{-j\omega\tau}$	$\tau_1 = 0$ $K = \omega_c$ $\tau_2 = 0$ $\tau = \tau_v$
$\frac{1}{[j\omega]^2}$	$j\omega \, \omega_c \, e^{-j\omega\tau}$	$\tau_1 >> 1$ $K = \omega_c/\tau_1$ $\tau_2 = 0$ $\tau = \tau_v$
$\frac{1}{[j\omega]^3}$	uncontrollable	--
$\frac{1}{1+j\omega\theta}$	$\omega c \frac{1+j\omega\theta}{j\omega} e^{-j\omega\tau}$	$\tau_1 = \theta$ $K = \omega_c \tau_2$

In formulating a decision model, Elkind made the following three assumptions:

- The human operator is well trained, that is, he possesses a good internal representation of each of the subsystems.
- The human data processing is considered to occur at discrete moments, with a constant sample interval of T seconds.
- The control strategy of the human operator can be characterized by three phases ranked in order: (1) Detection phase, (2) Identification Phase, and (3) Selection phase.

The human operator's behavior can now be explained by modeling each of the three phases. The model for the first two phases is based on statistical decision theory, the last phase will be described by a simple deterministic model. The model was validated with an intensive set of laboratory experiments (Kok, 1973), where Kok also showed that the results could be explained on the basis of the Cross-Over Model.

The Cross-Over Model is particularly applicable to the manual control of single input/single output linear time-invariant systems. However, in practice the systems to be controlled often are multiple input/multiple output and time-varying, so one has to search for other methods. Therefore, a model has been proposed in the time domain, which is based on modern control theory. The model is based on the following philosophy: The human operator will behave as an optimal controller, taking into account inherent human limitations such as processing time, inaccurate observation, inaccurate generation of system output, and limb dynamics. The limitations can be modeled by means of a time delay, an observation noise, a motor noise, and a neuromuscular system, respectively. By assuming, for modeling purposes only, that the system to be controlled is linear, that the cost criteria involved are quadratic, and that the noises are white and Gaussian (hence that the separation theorem can be applied - Kwakernaak and Sivan, 1972), optimal control behavior can be modeled by a Kalman filter, an optimal predictor and an optimal controller (Kleinman, Baron, and Levison, 1971).

In Fig. 3, the five subsystems of the Optimal Control Model are depicted. On the basis of a τ-seconds delayed and by observation noise $v_o(t)$ disturbed output $z(t-\tau)$, the Kalman-Bucy filter estimates the delayed state of the system to be controlled $\hat{x}(t-\tau)$. The predictor then generates the actual state vector $\hat{x}(t)$, on the basis of which the optimal controller defines a control signal $u_c(t)$ which after passing the neuromuscular system and adding the motor noise $v_m(t)$ will result in the system input $u(t)$. The control action has to be optimal according to a quadratic cost criterion. It has to compensate for the observation noise, the motor noise, and the time delay. It then follows that

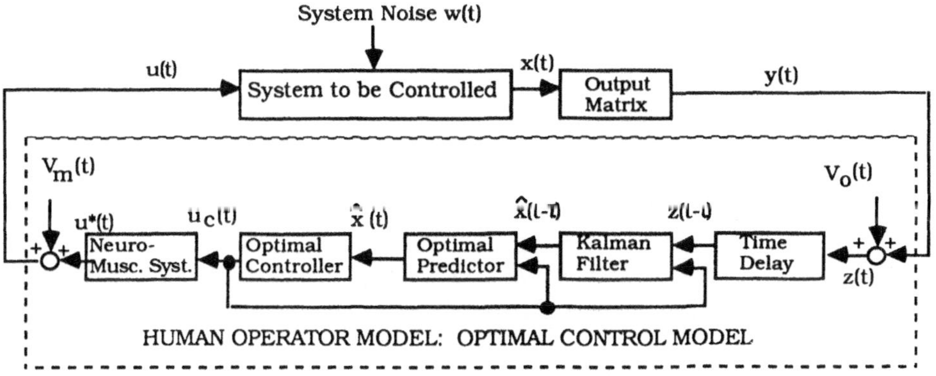

Figure 3. The Optimal Control Model

$$z(t-\tau) = y(t-\tau) + v_o(t-\tau); \qquad (5)$$

$$u(t) = u^*(t) = v_m(t); \qquad (6)$$

$$\tau_N \dot{u}^*(t) + u^*(t) = u_c(t), \qquad (7)$$

where the parameter τ_N is the neuromuscular time-constraint. Consider now the linear system, as given by the Eqs. (8) and (9):

$$\dot{x}(t) = Ax(t) + Bu(t) + G\,w(t); \qquad (8)$$

$$y(t) = Cx(t), \qquad (9)$$

where $x(t)$ means the state vector of the controlled system, $y(t)$ the output and $u(t)$ the control input, $w(t)$ the system noise, and where A is the system matrix, B the input matrix, C the output matrix, and G the noise input matrix. The vector $z(t-\tau)$ will then be the input for the Kalman filter.

Without any loss in generality, one may assume that:

- The mean values of $x(t)$, $y(t)$, $z(t)$, and $u(t)$ are zero.
- The observation noise $v_o(t)$ is white, hence:

$$E\{v_o(t_1)v_o^T(t_2)\} = V_o\delta(t_1 - t_2). \qquad (10)$$

- The system noise $w(t)$ and motor noise $v_m(t)$ can be combined, $v(t)$, and may be regarded as white:

$$E\{v(t_1)v^T(t_2)\} = V\delta(t_1 - t_2). \qquad (11)$$

The cost criterion to be optimized is defined as follows:

$$J(u_c;t) = E\{\frac{1}{T-t_o}\int_{t_o}^{T}[x^T(t)Qx(t) + u_c^T(t)Ru_c(t)]dt\} \quad . \qquad (12)$$

Optimization of Eq. (12) yields the control vector $u_c(t)$. This is achieved by:

- A Kalman-Bucy filter, by which from the observations $z(t-\tau)$, with knowledge of the control vector $u_c(t)$, as well as the matrices A, B, C and G and the intensities V_0 and V, a minimum variance estimate $\hat{x}(t-\tau)$ of the state $x(t-\tau)$ is made (Sage and Melsa, 1971).
- An optimal predictor, by which starting with $\hat{x}(t-\tau)$ and with knowledge of $u_c(t)$, A and B, an estimate $\hat{x}(t)$ of $x(t)$ is achieved (Sage and Melsa, 1971).

- An optimal control law, by which from $\hat{x}(t)$, and with knowledge of A, B, Q and R the control actions $u_c(t)$ are generated (Sage, 1968).

Some very important remarks should be made:

- Besides the fact that the Optimal Control Model is applicable to a much broader field, a fundamental difference exists with the Cross-Over Model. The Cross-Over Model was based on stability considerations, whereas the Optimal Control Model is based on the assumption that the human operator possesses perfect knowledge of the system dynamics and the disturbance statistics. This knowledge is modeled by an internal model, i.e., the matrices A,B,C and G.

- The validation of the model is based on a comparison of the power spectra of the human input and output with those of the model; in this way, a unique solution is not achieved. A much more fundamental approach is given by Kok and van Wijk (1978) and by Kok and Stassen (1980). Starting with the signals u(t) and y(t), and given the matrices A, B, C and G, it is possible to estimate the Kalman-Bucy gain factor, the predictor, and the control law directly. From these, the class of observation noise intensity Vo, the time delay τ and the class of weighting matrices Q and R can be determined. The last problem is known as the inverse control problem (Jameson and Kreindler, 1973; Molinari, 1973). The first one may be introduced as the inverse filter problem (Kok and van Wijk, 1978). The interesting feature of this approach is that it may provide insight into the way the human operator is optimizing. To a certain extent this approach makes an integration with experimental psychology possible. The Optimal Control Model has been applied mainly in the field of astronautics and aviation (Kleinman, Baron and Levison, 1971).

SUPERVISORY CONTROL: FINAL DISCUSSION

Stimulated by the successful attempts to model manual control behavior, several studies at the RBB level have been aimed at supervisory control behavior. Although manual and supervisory control tasks differ substantially, positive results were expected. In supervisory control, the overall task consists of a number of subtasks, such as monitoring, interpreting, set-point correction, planning, fault management, and intervention, whereas in manual control just direct closed-loop actions are involved. Hence, the supervisory task is more or less vaguely and globally defined, leaving the operator a lot of freedom to choose how to reach the goal. As a consequence, difficulties in modeling human supervisory behavior may arise, particularly with respect to the decision-making aspects. Moreover, for those cases where a model will fit the data well, such a model is burdened by a large number of parameters, complicated closed-loop identification techniques, relatively few measurements, and a lack of general decision making mechanisms.

Based on estimation and control theory, probably the two most complete models of supervisory control behavior during stationary process conditions are the Observer Controller Decision Model (Kok and Stassen, 1980) and PROCRU (Baron, Muralidharan, Lancraft, and Zacharias, 1980). The former is derived from the Optimal Control Model, assuming that for slowly responding processes the prediction for time delay, the time delay itself, and the neuromuscular system can be neglected. To describe the monitoring and decision making processes, a decision making system was added. To a certain extent, the model gave satisfactory results (Stassen, et al., 1986), but only during stationary conditions. PROCRU models the behavior of the crew of an aircraft. The state variables describing the aircraft are handled by the Optimal Control Model, whereas verbal messages from air traffic control or between crew members are handled by a rule based pattern recognition system. The output of the state estimation can either be used to drive SBB control actions or, by means of pattern recognition, to provide the left hand side of production rules for RBB. The model has been validated for the approach and landing phases of flight, and it has been suggested (but not tested) that it could be extended to full-scale industrial process control.

A promising new development is based on the Fuzzy Set theory (Zadeh, 1965). The basic idea is to extend classical set theory, where an element belongs, or not, to a set, and to introduce the membership function which describes to what extent an element belongs to a set. This allows one to represent the operator's knowledge in a way which is readily compatible with the linguistic forms used by operators when describing how they perform tasks. A nice example is the work in modeling the navigator's behavior in the maneuvering of tankers in narrow, shallow waterways with heavy traffic (Papenhuijzen and Stassen, 1987; Salski, Noback and Stassen, 1987). In the maneuvering of large vessels, the helmsman executes a manual control task of following the course or controlling the rudder, ordered by the navigator. The navigator in turn defines a desired track; his task includes the determination of the position and speed of the vessel on the basis of information such as boundaries of the fairway, obstacles and other traffic, current and wind, buoys and beacon lights. The task of the navigator requires an internal representation of the environment and the ship dynamics. The authors suggest a model with three subsystems which describe:

- The estimation of vessel position.
- The decision-making process to come to the desired track, including the fairway safety levels, on the basis of the travel plan and environmental data.
- The desired track-following process consisting of a prediction process and a process for following the desired track.

The position estimation is modeled by a Kalman filter which provides, combined with environmental data, the information for the fuzzy decision-making block generating the desired track. Next, a fuzzy predicting process estimates the future dynamic state of the vessel, and finally the fuzzy controller generates the desired heading and the main engine setting.

Tasks to be performed at the KBB level require the operator's creativity and intelligence, hence modeling KBB sounds to be contradictory. However, in recent years new modeling tools have become available. Concepts such as production systems and expert systems borrowed from artificial intelligence research make it possible to describe the internal representations of operators in terms of linguistic production rules. In this way, very complex decision making processes can be formulated on the basis of an internal model (Feigenbaum, 1979). As a consequence, a fundamental question arises: Is this kind of modeling a quantitative description of KBB, or is it just a way to transfer or to reduce (parts of) KBB to RBB?

REFERENCES

Allen, R.W., Jex, H.R., McRuer, D.T., and DiMarco, R.J. (1974). Alcohol effects on driving behavior and performance in a car simulator. In: Proc. 10th NASA-Univ. Conf. Manual Control, Dayton. 341-365.

Baron, S., Muralidharan, R., Lancraft, R., and Zacharias, G. (1980). PROCRU: A model for analyzing crew procedures in approach to landing. Tech Rep. NASA-10035, NASA-Ames, CA.

Borys, B.B., Johannsen, G., Hansel, H.G., and Schmidt, J. (1987). Task and knowledge analysis in coal-fired power plants. IEEE Control Systems Magazine, 7 (3), 26-30.

Conant, R. and Ashby, W.R. (1970). Every good regulator of a system must be a model of that system. Int. J. of Systems Science, 1, 89-97.

Costello, R.C. and Higgins, T.J. (1966) An inclusive classified bibliography pertaining to modeling the human operator as an element in an automatic control system. IEEE-Trans. on Human Factors in Electr. Vol. HFE-7, No. 4, 174-181.

Crossman, E.R.F.W. (1971). "Manual control" models of industrial management. In: Proc. 7th NASA-Univ. Conf. Manual Control, Los Angeles, CA, NASA SP-281, 89-103.

Elkind, G.G. and Miller, D.C. (1966). On the process of adaptation by the human controller. In: Automatic and Remote Control III. 3rd IFAC Congress, London, Vol. 1, Book 2, Paper 30A. 13 pp.

Elkind, J.I. and Miller, D.C. (1968). Adaptive characteristics of the human controller of time-varying systems. Springfield, NTIS. 191 pp. AD-665455.

Feigenbaum, E.A. (1979). Themes and case studies of knowledge engineering. In: D. Michie (Ed.) Expert Systems in the Microelectronic Age. Edinburgh University Press (Scotland).

Francis, B.A. and Wonham, W.M. (1975). The internal model principle of linear control theory. Proc., IFAC 6th World Congress. Boston MA, Paper 43.5.

Gai, E.G. and Curry, R.E. (1976). A model of the human observer in failure detection tasks. IEEE Trans. on SMC-6, (2) 85-91.

Jameson, A. and Kreindler, E. (1973). Inverse problem of linear optimal control. SIAM J. Control, 11, No. 1 (Feb), 1-19.

Jex, H.R. (1971). Problems in modeling man-machine control behavior in biodynamic environments. In: Proc. 7th NASA-Univ. Conf. Manual Control, Los Angeles, CA, NASA SP-281, 3-15.

Jex, H.R. (1988). Critical-instability tracking task - its background, development and application. In: Rouse, W.B., (Ed.) Advances in Man-Machine Systems Research, Vol. 5. Georgia Inst. of Technology, Atlanta, GA. To be published.

Jex, H.R., McDonnel, J.D., and Phatak, A.V. (1966). A "critical" tracking task for manual control research. IEEE Trans. on Human Factors in Electr. HFE-7, No. 4., 138-145.

Johannsen, G. (1984). Categories of human operator behavior in fault management situations. In Proc. IEEE Int. Conf. on Systems, Man, and Cybernetics, (Bombay/New Delhi), 884-889.

Johannsen, G. and Govindaraj, T. (1980). Optimal control model predictions of system performance and attention allocation and their experimental validation in a display design study. IEEE Trans. System, Man, Cybernetics, SMC-10, 249-261.

Kalman, R.E. and Bucy, R.S. (1961). New results in linear filtering and prediction theory. Trans. on ASME, J. of Basic Eng., 83, 95-107.

Kleinman, D.L., Baron, S., and Levison, W.H. (1971). A control theoretic approach to manned-vehicle systems analysis. IEEE Trans. on Automatic Control, AC-16, 824-832.

Kok, J.J. (1973). Human adaptive control: decision models. In: Stassen, H.G., et al., Progress Report Jan 1970 to Jan 1973 of the Man-Machines Systems Group. Report: Delft Univ. Tech., Dept. of Mech. Eng. 161-186, WTHD 55.

Kok, J.J. and van Wijk, R.A. (1978). Evaluation of models describing human operator control of slowly responding complex systems. Delft (The Netherlands) Delft Univ. Tech., 235 pp., Ph.D. Thesis.

Kok, J.J. and Stassen, H.G. (1980). Human operator control of slowly responding complex system: Supervisory control. J. of Cyb and Info. Sc. Special Iss. on MMS,3 (Nos. 1-4), 123-174.

Kwakernaak, H. and Sivan, R. (1972). Linear Optimal Control Systems. New York: Wiley-Interscience, 575 pp.

Levison, W.H. and Elkind, J.I. (1967). Two-dimensional manual control systems with separated displays. IEEE Trans. on Human Factors in Electr. HPB-8, No. 3., 202-209.

McRuer, D.T., Graham, D., Krendel, E.S, and Reisener, W.C. (1965). Human pilot dynamics in compensatory systems - theory, models and experiments with controlled element and forcing function variations. Tech Report: Dayton, 194 pp. AFFDL-TR-65-15.

McRuer, D.T., Hofman, L.G., Jex, H.R., Moore, G.P., Phatak, A.V., Weir, D.H. and Wolkovitch, J. (1968). New approaches to human-pilot/vehicle dynamic analysis. Tech Report: Dayton. 188 pp., AFFDL-TR-67-150.

McRuer, D.T., and Jex, H.R. (1967). A review of quasi-linear pilot models. IEEE Trans. on Human Factors in Electronics, HFE-8, (3), 231-249.

McRuer, D.T., and Krendel, E.S. (1959). The human operator as a servo system element. J. Franklin Inst., 267, 381-403 and 511-536.

McRuer, D.T., and Krendel, E.S. (1974). Mathematical models of human pilot behavior. Report: NATO, AGARD-AG-188, 72 pp.

McRuer, D.T., Magdaleno, R.E., and Moore, G.P. (1968). A neuromuscular actuation system model. In: Proc. IFAC Symposium on Technical and Biological Problems in Cybernetics, Yerevac. 13 pp.

McRuer, D.T., and Weir, D.H. (1969). Theory of manual vehicular control, Ergonomics 12, No.4, 599-633.

Molinari, B.P. (1973). The stable regulator problem and its inverse. IEEE Trans. on Aut. Control, Vol. AC-18, No. 5, 454-459.

Morris, N.M. (1982). The human operation in process control: A review and evaluation of the literature. Report 82-1, Dept of Ind Eng, Georgia Institute of Technology, Atlanta GA.

Papenhuijzen R. and Stassen, H.G. (1987). On the modeling of the behaviour of a navigator. Proc., 8th Ship Control Systems Symposium. The Hague (The Netherlands).

Rasmussen, J. (1983). Skills, rules and knowledge; signals, signs and symbols; and other distinctions in human performance models. IEEE Trans. on SMC, SMC-13, (3), 257-266.

Rasmussen, J. (1986). Information Processing and Human Machine Interaction. New York, North-Holland, 215 pp.

Repa, B.S., Albers, J.W., Rotvin, A.R., and Tourtelotte, W.W. (1971). The use of a battery of tracking tests in the quantitative evaluation of neurological functions. In: Proc. 7th NASA-Univ. Conf. Manual Control, Los Angeles, CA, NASA SP-281, 119-125.

Sage, A.P. (1968). Optimum Systems Control. Englewood Cliffs: Prentice-Hall, 526 pp.

Sage, A.P. and Melsa, J.L. (1971). Estimation Theory With Application to Communications and Control. New York: McGraw-Hill, 529 pp.

Salski, A., Noback, H. and Stassen, H.G. (1987). A model of the navigator's behavior based on fuzzy set theory. In: J. Patrick and K. Duncan (Eds.), Human Decision Making and Control. North Holland Publ. Co., 18 pp.

Sheridan, T.B and Ferrell, W.R. (1974). Man-Machine Systems: Information, Control and Decision Models of Human Performance. Cambridge (USA), MIT Press. 440 pp.

Stassen, H.G. (1976). Man as controller. In: K.P. Kraiss and J. Moraal (Eds.), Introduction to Human Engineering. Koln (FRG), Verlag TUV Rheinland, 1976, 61-81.

Stassen, H.G., Johannsen, G., and Moray, N.P. (1988) Internal representation, internal model, human performance and mental workload. In: Proc. of 3rd IFAC/IFIP/IFORS/IEA Conf. on Man-Machine Systems, Oulu (Finland), Univ. of Oulu, 10 pp.

Stassen, H.G., Kok, J.J., van der Veldt, R.J., and Heslinga, G. (1986). Modeling human operator performance, possibilities and limitations. Proc. of 2nd IFAC/IFIP/IFORS/IEA Conf. on Analysis, Design and Evaluation of Man-Machine Systems. Varese, Italy, 101-106.

Stassen, H.G., van Dieten, J.S.M.J., and Soede, M. (1975). On the mental load in relation to the acceptance of arm prostheses. In: Proc. 5th IFAC, Cambridge, MIT. 11 pp.

Stassen, H.G., Meijer, A.W.A., and van Lunteren, A. (1970). On the possibilities of tactile information transmission for the use in arm prostheses. In: Proc. 6th Conf. Manual Control, Dayton, 481-513.

Summers, L.G. and Ziedman, K. (1964). A study of manual control methodology with annotated bibliography. Report: NASA CR-125, 198 pp.

van Lunteren, A. (1979). Identification of human operator describing function models with one or two inputs in closed loop systems. Delft (The Netherlands), Delft Univ. Tech., 157 pp., Ph.D. Thesis.

van Lunteren, A. and Stassen, H.G. (1967). Investigations on the characteristics of the human operator stabilizing a bicycle model. In: Proc. Intern. Symp. Ergonomics in Machine Design, Prague, 349-370.

van Lunteren, A. and Stassen, H.G. (1970). On the influence of drugs on the behavior of a bicycle rider. In: Proc. 6th Conf. Manual Control, Dayton, 419-439.

Veldhuyzen, W. and Stassen, H.G. (1977). The internal model concept: An application to modeling human control of large ships. Human Factors, 19 (4), 367-380.

Weir, D.H., Heffley, R.K., and Ringland, R.F. (1972). Simulation investigation of driver/vehicle performance in a highway gust environment. In: Proc. 8th Conf. Manual Control, Dayton, 449-471, AFFDL-TR-72-92.

Zadeh, L.A. (1965). Fuzzy sets. Inf. Control 8, 339-353.

MODELS OF HUMAN MONITORING AND DECISION MAKING

IN VEHICLE AND PROCESS CONTROL

Willi Stein

Research Institute for Human Engineering (FAT)
5307 Wachtberg-Werthhoven
Federal Republic of Germany

INTRODUCTION

The growing interest in monitoring and decision making and the related models depends on many factors. Due to increasing automation and the use of advanced information technology, the human's function in vehicle and process control is shifting from a direct and continuously active involvement towards a supervisory control structure, where man-machine interaction is exercised through the mediation of a computer. Supervisory control indicates a hierarchy or coordinated set of human activities that includes initiating, monitoring, detecting events, recognizing, diagnosing, adjusting, and optimizing processes in systems that are otherwise automatically controlled. Thus, the spectrum of monitoring and decision making includes primary task components of the human operator in vehicle and process control. In this paper, a survey of models and related experimental studies of human monitoring and decision making is presented. Fitts et al. (1950) started with the empirical study of monitoring and Senders (1964) pioneered in the related model development (Moray, 1986). Special emphasis is given to the control theory models of monitoring and decision making which are based on the information processing structure of the optimal control model (OCM) developed by Kleinman, Baron, and Levison (see Baron (1984), Rouse (1980), and Sheridan and Ferrell (1974)). The experimental studies presented include a variety of multivariable monitoring and decision making tasks. Characteristic factors are, for example, the number of displayed dynamic processes, bandwidths, event probabilities and correlation among abnormal events and unfailed processes. The models predict several measures of monitoring and decision making behaviour, the decision speed/accuracy trade-off and the attentional characteristics, including the time requirements of effective instrument fixations and eye movements. Most of the predictions are based on a few free model parameters only. The considerable level of overall agreement between the models and the experimental results provides the predictive potential for the analysis, design, and evaluation of man-machine interfaces.

CHARACTERISTICS OF MONITORING AND DECISION MAKING

According to Moray (1986), a human operator is monitoring a system when he or she scans an array of displayed information without taking any action to change the system state. The purpose of monitoring is to update the operator's knowledge and so to permit appropriate decisions. Monitoring is normally dominated by vision, but auditory signals and communication may be involved, especially when coordinating action. Human decision making as a discrete control activity in the context of man-machine systems may be defined as the process of selecting an appropriate alternative from a set of possible alternatives, based on the perception of actual system states and other sources of information (Sheridan and Ferrell, 1974). Many types of human activity in man-machine systems have decision making as an implicit component, although they would be more commonly considered as sensory, sensory-motor, or even cognitive tasks. Deeper insights into the complexity of human decision making are given by Moray (1986), Rasmussen and Rouse (1981), Schrenk (1969), and Sheridan and Ferrell (1974). Mathematical theories of stochastic estimation, statistical signal processing, failure detection, and diagnosis are presented by Pau (1981), Sage and Melsa (1971), and Srinath and Rajasekaran (1979). Following these definitions of human monitoring and decision making, a model-oriented classification of related operator tasks can be proposed:

1) Pure monitoring (i.e., observing without acting).
2) Independent decisions based on non-sequential observations.
3) Independent decisions based on sequential observations.
4) Dependent (or dynamic) decisions based on heterogeneous types of observations.
5) Heterogeneous types of observations and decisions, often embedded in sequences of other task components (e.g., supervisory control situations).

During phases of pure monitoring or scanning, the nature of actions executed as a result of information obtained by scanning cannot be taken into account.

A particular type of task including independent decisions and non-sequential observation is the so-called tolerance-band monitoring (TBM) which has been extensively studied and modeled by Stein and Wewerinke (1983) and Wewerinke (1976). It is interesting to see that the experimental situations of Senders (1964, 1983), Levison (1971), and other researchers (see Moray, 1986) can be characterized as tolerance-band monitoring. TBM tasks involve observing a dynamic process (which can include stochastic and deterministic components) in respect to exceeding an explicitly indicated tolerance band. In a binary case, performing a TBM task can be characterized as making a sequence of independent binary decisions, where each single decision can be based on a single observation testing a pair of hypotheses.

A particular type of task including independent decisions and sequential observations is the so-called failure detection (FD) which has been extensively studied and modeled by Gai and Curry (1976), Kleinman and Curry (1977), Stein and Wewerinke (1983), and Wewerinke (1983). FD tasks involve observing a dynamic process (which can include stochastic and deterministic components) for the potential occurrence of an abnormal event (e.g., a failure), where an event is defined as a change in the statistics of the displayed process. This change may be constituted for example by changes in mean, standard deviation, dynamic properties, and other characteristics. Performance measures of FD tasks can include speed and accuracy data with related trade-offs. The detection time denotes the interval between occurrence and detection of a system failure. The accuracy of detecting failures is described by false alarm and miss probabilities. Optimally detecting events or failures with a given accuracy requires sequential observations, i.e., the number of subsequent observations used as input information for making decisions is not fixed, but greater than one (Pau, 1981; Sage and Melsa, 1971; Srinath and Rajasekaran, 1979). Thus, detecting a failure by observing a displayed dynamic process is an unstationary binary task that includes testing a pair of hypotheses.

A dependent (or dynamic) decision making task is given, if factors in subsequent decisions are influenced by a choice made earlier (Sheridan and Ferrell, 1974). Particular types of dynamic decision making tasks have been experimentally studied (i.e., based on specifically developed laboratory paradigms) and modeled by Tulga and Sheridan (1980) and Pattipati et al. (1983).

Tasks including heterogeneous types of observations and decisions are given, when the human operator becomes more of a supervisor within a man-machine system, where his primary responsibilities for example are state monitoring, situation assessment, plan execution, and failure management. Studying and modelling complex task situations like supervisory control has proven to be extremely difficult. Issues of human supervisory control have been considered by Baron (1984), Kok and van Wijk (1978), Moray (1986), Rouse (1980, 1981), Sheridan and Johannsen (1976), and Sheridan (1987).

BASIC MODEL CONCEPTS

The beginning of investigating human monitoring behaviour (see figure 1) is marked by the work of P.M. Fitts and his group (summarized by Fitts et al. (1950), reviewed by Moray (1986)). These studies included instrument landing approaches of 40 pilots. The point of fixation was identified by means of camera recordings, but the state of the cockpit instruments was not recorded simultaneously. Figure 1a) shows the positions of the cockpit instruments together with two statistics: FPM is the number of fixations per minute made on that instrument; the second number is the percentage of time during the flight that the pilot spent looking at the instrument. The eye movement link values shown in figure 1b) are not true transition probabilities. They are the fraction of all eye transitions that went between the instruments indicated. Values less than 2 per cent are omitted. The results of these studies show for example (see Moray 1986) that the mean fixation duration was 0.6 sec, with a standard devi-

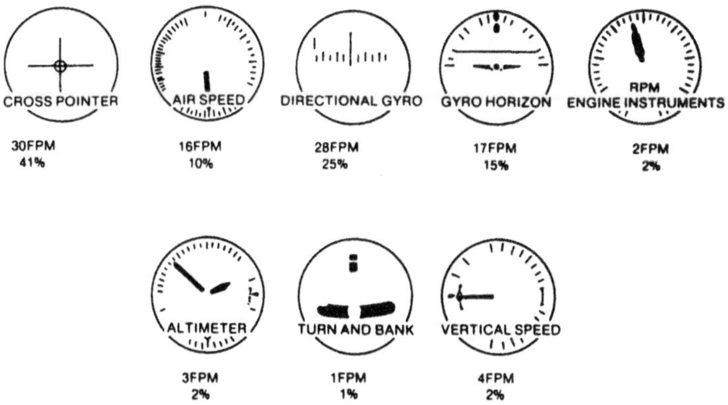

a) Cockpit instrument array (fixations per minute (FPM) and percentage of time).

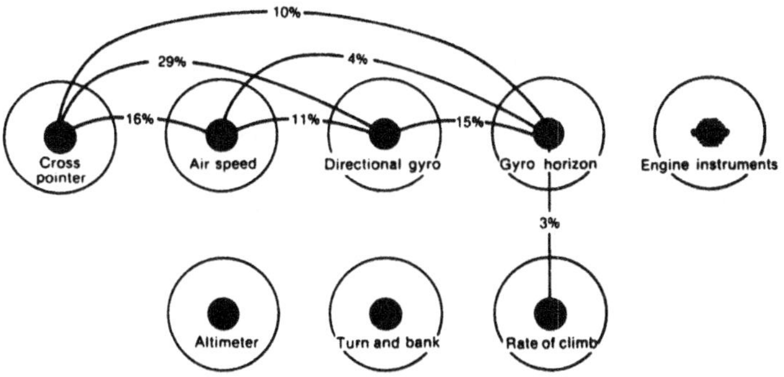

b) Eye movement link values (percentage of transitions, values < 2 % omitted).

Figure 1. Instrument fixations and eye movements of a pilot during a landing approach (Source: Moray (1986), adapted from Fitts et al. (1950)).

ation of 0.12. The range of fixation duration was 0.4 - 1.4 sec and depended both on the individual pilot and on the instrument that was fixated.

The first monitoring model (Senders, 1964), developed in the light of the work of Fitts' group on the eye movements of pilots in cockpits (see Moray, 1986), is very simple and assumes that the spectral characteristics of the displayed signal (e.g. represented in terms of the bandwidths) are the only determinant of the monitor's behaviour. The subsequent models are increasingly complex and take into account more and more factors of real-world tasks such as subjective value and the meaning of the displayed variables.

Visual Sampling Model of Senders

The pioneering approach to develop quantitative analytic models of human monitoring was made by Senders (1964). This model, based upon information theoretic concepts developed by Shannon and others, describes the way in which an operator devides his attention among a number of instruments while he monitors them. Fundamentally, it assumes that a human operator's fixation frequency for a particular instrument depends upon its information generation rate,

$$\dot{H}_i = W_i \log_2\left(A_i^2 / E_i^2 \right) \text{ bits/sec,} \tag{1}$$

where W_i is the bandwidth of the i-th displayed signal, A_i is its amplitude, and E_i is the permissible rms reading error of instrument i. For an observer to reconstruct the signal, the sampling theorem requires that his sampling rate F_i be at least 2 W_i. If $F_i = 2\,W_i$, then the average information to be assimilated by the operator at each fixation is

$$\dot{H}_i / \left(2W_i \right) = H_i = \log_2\left(A_i / E_i \right) \text{ bits.} \tag{2}$$

For an observer with a fixed channel capacity, who must share his attention among several displays presenting uncorrelated stochastic processes with known information rates, the attentional demand of a particular instrument is calculated to be where

$$T_i = 2K\ W_i\ \log_2\left(A_i/E_i\right) + 2\ W_i\ \ C\ \text{sec/sec}, \tag{3}$$

is the percentage of total time that must be devoted to displayed process i, K is a constant with dimensions of time per bit, and C (with dimensions of time per fixation) is a constant that accounts for movement time and minimum fixation time. Hence, the duration of a sample is given by

$$D_i = K\ \log_2\left(A_i/E_i\right) + C\ \text{sec}. \tag{4}$$

Validation studies, using an array of instruments displaying signals of various amplitudes and bandwidths, showed a good agreement with the model predictions. Values for the constants K and C must be extracted from data collected in a specific context. The model yields estimates of the fixation frequencies and durations for each signal, and for the probabilities of transitions between signals. Limitations of the model became apparent, when an attempt was made to take into account (1) correlations between displayed signals and (2) the interactions between control behaviour and visual sampling.

In a recent monograph, Senders (1983) developed his pioneering approach in several ways, provided new and simpler mathematical derivations, and proposed various models depending upon diverse definitions of the goals of the human monitor. The most elementary goal and perhaps the most unreal is that of signal reconstruction. An equally unreal goal is that of sampling in the way of pure random choice. Based on these considerations, the following strategies in visual sampling behaviour have been identified that should be regarded as complementary rather than alternatives:

1) Periodic sampling.
2) Random constrained sampling.
3) Conditional sampling:
   a) Sampling when probability is maximum.
   b) Sampling when probability exceeds a threshold.
   c) Variable Nyquist sampling.
4) Signal reconstruction with imperfect memory.

The various models predict different variances for the data. Periodic sampling yields no variability at all, a clearly unrealistic prediction. In the case of random constrained sampling, the variance is a direct consequence of the sampling process. In the case of conditional sampling, the interval is assumed to be a certain mathematical function of the previously observed value.

Although each strategy in visual sampling behaviour generates a different mathematical analysis, the assumptions about the signals will be the same:

1) The signals displayed are random, band-limited time functions with Gaussian amplitude density distributions.
2) The signals which drive the instruments in an array are assumed to be statistically independent and uncorrelated with one another.
3) There are always assumed to be three or more instruments in the array.
4) The different signals displayed do not differ in value.

Instrument Monitoring Model of Smallwood

The instrument monitoring model of Smallwood (1967) grew out of previous studies involving Senders, Carbonell et al. (1968). The task under consideration involves a human operator monitoring the readings of a given number of instruments (e.g. four instruments), which are driven by signals of different amplitudes and bandwidths, and signalling whenever any instrument exceeds a certain threshold.

The model assumes that the human operator constructs an internal model of the processes being monitored. The model further assumes (1) that a dead time of about 0.1 second is required to shift attention between two instruments and (2) that the time required to read an instrument is inversely related to its distance from the threshold. The predictions of the model include:

- Relative fixation frequency for each instrument.
- Duration of fixation for each instrument.
- Average transition probabilities between the instruments.

The concept of the internal model that has been used by Smallwood in an explicit form, plays an important role in the discussion of human operator behaviour. Indeed human operator models imply an internal model, even when they do not mention it.

Carbonell's Queueing Model

Carbonell (1966) introduced queueing theory concepts to human operator modelling and emphasized the importance of considering the operator's actions. Thus, he moved from the abstraction of Senders's models to more realistic tasks. The human monitor is modeled as a single-channel server that can attend to only one instrument at a time. It is assumed that at each step in his sampling process, the monitor determines for each instrument a subjectively expected cost for not observing it next, and then chooses to observe the instrument with the highest cost of being ignored. An additional assumption is made that the time involved in reading an instrument is constant (approximately 0.33 sec.). According to Pew et al. (1977), the total cost of not looking at any instrument is defined by

$$C(t) = \sum_{i=1}^{M} \frac{c_i P_i(t)}{1 - P_i(t)}, \tag{5}$$

where M is the total number of instruments, c_i is the cost associated with instrument i exceeding its allowed limit, and $P_i(t)$ is the probability that instrument i will exceed its threshold at time t. Thus, the total cost of looking at instrument j at time t is

$$C'_j(t) = C(t) - c_j P_j(t) \tag{6}$$

and the aim of the human monitor will be to choose the instrument j that will minimize $C'_j(t)$ at any time.

Carbonell's model was compared with Senders' model for eye movement data from realistic landings in an instrument flight simulator using pilots as subjects. It was found to be considerably more accurate than the simpler model but it had to be tuned to each pilot, using his individual estimates of costs, tolerances, and action thresholds. This need to fit the models to individuals emphasizes the concept that the operator has his own internal model of the process he is monitoring or controlling.

Using the model, one must specify the statistical characteristics of each displayed signal, the costs of exceeding given thresholds on each display, and the thresholds below which each instrument reading is ignored. Then the model yields a time sequence of instrument fixations which may be analyzed to get visual sampling parameters of interest. A significant feature of the model is that the displayed signals are not assumed to be Gaussian with zero mean.

Thus, the model represents a significant advance in modelling human sampling behaviour, although it does not attempt to take into consideration cross-coupling among instruments. Referring to Pew et al. (1977), the flexibility and power of this model is obtained at the cost of considerable analytical complexity.

CONTROL THEORY MODELS

Several different monitoring and decision making models have been derived using the information processing structure (i.e., the stages including perceptual limitations, delay, estimator and predictor of figure 3) of the optimal control model (OCM) outlined by Baron (1984), Sheridan and Ferrell (1974), and in other sources. It is apparent that the estimator-predictor combination produces outputs that can be used for assessing system states and detecting events (the field of control theory models and related experimental situations is illustrated in figure 2 and 3). Theoretical aspects of these OCM-based monitoring and decision making models are discussed by Phatak and Kleinman (1972) and Kleinman and Curry (1977); there also the possibility of using the residual sequence generated by the OCM information processing structure to model human failure detection is discussed.

Monitoring and Decision Making Model of Levison

According to Baron (1984), the first use of the OCM information processing structure in modelling monitoring and decision making was by Levison (1971). He studied the problem of how well subjects could determine whether a signal embedded in added noise was within specified tolerances. It is assumed that the operator perceives a noisy, delayed version of the displayed processes. The perceived data are then processed, via an optimal estimator-predictor combination, to generate (1) a maximum variance estimate of the system state vector and (2) the covariance of the error in that estimate. This estimator-predictor yield is a sufficient statistic for testing hypotheses about the state of the system. The model assumes that the operator is an optimal decision-maker in the sense of maximizing expected utility. This strategy is then applied to the problem of deciding whether or not a signal corrupted by noise, is within certain prescribed tolerances. For equal penalties on missed detections and false alarms, this rule reduces to one of minimizing the expected decision error. The resulting decision rule is that of a Bayesian decision maker using a likelihood ratio test. Experimental results have been compared with model predictions for the following task situations: (1) monitoring a single displayed process, (2) monitoring two processes and (3) concurrent manual control and monitoring task. Using fixed values for model parameters, model predictions of single-task and two-task decision performance are within an accuracy of 10 per cent.

Failure Detection Model of Gai and Curry

Based on the OCM information processing structure, a failure detection model has been developed by Gai and Curry (1976). They have tested that model in a simple laboratory task and in an experiment simulating pilot monitoring of an automatic approach. In both cases, step or ramp failures were added to an observed signal at a random time to simulate a failure. This produced a non-zero mean value for the signal and for the residual; failure detection consisted of testing an hypothesis concerning the mean of the distribution of the residuals. Sequential analysis was used to perform the hypothesis test. By summing the residuals, a likelihood ratio can be calculated and used to arrive at the decision. Gai and Curry modified classical sequential analysis to account for the fact that a failure detection problem is characterized by a transition from one mode of operation to another at a random time, whereas the classical analysis is based on the assumption that the same mode of operation exists during the entire observation interval. They reported good agreement between predicted and observed detection times for both the simple and more realistic situations. In later experiments, the model was used in a multi-instrument monitoring task and accounted for attention sharing and cross-checking of instruments to confirm a failure. A significant result of the experiments was that the property of integration of the residuals appeared to be confirmed for both step and ramp type failures.

Experimental Paradigm and Control Theory Models of Stein and Wewerinke

A laboratory paradigm has been developed by Stein and Wewerinke (1983) as an experimental basis for model-oriented research on various types of monitoring and decision making tasks including eye-movement studies. The corresponding model shown in figure 3, derived from the OCM information processing structure (Baron, 1984), has been developed by Wewerinke (1976, 1983). By using different decision rules the model can be adapted to different types of tasks:

1) In the case of independent decisions based on non-sequential observations (e.g., tolerance-band monitoring tasks, TBM), an optimal Bayesian decision rule (Sage and Melsa, 1971; Srinath and Rajasekaran, 1979) is involved in the model (Wewerinke, 1976).
2) In the case of independent decisions based on sequential observations (e.g., failure detection tasks, FD), a sequential decision rule based on a generalized likelihood ratio test (Sage and Melsa, 1971; Srinath and Rajasekaran, 1979) is involved in the model (Wewerinke, 1983).

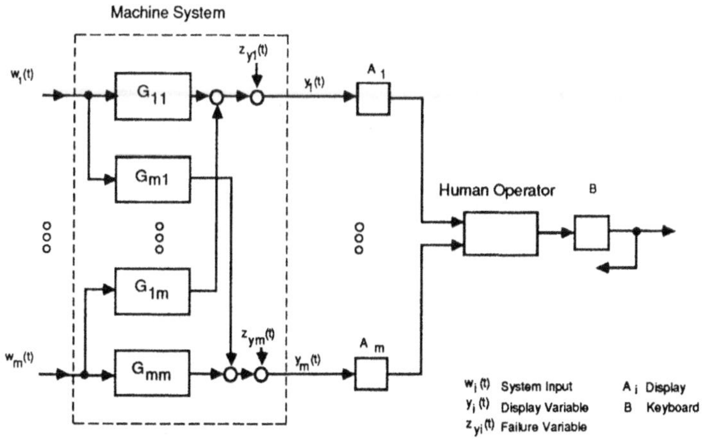

Figure 2. Experimental situation of monitoring and decision making.

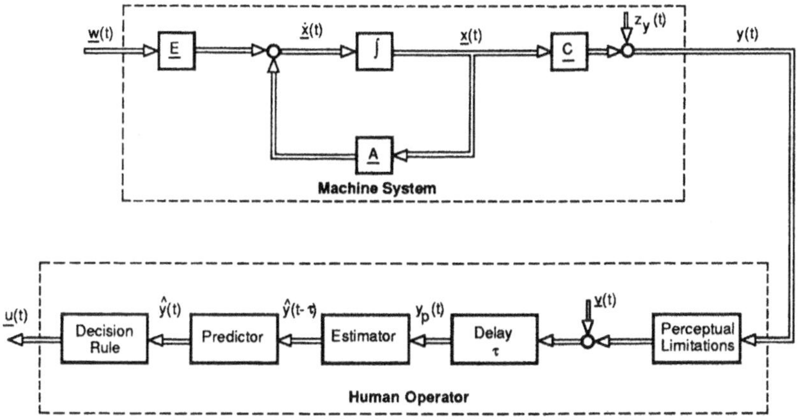

Figure 3. Control theory model of monitoring and decision making.

Based on the laboratory paradigm and the related model of figure 2 and 3, the following factors of tolerance-band monitoring (TBM) and failure detection (FD) have been experimentally studied including eye-movement recordings and have been modeled by Stein and Wewerinke (1983) and Wewerinke (1983):

1) Number of displayed processes.
2) Bandwidths of displayed processes.
3) Amplitude of processes or probability of resulting events.
4) Type and intensity of failures embedded in displayed processes (e.g., step, ramp, dynamic properties, etc.).
5) Couplings among unfailed displayed processes.
6) Couplings among displayed failures.
7) Size of display array and operator's field of view.

An overview of the results of tolerance-band monitoring is given in figure 4 (Stein and Wewerinke, 1983). The correspondence of data and model is high. Human time delay is assumed to be constant at 0.2 sec. Considering a given task situation with a constant process bandwidth, monitoring performance in terms of the decision error is very sensitive to the observation noise ratio or fraction of human attention devoted to the displayed process; the observation noise ratio increases, when attention is devoted to several processes. The decision error represents the cumulated time fractions of false alarm and missed tolerance-band exceedance. The decision error increases monotonically with bandwidth; the increase begins linearly and becomes progressively nonlinear as a function of both bandwidth and observation noise.

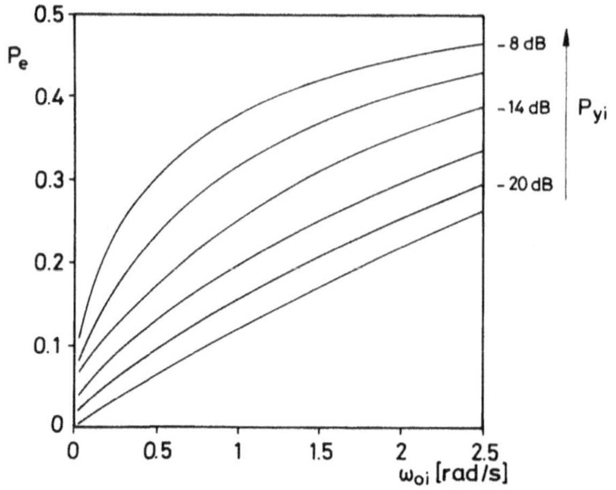

Figure 4. Decision error P_e vs bandwidth ω_{oi} as function of observation noise ratio P_{yi}.

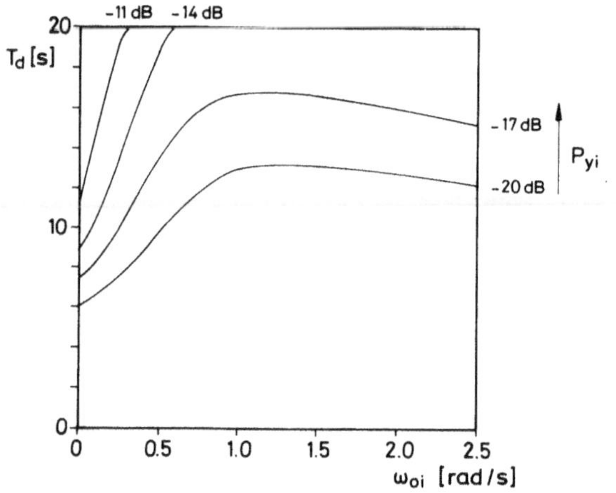

Figure 5. Detection time T_d vs bandwidth ω_{oi} as function of observation noise ratio P_{yi}.

An overview of the results of failure detection is given in figure 5 (Stein and Wewerinke, 1983). These results are restricted to situations with ramp failures. The accuracy of failure detection in terms of false alarm probability is assumed to be constant on a level of 0.05. The detection time increases with observation noise ratio, e.g., when human attention is devoted to several displayed processes. Compared with tolerance-band monitoring, process bandwidth is a factor of reduced influence in failure detection tasks. The predictor portion of the model may be dropped, if human time delay (e.g., 0.2 sec) is small in comparison with detection time.

MODELS OF MULTI-TASK SITUATIONS

Dynamic Decision-Making Model of Tulga

A model of human dynamic decision making in multi-task situations has been proposed and validated on the basis of a laboratory paradigm by Tulga and Sheridan (1980). Tulga modeled the monitoring and control strategy of the human operator as a dynamic queueing system based on the supervisory sampling concept of Sheridan (1970) trying to maximize value over a limited planning horizon. The tasks are characterized in terms of the distribution of times between arrivals, the distribution of times to perform tasks (i.e., the service time), and the priority structure among tasks. As a particular result, the model produces predictions of the sequence of tasks likely to be chosen by the human.

Dynamic Decision-Making Model of Pattipati

The dynamic decision making model (DDM) developed by Pattipati et al. (1983) uses the information processing structure of the OCM, extends the control theory approach to dynamic decision making and particularly addresses the problem of task selection in a dynamic multi-task environment. The experimental paradigm of Tulga and Sheridan (1980) was modified to provide an appropriate laboratory task for validating the DDM. In case of a situation with N independent tasks, the DDM includes a set of N independent estimator-predictor combinations shown in figure 3.

The DDM can predict various measures related to decision-making performance, including task completion probability and error probability. As in various other models, situation assessment in the DDM involves estimation of the time available and the time required for task completion. A major contribution of the DDM work is the experimental validation of the model. By constraining the paradigm to a situation that can be treated carefully in an experimental environment, it becomes possible to test model hypotheses with reasonable cost and control.

Decision, Monitoring, and Control Model

A decision, monitoring, and control model (DEMON) has been developed by Muralidharan and Baron (see Baron, 1984) for analyzing the operational control of a set of N remotely piloted vehicles. The operator's task is to monitor the trajectories and estimated times of arrival, to decide if deviations from desired flight paths and time errors exceed some tolerance thresholds and to correct the paths by appropriate control commands. Display information is assumed to be updated at discrete times. According to Baron (1984), the essence of the DEMON model is to characterize the operator limitations and the mission goals and to predict operator strategies and overall system performance. The DEMON model is an important integrative step in the development of supervisory control models.

Procedure-Oriented Crew Model

The supervisory control model PROCRU (procedure-oriented crew model) has been developed by Baron et al. (see Baron, 1984) for analyzing flight crew procedures in commercial aircraft control situations. PROCRU incorporates both "by the book" procedures and more unconstrained forms of monitoring and control behaviour. It models continuous tasks and also accounts for the effects of discrete control tasks and for the time to perform them. PROCRU can be extended to a full range multi-operator model. The monitoring and information processing portions of PROCRU are not unlike those of the optimal control model (OCM), though they have some novel features and extensions.

The supervisory control model PROCRU includes representations of a series of operator tasks and processes: (1) monitoring displays, (2) situation assessment, (3) decision to act - or not to act - based on that assessment, and (4) action to implement the decision. These functions are implemented by various processors:

1. A display processor selects an appropriate displayed quantity and accounts for sensory/processing limitations in observation.

2. An information processor includes a mental model of the plant from which is derived a predict/correct logic for state estimation and prediction.

3. A situation assessor provides a template matching scheme which checks symptoms against a template which is part of a procedure residing either in a manual or in memory.

4. A procedure selector includes major decision making at several levels. Choices are made on the basis of utility theory.

5. A procedure effector permits three types of actions: control, observation, and communication. Time is associated with each action.

According to Baron (1984), PROCRU has been used in model studies but is conceptual in that some of its features have yet to be implemented.

MODEL APPLICATIONS

The usefulness of monitoring and decision making models to the design and evaluation of display arrays and man-machine interfaces can be illustrated in the following way. Given the mathematical equations for the vehicle and process dynamics, the statistical properties of disturbance, and the performance tolerances of each display then the model user is enabled to calculate the fraction of time that the human operator will spend looking at each display, as well as likely transitions among displays. Thus, the model user can determine valuable indices of human operator behaviour. For example, displays that require a relatively large time fraction of looking should be placed near each other, or perhaps be integrated into a single display.

A design procedure to determine the optimal layout of multiple-instrument displays has been proposed by Freund and Sadosky (1967). The design has been formulated as an optimization problem using linear programming and the type of information produced by human sampling models (e.g., fixation durations and matrices of eye transitions between displays).

The first design approach based on a human performance model is the display design procedure for manual control situations developed by Clement et al. (1972). This approach utilizes a quasilinear pilot model, a rationale based on visual scanning results, and an empirical workload metric.

According to Baron and Levison (discussed by Rouse, 1980), the following general display design issues can be addressed using the optimal control model and its derivations:

- Is status information acceptable?
- Will additional information degrade performance due to interference and/or high workload?
- Do the advantages of display integration outweigh the improved scaling possible with separate displays?
- Does command information integrate status effectively and, if not, how should it be done?
- What performance and workload levels can be achieved with a perfectly integrated and scaled display?
- Will quickening, prediction, or preview displays improve performance?
- What format should such displays have?

The advantage of design and evaluation approaches based on the optimal control model and its derivations stems from (1) the model structure composed of modules for separate human functions (e.g. visual perception, central processing, motor response), (2) the flexible information structure suited for multivariable, multiple process and/or multitask situations, (3) the unique performance/workload or performance/attention metric, (4) the comparably high level of model validation, and (5) the underlying normative modeling perspective. The OCM-approaches are highly developed and seem to be very attractive.

CONCLUSIONS

In this paper a survey of human monitoring and decision making models has been given. The task of modeling the human operator is not an easy one, as might be imagined. Thus, the number of models developed so far is impressive and might still increase. Currently, the most highly developed theories of supervisory control and there especially of the field of monitoring and decision making are extensions of optimal estimation and control theory (Baron, 1984; Moray, 1986). While the usefulness of models in terms of quantitative predictions is very important, it should be stressed that models have other important uses, such as providing an organized means for thinking about a system design problem; these qualitative uses of models are likely to become more important (Rouse, 1981).

REFERENCES

Baron, S. (1984). A Control Theoretic Approach to Modelling Human Supervisory Control of Dynamic Systems. In: W.B. Rouse (Ed.). Advances in Man-Maschine Systems Research. Greenwich, Conn.: JAI Press.

Carbonell, J.R. (1966). A Queueing Model of Many-Instrument Visual Sampling. IEEE Trans. Human Factors in Electron., Vol. 7, No. 4, p. 157-164.

Carbonell, J.R., Ward, J.L., and Senders, J.W. (1968). A Queueing Model of Visual Sampling Experimental Validation. IEEE Trans. Human Factors in Electron., Vol. 9, No. 3, p. 82-87.

Clement, W.F., McRuer, D.T., and Klein, R.H. (1972). Systematic Manual Control Display Design. Proceedings, Guidance and Control Displays, CP-96. Neuilly sur Seine, France: AGARD.

Curry, R.E., Kleinman, D.L., and Hoffman, W.C.(1977). A Design Procedure for Control/Display Systems. Human Factors, Vol. 19, No. 5, p. 421-436.

Fitts, P.M., Jones, R.E., and Milton, J.L. (1950). Eye movements of aircraft pilots during instrument landing approaches. Aeronautical Engineering Review, Vol. 9, p. 1-5.

Freund, L.E., and Sadosky, T.L. (1967). Linear Programming Applied to Optimization of Instrument Panel and Workplace Layout. Human Factors, Vol. 9, No. 4, p. 295-300.

Gai, E.G., and Curry, R.E. (1976). A Model of the Human Observer in Failure Detection Tasks. IEEE Trans. Syst., Man, and Cybern., Vol. 6, No.2, p. 85-94.

Gould, J.D. (1968). Visual Factors in the Design of Computer-Controlled CRT Displays. Human Factors, Vol. 10, No. 4, p. 359-376.

Jones, J.C. (1981). Design Methods. Chichester, UK: John Wiley.

Kleinman, D.L., and Curry, R.E. (1977). Some New Control Theoretic Models of Human Operator Display Monitoring. IEEE Trans. Syst., Man, and Cybern., Vol. 7, No. 11, p. 778-784.

Kok, J., and van Wijk, R. (1978). Evaluation of Models Describing Human Operator Control of Slowly Responding Complex Systems. Delft, Netherlands: Delft University Press.

Levison, W.H. (1971). A Control-Theory Model for Human Decision Making. Seventh Annual Conf. Manual Control, NASA SP-281. Washington, D.C.: National Aeronautics and Space Administration.

Moray, N. (1985). Monitoring Behavior and Supervisory Control. In: K.R. Boff, L. Kaufman, J.-P. Thomas (Eds.), Handbook of Perception and Human Performance. New York: John Wiley.

Nadler, G. (1985). Systems Methodology and Design. IEEE Trans. Syst., Man, and Cybern., Vol. 15, No.6, p. 685-697.

Pattipati, K.R., Kleinman, D.L., and Ephrath, A.R. (1983). A Dynamic Decision Model of Human Task Selection Performance. IEEE Trans. Syst., Man, and Cybern., Vol. 13, No. 3, p. 145-166.

Pau, L.F. (1981). Failure Diagnosis and Performance Monitoring. New York: Marcel Dekker.

Pew, R.W., Baron, S., Feehrer, C.E., and Miller, D.C.(1977). Critical Review and Analysis of Performance Models Applicable to Man-Machine Systems Evaluation. BBN Report 3446. Cambridge, Mass.: BBN Laboratories.

Pew, R.W., and Baron, S. (1983). Perspectives on Human Performance Modelling. Automatica, Vol. 19, No. 6, p. 663-676.

Phatak, A.V., and Kleinman, D.L. (1972). Current status of models for the human operator as a controller and decision maker in manned aerospace systems. Proceedings, Automation in Manned Aerospace Systems, CP-114. Neuilly-sur-Seine, France: AGARD.

Rasmussen, J., and Rouse, W.B.(1981). Human Detection and Diagnosis of System Failures. New York: Plenum Press.

Rouse, W.B. (1980). Systems Engineering Models of Human-Machine Interaction. New York: North Holland.

Rouse, W.B. (1981). Human-Computer Interaction in the Control of Dynamic Systems. Computing Surveys, Vol. 13, No. 1, p. 71-99.

Rouse, W.B., and Boff, K.R. (Eds.)(1987). System Design - Psychological Aspects. New York: North-Holland.

Sage, A.P., and Melsa, J.L. (1971). Estimation Theory with Applications to Communications and Control. New York: McGraw-Hill.

Schrenk, L.P. (1969). Aiding the Decision Maker: A Decision Process Model. Ergonomics, Vol. 12, p. 543-557.

Senders, J.W. (1964). The Human Operator as a Monitor and Controller of Multidegree of Freedom Systems. IEEE Trans. Human Factors in Electron., Vol. 5, No. 1, p. 2-5.

Senders, J.W. (1983). Visual Scanning Processes. Tilburg, Netherlands: Tilburg University Press.

Sheridan, T.B. (1970). On how often the supervisor should sample. IEEE Trans. Syst., Sci., and Cybern., Vol. 6, p. 140-145.

Sheridan, T.B., and Ferrell, W.R.(1974). Man-Machine Systems: Information, Control, and Decision Models of Human Performance. Cambridge: The MIT Press.

Sheridan, T.B., and Johannsen, G. (Eds.)(1976). Monitoring Behavior and Supervisory Control. New York: Plenum Press.

Sheridan, T.B. (1987). Supervisory Control. In: G. Salvendy (Ed.). Handbook of Human Factors. New York: John Wiley.

Smallwood, R.D. (1967). Internal Models and the Human Instrument Monitor. IEEE Trans. Human Factors in Electron., Vol. 8, No. 3, p. 181-187.

Srinath, M.D., and Rajasekaran, R.K. (1979). An Introduction to Statistical Signal Processing with Applications. New York: John Wiley.

Stein, W. (1981): A Monitoring and Decision Making Paradigm: Experiments and Human Operator Modelling. Proceedings, First European Annual Conference on Human Decision Making and Manual Control. Laboratory for Measurement and Control. University of Technology, Delft, The Netherlands.

Stein, W., and Wewerinke, P.H. (1983). Human Display Monitoring and Failure Detection: Control Theoretic Models and Experiments. Automatica, Vol. 19, No. 6, p. 711-718.

Tulga, M.K., and Sheridan, T.B. (1980). Dynamic Decisions and Work Load in Multitask Supervisory Control. IEEE Trans. Systems, Man, and Cybern., Vol. 10, No. 5, p. 217-232.

Wewerinke, P.H. (1976). Human control and monitoring - models and experiments. Proceedings, 12th Ann. Conf. Manual Control. Springfield, Virg.: National Technical Information Service.

Wewerinke, P.H. (1983): Model of the Human Observer and Decision Maker - Theory and Validation. Automatica, Vol. 19, No. 6, p. 693-696.

Wierwille, W.W. (1981). Statistical Techniques for Instrument Panel Arrangement. In: J. Moraal, and K.-F. Kraiss (Eds.). Manned Systems Design. New York: Plenum Press.

TARGET ACQUISITION: PERFORMANCE MEASURES, PROCESS MODELS, AND DESIGN IMPLICATIONS

Richard J. Jagacinski

Ohio State University
Department of Psychology
Columbus, Ohio 43210

INTRODUCTION

Two types of movement tasks that have been studied from an engineering psychology perspective are continuous tracking and target acquisition. These two types of tasks can be distinguished by their goals. The goal of a continuous tracking task is typically to minimize the time-averaged discrepancy between a target and the output of some dynamic system. Keeping an aircraft on a predetermined flight path or regulating the headway between one's own automobile and another vehicle are examples of continuous tracking tasks. In contrast, the goal of a target acquisition task is to bring the output of a dynamic system to some desired terminal state. Moving one's finger to a numerical entry key on a calculator or positioning one's automobile in a parking space are examples of target acquisition tasks. Often continuous tracking is preceded by target acquisition in order to bring the target within close range of the tracking system. This chapter will discuss target acquisition tasks. A discussion of continuous tracking tasks can be found in the chapters by Stassen and by Allen, McRuer, and Thompson.

Two very important factors influencing the performance of target acquisition tasks are speed-accuracy tradeoffs and stimulus-response compatibility. In evaluating systems for performing target acquisitions, it is important to realize that a human operator is capable of exhibiting a range of performance over which increased speed can be traded for decreased accuracy. If only a single point on this tradeoff is sampled for each of two systems, it may be difficult to decide which system is superior, especially if performance on one system is faster, and performance on the other system is more accurate. Under such circumstances it becomes important to consider the speed-accuracy tradeoff function.

The concept of stimulus-response compatibility refers to the correspondence between the structure of stimulus information and the structure of the required movements. A simple correspondence between these structures can often facilitate performance. The practical implication is that the effectiveness of a movement system cannot be considered in isolation of the stimulus display with which it is associated. Furthermore, if the perceptual and motor structure of the human operator are organized at different levels of abstraction, the problem of designing simple correspondences between these different levels may be multi-facetted.

Although these two concepts regarding human performance have primarily been applied to position control systems, the relative efficiency of higher order dynamic systems for performing target acquisition tasks can also be understood in terms of these factors. The present chapter discusses these issues.

SPEED-ACCURACY TRADEOFFS IN ACQUIRING STATIONARY TARGETS

There has been considerable experimental study of movements from an initial stationary position to a stationary target. In an early study by Fitts and Peterson (1964), subjects held a sharp stylus in their hand and initially positioned the stylus directly in front of them. Two targets, each of width $\underline{W}$, were positioned a distance $\underline{A}$ to the right and left of the stylus. Subjects were instructed to wait until a right- or left-hand light was turned on, and then they were to move the stylus to the corresponding target as quickly as possible. There were two temporal measures of performance, reaction time and movement time. Reaction time was defined as the interval between the light turning on and the subject lifting the stylus from the starting position. Movement time was defined as the interval from the end of the reaction time interval until the stylus touched the target. Fitts and Peterson (1964) found that movement time increased linearly as a function of $\log_2(2\underline{A}/\underline{W})$, which they called the Index of Difficulty. As the Index of Difficulty ranged from roughly 3 to 8, movement time ranged from about 150 ms to 500 ms, and the linear correlation between these two variables was a remarkably high .99 . In contrast, reaction time changed much less, ranging from about 275 ms to 305 ms, and the linear correlation between reaction time and the Index of Difficulty was .79 . Subsequent research by other investigators has not always replicated the positively increasing effect of Index of Difficulty on reaction time (e.g., Hartzell, Dunbar, Beveridge, & Cortilla, 1982; Jagacinski & Monk, 1985). However, the effect of Index of Difficulty on movement time is a very large, highly replicable, and remarkably linear effect that has attracted considerable interest.

The Index of Difficulty can be considered as a logarithmic scale of the relative accuracy of a movement. Namely, $(\underline{W}/2)$ is how close the movement termination must come to the center of the target; it is an absolute error tolerance. $(\underline{W}/2)/\underline{A}$ represents this tolerance as a proportion of the movement length. The inverse of this measure, $\underline{A}/(\underline{W}/2)$ or $2\underline{A}/\underline{W}$, can be considered the required relative accuracy. A more strict proportional error tolerance corresponds to a higher relative accuracy. The Index of Difficulty simply converts this relative accuracy measure to a logarithmic scale. An Index of Difficulty of 3 thus corresponds to a proportional error tolerance of $1/2^3$ or 1/8. An Index of Difficulty of 8 corresponds to a proportional error tolerance of $1/2^8$ or 1/256. Movement time is an increasing linear function of the logarithm of relative accuracy, and this relation is referred to as Fitts' Law.

The slope of the linear relation between movement time and Index of Difficulty in the Fitts and Peterson (1964) experiment was 74 ms per bit, where a "bit" is a unit change in the Index of Difficulty. This number is the increment in movement time for either a doubling of the movement amplitude or a halving of the target width. From an information theory perspective, the slope can be interpreted as the incremental rate of information transmission in choosing one stopping region of width $\underline{W}$ from a set of $2\underline{A}/\underline{W}$ possible stopping regions (Fitts, 1954). The slope is also somewhat analogous to the way the time constant characterizes the response of a first-order linear system to a step input (Crossman & Goodeve, 1963; Langolf, Chaffin, & Foulke, 1976). The time constant represents the increment in time for a first-order linear system to decrease the discrepancy between its output and the target level by a factor of *e*, i.e., to 0.37 of the preceding discrepancy. The slope of the Fitts' Law relation represents the increment in movement time for the human performer (a generally nonlinear system) to reduce the terminal discrepancy from the target center by a factor of two.

For manual movements that require very little accuracy, e.g., Fitts' Index of Difficulty around 2 or less, relative accuracy is no longer a good predictor of movement time (Crossman & Goodeve, 1963). These movements are typically very rapid (around 200 ms or less). They are apparently not strongly dependent on visual feedback (Klapp, 1975) and have been termed "ballistic." Gan and Hoffman (in press) have noted that the duration of such ballistic movements is proportional to the square root of movement amplitude.

Fitts' Law has been found to hold in a wide variety of situations. A comprehensive review is beyond the scope of the present discussion (see Keele, 1986; Anderson, 1987). Instead, the effects of three factors will be considered: the limb performing the movement, the physical form of the movement device, and the dynamic response of the movement device.

Limbs

Fitts' Law has been found to hold for arm movements (A = 5 - 30 cm), wrist movements (A = 1.3 cm), and finger movements (A = 0.25 cm) by Langolf et al. (1976). The arm movements involved a reciprocal tapping task, and the movement time slope was 106 ms/bit. The wrist and finger movements involved a peg insertion task, in which the effective target width was taken to be the difference between the peg diameter and the hole diameter. Peg insertion was performed under a seven power stereoscopic microscope to mimic conditions that might occur in an industrial assembly task. Movement time slopes were 43 ms/bit (wrist) and 26 ms/bit (finger).

Jagacinski and Monk (1985) found Fitts' Law to hold for two-dimensional hand movements and head movements. The hand movements, which involved the fingers, wrist, and forearm to some degree, were performed with a two-dimensional joystick. The head movements were performed with a helmet-mounted sight. In both cases, the target and cursor were displayed on a cathode-ray tube (CRT) display. Movement time slopes were 199 ms/bit for both hand and head movements. Movements along diagonal directions were 7 - 9% slower than vertical and horizontal movements.

Drury (1975) found Fitts' Law to hold for foot positioning movements primarily involving the lower leg. Subjects had to tap their foot back and forth between two foot pedals. The effective target width was taken to be the width of the pedal plus the width of the subjects' shoe, because subjects could touch the pedal with any part of their shoe. The movement time slope was found to be 55 ms/bit for values of the Index of Difficulty ranging from 1 to 3 . A modified Index of Difficulty proposed by Welford (1968), $\log_2[(A/W) + 0.5]$, provided a slightly better fit to these data. It should be noted that the movement times associated with reciprocal tapping movements are generally longer than the movement times for single discrete movements (Fitts & Peterson, 1964).

In summary, Fitts' Law has been found to describe the speed-accuracy tradeoff for positioning movements involving the arm, wrist, fingers, head, and foot. It is difficult to isolate the effects of different limbs from these data, because other aspects of the movement tasks have also been varied across these experiments. Nevertheless, these results do provide a sampling of speed-accuracy tradeoff functions.

System Dynamics

In the above examples, the position of a limb directly controlled the position of some device, which had to be placed in the target region. These are examples of position control systems. It is also possible to have target acquisition systems in which the limb position determines the velocity (velocity control system) or acceleration (acceleration control system) of the system output, and it is this output that must be placed in the target region. For example, Jagacinski, Hartzell, Ward, and Bishop (1978) had subjects manipulate a single-dimensional joystick to capture targets that appeared on a CRT screen. Fitts' Law was found to hold when the joystick controlled cursor position or cursor velocity. The slope was steeper for the velocity control system (200 ms/bit) than for the position control system (113 ms/bit).

Hartzell et al. (1982) found that Fitts' Law held for step changes in altitude and airspeed in a simulated helicopter. With their left hand subjects manipulated a collective (joystick) which determined altitude through a lagged velocity control system. With their right hand subjects manipulated a cyclic (joystick) which determined airspeed through an approximate lagged acceleration control. Target altitude and airspeed were indicated on a CRT display, and only one of these variables changed on any given trial. The movement time slopes were 498 ms/bit for collective (altitude) control and 1,387 ms/bit for cyclic (airspeed) control for displays that simulated a typical helicopter. The higher slopes were thus associated with the higher order dynamics.

That Fitts' Law should hold for different orders of dynamic systems is remarkable when one considers that the basic spatio-temporal shape of the movement is being changed. Namely, to capture a stationary target with a position control, the time optimal movement pattern is simply a corresponding step change in the position of the control device. However, with a velocity control, the time optimal movement pattern is a pulse. The leading edge of the pulse moves the control device from its starting neutral position to its limit, and hence the cursor moves at maximum velocity toward the target. Once the cursor is over the target, the cursor is stopped by the return of the control device to its neutral (center) position. This return movement is the trailing edge of the overall pulse-like movement pattern. With an acceleration control, the time optimal movement pattern of the control device is a double-pulse. The first pulse accelerates the cursor, and the second pulse decelerates the cursor so it stops over the target (see Wickens, 1986).

Given these radically different movement patterns, it is surprising that the straight line form of the Fitts' Law relation is invariant across these situations. In a similar manner, in continuous tracking tasks it is remarkable that an invariant relation holds across changes in system dynamics. Namely, the McRuer Crossover Model (McRuer & Jex, 1967) in its simplest form states that in compensatory tracking of random inputs the rate of change of system output is proportional to the magnitude of displayed error (delayed by a dynamic reaction time). The form of the describing function of the human operator alone (the approximate linear relationship between the displayed error and the movement of the control device) changes with position, velocity, or acceleration controls. However, the form of the describing function of the human operator plus system dynamics maintains an approximate invariant proportional relation between error and the output rate of the overall system. In the case of stationary target capture, Fitts' Law specifies an invariant relationship between relative error tolerance, $(\underline{W}/2)/\underline{A}$, and movement time for the overall system.

Manipulators

A third factor affecting the speed-accuracy tradeoff is the particular movement device or manipulator that a person controls. For example, Card, English, and Burr (1978) compared four devices for selecting text on a CRT display: a mouse, a nonlinear, isometric, velocity-control joystick, step keys (home; right, left, up, and down, all with automatic repeat) and text keys (paragraph, line, word, and character, reverse, and all but paragraph with automatic repeat). Fitts' Law was found to hold approximately for the mouse (slope = 96 ms/bit) and joystick (slope = 220 ms/bit) using the modified Index of Difficulty due to Welford (1968). For the two sets of keys, movement time was proportional to the required number of keystrokes. Overall, the mouse (a continuous position control) was superior to the joystick (a continuous velocity control) and to the two sets of keys (discrete controls). English, Engelbart, and Berman (1967) compared the mouse and joystick with both as position control systems. Their data suggests that the mouse is superior for text selection. However, no statistical tests were reported. Greenstein and Arnaut (1987) provide a short review of these and other manual computer input devices, and note some contradictory results in the experimental literature.

Implications for System Evaluation

In summary, Fitts' Law has been found to describe the speed-accuracy tradeoff for movements to a stationary target in a wide variety of applications: moving one's foot to a pedal, selecting text on a CRT screen, performing a microassembly task, and changing the altitude and airspeed of a helicopter. One practical implication is that in empirically comparing target acquisition systems that differ in the limbs used, the manipulation devices, the displays, and the dynamic coupling between them, it is important to control the relative accuracy of the movements. It is difficult to reach any conclusions when one system has exhibited faster target acquisitions, but has done so with less accuracy in comparison with another system. Both systems might have the same speed-accuracy function, and the experimental evaluation might have simply sampled different points along this common function. Both systems would thus be capable of generating the same range of behaviors, given appropriate instructions and motivation to the human performers. On the other hand, the two systems might have distinct speed-accuracy functions that would recommend the use of one system over the other.

One approach to this problem is to try experimentally to keep movement accuracy constant, so that differences between systems are restricted to movement time differences. Using a single required movement amplitude and target width for all the movement systems being evaluated may be an inadequate experimental procedure for achieving this goal, because the "effective target width" used by the subjects may not correspond to the target width intended by the experimenter. In movement tasks in which the point of "touching down" and terminating the movement can be measured, the spread in these points can be used to infer an "effective target width." This width can be used in specifying the effective Index of Difficulty (Welford, 1968). This approach is useful for mapping out a speed-accuracy tradeoff function. However, if only a single target is used for comparing systems, there still may be difficulty in making comparisons if the effective target width varies. One may still have differences in both speed and accuracy across systems.

Another technique that is useful when the target appears on an electronically controlled video display is to define target capture in terms of staying within the target region for some duration. The measurement of movement time continues until the capture criterion is met, at which point the target disappears as feedback to the performer (Jagacinski, Repperger, Moran, Ward, & Glass, 1980a). There are no missed targets with this methodology. The effective target width cannot therefore exceed the experimentally specified target width, although it could be smaller. This characteristic of continuing the target acquisition process until the target is captured is typical of many practical target acquisition tasks.

Another important point is that in comparing target acquisition systems, it may not be sufficient to evaluate performance at a single value of relative accuracy, even if this value is equated across systems. The superiority of one system over another may vary as a function of the relative accuracy of the movement. A prudent approach to target acquisition system evaluation is therefore to use the speed-accuracy tradeoff function as the basis for comparing systems. The need to look at performance tradeoff characteristics in evaluating systems is not a unique feature of movement systems. Speed-accuracy tradeoffs are also a major source of concern when reaction time is used to assess cognitive tasks (Pachella, 1974). Similarly, signal detection performance is typically evaluated in terms the tradeoff between correct detections and false alarms (Swets, 1973).

There is a tendency in some of the human factors literature for authors to emphasize the physical device involved in a target acquisition system and deemphasize the particular dynamics. Thus, for example, in summarizing the research by Card et al. (1978), reviewers may over-generalize and refer to a comparison between a position control mouse and a velocity control joystick as a comparison between a mouse and a joystick. System dynamics are very strongly emphasized in descriptions of continuous tracking systems. They are also an important aspect of target acquisition systems that should not be overlooked in trying to reach generalizations.

Implications for System Design

In choosing a particular target acquisition system, i.e., a manipulation device, display, and coupling dynamics, a designer is choosing a particular speed-accuracy tradeoff function. For many systems, movement time is a monotonically increasing function of the relative accuracy of the movement. In order to decrease movement time, a designer should make the relative accuracy of the required movement low by decreasing *A* or increasing *W* where possible. When the shape of the speed-accuracy tradeoff is a straight line (Fitts' Law), the slope is a measure of the sensitivity of movement time to relative accuracy (e.g., see Drury, 1975). This measure can vary drastically across different target acquisition systems. With this measure, the system designer can better judge the tradeoffs involved in requiring movements of higher relative accuracy in order to achieve other design objectives, e.g., making the target small to minimize production costs. Of course, other aspects of the overall task environment will affect what the designer considers to be a reasonable tradeoff. The cost associated with increased target acquisition time may vary greatly across tasks, e.g., word processing vs. nap-of-the-earth helicopter maneuvers (Hartzell et al., 1982).

It is probably the marked linearity of the Fitts' Law relation that has attracted much theoretical interest on the part of behavioral scientists. However, there are some examples of target acquisition tasks in which the logarithm of movement time was linearly related to the Fitts' Index of Difficulty. Two examples are tasks involving remote manipulators with appreciable time delays (Sheridan & Ferrell, 1963; Ferrell, 1965) and movements performed under a stereoscopic microscope to grasp an object with tweezers (Hancock, Langolf, & Clark, 1973). The more qualitative property that movement time is a monotonically increasing function of relative accuracy still applies to these latter examples and is probably a more important point for the system designer. This qualitative property implies that in cases where the width of the cursor or the tip of the movement device is negligible, movement time will not be changed by magnifying or minifying a target and the surrounding movement region (within limits such that the limb executing the movement does not change). Namely, scaling up the size of both the movement amplitude and target width by the same factor will not change relative accuracy, and hence movement time will not change. Thus equivalence classes of targets become apparent through this qualitative principle. When the width of the cursor or the tip of the movement device must be added (e.g., Drury, 1975) or subtracted (e.g., Langolf et al., 1976) from W to specify the effective target width, this factor must also be taken into account in specifying target equivalence classes.

Information regarding speed-accuracy tradeoffs has been incorporated into more comprehensive systems for predicting human performance. For example, the information processing model by Card, Moran, and Newell (1983) for predicting the time to perform various text editing tasks on a computer system uses Fitts' Law to predict movement times. Another example is MTM-M, a Methods-Time Measurement system for establishing the time an industrial worker should take to perform microassembly work involving a stereoscopic microscope (Karger & Hancock, 1982). This system uses relative accuracy as a key variable. In the research leading to this latter system, the Fitts' Index of Difficulty was found to be linearly related to movement time for some classes of movements, and linearly related to the logarithm of movement time for others (Hancock et al., 1973).

STIMULUS-RESPONSE COMPATIBILITY

Stimulus-response compatibility refers to the correspondence between a set of stimuli and a set of responses. One aspect of compatibility involves the spatial isomorphism between stimuli and responses. In the case of target acquisitions, this aspect must be expanded to spatio-temporal isomorphism because the response is extended over time. If this isomorphism is simple, then the correspondence is said to be highly compatible. Responses tend to be faster and more accurate under conditions of high compatibility. Other aspects of stimulus-response compatibility involve population stereotypes or expectations based on past experience and also relations between stimulus and response modalities. Wickens (1984) provides a review of these latter aspects, which are beyond the scope of the present chapter.

Spatial Correspondence

In an early experiment on stimulus-response compatibility by Fitts and Seeger (1953), eight different light stimuli had to be responded to with eight different patterns of stylus movement. When there was a simple spatial isomorphism between the pattern of lights and the required movements, reaction times were faster and responses were more accurate. The superiority of one form of response organization over another was conditional on its being paired with a spatially compatible set of stimuli.

A simpler example comes from part of a larger experiment by Nicoletti, Anzola, Luppino, Rizzolatti, and Umilta (1982). In this example, there were two light stimuli, a "left" one and a "right" one, and two response buttons, one for the left hand and one for the right hand. Both lights appeared in the same visual field, so that "left" was relative to the other stimulus. When the left and right stimulus lights were respectively mapped to the left and right response buttons, reaction times were faster than when the mapping was crossed (left stimulus light to right response button). A simple spatial isomorphism thus enhanced performance.

In many helicopters, the collective control stick is on the pilot's left, and it is the primary control for altitude; however, the altitude indicator is on the right of the pilot's line of sight. Similarly, the cyclic control stick is on the pilot's right and is the primary control for airspeed; however, the airspeed indicator is on the left of the pilot's line of sight. This crossed spatial placement of controls and displays is similar to the incompatible condition in the experiment by Nicoletti et al. (1982), and one might expect such placement to be detrimental to the target acquisition behavior involved in changing airspeed or altitude. In tasks requiring subjects to make step changes in airspeed or altitude as quickly as possible, Hartzell et al. (1982) found reaction times to be about 74 ms longer when the displays were on the opposite sides from their controls. Also the slope of movement time vs. Index of Difficulty was 20% steeper for step changes in airspeed with the crossed display. At an Index of Difficulty of 6.3 , this slope difference corresponded to a 1.7 s difference in movement time. These data were based on about 40 hours of practice with separate groups of subjects assigned to each display placement. Hartzell et al. (1982) argue that for nap-of-the-earth flight, even small reductions in reaction time or movement time can be very important.

Correspondence with System Dynamics

A second finding that can be interpreted in terms of stimulus-response compatibility is that acquisition of stationary targets with a position control is generally superior to acquisition with a velocity or acceleration control (Poulton, 1974). The time optimal control movement with a position control system is a step, whereas a velocity control system requires a pulse movement to capture a stationary target. An acceleration control system requires two pulses of equal duration, but in opposite directions. If one regards the primary stimulus dimension to be target position, and the primary response dimension to be control stick position, then the position control system permits a very simple spatio-temporal correspondence between these dimensions. A step change in stimulus position requires a step change in stick position. This simple structural isomorphism can be regarded as stimulus-response compatibility.

The slope of the Fitts' Law relation is considerably steeper for velocity control joysticks than position control joysticks (Jagacinski, 1980a). Therefore, the advantage of the position control is especially great for movements of high relative accuracy. Foley and van Dam (1982) note that unsteadiness can be a problem with a sensitive position control. Although they recommend a velocity control to overcome this problem, Gibbs' (1962) data for acquisition of a stationary target at various joystick sensitivities suggest that better performance will be obtained with a position control system. The sensitivity of the position control should be set very low for a joystick control (Gibbs, 1962; Jenkins & Olson, 1952) and at empirically determined intermediate levels for rotary controls (Jenkins & Connor, 1949). Poulton (1974, Ch. 5) and Wickens (1986) provide reviews of data on optimal sensitivity.

Velocity controls, while not as good as position controls for stationary target acquisition, permit faster movement times than acceleration controls. Analogous to the comparison of position and velocity controls, the differences in target acquisition times between velocity and acceleration controls tend to be largest for movements of high relative accuracy (Hartzell et al., 1982). The sensitivity of a joystick velocity control should be set at an empirically determined intermediate level. Sensitivities around 1-3°/s of cursor movement per 1° of joystick displacement are suggested by the results of Gibbs (1962) and Hammerton (1962).

The idea of describing the speed-accuracy tradeoff for movement time with an Index of Difficulty can be generalized to moving targets. Jagacinski, Repperger, Ward, and Moran (1980b) had subjects use a joystick to capture single-dimensional moving targets displayed on a CRT screen. A "capture" was defined as holding a cursor over the target for 350 ms. The time that it took subjects to capture targets moving at a constant velocity, V, was approximated by the equation

$$CT = c + dA + e(V+1)[(1/W) - 1] \qquad (1)$$

CT is capture time (reaction time + movement time), and A, V, and W are specified in degrees of visual angle. c, d, and e are regression coefficients. An important feature of this equation is that target velocity amplifies the effects of small target widths in lengthening the capture time. This equation described the speed-accuracy tradeoff for both position and velocity control systems for a mixture of stationary and moving targets. In contrast, the Fitts' Index of Difficulty did poorly for moving targets captured with a position control system. Epps (1986) found both the Fitts' Index of Difficulty and Equation 1 to be good predictors of the speed-accuracy tradeoff in acquiring stationary targets with a variety of devices used for text editing.

In the above experiment by Jagacinski et al. (1980b), the velocity control system was superior for capturing narrow ($W = 0.30^{o}$), fast moving targets ($V = 3.14^{o}/s$). Can one interpret this superiority in terms of stimulus-response compatibility? Capturing a moving target generally requires both position and velocity corrections. With a position control system, time optimal capture of a constant velocity target requires an initial step movement to null out the position error, followed by a constant velocity or "ramp" movement to continue to match the target position and velocity. With a velocity control system, a pulse movement is required to null out the position error, and the trailing edge of the pulse must stop at a constant, non-zero displacement of the joystick just sufficient to match target velocity. A velocity control system allows a person to generate a constant velocity pattern with a constant displacement of the joystick, whereas a position control requires continuous movement. If target velocity is the primary stimulus dimension and control stick position is the primary response dimension, then a velocity control system affords a simple correspondence between these dimensions. Matching cursor velocity to target velocity can be considered a primary subgoal that is a prerequisite for achieving target capture. Under these assumptions, the superiority of the velocity control system in capturing difficult moving targets may be considered an example of stimulus-response compatibility. Converging evidence is necessary to establish the behavioral validity of this interpretation.

A system that is intermediate between the position and velocity control systems is a "rate-aided" control, which has been found to be superior for continuous tracking of some constant velocity ramps and irregularly moving targets (Poulton, 1974; Wickens, 1986). The author is unaware of empirical tests of a rate-aided control for the acquisition of moving targets. Based on the empirical findings for position and velocity control systems, one would expect the rate-aided control to generate a speed-accuracy function of the form indicated by Equation 1.

Levels of Abstraction

Another important issue in considering stimulus-response compatibility concerns different levels of abstraction in characterizing movement behavior. For example, the task of "getting" an object might involve a sequence of qualitatively different movements such as moving one's hand to the object, orienting the hand with respect to the object, grasping the object with sufficient force to hold it without crushing it, and transporting the object toward the individual. Within each of these different movements, there may in turn be a sequence of episodes of acceleration, coasting, and deceleration for each of several physical axes of control in order to generate the desired trajectory. Further, the dynamic coupling among these axes may change in an episodic manner. If such a task is to be performed by a robotic arm, one might give an operator various levels of control over the task, from simply specifying the command to "get an object" to a more detailed level of trajectory control. The type of person-machine interface that is most appropriate may vary as a function of the level of control. For example, because of the discrete serial structure of language, voice control may be most appropriately used for higher level discrete commands for supervisory control of automated movement processes (e.g., see Leifer, Michalowski, & Van der Loos, 1986). Examples include naming a task, a goal, or a movement mode. Continuous movement may be more appropriate for directly controlling the lower level detailed trajectories of particular movement episodes, especially if they require finely graded control and/or simultaneous coordination of multiple degrees of freedom (Kwee, 1986). Much more experimental work is needed on this issue.

Although stimulus-response compatibility is important for efficient performance, there may sometimes be other design constraints that take precedence. For example, a recent design of a robot arm to assist quadriplegic individuals utilizes the individual's head movements to control the position of the robotic hand (Kwee, 1986). It is easier for people to learn to control this system if their fore-aft head movements result in corresponding fore-aft movements of the robot hand. However, Kwee (1986) reports that this simple compatible relationship was not used because of safety considerations. Namely, suppose the quadriplegic individual moves his (or her) head backwards to bring an object to his mouth. If the individual overshoots the approach to the mouth, the robot arm could push his head backwards, which in turn would make the robot arm push even harder against the head. In other words, a positive feedback loop might be established with very unsafe consequences. To avoid such situations, a mirror image control-display relationship was used. Forward head movements bring the robot hand toward the head. Hence, if there is any overshooting by the robot hand, the resulting backward movement of the head will also move the robot hand away from the person's face rather than more forcefully toward it. This control scheme is more difficult for individuals to learn, but is safer (Kwee, 1986).

This example points out that there may be situations in which a designer wishes to sacrifice stimulus-response compatibility to some other goal such a safety. However, in general it is probably more often the case that stimulus-response compatibility is not given sufficient emphasis in the design process. Ease of construction is often considered first, and ease of learning and use considered only secondarily, much to the detriment of the overall product (e.g., see Norman, 1988).

PROCESS MODELS

The indices of difficulty for stationary and moving targets are useful for estimating how movement time will vary as a function of amplitude, target width, and target velocity. The indices of difficulty are basically regression equations that summarize speed-accuracy tradeoff functions. The equations do not, however, offer very much insight into the underlying movement generation processes. A similar issue has been raised in the decision making literature, where regression equations provide descriptions of decision policies, but do not necessarily bear any close relationship to the way an individual cognitively combines information (Hoffman, 1960; Einhorn, Kleinmuntz, & Kleinmuntz, 1979).

If instead of using movement time as the primary measure of performance, one examines the details of movement trajectories, there is a wealth of additional information regarding target acquisition. First of all, there is a good deal of trial-to-trial variation in the trajectory. This observation has led to a number of models relating variability in force generation to variability in the movement trajectory. For example, Meyer, Smith, and Wright (1982) and Schmidt, Sherwood, Zelaznik, and Leikind (1985) review evidence that people program ballistic impulses of force to move their limbs toward a target point in a brief *prespecified* time (typically less than 200 ms). The impulse has a brief period of acceleration followed by a brief period of deceleration. The intensity and duration of the force impulse are both assumed to have variabilities that increase with their magnitudes. When filtered through the dynamics of the limb, these variabilities produce variability in movement amplitude, W_e, that is proportional to the ratio of the distance moved to the duration of the impulse.

$$W_e = K \times A/MT \qquad (2)$$

K is a constant, and the duration of the impulse is equivalent to the movement time, MT (Schmidt, Zelaznik, Hawkins, Frank, & Quinn, 1979). This relation, referred to as Schmidt's Law, is equivalent to movement duration being linearly proportional to relative accuracy for brief force impulses. Temporally patterned force generation is an inherently noisy process according to this view.

A second observation is that movement to a stationary target often consists of a sequence of episodes of increasing and decreasing velocity. These episodes, termed submovements, are especially evident for wrist movements (Crossman & Goodeve, 1963; Jagacinski et al., 1980a), and less evident for arm movements (Crossman & Goodeve, 1963). They are most evident for movements of high relative accuracy. The first episode is generally slower and more accurate than the subsequent episodes for both position and velocity control systems (Jagacinski et al., 1980a). Meyer et al. (1982) postulated that this initial submovement might consist of an overlapping sequence of force impulses. In formulating a different theory to account for variations in the initial submovement, Abrams, Kornblum, Meyer, and Wright (in press) have shown that movements consisting of only one or two episodes can generate a speed-accuracy tradeoff function consistent with Fitts' Law. The model assumes that each submovement has an endpoint variability that is characterized by Equation 2. The intensity of the initial force impulse is programmed by the performer to make the distribution of initial submovement endpoints lie directly over the target center. The performer effectively assumes that a second submovement may be necessary to reach the target, and programs the average duration of the initial submovement to minimize the overall movement time. This theory characterizes the human performer as a problem solver who deals with the inherent noisiness of force generation by optimally programming the duration of the first submovement.

Other investigators have noted considerably more that two movement episodes in approaching a target (Crossman & Goodeve, 1963; Jagacinski et al., 1980a). These observations have led to an iterative corrections model that emphasizes successive, feedback-based submovements that eventually bring the cursor to the target (Crossman & Goodeve, 1963; Keele, 1968). Each submovement has a relative accuracy and a duration. The durations of successive submovements add up to the total duration of the movement. Similarly, the relative accuracies of successive submovements multiply to determine the relative accuracy of the overall movement. For example, if two successive submovements each reduce the remaining distance to the target center by a factor of two, together they will reduce the distance to the target center by a factor of four. The combination of relative accuracies becomes additive on a logarithmic scale. Hence, each submovement provides an additive increment of logarithmically scaled relative accuracy and an additive increment of movement time. Some versions of the iterative corrections model assume that relative accuracy and duration are each constant across submovements. However, this assumption is not strictly necessary. If the ratio of duration to the logarithm of relative accuracy is constant across submovements, then Fitts' Law will result (Crossman & Goodeve, 1963; Keele, 1968; Jagacinski et al., 1980a). In other words, Fitts' Law corresponds to a constant ratio of incremental duration to incremental logarithmic relative accuracy.

This iterative corrections model provides a logical connection between the concepts of stimulus-response compatibility and speed-accuracy tradeoffs for target acquisition. Namely, assume that the superiority of a highly compatible position control system over a velocity control system is evidenced in each corrective movement episode (Jagacinski et al., 1980a). Then the ratio of submovement duration to the logarithm of relative accuracy will be greater for the less compatible system. In other words, the slope of the Fitts' Law relation will be steeper for the velocity control system. The difference between systems will be magnified for movements requiring numerous corrective submovements, namely movements with high values of Index of Difficulty.

In the capture of a reactive moving target, Jagacinski, Plamondon, and Miller (1987) noted episodes of qualitatively different styles of movement behavior: "fast acquisitions," in which the cursor velocity greatly exceeded target velocity; a "predictive mode," in which the cursor moved very slowly relative to the target; and "close following," in which cursor movement closely mimicked target movement. Each of these styles of movement appeared to have a different subgoal: fast acquisitions were used to null large discrepancies between cursor and target; the predictive mode was used to dampen large oscillations of the target; close following was used to achieve the capture criterion of being within 1 mm of the target for 400 ms. The overall description of target acquisition in this very demanding task was that different movement styles were used to achieve a sequence of subgoals, the final goal being the actual target capture. From this perspective, capturing an elusive target can be considered a form of problem solving. Modeling of these qualitatively different episodes involved specifying both the control laws used to generate the different spatio-temporal trajectories and also specifying a set of rules for triggering transitions among movement styles. While subjects in this experiment used an isometric position control, it may be that different control systems are advantageous for the different styles of movement, and that a more adaptive control system would facilitate target acquisition of very elusive targets. Designing such a target acquisition system then becomes a matter of providing the performer with the capability of generating a set of different movement styles in order to engage in the problem-solving process.

In summary, process models of target acquisition have emphasized the inherent noisiness of brief impulses of force generated by human subjects. This undesirable variability is overcome by optimally programming submovement duration and/or using a succession of iterative corrections to eventually capture the target. For very elusive targets, the performer may use qualitatively different styles of movement to achieve a sequence of subgoals that eventuates in target capture. All of these models view movement as a kind of problem solving. They provide a qualitative conceptual framework that is useful for organizing and integrating the various parametric equations and heuristics that form the data base on target acquisition. Without such a conceptual framework, even a very extensive data base will be less helpful than it might be.

SUMMARY

The present chapter has discussed parametric descriptions of speed-accuracy tradeoffs. For stationary targets, movement time increases monotonically with relative accuracy. In many cases, movement time is proportional to the logarithm of relative accuracy, although this is not a universal finding. For constant velocity moving targets, Equation 1 indicates that target velocity strongly interacts with target width to determine the capture time. Regarding stimulus-response compatibility, it is important to have a simple correspondence between the primary stimulus dimensions and the required control movements. Position control systems are superior for stationary targets, and velocity control systems are superior for small, fast moving targets. It also is important to realize that movements may be behaviorally organized at multiple levels of abstraction, and that the kind of control that is appropriate for one level may not be appropriate for another level. Process models of performance emphasize the conception of movement as problem solving. The emphasis in this chapter has been on relatively direct control by the human operator. For a discussion of target acquisition in more highly automated systems, the reader is referred to articles by Bejczy (1980) and Sheridan (1984).

ACKNOWLEDGMENT

The author wishes to thank Carroll Day, Jim Hartzell, and Grant McMillan for their support and encouragement, and Alex Kirlik, Al Miller, and Harvey Shulman for helpful comments.

REFERENCES

Abrams, R. A., Kornblum, S., Meyer, D. E., and Wright, C. E., in press, Fitts' Law: Optimization of initial ballistic impulses for aimed movements, Psychological Review.

Anderson, N. S., 1987, The pervasiveness of Fitts' Law, Presidential Address to the Division of Applied Experimental and Engineering Psychologists, Ninety-fifth Annual Convention of the American Psychological Association, New York, New York.

Bejczy, A. K., 1980, Sensors, controls, and man-machine interface for advanced teleoperation, Science, 208:1327-1335.

Card, S. K., English, W. K., and Burr, B. J., 1978, Evaluation of mouse, rate-controlled joystick, step keys, and text keys for text selection, Ergonomics, 21:601-613.

Card, S. K., Moran, T. P., and Newell, A., 1983, "The Psychology of Human Computer Interaction," Erlbaum, Hillsdale, New Jersey.

Crossman, E. R. F. W., and Goodeve, P. J., 1963, Feedback control of hand-movement and Fitts' Law. Paper presented at the meeting of the Experimental Psychology Society, Oxford, July, 1963, in: Quarterly Journal of Experimental Psychology, 1983, 35A:251-278.

Drury, C. G., 1975, Application of Fitts' Law to foot-pedal design, Human Factors, 17:368-373.

Einhorn, H. J., Kleinmuntz, D. N., and Kleinmuntz, B., 1979, Linear regression and process-tracing models of judgment, Psychological Review, 86:465-485.

English, W. K., Engelbart, D. C., and Berman, M. L., 1967, Display-selection techniques for text manipulation, IEEE Transactions on Human Factors in Electronics, HFE-8:5-15.

Epps, B. W., 1986, Comparison of six cursor control devices based on Fitts' Law models, Proceedings of the Human Factors Society Thirtieth Annual Meeting, Dayton, Ohio, 327-331.
Ferrell, W. R., 1965, Remote manipulation with transmission delay, IEEE Transactions on Human Factors in Electronics, 6:24-32.
Foley, J. D., and van Dam, A., 1982, "Fundamentals of Interactive Computer Graphics," Addison-Wesley, Reading, Massachusetts.
Fitts, P. M., 1954, The information capacity of the human motor system in controlling the amplitude of movement, Journal of Experimental Psychology, 47:381-391.
Fitts, P. M., and Peterson, J. R., 1964, Information capacity of discrete motor responses, Journal of Experimental Psychology, 67:103-112.
Fitts, P. M., and Seeger, C. M. S-R compatibility: Spatial characteristics of stimulus and response codes, Journal of Experimental Psychology, 46:199-210.
Gan, K., and Hoffmann, E. R., in press, Geometrical conditions for ballistic and visually controlled movements, Ergonomics.
Gibbs, C. B., 1962, Controller design: Interactions of controlling limbs, time-lags and gains in positional and velocity systems, Ergonomics, 5:385-402.
Greenstein, J. S., and Arnaut, L. Y., 1987, Human factors aspects of manual computer input devices, in: "Handbook of Human Factors," G. Salvendy, ed., Wiley, New York.
Hammerton, M., 1962, An investigation into the optimal gain of a velocity control system, Ergonomics, 5:539-543.
Hancock, W. M., Langolf, G., and Clark, D. O., 1973, Development of standard data for stereoscopic microscope work, AIIE Transactions, 5:113-118.
Hartzell, E. J., Dunbar, S., Beveridge, R., and Cortilla, R., 1982, Helicopter pilot response latency as a function of the spatial arrangement of instruments and controls. Proceedings of the Eighteenth Annual Conference on Manual Control, Dayton, Ohio, 345-364.
Hoffman, P. J., 1960, The paramorphic representation of clinical judgment, Psychological Bulletin, 57:116-131.
Jagacinski, R. J., Hartzell, E. J., Ward, S., and Bishop, K., 1978, Fitts' Law as a function of system dynamics and target uncertainty, Journal of Motor Behavior, 10:123-131.
Jagacinski, R. J. and Monk, D. L.,1985, Fitts' Law in two dimensions with hand and head movements, Journal of Motor Behavior, 17:77-95.
Jagacinski, R. J., Plamondon, B. D., and Miller, R. A., 1987, Describing movement control at two levels of abstraction, in: "Human Factors Psychology," P. A. Hancock, ed., North-Holland, Amsterdam.
Jagacinski, R. J., Repperger, D. W., Moran, M. S., Ward, S. L., and Glass, B., 1980a, Fitts' Law and the microstructure of rapid discrete movements, Journal of Experimental Psychology: Human Perception and Performance, 6:309-320.
Jagacinski, R. J., Repperger, D. W., Ward, S. L., and Moran, M. S., 1980b, A test of Fitts' Law with moving targets, Human Factors, 22:225-233.
Jenkins, W. L., and Connor, M. B., 1949, Some design factors in making settings on a linear scale, Journal of Applied Psychology, 33:395-409.
Jenkins, W. L., and Olson, M. W., 1952, The use of levers in making settings on a linear scale, Journal of Applied Psychology, 36:269-271.
Karger, D. W., and Hancock, W. M., 1982, "Advanced Work Measurement," Industrial, New York.
Keele, S. W., 1968, Movement control in skilled motor performance, Psychological Bulletin, 70:387-403.
Keele, S. W., 1986, Motor control, in: "Handbook of Perception and Human Performance," Volume 2, K. R. Boff, L. Kaufman, and J. P. Thomas, eds., Wiley, New York.
Klapp, S. T., 1975, Feedback versus motor programming in the control of aimed movements, Journal of Experimental Psychology: Human Perception and Performance, 104:147-153.

Kwee, H. H., 1986, Spartacus and Manus: Telethesis developments in France and the Netherlands, in: "Interactive Robotic Aids -- One Option for Independent Living," R. Foulds, ed., World Rehabilitation Fund, New York.

Langolf, G. D., Chaffin, D. B., and Foulke, J. A., 1976, An investigation of Fitts' Law using a wide range of movement amplitudes, Journal of Motor Behavior, 8:113-128.

Leifer, L. J., Michalowski, S. J., and Van der Loos, H. F. M., 1986, Development of an advanced robotic aid: From feasibility to utility, in, "Interactive Robotic Aids -- One Option for Independent Living," R. Foulds, ed., World Rehabilitation Fund, New York.

McRuer, D. T., and Jex, H. R., 1967, A review of quasi-linear pilot models, IEEE Transactions on Human Factors in Electronics, 8:231-249.

Meyer, D. E., Smith, J. E. K., and Wright, C. E. , 1982, Models for the speed and accuracy of aimed limb movements, Psychological Review, 89:449-482.

Nicoletti, R., Anzola, G. P., Luppino, G., Rizzolatti, G., and Umilta, C., 1982, Spatial compatibility effects on the same side of the body midline, Journal of Experimental Psychology: Human Perception and Performance, 8:664-673.

Norman, D. A., 1988, Infuriating by design, Psychology Today, March, 1988, 53-56.

Pachella, R. G., 1974, The interpretation of reaction time in information processing research, in: "Human Information Processing," B. H. Kantowitz, ed., Erlbaum, New York.

Poulton, E. C., 1974, "Tracking Skill and Manual Control," Academic, New York.

Schmidt, R. A., Sherwood, D. E., Zelaznik, H. N., and Leikind, B. J., 1985, Speed-accuracy trade-offs in motor behavior: Theories of impulse variability, in: "Motor Behavior: Programming, Control, and Acquisition," H. Heuer, U. Kleinbeck, and K. H. Schmidt, eds., Springer-Verlag, Berlin.

Schmidt, R. A., Zelaznik, H., Hawkins, B., Frank, J. S. and Quinn, J. T., Jr., 1979, Motor-output variability: A theory for the accuracy of rapid motor acts, Psychological Review, 86:415-451.

Sheridan, T. B., 1984, Supervisory control of remote manipulators, vehicles and dynamic processes, in: "Advances in Man-Machine Systems Research," Volume 1, W. B. Rouse, ed., JAI, Greenwich, Connecticut.

Sheridan, T. B., and Ferrell, W. R., 1963, Remote manipulative control with transmission delay, IEEE Transactions on Human Factors in Electronics, 4:25-29.

Swets, J. A., 1973, The relative operating characteristic in psychology, Science, 182:990-1000.

Welford, A. T., 1968, "Fundamentals of Skill," Methuen, London.

Wickens, C. D., 1984, "Engineering Psychology and Human Performance," Merrill, Columbus, Ohio.

Wickens, C. D., 1986, The effects of control dynamics on performance, in: "Handbook of Perception and Human Performance," Volume 2, K. R. Boff, L. Kaufman, and J. P. Thomas, eds., Wiley, New York.

MODELLING OF AUDITORY MASKED THRESHOLDS IN HUMANS

G.M. Rood, Royal Aerospace Establishment, Farnborough, UK

R.D. Patterson, Applied Psychology Unit, Cambridge

M.C. Lower, ISVR, University of Southampton

INTRODUCTION

In order to be able to predict the probability of detection of an acoustic sound in noise, it is necessary to be able to define the masked threshold of that noise along with the level of the sound required to be detected. The phenomenon of obscuring the detection of one sound by another is defined as auditory masking and is one of the more classical problems of the detecting and classifying of signals in noise. Auditory masking is particularly important when threshold listening is involved, that is listening for low level signals in conditions of high noise, and from a military viewpoint it is of great importance when attempting to complete auditory monitoring tasks of sonar signals or electronic warfare returns. Masking, however, is not solely a military problem, but is important wherever a human is required to listen, detect and classify.

The origin of this research grew from the high noise levels in helicopters and the necessity for the human operator to listen for low level sonar returns from helicopter-borne sonar systems. The returns could be active or passive, but relied upon the operator aiding the sonar processing equipment both visually and aurally. As in most systems which use a human operator in the loop, the system should be designed such that human operator and the processing equipment both complement each other and interact such that the combination provides an improved performance over the two constituent parts. For instance, in sonar operations, the human provides a better detector of transient sounds and the classification of those sounds than the processor, whilst the processor provides other benefits - mainly in long term detection of low level sounds in noise - at which the human is less efficient.

Whilst airborne sonar operators are protected from the high levels of noise by a flying helmet, such helmets are unable to prevent some noise reaching the ear, and it is these noise levels which are liable to mask the incoming detection signals. Fig. 1 shows typical cabin noise in an airborne helicopter and clearly shows the low frequency content of the noise spectrum. Generally the helicopter rotor noise is at the lower frequency end of the spectrum, typically 16 Hz to 20 Hz, and at levels of up to 125 dB SPL, whilst gearbox and associated mechanical noise is in the region of 400 Hz to 600 Hz at slightly lower levels (Lucas 1982, 1984). Due to the inability of the flying helmet to attenuate the cabin noise to any great extent at frequencies below 300 Hz or so (Rood 1978), the lower frequency noise levels experienced at the ear are high. Fig. 2 shows a helicopter cabin noise spectrum with the noise at the ear overplotted. Quite clearly high levels of low frequency noise are apparent.

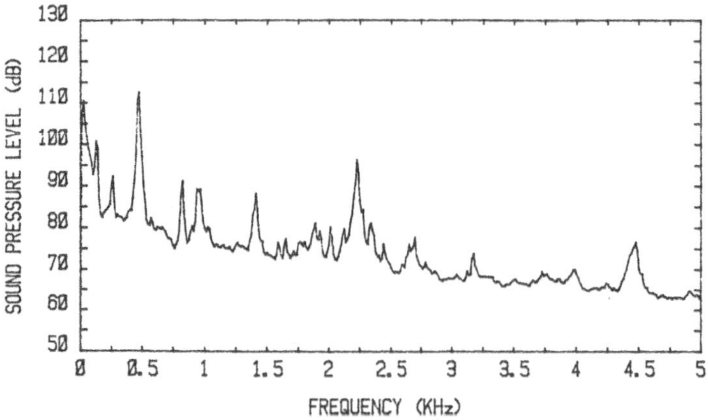

Fig. 1. Helicopter cabin noise spectrum (12.5 Hz resolution)

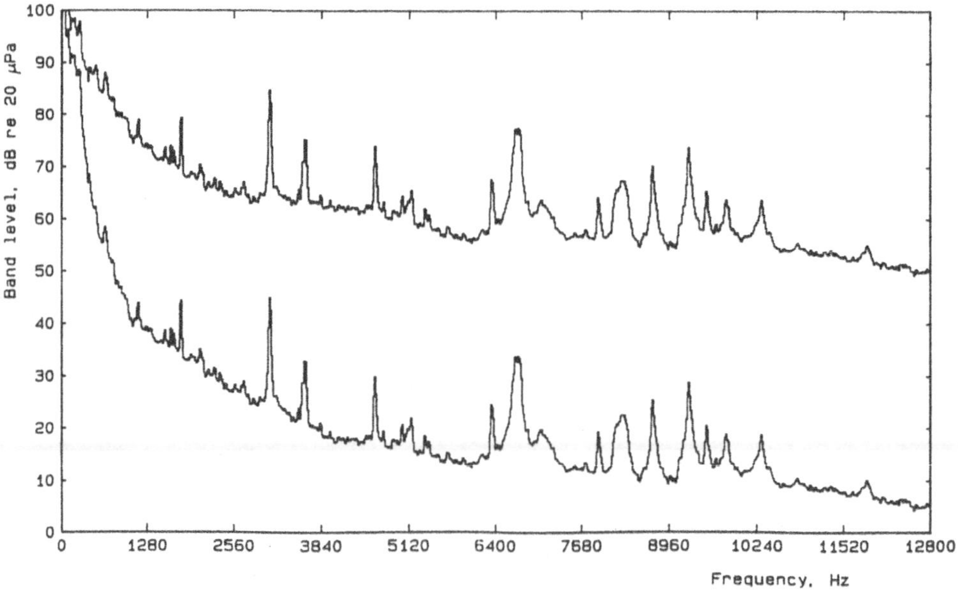

Fig. 2. Helicopter cabin noise spectrum and noise levels at the ear measured under a flying helmet (16 Hz resolution)

Whilst masking is caused by all noise levels, it is the combination of the masking noise and the masked detection signal and their relative positions in the frequency spectrum that are important. Masking is essentially present in three forms; direct, upward and downward. Direct masking occurs when the masker is at the same frequency as the signal, and both upward and downward masking are caused by noise at a given frequency masking signals of higher or lower frequencies (upward or downward, respectively).

Thus, in order to optimise the positioning of signal returns in the frequency domain and to be able to transmit signals at levels that are adequate for a 100% detection rate, but no louder; or to predict the auditory detection rate for a given signal strength in noise, a research programme was initiated by MOD(PE) and the Royal Aerospace Establishment, Farnborough to develop a mathematical model of auditory masking. The project was structured to become a joint project between Human Engineering Division of RAE Mission Management Department, MRC/Applied Psychology Unit, Cambridge - whose auditory model was utilised and developed - and Dr Mike Lower of the Institute of Sound and Vibration Research, Southampton University with funding from MOD(PE)DA Radio in London. Experimental work was carried out at all of these locations and this paper describes the validation of the model in the RAE Helicopter Noise Simulator.

THE AUDITORY FILTER MODEL

Patterson and Nimmo-Smith (1980) published data on experiments concerned with the auditory filter shape and the asymmetry of such filter shapes, based on previous research by Patterson (1974, 1976) and Patterson and Henning (1977).

In essence, Patterson argued that since the rise and fall times of the auditory systems are short with respect to the duration of speech sound or signals, and since the relative phase of the spectral components has essentially no effect on masking levels, it is possible to predict auditory thresholds in noise using a model of auditory masking in which the stimuli are represented by their long term power spectra and the selectivity of the auditory system is represented by an auditory filter (Patterson 1982). An assumption is made that if the listener is asked to detect a signal in the presence of a noise background he listens for a signal through an auditory filter centred near the peak of the signal spectrum.

Patterson quantified this model in the form

$$P_S = K \int_{-\infty}^{\infty} N(f)\, W(f)\, df \qquad (1)$$

In other words, the power of the signal at threshold, *Ps,* is some constant proportion *K,* of the integral of the noise spectrum, *N(f),* times the auditory filter characteristics, *W(f).* This auditory filter's characteristics were determined in a series of experiments which showed that the passband of the filter is virtually symmetric when plotted on a linear frequency scale (Patterson 1974, 1976). The filter has a passband with skirts that fall at around 100 dB/octave, with the passband having a dynamic range in the region of 40 dB. Outside the passband the slope of the filter shape drops rapidly to about 15 dB/octave. An equivalent rectangular bandwidth of the filter, *ERB,* may be determined, which changes marginally with age and frequency but which for practical purposes may be defined from the equation:

$$ERB = 6.23f^2 + 93.39f + 28.52 \qquad \text{(f in kHz)}$$

The equation is an estimate of how the filter width increases with centre frequency

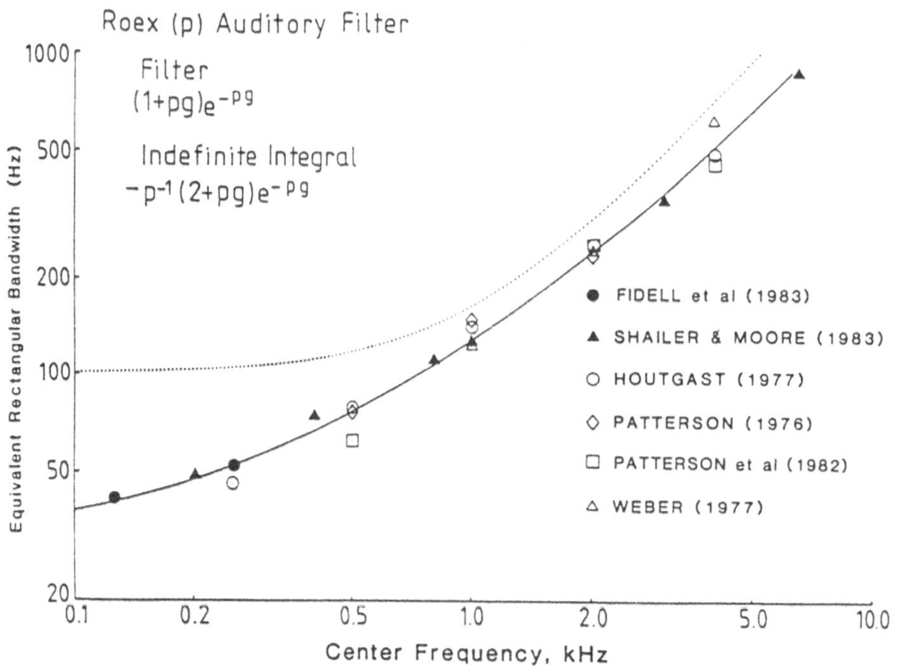

Fig. 3. Increase of ERB filter bandwidth with centre frequency

(Moore and Glasberg 1983) and the curve of the equation is shown in Fig. 3. From these data the passband may be approximated by a pair of back-to-back exponential functions, and since the filter is roughly symmetric, only one exponential parameter, *p*, is required.

An adequate first approximation to the filter passband is provided by

$$W(g) = (1 + pg)e^{-pg} \tag{2}$$

where g is the normalised separation from the centre of the filter, *fo*, to the evaluation point, *f*.

The parameter, *p*, determines the width of the passband of the filter, and the term, $(1 + pg)$ rounds off the peaked top of the double exponential. This rounded exponential (ro-ex) is referred to as the ROEX(p) filter. Further, a dynamic range restriction, *r*, may be introduced since the auditory filter does have limited skirts.

A useful approximation to the entire filter is then provided by

$$W(g) = (1 - r)(1 + pg)e^{-pg} + r$$

The factor $(1 - r)$ is introduced to ensure that the value of the filter remains at unity at its maximum point of sensitivity. This is referred to as the ROEX *(p, r)* filter.

This filter shape may now be substituted in the general masking equation, (1), to provide an expression for calculation of thresholds from an arbitrary noise spectrum. The proportionately constant, k, can be assumed to have a value of 1.0 for practical purposes.

Thus the general expression for threshold becomes

$$P_S = fo\int_0^{0.8} N(g)[(1 - r)(1 + pg)e^{-pg} + r]\,dg$$

the constant, fo, is used to convert from the normalised frequency domain to physical power and since the dynamic range limitation is implemented, the integration is restricted in frequency to 0.8. Patterson (1982) notes that this expression may be used in the prediction of threshold when the total noise levels do not exceed 95 dB(A), since above this level the auditory filter broadens and corrections must be included.

This, then, was the basic auditory filter model which was to be used in determining the masked auditory threshold from experiments carried out in the helicopter noise simulator.

MODEL VALIDATION

The experimental work was carried out in the RAE Helicopter Noise Simulator, which accurately reproduces the noise field experienced in real flight, including the temporal and frequency variations. One advantage of such a simulator, utilised fully in this experiment, is the ability to rapidly change the noise spectrum from one helicopter to another. For this particular series of masking experiments, three helicopter noise spectra were chosen, all of which have differing spectral characteristics, the spread of which would adequately test and validate the masking model.

The basis of the experiment was to measure the auditory threshold of a number of subjects to a range of pure tone frequencies whilst exposing the subjects in the helicopter noise simulator to "real-time" noise conditions. Comparison would then be made with the calculated threshold data from the mathematical model. Ten listeners were used with a series of 17 pulsed pure tones, spaced over the frequency spectrum from 100 Hz to 4.5 Hz. In addition, more complex "real-time" electronic warfare returns were used although the results are not included in this paper. Noise spectra from the Chinook, Sea King and Lynx helicopters were used.

Each subject, whose hearing was normal to ISO standards, wore a Mk 4 flying helmet for the duration of the exercise, fitted with experimental PVDF headphones which have a low-frequency response which allows the lower frequency signals to be clearly perceived. A passive acoustic attenuation measurement was made on each subject, using the standard RAE method with miniature microphones, to ensure that helmet fit was acceptable and within normal limits. In addition to the miniature microphone at each ear, placed over the external meatus, a microphone was placed on each side of the helmet to monitor the external noise field. To ensure that any variance due to differences in cabin noise fields between subjects was minimised, each subject adjusted the helicopter seat until his head was in a particular position fixed by sets of cross wires.

Whilst being exposed to the noise, which was measured at both ears, the listener used a Bekesy tracking procedure to measure the thresholds at each of the discrete frequencies, each frequency being exposed for 30 seconds, allowing about 10-12 turnarounds in that time. Prior to these detections the audiometer had been calibrated against the sound pressure levels measured at the ear for each frequency. Similarly the audiometer had also been calibrated against artificial ear (Bruel and Kjaer Type 4153) measurements.

From the measurements of the noise levels at each ear the predicted masked thresholds could be calculated, which were then compared with the measured thresholds. The threshold curves for different listeners had very similar shapes, and whilst one listener may be consistently above or below another, indicating a broader or narrower filter, all of the functions followed the spectrum quite closely and the mean data are thus considered relevant.

During the whole experiment, one of the major concerns was predictive efficiency; that is the final model was required to be as complex as necessary for predicting threshold in helicopters, but theoretical complications were not required which would increase the computation time without increasing the predictive accuracy. As a starting point, the simplest of the theoretical models was used, the Rounded Exponential Filter, having only a single parameter, filter bandwidth. To initially maintain the simple approach, aspects of off frequency listening, broadening of the filter shape at high levels and localised reductions in masked variability were ignored.

The results of the experiment are shown in Fig. 4 for Lynx cabin noise, Fig. 5 for Chinook noise and Fig. 6 for Sea King noise. Each plot shows the subjectively measured and objectively calculated threshold from the noise level measured at the ear. The solid line through the data is the average of ten predicted threshold functions. In the upper of the two plots the star shows the average of the ten listeners and the corresponding standard deviations, whilst the lower plot shows each of the ten individual points.

It is, incidentally, well worth noting that this calculated threshold represents a true prediction of the data, rather than a fit to the data, in the sense that the parameter values were taken from classical literature rather than being estimated from the experimental data - the values of 'p' being obtained from classical critical ratio research. To obtain this calculated average, each of the 10 subjects' threshold was calculated from the noise level measured at the ear during the course of the experiment. The noise was in fact measured twice, once at the start of the experiment and once at the finish, and measured at both ears. Fig. 7 shows the calculated threshold for one subject and two helicopter noise spectra indicating the variability of noise levels at each ear due predominantly to helmet fit. Fig. 8 shows one more consistent set of threshold curves again for two different helicopters. It must be strongly emphasised, however, that these differences are NOT solely due to error variance but are a correct indication of the variance found not only in experiments of this type but during in-flight measurements, with error variance contributing only minimally to the overall variance figure.

From the data an assumption was made that for detection at each discrete frequency, the listener would use the ear which provides the best signal-to-noise ratio. Thus each of the two left and right ear thresholds was averaged to give a mean left and mean right threshold, and the lower of these (which would give the best signal-to-noise ratio) was used as the threshold for that particular listener in that particular helicopter noise.

Fig. 9 and Fig. 10 show plots of the difference between the calculated and measured mean threshold for the Lynx and Chinook helicopters respectively, based on the above assumption. Table 1 summarises the differences between the measured and predicted mean values of auditory threshold for all three helicopter noise spectra. The corresponding individual differences across subjects are shown for Lynx, Chinook and Sea King in Tables 2, 3 and 4 respectively in association with the corresponding standard deviations.

The measured data were taken from the audiometry and it is clear from a comparison of the measured and calculated data that whilst the mean values are surprisingly good, the differences in variance are significant, particularly so in Lynx noise at the higher frequencies (>3 kHz). There are two predominant reasons for this, the first being an experimental factor that is only apparent in Lynx noise and the second being valid across all helicopter noise spectra. The experimental factor concerns the wide dynamic range of SPLs at the ear when measuring in helicopter noise under a flying helmet. In Chinook, for example, the dynamic range may well be over 100 dB, which is

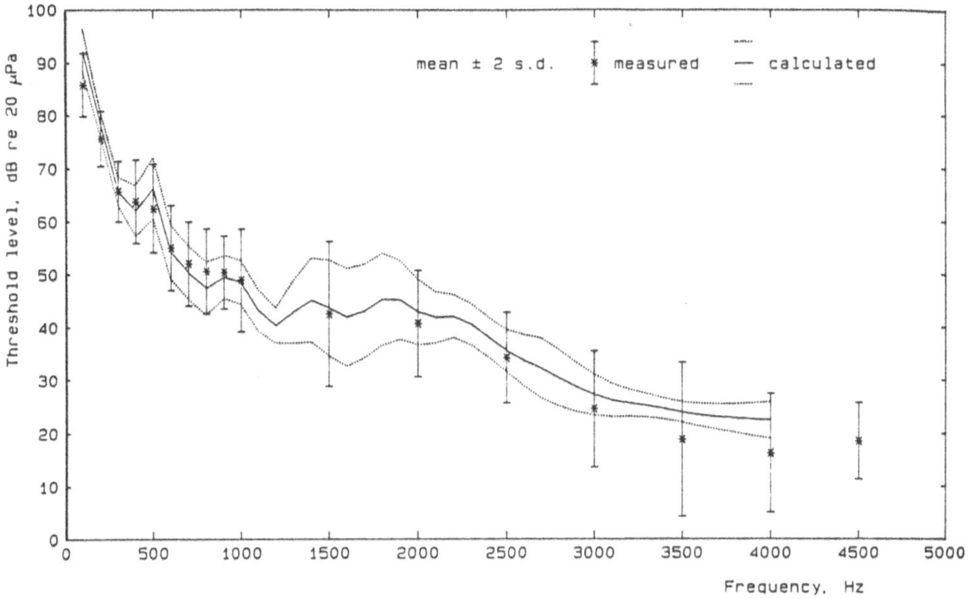

Measured values compared with calculated mean ± 2 s.d.; Lynx noise.

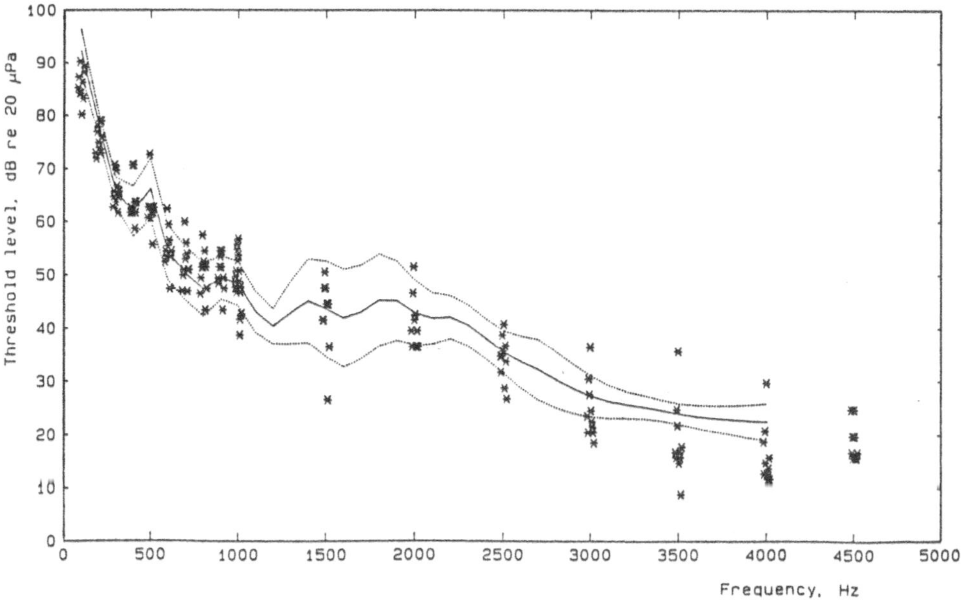

Fig. 4. A comparison of the measured and predicted auditory thresholds showing the mean data (upper figure) and the individual measured data points for 10 subjects (lower figure), both against predicted values: Lynx noise

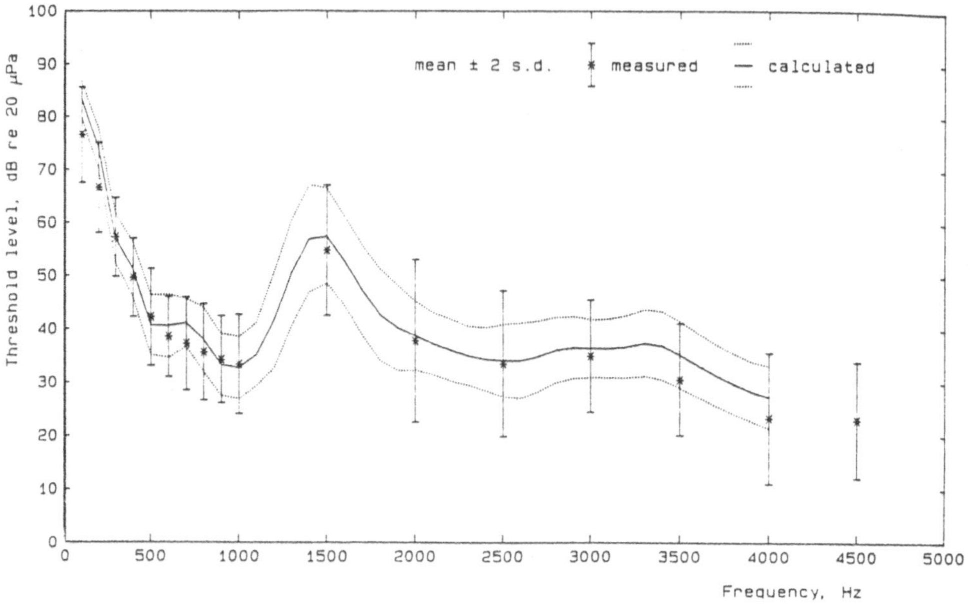

Measured values compared with calculated mean ± 2 s.d.; Chinook noise.

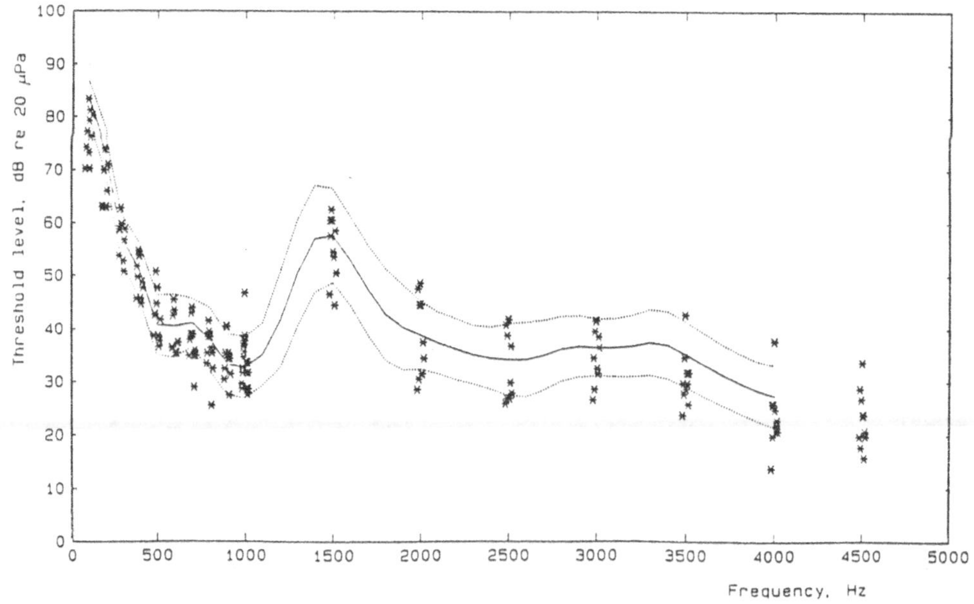

Fig. 5. A comparison of the measured and predicted auditory thresholds showing the mean measured data (upper figure) and the individual measured data points for 10 subjects (lower figure), both against predicted values: Chinook noise

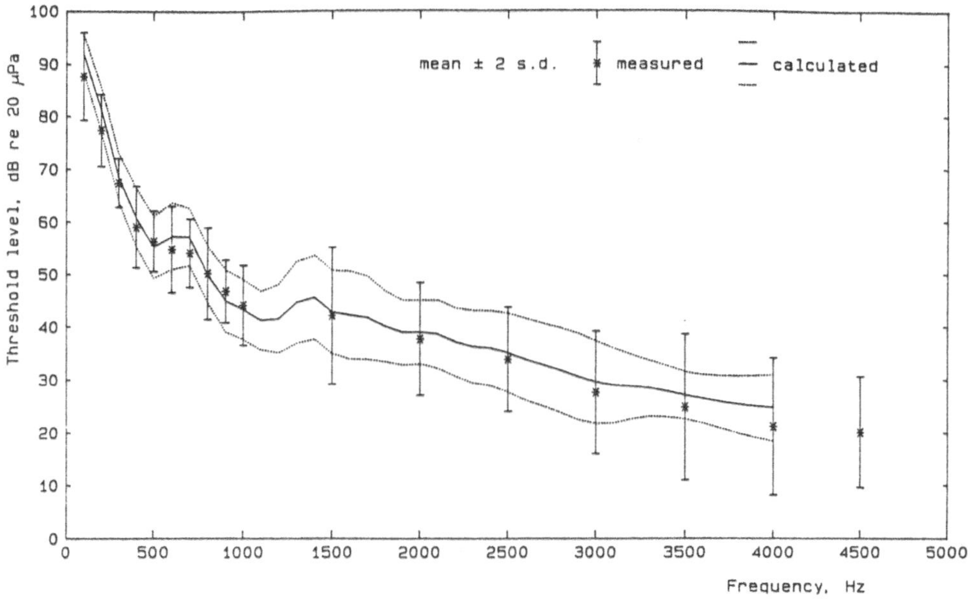

Measured values compared with calculated mean ± 2 s.d.; Sea King noise.

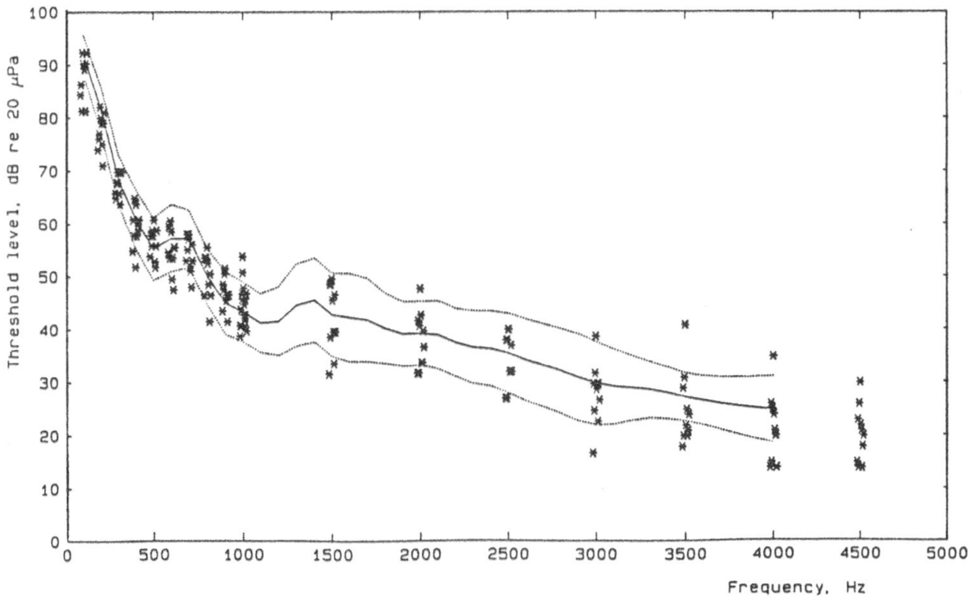

Fig. 6. A comparison of the measured and predicted auditory thresholds showing the mean measured data (upper figure) and the individual measured data points for 10 subjects (lower figure), both against predicted values: Sea King noise

GR (Artificial ear calibration) Sea King noise

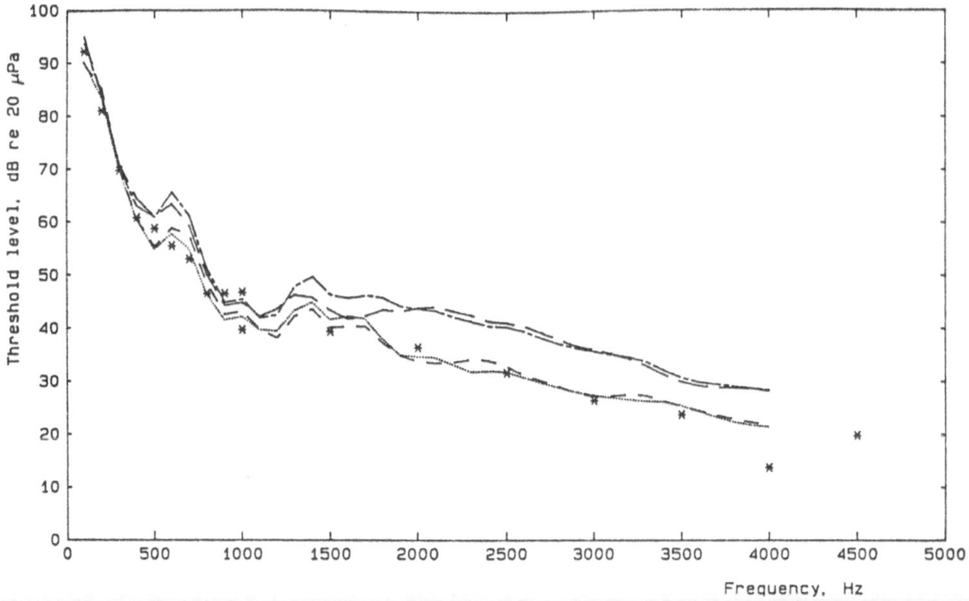

GR (Artificial ear calibration) Chinook

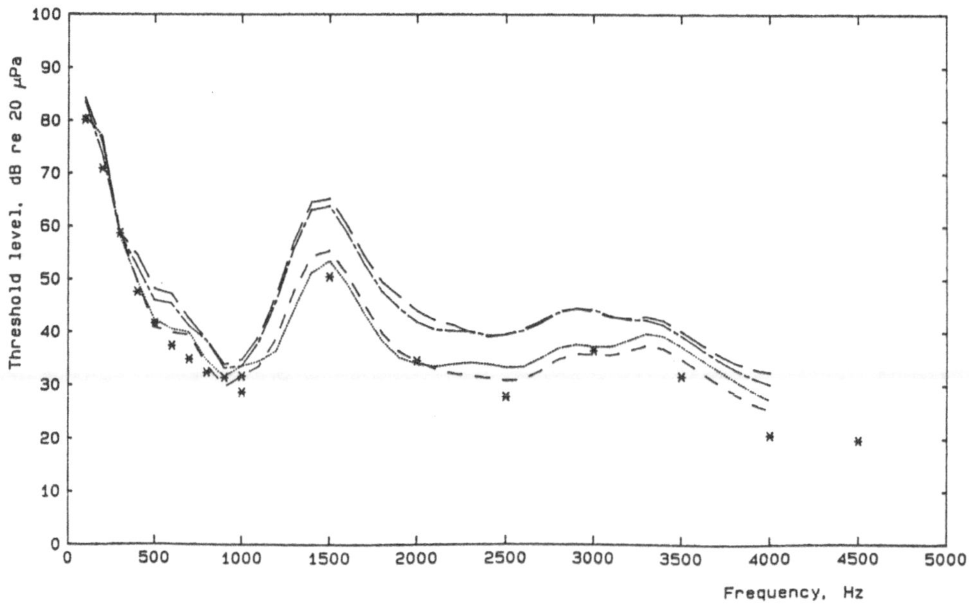

Fig. 7. A comparison of measured and predicted thresholds (left and right ears) for an individual subject. The graphs highlight the differences in helmet fit between left and right ears

DK (Artificial ear calibration) Chinook

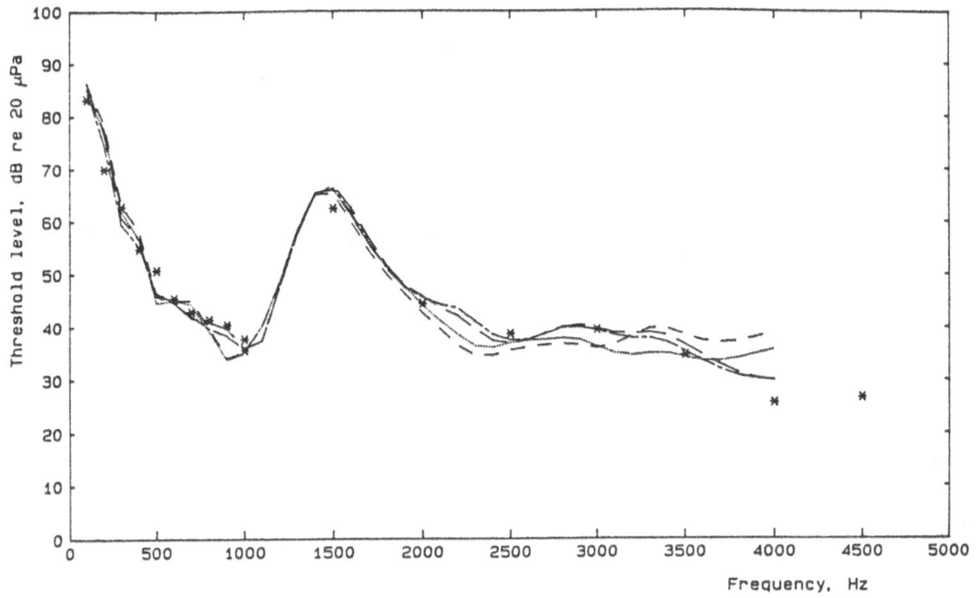

DK (Artificial ear calibration) Lynx

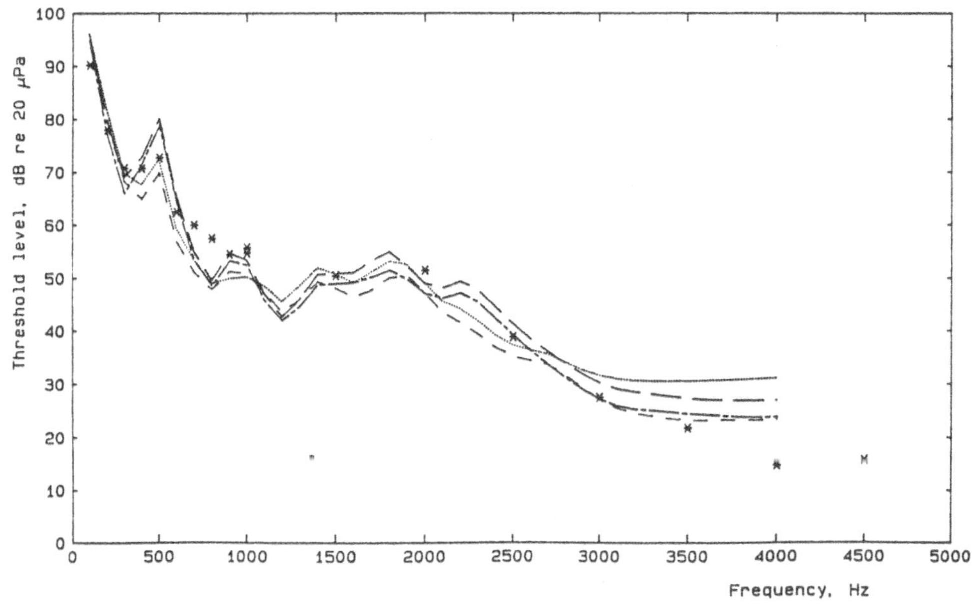

Fig. 8. A comparison of measured and predicted thresholds (left and right ears) for an individual subject with consistent helmet fit for left and right ears (compare with Fig. 7)

Table 1. A comparison of measured and predicted mean thresholds for all three helicopter noise spectra

Freq	Lynx		Chinook		Sea King	
Hz	Meas	Calc	Meas	Calc	Meas	Calc
100	85.85	92.30	76.55	83.14	87.65	92.05
200	75.60	78.33	66.60	73.89	77.40	81.57
300	65.65	65.69	57.25	57.01	59.05	60.92
400	63.75	61.96	49.65	51.05	59.05	60.92
500	62.45	66.26	42.25	40.71	56.35	55.23
600	55.00	54.21	38.60	40.53	54.80	57.34
700	52.00	50.25	37.30	41.17	54.10	57.15
800	50.60	47.31	35.70	38.01	50.20	50.03
900	50.40	49.50	34.30	33.30	46.80	44.95
1000	48.90	48.39	33.45	32.74	44.15	43.32
1500	42.50	43.59	54.90	57.60	42.10	42.74
2000	40.70	42.85	37.90	38.80	37.80	39.12
2500	34.25	35.57	33.65	34.17	34.05	35.25
3000	24.60	27.24	35.10	36.52	27.70	29.65
3500	18.85	23.91	30.65	35.28	24.85	27.09
4000	16.25	22.42	23.35	27.27	21.15	24.66

Table 2. Individual differences between measured and predicted thresholds: Sea King noise

Mean threshold of ear with lower threshold - measured threshold

Frequency	Subjects											
Hz	1	2	3	4	5	6	7	8	9	10	mean	s.d.
1000	2.9	.8	-1.4	-4.4	-8.2	-10.7	-2.6	7.6	2.2	4.4	-.93	5.71
1500	1.5	-.6	5.5	-2.7	-2.0	-3.1	.2	6.5	-5.6	6.6	.64	4.32
2000	-2.2	2.6	4.2	3.3	-3.9	-2.9	-.1	6.5	-.1	5.8	1.32	3.68
2500	.6	6.1	.8	-.7	-3.1	-3.0	-1.8	6.5	1.4	5.1	1.20	3.61
3000	.8	6.4	6.6	-4.0	4.9	-6.6	-1.6	2.0	3.5	7.3	1.95	4.75
3500	1.6	7.9	7.1	1.2	5.0	-11.1	-.7	4.7	.7	6.1	2.24	5.53
4000	7.8	5.5	4.2	1.9	.1	-5.0	5.8	7.6	-1.1	8.3	3.51	4.41
4500	thresholds not calculated											
1000	-4.1	3.8	-.4	-4.4	-3.2	-7.7	-1.6	4.6	-.8	6.4	-.73	4.49
900	-4.4	2.7	-1.1	-3.5	-2.2	-7.7	-4.3	.5	-3.3	4.9	-1.85	3.71
800	.9	3.0	5.5	.1	-2.5	-8.9	-1.0	-.8	-4.2	6.3	-.17	4.52
700	3.1	3.6	6.9	2.9	3.3	-3.8	3.2	4.5	-1.1	8.1	3.05	3.45
600	2.8	5.0	6.0	-2.3	6.5	-1.9	1.0	5.5	-2.2	5.1	2.54	3.61
500	-3.8	2.1	1.5	-3.7	-5.1	-2.7	1.6	-1.0	-3.4	3.1	-1.12	2.98
400	-.3	5.7	.4	-1.8	-1.3	9.4	1.2	1.9	-1.3	4.8	1.87	3.65
300	.4	1.3	3.4	-1.4	.6	.6	1.6	.4	1.3	2.8	1.09	1.36
200	3.0	4.8	11.0	1.9	3.3	1.8	1.1	4.2	3.2	7.4	4.17	2.99
100	-2.2	4.8	9.2	2.0	3.0	.4	2.3	9.2	6.7	8.6	4.40	3.95

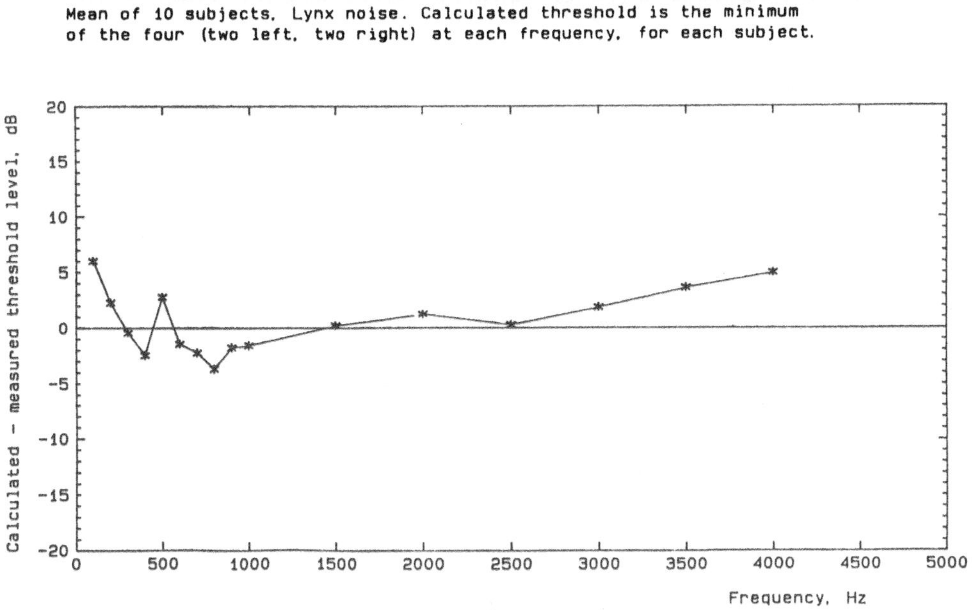

Fig. 9. A summary of the differences between calculated and measured mean auditory thresholds: Lynx noise

Table 3. Individual differences between calculated and measured auditory thresholds: Lynx noise

Mean threshold of ear with lower threshold - measured threshold

Frequency	Subjects											
Hz	1	2	3	4	5	6	7	8	9	10	mean	s.d.
1000	6.2	3.3	8.0	1.2	-7.0	-5.3	-4.2	-1.3	-5.8	-1.8	-.66	5.18
1500	5.0	4.2	14.5	-5.0	.7	-3.3	-1.0	2.0	-4.4	-1.4	1.09	5.80
2000	5.6	6.5	6.0	3.8	-2.8	-.6	-3.5	.8	3.8	1.9	2.15	3.62
2500	5.4	4.6	1.0	4.5	-5.4	-.8	-2.5	2.9	4.0	-.3	1.32	3.56
3000	6.4	9.6	7.7	1.9	2.3	-9.4	1.2	-1.4	5.4	3.1	2.65	5.37
3500	6.9	15.0	9.2	6.1	8.4	-11.9	4.1	-1.0	7.6	6.2	5.05	7.19
4000	12.9	7.2	9.5	10.6	9.4	-9.0	10.8	.8	8.6	1.1	6.18	6.67
4500	thresholds not calculated											
1000	6.2	2.3	5.0	-.8	-4.0	-5.3	-5.2	2.8	-3.8	-.8	-.36	4.22
900	2.8	1.8	5.9	-3.7	-.1	-6.5	-3.8	-.5	-2.8	-2.0	-.91	3.66
800	-2.9	.6	5.8	-8.0	-3.8	-7.6	-9.0	-6.4	-3.0	1.5	-3.29	4.76
700	-1.6	3.0	4.9	-8.5	-3.8	-5.5	-7.6	-.6	-.3	2.5	-1.75	4.56
600	.8	3.0	8.3	-7.8	-4.9	-2.3	-4.4	-.3	-1.2	.9	-.79	4.53
500	6.0	4.9	12.7	-.9	3.2	6.2	-1.5	2.4	2.3	2.8	3.82	4.02
400	-1.0	2.3	4.3	-6.3	-9.0	-.4	-4.5	-2.3	.6	-1.8	-1.80	3.95
300	-.5	2.3	3.3	-3.3	-2.8	2.9	-3.3	.2	-.5	1.8	.03	2.55
200	2.4	1.2	4.0	-1.1	4.3	4.8	.3	2.6	4.8	4.3	2.74	2.06
100	-.3	4.2	10.3	5.2	13.5	6.2	4.2	5.6	7.2	8.5	6.45	3.76

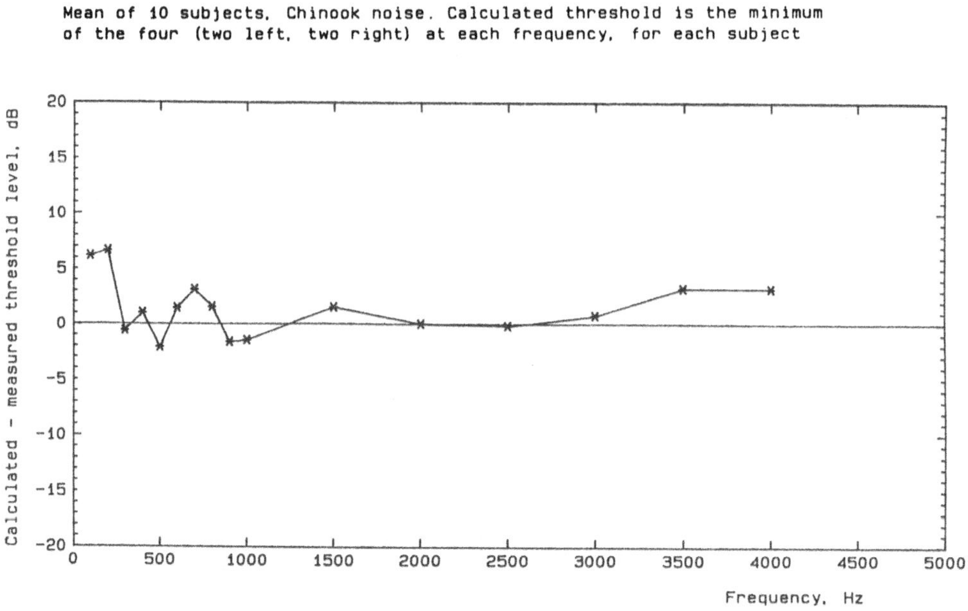

Fig. 10. A summary of the differences between calculated and measured mean auditory thresholds: Chinook noise

Table 4. Individual differences between calculated and measured auditory thresholds: Chinook noise

Mean threshold of ear with lower threshold - measured threshold

Frequency	Subjects											
Hz	1	2	3	4	5	6	7	8	9	10	mean	s.d.
1000	1.0	8.3	1.2	-5.3	-6.5	-16.0	-.3	.5	-1.5	5.6	-1.30	6.80
1500	3.9	5.0	10.4	-2.4	.3	0.0	3.2	-.5	-4.5	11.5	2.69	5.21
2000	-.5	6.1	5.8	3.3	-6.5	-7.1	-1.0	8.4	-9.5	10.0	.70	6.91
2500	4.1	3.5	2.6	2.4	-5.0	-5.5	-2.3	4.8	-4.9	5.5	.53	4.45
3000	-.1	3.8	6.5	.7	-5.5	-4.0	-3.4	7.3	2.9	5.9	1.42	4.61
3500	4.3	16.8	3.1	2.2	3.5	-4.8	1.3	6.4	5.0	8.6	4.63	5.56
4000	5.5	9.1	3.3	6.1	.1	-5.8	4.2	5.9	.9	10.1	3.94	4.66
4500	thresholds not calculated											
1000	4.0	2.3	2.2	-3.3	-2.4	-8.0	-2.3	2.5	-.5	4.6	-.10	3.96
900	-.9	3.4	1.2	-4.5	-.7	-7.5	-6.5	1.1	-.7	5.3	-1.00	4.13
800	1.7	6.6	10.0	-1.2	2.9	-2.4	-2.0	2.8	-.5	5.4	2.32	4.06
700	4.6	8.6	11.7	1.9	1.7	-3.6	-1.0	5.2	2.3	7.4	3.86	4.59
600	2.7	5.8	2.9	-1.0	3.7	-3.3	-.8	-.8	3.4	6.9	1.94	3.30
500	-.1	6.1	1.7	-3.5	-6.9	-3.3	-5.3	-2.3	-3.6	2.1	-1.54	3.92
400	2.0	5.9	4.7	1.5	-2.5	-2.5	1.3	1.9	-1.7	3.7	1.41	2.91
300	-.3	3.0	5.2	-.5	-2.1	-2.7	-2.7	-1.3	-4.0	3.1	-.23	3.00
200	3.8	10.1	10.1	7.0	.1	4.0	6.0	12.2	8.8	11.1	7.29	3.86
100	.7	9.3	9.9	9.7	2.5	2.5	2.1	8.7	6.6	14.2	6.60	4.45

difficult to encompass in measuring equipment - although the ear itself has no problem! To reduce this problem the input spectrum was initially fed through an 'A' weighting filter, which reduced the dynamic range, but for the first 5 subjects there remained some problems of dynamic range above 3 kHz and measurements were running into the noise floor. Thus half of the Lynx data above 3 kHz are contaminated and in their final form will not be used in the calculations.

The other factor which causes these differences is that the measured and calculated values are obtained using different - but realistic - parameters. The measured threshold is from the noise level data at the ear and thus takes into account the helmet fit, the individual subject performance during the audiometry task, his particular criteria for deciding what is detectable as well as the individual differences in age, auditory filter width and characteristic, and off frequency listening - to list only a few differences. This is then a relatively true measure of detection. On the other hand the calculation is based solely on classical literature and the variability is only due to the sound pressure levels at the ear, which is then processed for the 'standard' human listener with no allowances made for either the variations found in real-life or individual differences.

At the lowest frequencies the predictive values are consistently above the measured data. This indicates that when the dominant masker component is at very low frequency, the subject is listening for the signal in the troughs between the peaks of the masker wave - and this is a factor which will be taken into account in the modifed version of the auditory model.

Both of these factors, at the high and low frequency end of the spectrum, can be seen in Fig. 11, which shows the correlation between the measured and calculated thresholds and the regression line. All 48 pairs of points are plotted and the correlation is across all three helicopters - since the correlation should be independent of noise spec-

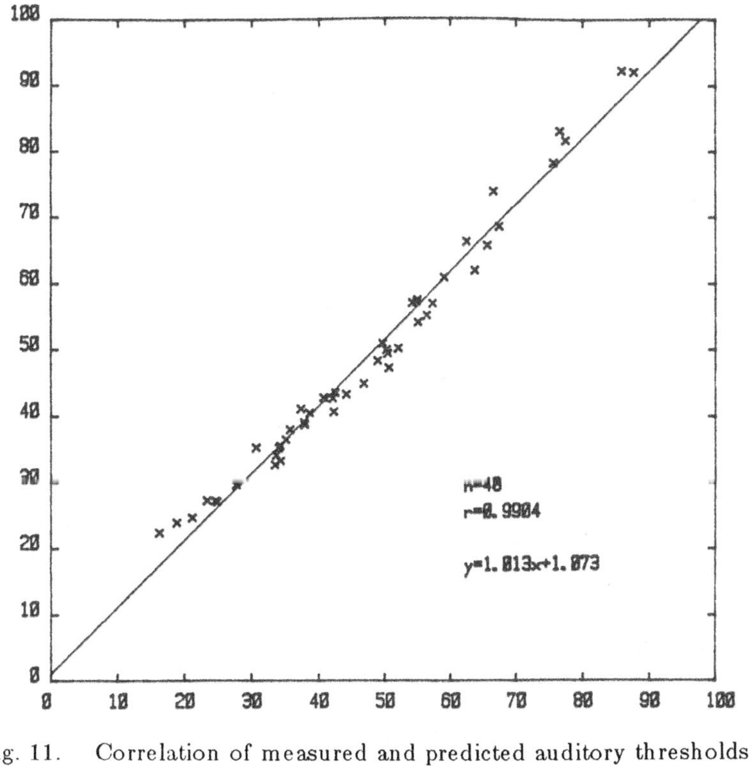

Fig. 11. Correlation of measured and predicted auditory thresholds

trum. The correlation coefficient is 0.990 ($p << 0.001$) and the equation of the regression line is $y = 1.013x + 1.073$ with neither the slope or intercept being significantly different ($p < 0.001$) from the theoretical $y = x$. The standard error of the estimate is calculated at 2.43 dB which gives a 95% confidence limit of 4.76 dB.

The slight variation at the extremes of the data points, at the low and high sound pressure levels are due to the noise floor and the inter-peak listening respectively (i.e., at high and low frequency). A minor change in the constants for low frequency listening will correct this minor discrepancy in the model.

Individual correlations for each particular helicopter give virtually identical results to the overall calculation with the correlation coefficients for Lynx, Chinook and Sea King being 0.989, 0.989 and 0.996 respectively - all highly significant ($p < 0.001$).

The general conclusion from these data is that the simple ROEX(p) auditory filter model provides an accurate enough model at present to determine the noise masked threshold in helicopters, with an accuracy which is well within the boundaries of individual differences. Minor modifications to the model to suit the low frequency aspects of helicopter use will enhance accuracy of prediction. Calculations involving more complex models indicated that, at present, no advantages are gained by using such models.

APPLICATIONS OF THE MODEL

Whilst the validation of the model has been carried out in helicopter noise levels, which provide a stringent test due to the high-level discrete frequency components present, the model will be valid for all other types of noise spectrum. Also the model is general purpose in that it may be used to predict the detectability of any type of sound (auditory warnings, auditory cues, etc.) in any noise. Currently the same scientific team are providing a set of auditory warning signals for helicopter use that are designed using the masking model, to be clearly detectable without being at such an acoustic level that startle may occur and are so aversive that the first action is to seek the cancel or mute button. Fig. 12 and Fig. 13 show such a procedure where the masked threshold is predicted, the 100% detection line drawn, a + 10 dB band added to this detection line to allow for the variations in detection levels (due to variations in cabin noise, helmet fit, individual variability, etc.) and the spectral characteristics of the auditory warning fitted to the aircraft spectra taking into account the limitations of the communications system through which the signals must be transmitted - including the headset or flying helmet.

A further current practical use is shown in Fig. 14 where the masking thresholds are plotted for a number of helicopters in order to be able to construct a single set of auditory warnings that may be used reliably across different helicopter types. In these cases it is important that the same auditory warning and its associated meaning sound subjectively the same to the aircrew in different types of helicopter with their different noise spectra. For instance, a fire warning must sound the same in all types of helicopter, irrespective of the many differences in noise spectrum.

The general conclusion from the experiments is that the auditory filter model provides a design accurate enough to be able to determine the noise masked threshold for aircrew in helicopter operations and may be used for noise fields of different spectral characteristics. A current limitation, although not yet critical in these cases, is that the model cannot yet run in real time. This would allow signal levels to be continually optimised as noise fields change, during manoeuvring flight for instance. This problem is being addressed by the same team in looking at, and producing, an auditory filter bank model which will allow not only real-time masking curves to be produced, but will allow the outputs from the cochlea and higher levels of processing to be determined and used in various models of pitch extraction, pulse streams of auditory nerve firing, etc.

Such models, as with most models of human performance, are often critical to further understanding of not only human systems themselves, but of numerous interactions between the human and the relevant machine interface.

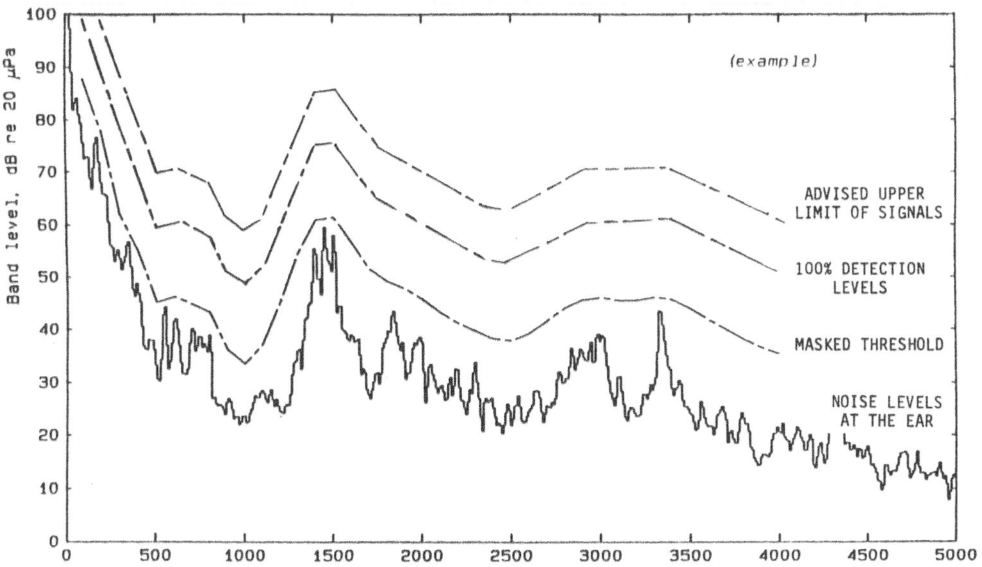

Fig. 12. An example of the use of the masked threshold prediction in setting 100% detection levels for warning signals

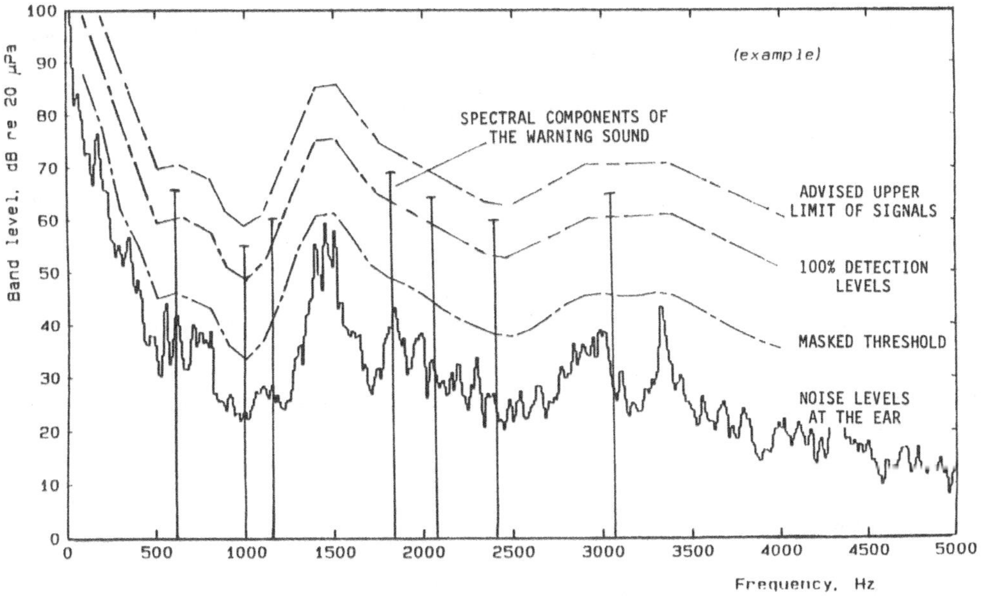

Fig. 13. Spectral content and levels of a warning sound set against 100% detection levels

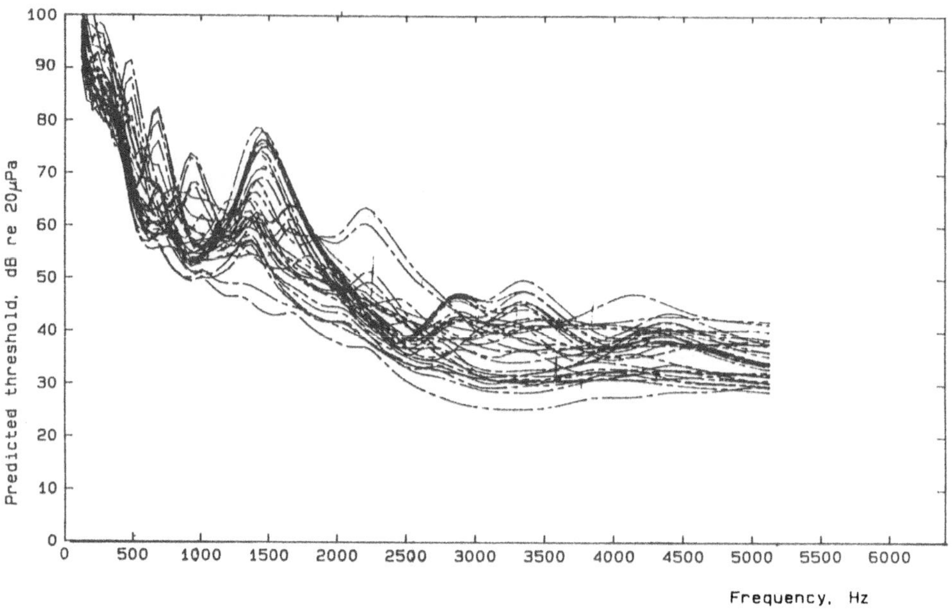

Fig. 14. Masked auditory thresholds of a number of helicopter noise spectra

REFERENCES

Lucas, S.H., 1982, Measurement of Cabin Noise in Lynx Helicopters Royal Aircraft Establishment Technical Memorandum FS(F) 498

Lucas, S.H., 1984, Measurement of Cabin Noise in Five Chinook Helicopters. Royal Aircraft Establishment Technical Report TR84101

Moore, B.C.J. and Glasberg, B.R., 1983, Suggested Formulae for Calculating Auditory-filter Bandwidths and Excitation Patterns. J. Acoust. Soc. Am. 74

Patterson, R.D. and Nimmo-Smith, I., 1980, Off-frequency Listening and Auditory-filter Asymmetry. J. Acoust. Soc. Am. 67, 229-245

Patterson, R.D., 1974, Auditory Filter Shape. J. Acoust. Soc. Am. 55, 802-809

Patterson, R.D., 1976, Auditory Filter Shapes Derived with Noise Stimuli. J. Acoust. Soc. Am. 59, 640-654

Patterson, R.D. and Henning, G.B., 1977, Stimulus Variability and Auditory Filter Shape. J. Acoust. Soc. Am. 62, 649-664

Patterson, 1982, Chapter 9 - Voice Communications in "Human Factors for Designers of Naval Equipment". MRC Royal Naval Personnel Research Committee: Operational Efficiency Sub-committee

Rood, G.M., 1978, The Acoustic Attenuation of the Mk 4 Flying Helmet Measured by Semi-objective Methods. Royal Aircraft Establishment Technical Memorandum FS 171

DYNAMIC SYSTEMS ANALYSIS PROGRAMS WITH CLASSICAL AND OPTIMAL CONTROL APPLICATIONS OF HUMAN PERFORMANCE MODELS

R. Wade Allen
Duane T. McRuer and
Peter M. Thompson

Systems Technology, Inc.
13766 Hawthorne Blvd.
Hawthorne, CA 90250

OVERVIEW

This paper describes the application of two general purpose dynamic systems analysis programs that have been set up to analyze human dynamic performance models. One program is designed to handle very large dynamic systems (up to 102 states). The program interfaces are set up to make it convenient to specify and check out large models, and the example application involves pilot longitudinal control of a CTOL aircraft. This program, referred to as the Linear Systems Modeling Program (LSMP), permits the user to easily change model parameters and loop structure, and permits transfer functions and transient responses to be specified between any combination of system inputs and outputs. LSMP also has a feature that allows transfer function files to be converted to a form that can be analyzed in a general control systems analysis program (Program CC) that permits a wide range of classical and modern analysis procedures.

Program CC, as use in the second application example, provides an optimal solution for a manual control system problem. The example application involves not only obtaining optimal control solutions, but also the analysis of the solution in order to provide some insight into the behavior of the resulting system. This second step, which is ordinarily not provided by other programs, can be carried out conveniently because of the wide range of analysis modules provided in Program CC, and the convenient features that permit passing system models from one procedure to the next.

INTRODUCTION

The human performance models discussed in this paper are designed for the analysis of man-machine system dynamic response including stability and control considerations. The models are implemented with general purpose microcomputer programs running under MS-DOS. The programs are designed for general linear dynamic systems analysis, and can be used to model and analyze various system components including the human operator, vehicle equations of motion, stability augmentation systems, fire control systems, etc. The programs are designed with convenient user interfaces that permit system dynamic models to be easily specified and analyzed with various procedures.

The models and procedures presented here are suitable for analyzing general vehicle control systems such as illustrated in Figure 1. Manual vehicle control (both direct and supervisory) can be described in terms of three major elements: vehicle equations of motion (including stability augmentation), human operator/supervisor dynamics, and kinematic and dynamic equations describing the operator's task (including display presentation). Inputs to the system include disturbances to vehicle motions (e.g. wind gusts, tire/surface disturbances for ground handling), task input commands (e.g. target maneuvering for air to air combat, path curvature for ground vehicles), and task induced remnant or noise generated by the human operator (i.e. responses not directly correlated with other system inputs).

The general modeling approach summarized above, including microcomputer software implementation has been described in detail elsewhere for aeronautical and ground vehicles (Refs. 1,2). Optimal control implementation of single input, single output models (SISO) in a microcomputer environment as discussed herein has also been documented previously (Ref. 3). The purpose of this paper is to summarize computer model capabilities and user interaction.

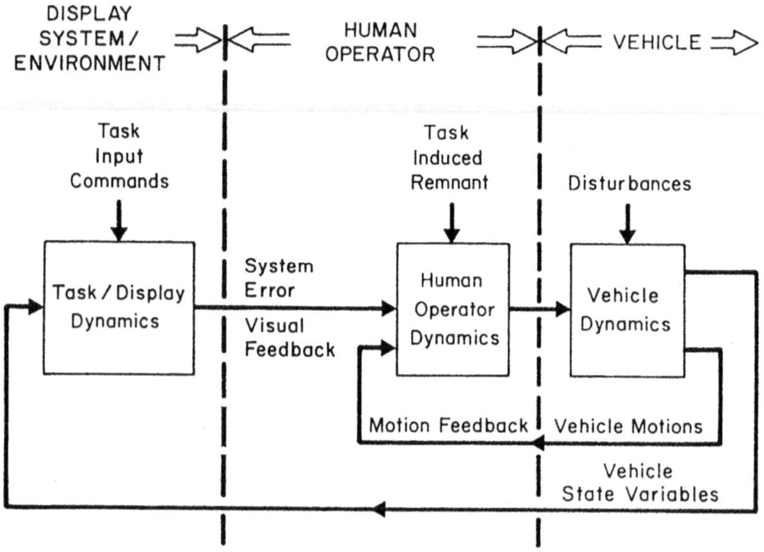

Figure 1. Generic Manual Vehicle Control System

BACKGROUND

The two human performance modeling approaches and associated software described here are related to the structural and algorithmic models described in Refs. 4,5. The structural model has component elements that are isomorphic with more or less identifiable human operator components. This approach is implemented in software that is designed to simplify the specification and analysis of large structural (i.e. physically oriented) dynamic models (Ref. 6). The algorithmic model is based on linear-quadratic-gaussian optimal control theory modified to permit the inclusion of the human operator's time delay and remnant, and derives from the work described in Refs. 7,8. This optimal control model (OCM) is realized with software that includes modern control systems analysis procedures (Ref. 9).

As examples of typical manual control problems that can be analyzed with the above procedures consider the vehicle control models summarized in Fig. 2. The overall structure of the system dynamics are determined through mission and task analysis. In both of the Fig. 2 systems the task involves controlling the vehicle relative to some look ahead reference point. Given the task kinematics and dynamics, vehicle dynamic models must be developed from force and moment equations and augmented with active control systems if required. Fairly complicated vehicle models are available, but the model form selected should be as simple as possible and still meet the objectives of the desired man-machine systems analysis. In the Fig. 2 examples most vehicle characteristics, even including active augmentation, can be adequately approximated as two or three degree of freedom systems for the purpose of overall system dynamic performance analysis.

The form of the human operator model and parameter values depend on task specifics. The Fig. 2 examples assume a relatively simple form, based on visual feedback. The human operator's compensation and time delay penalty depend on task specifics (Refs. 1,4). The human operator optimizes system performance by variations in compensation to meet stability and control requirements, subject to the constraints imposed by visual and motor time delays and remnant. The structural model approach discussed next is optimized by meeting robustness criteria placed on the open loop system dynamic response, including phase and gain margins. The optimal control model discussed subsequently is optimized by algorithms that are designed to minimize a specified mean square performance criterion.

It is useful here to briefly consider the connections between the structural and optimal control modeling approaches. For this purpose a simple crossover law performance model with processing (display) remnant proportional to mean square error has been developed and analyzed (Refs. 1,5). Fig. 3 shows normalized results for a particular input spectrum as a function of an attentional demand parameter that determines the intensity of injected processing remnant. Other conditions are analyzed in detail in Ref. 1. The main point here is to note the relationship between mean square error performance and phase margin. Performance blows up (referred to as instability in the mean square sense) at larger and larger phase margins as the human operator devotes less attention to the control task. Furthermore, the regions of near-minimum mean squared error are fairly broad. Phase margins on the order of 30 to 50 degrees can be justified depending on the attention the human operator devotes to the task. Thus, for divided attention tasks, the specification of phase margin in the structural model is related to achieving minimum mean squared error with an OCM model.

CLASSICAL CONTROL ANALYSIS

Structural human operator dynamic performance models are implemented and analyzed by classical dynamic systems analysis procedures. The software discussed here for this analysis, the Linear Systems Modeling Program (LSMP, Ref. 6), is designed to solve large sets of simultaneous ordinary differential equations using Laplace transform techniques. The model input portion of LSMP is arranged to simplify the specification of dynamic

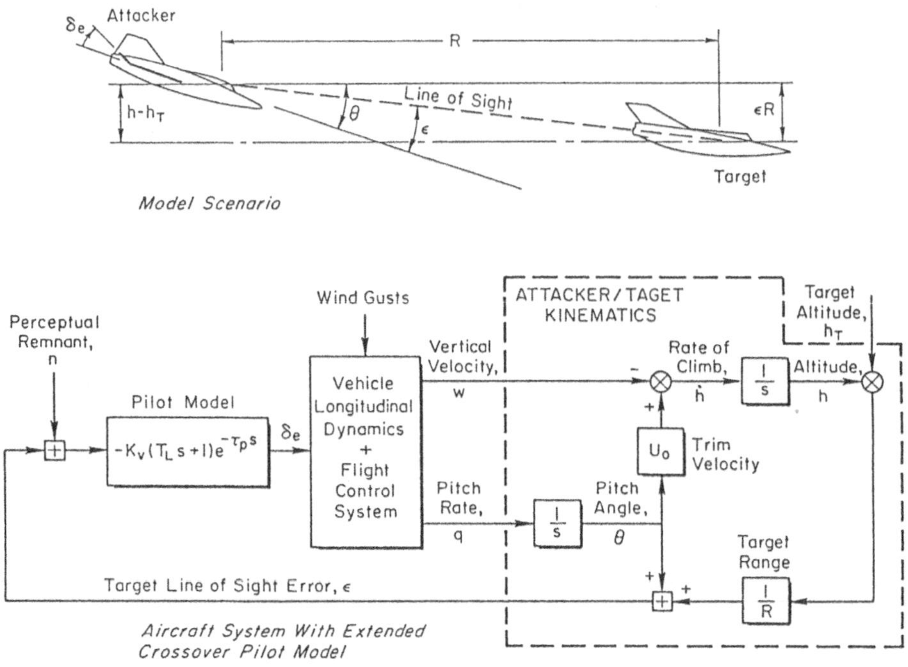

a) System Model for Air-to-Air Target Tracking

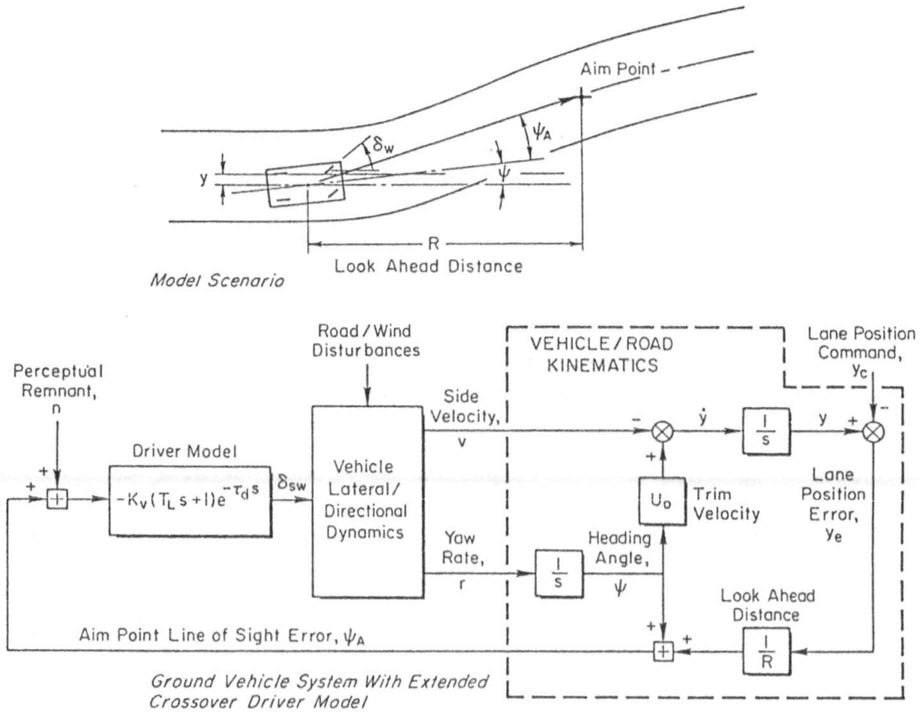

Figure 2. Example Man/Machine System Dynamic Models for Vehicle Control Problems

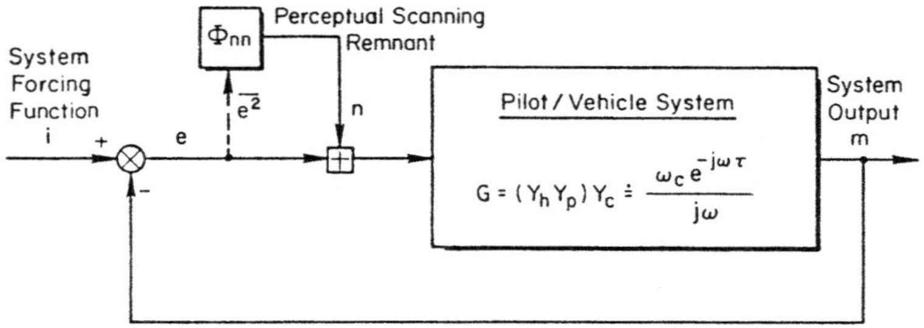

Y_c = Effective dynamics of vehicle (e.g. aircraft plus stability augmentation plus displays)

Y_p = Full attention pilot describing function

Y_h = Perceptual describing function to account for divided attention

ω_c = System crossover frequency

τ = Overall pilot-vehicle system effective latency

PM = System phase margin ($\pi/2 - \tau\omega_c$)

η = T_d/T = Attentional Dwell Fraction

Φ_{nn} = Processing Remnant Spectrum ($\overline{n^2} \sim \overline{e^2}$)

a) Crossover Model

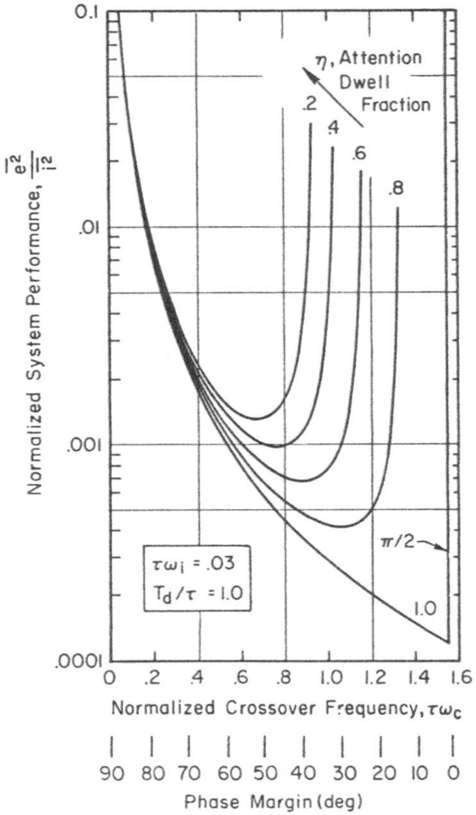

b) Normalized System Performance

Figure 3. Crossover Model Performance Analysis

system equations. The model is expressed as elements of the system transform matrix equation

$$[L(s)][y(s)] = [R(s)][x(s)]$$

where y(s) represents the system output or dependent variables and x(s) represent the system input variables. The system dynamics are defined by the left and right hand matrices L(s) and R(s) respectively.

A typical manual control system model that can be setup in LSMP is illustrated in block diagram form in Fig. 4. This longitudinal aircraft control model has been previously described and analyzed elsewhere (Refs. 1,10). An additional front end program has been prepared to simplify the set up of this generic manual control system. Model parameters are first defined in an annotated file as indicated in Table 1. This file is then submitted to the front end program LONMOD (LONgitudinal MODel) which in turn specifies the L(s) and R(s) matrices of the LSMP model.

The Fig. 4 longitudinal control model basically represents a two degree of freedom manual control task involving vehicle pitch angle, θ, and altitude, h. The system dynamics are further dependent on the task alternatives defined in Fig. 5. Given specification of the control task kinematics/dynamics, the formulation of the overall pilot/vehicle/task model is complete. The next step is to optimize the various vehicle stability augmentation characteristics and pilot parameters in order to obtain a robust dynamic response for the overall system.

Parameter optimization with the structural model of Fig. 4 proceeds in three steps as described in Refs. 1 and 10. The first step is to obtain a good stability augmentation design for the vehicle, which can be accomplished using superaugmentation concepts (Ref. 11). As discussed in Ref. 10 this procedure was carried out to insure that the augmented vehicle dynamic modes had good closed loop damping, met appropriate flying qualities criteria, etc. The second step involves closing the pilot's motion feedback loop which has a similar effect to the rate gyro loop closure in the flight control system. The final stage involves closing the pilot's visual loop with appropriate compensation parameters depending on the selected task.

Given that the flight control system and pilot motion feedback loops in Fig. 4 have been properly closed, an equivalent open loop pilot/vehicle transfer function can be used as discussed in Refs. 1 and 10 in closing the pilot's visual loop. An example of this transfer function as generated by LSMP is illustrated in Fig. 6. Using this approach the pilot/vehicle dynamics can be partitioned into low, middle, and high frequency components. The high frequency dynamics are associated with the augmented aircraft dynamics and the pilot's neuromuscular dynamics and response to motion cues. The mid frequency dynamics are associated with the pilot's crossover frequency compensation. The low frequency effects arise from the pilot's trim behavior and the kinematics of the target tracking task.

Optimization of the structural pilot model is carried out as follows. First consider the Fig. 6a Bode plot which has the pilot's trim and motion feedback set to zero and visual gain set to unity. The motion feedback (i.e. pitch rate) gain provides the equivalent of a lead time constant for pitch attitude which is the primary visual feedback variable. Set this lead to some value beyond crossover (i.e. > 3 r/s) in order to reduce high frequency phase lag since the net vehicle dynamic lags do not require offsetting pilot lead. Now introduce low frequency trim to obtain low frequency high gain (i.e. K'< 1.0 r/s). Now check the conditionally stable phase 'bell' to make sure adequate phase margin can be achieved (i.e. 50-60 deg). Finally, increase pilot visual loop gain in order to obtain

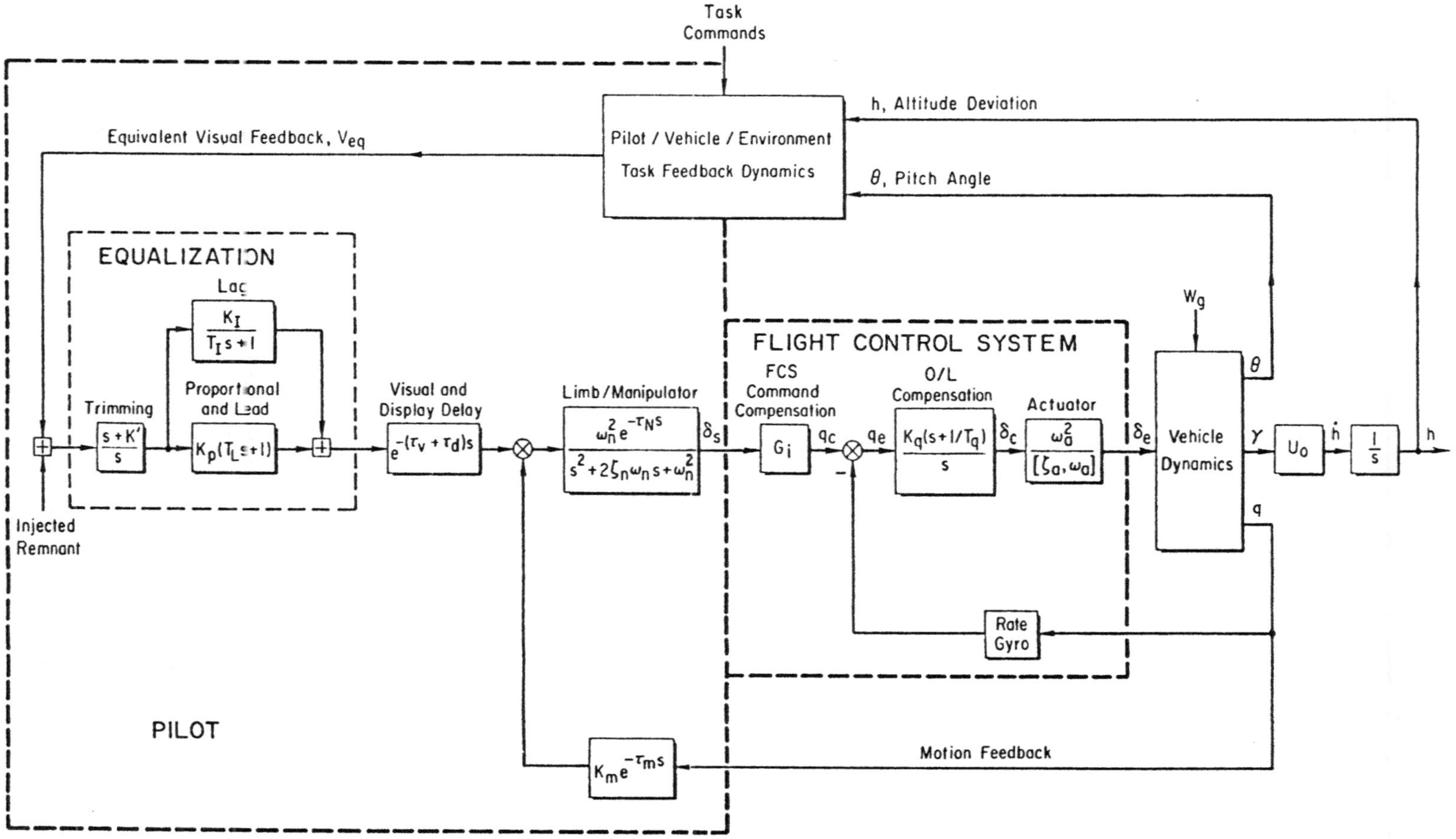

Figure 4. Pilot/Vehicle Model for Longitudinal Aircraft Control

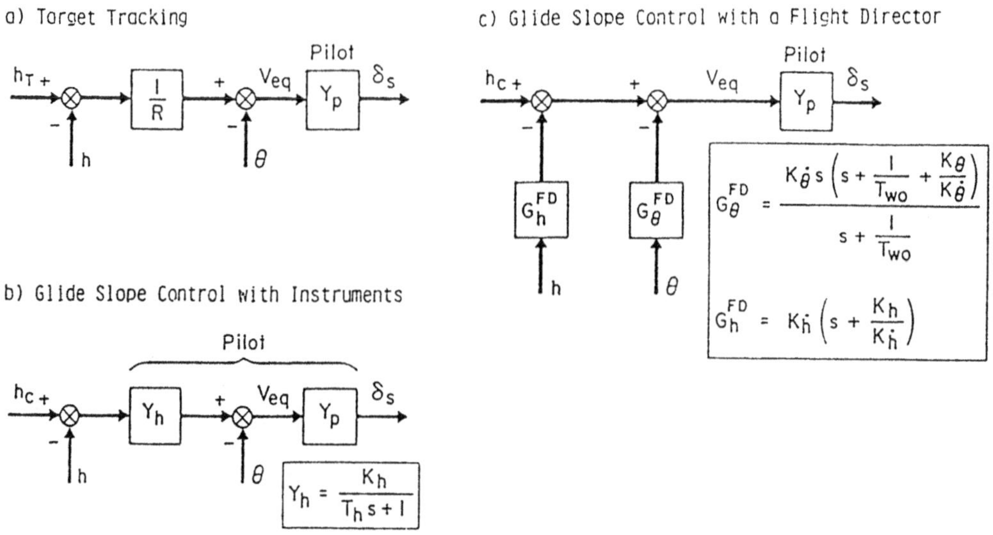

Figure 5. Kinematics/Dynamics for Different Control Tasks

TABLE 1. ANNOTATED FILE FOR DEFINING STRUCTURAL PILOT/VEHICLE MODEL PARAMETERS

```
High Performance Fighter
* VEHICLE DYNAMICS
830., UO   Airspeed
0.,   THETAO
-1.,  ZW
.014, MW
-.3E-3,    MWD
-.829,     MQ
1.,   LXA
-127.,     ZDE
-21., MDE
* SAS - ACTUATOR, COMPENSATION AND S
.707, ZA   Actuator dampin
30.,  WA   Actuator na
-.428,     K
6.,   WQ                       requency
0.,                            constant

                        ter Transfer Function

                   GA2*sq + GA1*s + GAO

      AY/PILOT    ( HA1*s + HAO )/( HB1*s + HBO )
*     Target    IFR  Flight Director
1.,   1.   HAO  1.
0.,   0.   HA1  0.
1.,   range     HBO  1.
0.,   0.   HB   0.
1000, 0.   0.   Ktheta
100,  0.   0.   Ktheta dot
.1,   0.   0.   washout (rad/sec)
1.,   0.   0.   Kh
1.,   0.   0.   Kh dot
^Z
```

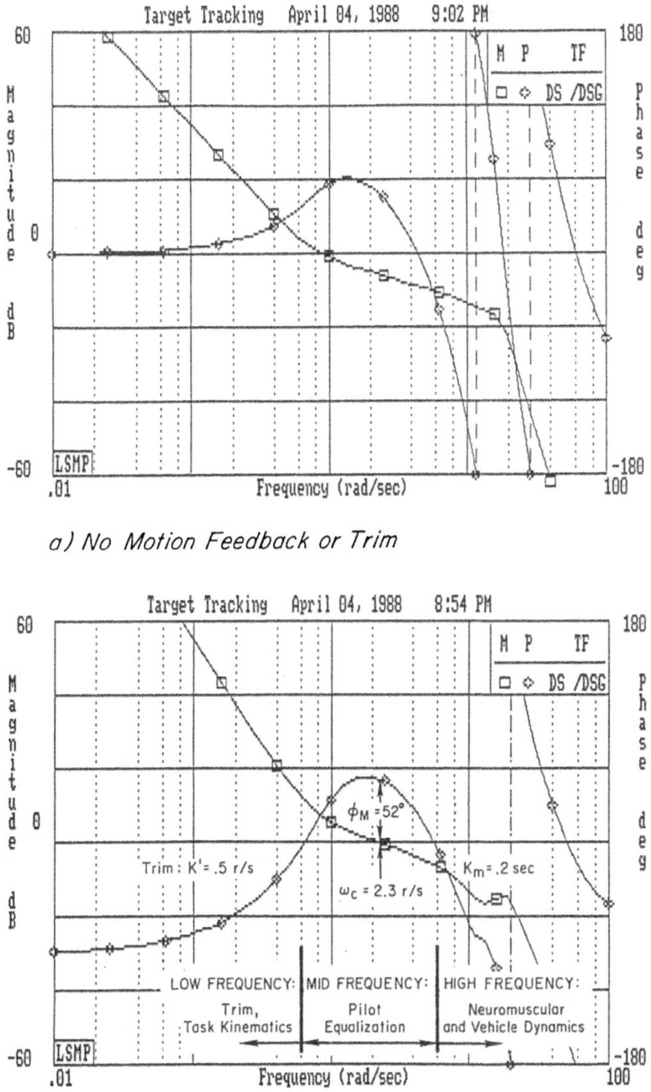

a) No Motion Feedback or Trim

b) Robust Pilot Feedbacks

Figure 6. Bode Plots for Equivalent Pilot/Vehicle Open Loop Response

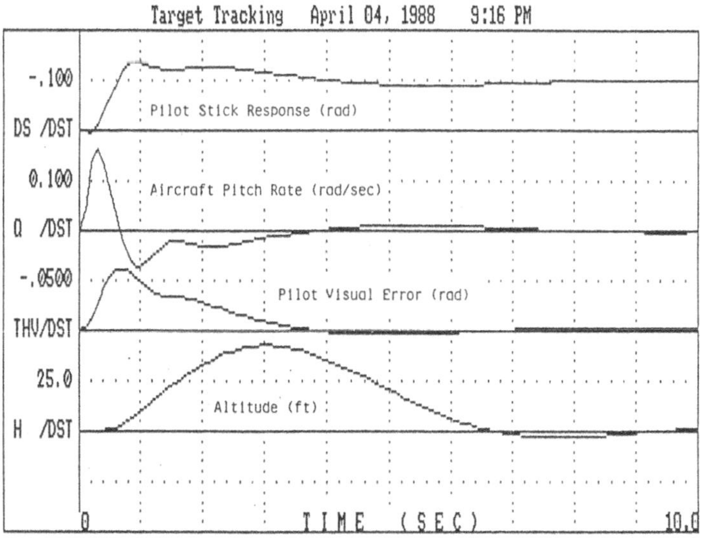

Figure 7. Wind Gust Disturbance to the Target Tracking Task with a Structured Pilot Model

as high a crossover frequency as possible with adequate phase margin. The resulting transfer function Bode plot is shown in Fig. 6b.

The Fig. 6b Bode plot phase lag function shows a typical conditionally stable range over which the system crossover frequency can be established by the pilot. As indicated in Fig. 6 the crossover frequency is set to give near maximum phase margin. Given this set of model parameters the time response of system variables to arbitrary inputs can also be computed and displayed in LSMP. Typical step responses of aircraft and pilot variables to a windgust disturbance are shown in Fig. 7. The transient response shows good damping in both the pitch and altitude degrees of freedom.

OPTIMAL CONTROL MODEL

The algorithmic model considered here is the BBN optimal control model (OCM) described in, for example Refs. 7 and 8. It is implemented with a general purpose control systems analysis program that runs on IBM-PC compatible computers (Program CC, Ref. 9). This OCM implementation has been described in detail elsewhere (Ref. 3). The objective of the OCM is to optimize the adaptive characteristics of the human operator, subject to the constraints presented by visual and motor time delays and remnant.

A block diagram of the OCM implementation in Program CC is shown in Fig. 8. The top path represents the machine dynamics, which are represented by linear state vector and display vector-matrix equations, and provide for an external system input w(t). As indicated in Fig. 8 the system input, considered to be Gaussian white noise, can be prefiltered and injected before, after, or at an intermediate stage in the vehicle dynamics in order to allow for the representation of a variety of system disturbances and commands.

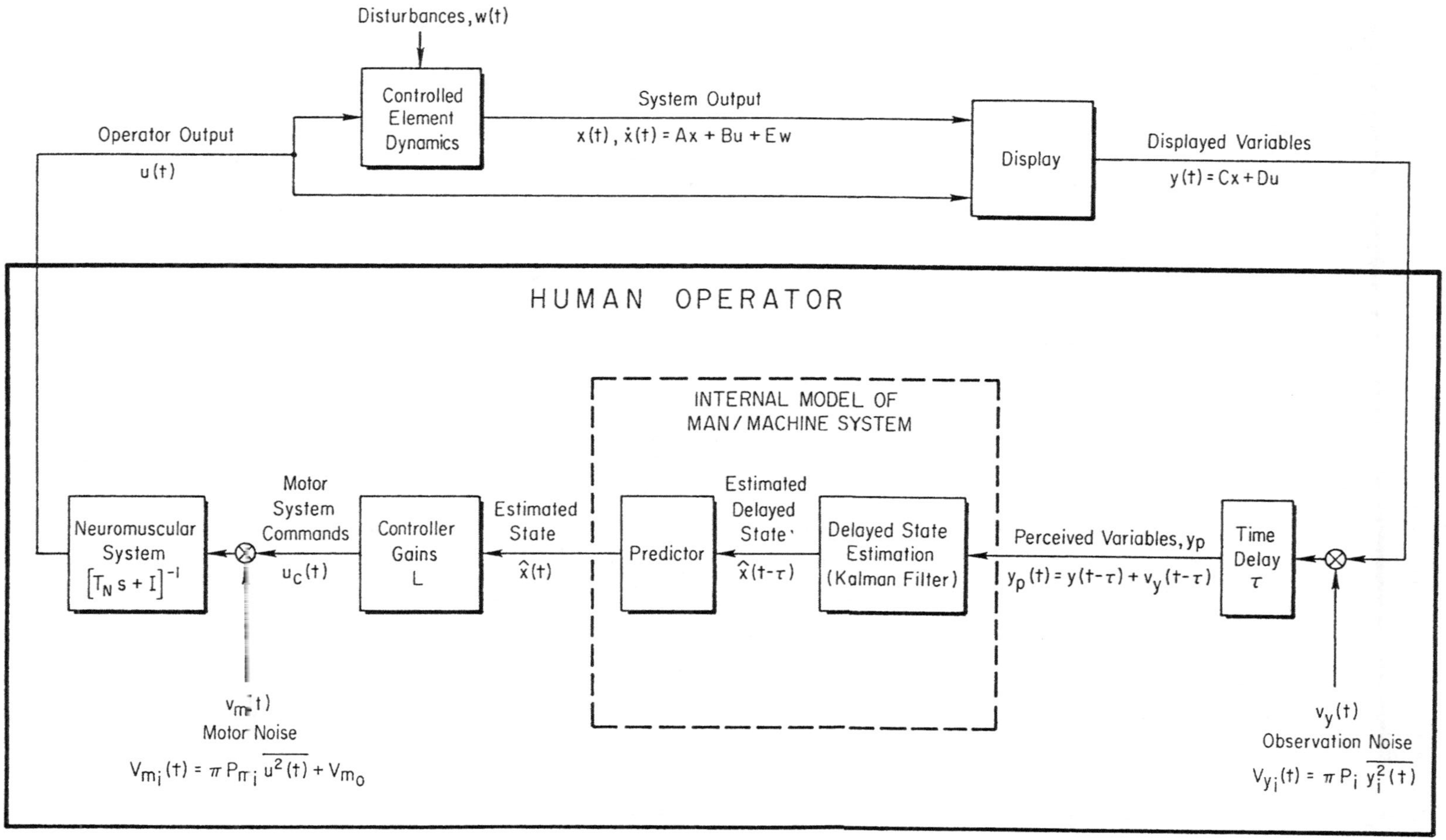

Figure 8. OCM Model Implemented in Program CC

The human operator is defined in the lower path of the Fig. 8 block diagram. The human operator's primary constraints or limitations are represented by an effective time delay, τ, and remnant sources including observation noise applied to the display vector and motor noise injected at the control output prior to the neuromuscular dynamics. Observation noise intensity for each displayed variable is considered to be proportional to the mean square value of the variable in a Weber law sense. Similarly, motor noise intensity is assumed proportional to the mean square value of the human operator's control output. The OCM user can specify both the time delay and noise proportionality constants.

The remaining human operator components in Fig. 8 are adaptive, and are adjusted in an optimal sense to minimize a mean square cost functional composed of display error and control output terms. A Kalman filter is used to estimate the delayed state vector, which is then fed to a least-mean-squared predictor to yield the estimated state vector. Finally the optimal gain matrix, L, is generated by solving the optimal regulator problem for the simple quadratic cost function

$$J = \lim_{T \to \infty} \left\{ \frac{1}{T} \int_0^T (y_e^2 + G\dot{u}^2)\, dt \right\}$$

The weighting parameter, G, is adjusted to provide an appropriate "neuro-muscular" time constant T_N (Refs. 7,8).

Program CC contains all of the algorithms required to solve the OCM problem as summarized above, and the solution method is described in some detail in Ref. 3. The direct and iterative portions of the OCM solution in program CC for a given set of inputs is carried out by a series of macro instructions which call appropriate numerical procedures as required. The user carries out an OCM solution in basically five steps as follows:

1) The human operator constraints consisting of the time delay and noise ratios are specified. These constraints remain constant over a wide range of conditions, but can change depending on control devices and display conditions.

2) The controlled element, Y_c (e.g. vehicle dynamics) and system input filter, Y_w are defined in Laplace transform notation.

3) A macro is called to convert Y_c and Y_w to state space form, based on which input path is chosen from the Fig. 9 options.

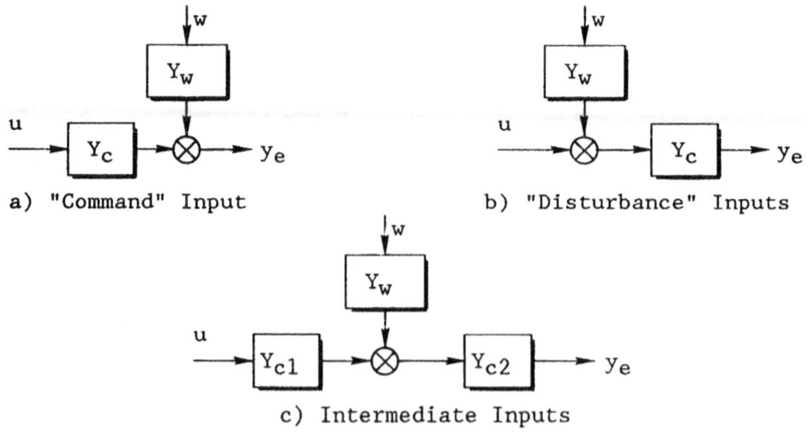

Figure 9. OCM Model Input Options

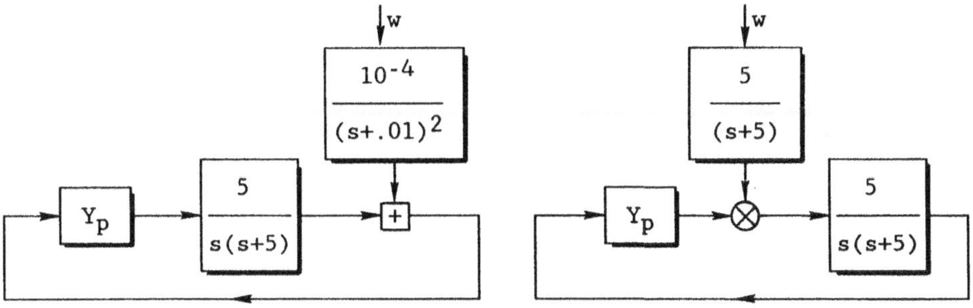

a) Command Input b) Disturbance Input

Figure 10. Example OCM Problems

4) A macro is called to carry out the iterative OCM solution. Mean square performance estimates are available at this point.

5) A macro is called to convert the state space realization of the optimal human operator dynamics, Y_p, to a convenient transfer function form for interpretation. As discussed in Refs. 3 and 5 some simplification and approximation are carried out at this point (e.g. cancellation of closely spaced poles and zeros).

As examples of OCM solutions consider the problems defined in Fig. 10. The controlled element dynamics are such that the human operator should provide some lead compensation. In one case the input is injected as though it were a disturbance such as wind gusts, the second example the input is applied as a very low bandwidth command. Bode plots of the OCM solution for the system open loop transfer function, Y_p*Y_c, are illustrated in Fig. 11. For the disturbance input case the transfer function shows a basically K/s form at and below crossover frequency, a neuromuscular-like resonant mode beyond crossover, and a high frequency effective time delay. For the command input case the region of crossover frequency and above are similar to the disturbance input case, but additional low frequency compensation has been added (gain is increased over a K/s function) in order to effectively deal with the low frequency power in the command input. The command input example is analogous to the target tracking example given previously for the structural model, while the disturbance input example would be analogous to control against wind gusts during landing approach.

As has been noted elsewhere (Refs. 3,5) the OCM gives solutions that compare qualitatively with the structural model on a transfer function basis, but dynamic modes are not necessarily equivalent. The OCM tends to have a significant number of exact and near cancelling poles and zeros. Among these cancellations is the original OCM neuromuscular mode (the exact cancellations are not shown in the Y_p forms given on Fig. 11). The resonant peak beyond crossover results from fortuitous closed loop shifts in the denominator term of the Pade approximation used for the human operator time delay. The recovered pure time delay of 0.17 seconds shown on Fig. 11 does not correspond with the 0.2 second assumed for τ at the start of the problem. However, if the high frequency modes shown in the Y_p approximations (i.e., [.229, 12.9] for command input and [.241, 13.1] for disturbance input) are replaced by their low-frequency approximation as a $e^{-0.03s}$ delay, the proper effective delay is recovered. Finally note that the low frequency compensation in the command input example is

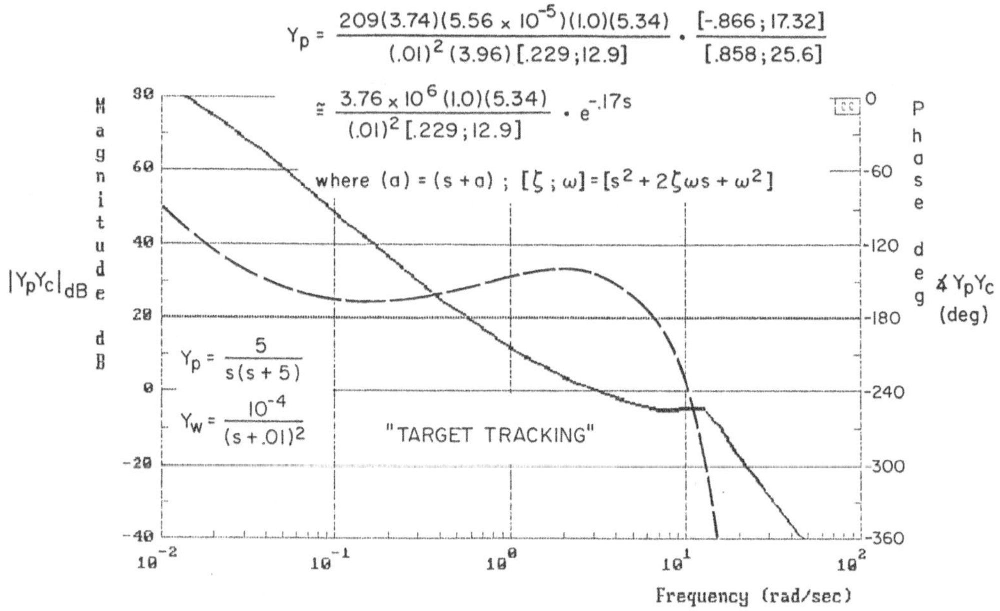

a) Command Input

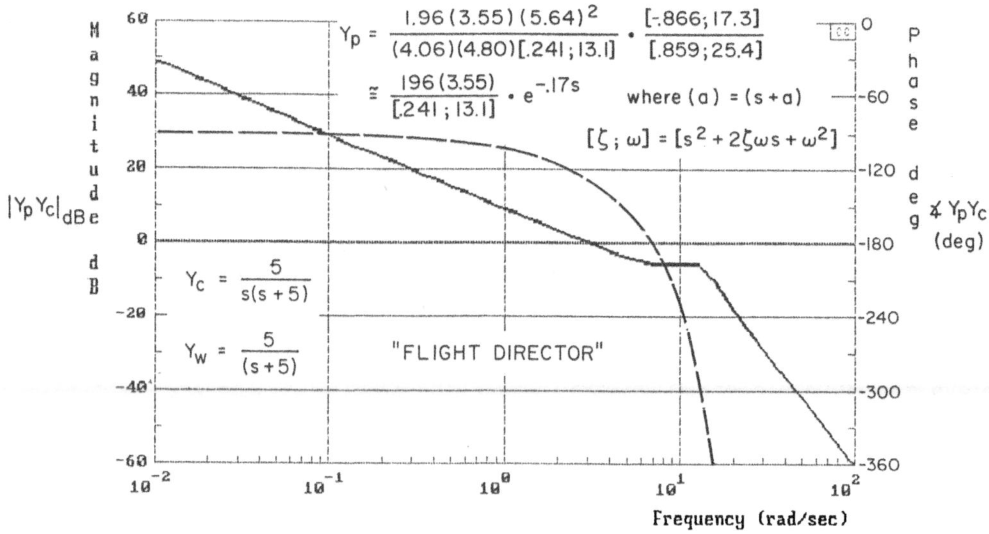

b) Disturbance Input

Figure 11. System Open Loop Transfer Function Solution for OCM Examples not a pure trim function as included in the structural target tracking model, but a series of low frequency poles and zeros which eventually return to a K/s slope at zero frequency. Nevertheless, qualitatively the OCM does adaptively set the overall system open loop characteristics to rational values, and significant agreement has been noted in the past with actual human operator measurements.

CONCLUDING REMARKS

The two man-machine system dynamic performance models discussed here both require the user to define the structure of a given problem, select model parameters, then carry out the computer-aided model analysis process. The OCM attempts to handle most of the analysis steps algorithmically, but solutions must still be simplified, analyzed in order to verify credibility, and should include assessment of parameter sensitivity. With the structural model approach parameter sensitivity is explored as part of the solution process, so that considerable insight is gained into the closed loop system dynamics as the analysis proceeds.

The two programs described in this paper have been designed to work in conjunction with each other, which is useful in developing insight as well as in simplifying the definition of controlled elements for the OCM. The structural modeling program LSMP has been designed to permit the formulation of transfer functions in the format required by Program CC. Thus, it is possible to prepare a structural model for the dynamics of a given vehicle in LSMP, then generate transfer functions to be used as controlled elements in the OCM.

The combination of microcomputer processing power, a convenient, graphically oriented user interface and efficient numerical analysis software make it convenient to carry out dynamic systems analysis with the programs discussed herein. Both programs have successfully solved IEEE benchmark systems analysis problems (Refs. 12,13), and have well defined methods for the analysis of man-machine dynamic systems problems. The analysis power of the programs discussed herein and computer-aided dynamic systems analysis software in general should continue to improve in the foreseeable future with advancements in microprocessor speed and power, memory, and operating systems software.

REFERENCES

1. Allen, R. W., D. T. McRuer, T. J. Rosenthal, et al., Computer-Aided Procedures for Analyzing Man-Machine System Dynamic Interactions, Vol. I: Methodology and Application Examples, Apr. 1987.

2. Allen, R. W., H. T. Szostak, and T. J. Rosenthal, "Analysis and Computer Simulation of Driver/Vehicle Interaction," SAE Paper No. 871086, May 1987.

3. Thompson, P. M., "Program CC's Implementation of the Human Optimal Control Model," AIAA Guidance, Navigation and Control Conference, Monterey, CA, Aug. 1987.

4. McRuer, D. T., "Human Dynamics in Man-Machine Systems," Automatica, Vol. 16, No. 3, pp. 237-253, May 1980.

5. McRuer, D. T., "Pilot Modeling," Advances in Flying Qualities, AGARD LS 157, Jan. 1988.

6. Allen, R. W., T. J. Rosenthal, and R.E. Magdaleno, "A Microcomputer Program for Linear Systems Modeling and Analysis (LSMP)," SCS Multiconference, San Diego, CA, Feb. 1988.

7. Kleinman, D. L., S. Baron, and W. H. Levison, "An Optimal Control Model of Human Response," Parts 1 and 2, Automatica, Vol. 9, No. 3, pp. 357-383, May 1973.

8. Kleinman, D. L., S. Baron, and W. H. Levison, "A Control Theoretic Approach to Manned-Vehicle Systems Analysis," IEEE Trans., Vol. A-C16, No. 6, Dec. 1971.

9. Thompson, P. M., Program CC Version 3 User's Guide, Systems Technology Inc., Hawthorne, CA, 1985.

10. Allen, R. W., D. T. McRuer, R. E. Magdaleno, et al., "Computer-Aided Procedures for Analyzing Pilot/Vehicle/System Interactions," NAECON '86, Dayton, Ohio, Systems Technology Inc., P-385, May 1986.

11. McRuer, D. T., D. E. Johnston, and T. T. Myers, "A Perspective on Superaugmented Flight Control Advantages and Problems," Journal of Guidance, Control & Dynamics, Vol. 9, No. 5, Sept/Oct 1986, pp. 530-540.

12. Thompson, P. M., and D. H. Klyde, "IEEE CACSD Benchmark Problem Number 2 Worked with Program CC," Systems Technology, Inc., WP-430-10, Jan. 1988.

13. Klyde, D. H., T. J. Rosenthal, R. W. Allen, et al., "IEEE CACSD Benchmark Problem Number 2 Worked with LSMP," Systems Technology, Inc., WP-440-1, Jan. 1988.

THE OPTIMAL CONTROL MODEL FOR

MANUALLY CONTROLLED SYSTEMS

William H. Levison

Experimental Psychology Department
BBN Laboratories Incorporated
Cambridge, MA 02238

INTRODUCTION

This objective of this paper is to provide an overview of the theoretical concepts underlying the optimal control model for manually-controlled systems, and to give the reader a feel for the software system that implements this model.

The optimal control model (OCM) is based on the assumption a well-trained and well-motivated human controller will act in a near-optimal manner, subject to certain internal constraints on the human's information-processing capabilities and subject to the operator's understanding of task objectives. This assumption is consistent with notions of human response behavior discussed in the psychological literature. What differentiates the optimal control model from other models of the human operator are the methods used to represent human limitations, the inclusion in the model of elements that compensate optimally for these limitations, explicit representation of the human's internalization ("internal model") of the control task, and the extensive use of state-space concepts and techniques of modern control theory.

The OCM has been implemented as the SSOCM (Steady-State Optimal Control Model) computer program. The SSOCM software system is used to predict operator/vehicle behavior in linear, steady-state control tasks (i.e., tasks where all external inputs are statistically stationary and all dynamical response elements can be modeled as linearly responding systems). A model for task interference and attentional workload is incorporated in the program, and perceptual limitations such as thresholds and resolution limitations can be accounted for. The major assumptions underlying use of this model are:

- The operator is sufficiently well-trained and motivated to perform in a near-optimal manner subject to system goals and human limitations.
- The system to be controlled can be approximated by a linear state-space dynamical representation.
- Effects of nonlinearities can be approximated by stochastic processes.
- The operator constructs an internalized representation ("internal model") of system dynamics that is mathematically equivalent to the true (linear) system dynamics.
- The performance objective can be represented by a quadratic performance index.
- Steady-state external disturbances and commands can be represented as linearly filtered white Gaussian noise processes.

To obtain a model solution, the user must describe the task in terms of the quadratic performance objective, matrices of constant coefficients to quantify system dynamical response, and magnitudes of external white-noise inputs; the operator's limitations are quantified in terms of response-related limitations, time delay, and "noise" variances to account for information-processing limitations.

The SSOCM software does not predict time histories; rather, it yields predictions of the closed-loop performance statistics of the operator/vehicle system. Program outputs include (1) the full state covariance matrix from which are derived standard deviations for all system variables; (2) covariances and standard deviations for the operator's estimates and estimation errors of system variables; (3) power spectral densities for all system variables; and (4) operator frequency response metrics.

The OCM has been applied as a predictive and diagnostic tool -- primarily with regard to problems relating to aircraft flight and flight simulators, and to a lesser extent to automobile driving. Potential areas of application include display design and evaluation, control design and evaluation, prediction of aircraft handling qualities, simulator design and evaluation, and design of simulation experiments. The reader is referred to Levison (1982) for a discussion of representative applications, and to Levison (1985) for examples of how the OCM is used to aid experiment design.

Two major sections follow. The first discusses the conceptual model underlying the SSOCM software; the second reviews implementation and operation of the software.

THE CONCEPTUAL MODEL: THEORY AND PARAMETERIZATION

The basic assumptions underlying the optimal control model have been stated above. We discuss here the organization and parameterization of the conceptual model. We first give separate treatment to the three major model elements (or "submodels") of (1) the task environment, (2) the operator's perceptual processes, and (3) the operator's control processes. We then discuss the conceptual integration of these submodels into a unified system model.

The following discussion deals primarily with model structure and is largely qualitative in nature. Comprehensive mathematical treatments of the algorithms involved in obtaining problem solutions are provided by Kleinman, Baron, and Levison (1970, 1971).

Model of the Task Environment

In general, two representations of the task environment should be considered: the representation of the actual or simulated physical environment, and the operator's understanding or "internal model" of the task environment. The operator's perceptual and control strategies are based largely on his internal model of the task environment; closed- loop operator/vehicle performance is influenced by both the "actual" and internal models of the task environment. These two representations will be identical for some control situations, and different for others. For the most part, the implementation of the OCM discussed in this paper requires the internal and actual system models to be identical.

There are two major aspects of the task environment, or problem description, that must be quantified: (1) the "task dynamics", i.e., the linear equations of motion describing all dynamical relationships among variables not directly related to the operator, and (2) a quadratic performance index or "cost function" to specify the performance goal(s) of the closed-loop system.

Elements included in the description of task dynamics are shown in Figure 1. The "plant dynamics" element describes the dynamical response of the real or simulated vehicle to be controlled by the operator. The task environment also includes dynamical response characteristics of external disturbance or command inputs, which are typically modeled as linear processes as discussed below. If the task environment includes panel instruments that have response lags, the dynamical equations of motion of such instruments become part of the task description. Also, dynamical response limitations associated directly with the operator's sensory mechanisms (e.g., sensing of whole-body motion via the vestibular apparatus) are included in this element of the "task description". (Other aspects of the operator's information processing are included in the submodels associated with perception and control.)

One or more external signals acting on the operator/vehicle system must be defined in order to have a meaningful problem for model analysis; that is, there must be some externally-imposed task load. This task load will generally consist of a disturbance force acting to perturb the vehicle from its desired path, or a path command that the operator is to follow. Command inputs may be imposed by the physical environment or they may be self-imposed by the operator. Command and disturbance inputs may be applied in combination. These inputs are represented as filtered noise processes. Each such input is modeled by its corresponding filter equations as implied by the "disturbance dynamics" or "command dynamics" blocks of Figure 1, and by the covariance of the corresponding zero-mean white Gaussian noise process.

As shown in Figure 1, the inputs to the task-environment submodel are the external inputs described above, plus the operator's control inputs. The outputs of this submodel are the "output" variables available to the operator.

Task dynamics are described by the following linear vector/matrix relationships:

$$\dot{x}(t) = Ax(t) + Bu(t) + Ew(t) \qquad (1)$$
$$y(t) = Cx(t) + Du(t)$$

where the vector **x**(t) represents the vehicle state, **u**(t) is the operator's control input, **w**(t) is the white Gaussian noise inputs, **y**(t) is the vector of system outputs, and **A**, **B**, **C**, **D**, and **E** are matrices of constant coefficients.

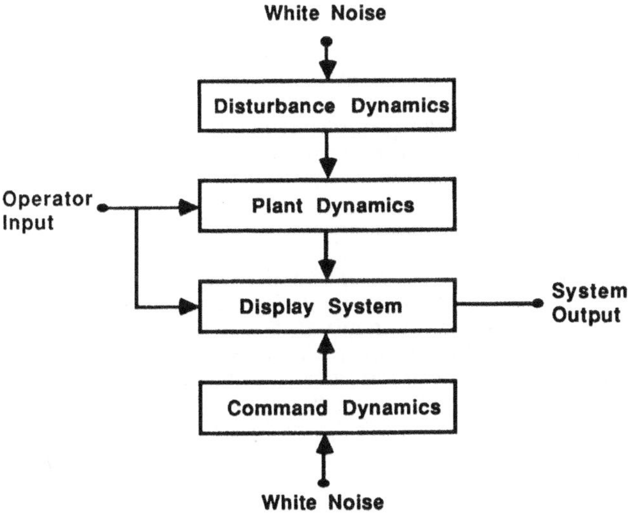

Figure 1. Submodel for the Task Environment

The time-dependent variables included in Equation 1 are assumed to be perturbational variables; i.e., variables whose desired values are all zero. If the problem is such that the operator is to maintain one or more variables at a desired state that is not numerically zero (e.g., maintain constant turn rate), the problem is often formulated in such a manner that problem variables are defined in terms of their excursions from the desired state. In this case, a value of zero reflects zero "error".

Both conceptually and implementationally, task dynamics are parameterized by specification of the number of state, control, noise, and output variables, and by the numerical entries in the five matrices shown in Equation 1. These matrices reflect vehicle, input, and display dynamics integrated into a single vector/matrix equation.

The other aspect of the task-environment -- task requirements -- is expressed by the following scalar quadratic performance index:

$$J = E\left\{\sum_i q_i\, y_i^2 + \sum_j r_j\, u_j^2 + \sum_j g_j\, \dot{u}_j^2\right\} \quad (2)$$

where J is the numerical performance index, "E" in this case is the expectation operator, q_i are the "cost weightings" associated with output quantities, and r_j and g_j are the weightings associated with the operator's control and rate-of-change of control, respectively. Thus, the operator's assumed performance objective is to minimize a weighted sum of mean-squared output and control quantities, where these quantities represent deviations from desired reference values.

To be consistent with the treatment of task dynamics, one could consider two performance indices -- one based on an objective analysis of task requirements, and one reflecting the operator's perception of these task requirements. Since only the operator's perception of task requirements influences operator and closed-loop performance, the conceptual and implementational models are concerned with a single performance index. Although classified as an element of the "task environment", the performance index will typically reflect objective task requirements as well as operator preferences and limitations.

Parameterization of the performance index requires specification of which variables are to be included in the performance index, along with values for the associated weighting terms. Application of the optimal control model has led to two philosophies for selecting parameters. In the case of a simple single-variable laboratory-type manual control experiment, the performance index contains two terms: tracking error and control rate. Tracking error is included to reflect the instructions usually given to the subject to minimize mean-squared error, and, as we discuss later in this section, control- rate is included both to reflect an assumed limitation on operator response capabilities, and to accommodate a mathematical requirement of the problem formulation.

In general, the overall scaling on the performance index is unimportant (it does not influence the optimal operator response behavior), and only the relative weightings are important. Thus, we typically assign a value of unity, and we (iteratively) assign a value to the weighting on control rate to satisfy assumed response constraints as discussed in the subsequent treatment of the control submodel.

A different philosophy is adopted for complex tasks that are more representative of land and air vehicle operations. In this case, cost weightings may be derived by first associating a maximum allowable value (or "limit") with each variable in the performance index, and then setting the corresponding cost weighting equal to the square of the reciprocal of the limit. Limits are generally determined from considerations of desired performance tolerances, hardware constraints, and operator preferences. This scheme has been used with apparent good results in analytical studies of aircraft operations (Hess and Wheat, 1976; Levison, 1978; Levison and Rickard, 1982).

The Perceptual Submodel

Important aspects of the perceptual submodel for the operator are shown in the block diagram of Figure 2. The various sources of operator response randomness are lumped into an equivalent "observation noise" process. In effect, we assume that each system output variable utilized by the operator is perturbed by a white Gaussian noise process that is linearly independent of all other such noise processes and of white noises associated with external task demands. Similarly, the pure transport delay associated with human operator response is lumped into an equivalent "time delay" at the perceptual end. All perceptual variables are assumed to be delayed by the same amount.

The association of response randomness and transport delay with the operator's perceptual input is a mathematical assumption rather than an assumption relating to human physiology. From the latter viewpoint, we would assume that these information-processing limitations are distributed throughout the perceptual, information-transformation, and motor-response processes. With little loss in generality, however, we gain considerable mathematical convenience by modeling response randomness and transport delay as largely perceptual limitations.

The conceptual model for the operator's perceptual input is expressed as:

$$y_p(t) = y(t-\tau) + v_y(t-\tau) \quad (3)$$

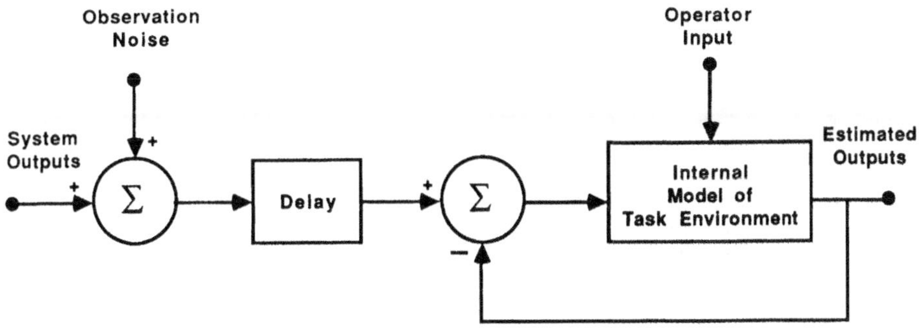

Figure 2. Submodel for the Operator's Perceptual Process

where $\mathbf{y}_p(t)$ is the vector of "perceived displays", $y(t-\tau)$ is the output vector delayed by the operator's transport delay τ, and $\mathbf{v}_y(t-\tau)$ is the vector of white observation noise processes. Thus, the operator is assumed to perceive a delayed noisy representation of each output quantity.

Because white observation noise processes are assumed, each such noise process (one per output variable utilized by the operator) is parameterized by a single quantity -- the autocovariance (equivalent to π times the power density level). To guide the user (and, the computerized model as well) in the selection of appropriate noise covariances, we adopt the following submodel for the observation noise:

$$V_y = \pi P \left(\sigma_y^2 + \sigma_o^2\right)/f_y \tag{4}$$

where V_y is the noise covariance, π is the scale factor that converts power density level to covariance, P is the observation noise-to-signal ratio, σ_y^2 is the expected squared deviation of the (noise-free) output variable "y", σ_o^2 is the variance of a "residual noise" process associated with perception of the variable "y", and f_y is the fraction of attention devoted to "y".

A separate relationship of this form is required for each output variable utilized by the operator. In general, then, the observation noise aspect of the model requires specification of an overall noise-to-signal ratio, plus a residual noise variance (or standard deviation) and a fraction of attention for each output variable. (The variable σ_y is computed as part of the problem solution and is therefore not an operator-related independent model parameter.)

This submodel is perhaps best appreciated by a review of the way in which it was developed. Early model application revealed that the observation noise covariance tended to scale with the variance of the corresponding output variable for single-axis laboratory tracking tasks utilizing optimal display formats and scalings (Levison, Baron, and Kleinman, 1969). These early results led to the notion of an observation noise/signal ratio, which is consistent with the results of psychophysical experiments that show the human's estimation error variance to scale in rough proportion to the squared magnitude of the physical stimulus. Subsequent analysis of laboratory tasks requiring concurrent monitoring and control of multiple display variables led to the model for attention in which the observation noise associated with a given perceptual variable scales inversely with the fraction of attention allocated to that variable (Levison, Elkind, and Ward, 1971; Levison, 1979). This representation is consistent with the notion that the operator has a fixed "channel capacity" that must be shared among the various task-relevant output variables.

Analysis of non-ideal display formats and scalings led to the need to include additional parameters to account for the effects of perceptual resolution limitations (and, in some cases, "indifference thresholds" representing a minimum acceptable error magnitude). For the treatment presented in this paper we use the residual-noise variable to reflect limitations of this sort. Additional discussion of the perceptual submodel, including alternative representations for perceptual limitations, is provided by Baron and Levison (1975, 1977).

For simple tracking tasks, then, the observation noise for all perceptual quantities is specified by a single parameter -- the observation noise-to-signal ratio (or, for short, the "observation noise ratio"). Typical values for simple tracking tasks are on the order of 0.01 (i.e., -20 dB).

For tasks more representative of operational control tasks, attentional allocations and residual noise will have to be specified as well. There are two basic philosophies for selecting fractional attentions. The more rigorous method, and the one that is consistent with the concept of optimal operator response, is to perform iterative model analysis until attention is allocated optimally among the available displays. A simpler method, and one that will usually yield results similar to the rigorous method, is to first reduce the set of output variables to those actually required for control, and then assume equal allocation of attention among the reduced output set.

The penalty associated with attention-sharing depends on assumptions made concerning the display environment specific to the problem at hand. If the operator is not able to process output variables concurrently, then the fractional attentions associated with the various outputs must sum to unity (or less than unity if one wishes to allow for "dead time" during eye movements). On the other hand, if the display elements are assumed to be well integrated, then no attention sharing penalty is assumed (i.e., all attentions may be unity). In general, the operator is assumed to perceive both displacement and rate information from a given display element without an attention-sharing penalty.

Selection of residual noise levels is highly problem-dependent. In principle, each physical display providing task-relevant information to the operator should be analyzed to determine which cues are useful, and, for these cues, a quantification of the perceptual limitations. Data obtained from previous experience with the model, plus the literature on perception, will often allow the user to assign reasonable values to the "residual noise" parameters.

The remaining operator-related independent model parameter associated with the perceptual submodel is time delay. Values on the order of 0.20 to 0.25 seconds are typical.

In addition to the noise and delay parameters, the perceptual submodel shown in Figure 2 embodies the notion of an "internal model" of the task environment that allows the operator to generate expectations ("estimated outputs") of the perceptual inputs. Differences between expected and actual perceptual inputs, along with the operator's perception of his own control input, are used to drive the internal model so as to update the estimated outputs. The internal model is also used to obtain estimates of the (undelayed) system state vector.

The Control Submodel

The conceptual model of the operator's control activity is diagrammed in Figure 3. The block labeled "optimal estimation and prediction", which includes the internal model concept as discussed above, yields estimates of the current vehicle state. These state estimates are then processed by a set of optimal gains to yield a commanded control input. This commanded control is assumed to be perturbed by a "motor noise" and then filtered by a first-order lag to yield the actual control (operator input) that is applied to the vehicle.

The motor noise is not intended to reflect a major source of response randomness -- that function is fulfilled by the observation noise vector discussed above. Rather, the motor noise is intended largely to reflect limitations on the operator's ability to predict the effects of his control input (e.g., prevent the operator from perfectly predicting the response of the vehicle to his control input).

Conceptually, there are two motor noise parameters for each control input: an "actual" motor noise and an "internal" motor noise, each of which is modeled as a white noise process that scales with control-rate variance. The internal motor noise fulfills the function described above. Values of -60 to -50 dB for the internal motor noise/signal ratio have been typically used in previous applications of the optimal control model. As the name implies, this noise influences the operator's response strategy, but it does not represent a physical noise process.

The "actual" motor noise is included for completeness to represent a motor-related noise process that is applied directly to the vehicle. Unless the vehicle control gain is overly sensitive, or the

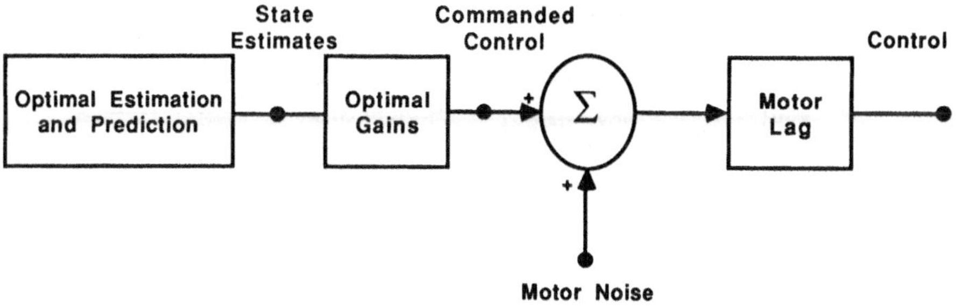

Figure 3. Submodel of the Operator's Control Process

operator has significant neuro-motor deficits, this parameter is usually set to a negligibly small level (e.g., a motor noise/signal ratio of -90 dB).

Early applications of the optimal control model to laboratory tracking tasks suggested that better and more consistent matches to observed manual control response could be obtained by including a performance penalty on mean-squared control-rate activity as opposed to mean-squared control force or displacement. A consequence of penalizing control rate is to induce a first-order lag in the operator's control response, the time constant of which is called the "motor time constant". (If the problem contains more than a single control input, this lag is a matrix quantity.) One might also argue from physiological grounds that some form of filtering is required at the motor end to reflect the human operator's inherent bandwidth constraints.

Mathematically, the motor time constant is an aspect of the problem solution and is therefore not an independent model input. Conceptually, however, the motor time constant may be treated as an independent operator parameter for wide-bandwidth control tasks in which operator response capabilities -- rather than vehicle response bandwidth -- is presumed to be the limiting factor. In this case, the penalty associated with control-rate is (automatically) iterated until the desired motor time constant is achieved. Motor time constants in the range of 0.09 to 0.12 appear to be typical of wide-bandwidth manual control tasks using force controllers. Larger values are to be expected for manipulators having significant displacement.

For relatively low-bandwidth tasks, the rate of control is more likely to be limited by physical constraints or by operator preferences, rather than by operator response bandwidth. In this case, the control-rate coefficient is chosen on the basis of a maximum allowable value, as are other cost weightings. The resulting motor time constant then serves as a check that the user has not required an excessive operator response bandwidth.

Integrated Operator/Task Model

The conceptual model developed thus far is integrated and summarized in the flow diagram of Figure 4. The portion of the operator/task model associated specifically with the operator is enclosed by the dotted lines. The flow of information through the conceptual model is summarized as follows:

- External white noise processes **w** (designated as "disturbances" in the diagram) along with the operator's inputs **u** to perturb the system state vector **x** and the system output **y**.
- Observation noise and delay are added to each output variable to yield the vector of perceived variables $\mathbf{y_p}$.
- The adaptive portion of the operator's response operates on the perceived variables to generate the commanded control. This adaptive response is shown as three elements: (1) an optimal (Kalman) estimator that estimates the delayed state vector, (2) an optimal predictor to partially compensate for operator delay to yield the least-squared-error estimate of the current state, and (3) a set of optimal control gains (-L*) operating on the state estimate to generate the commanded control.

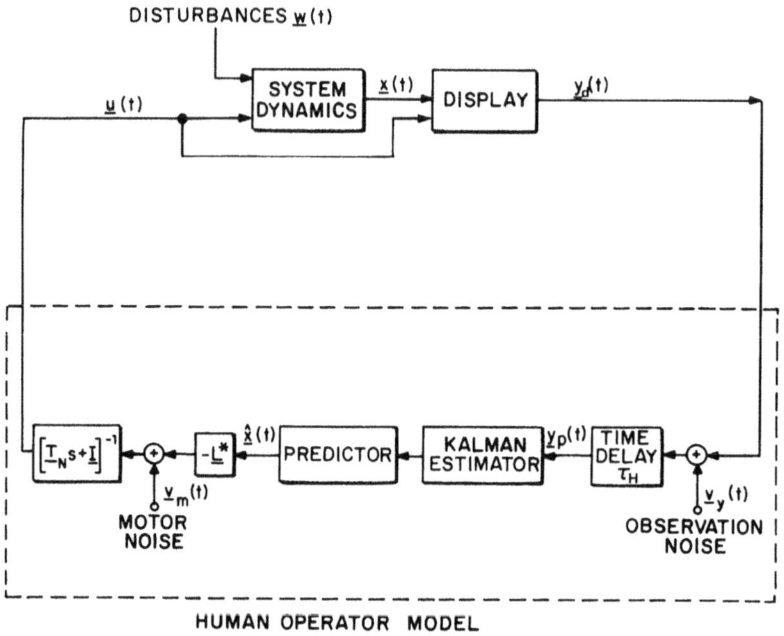

Figure 4. Flow Diagram of the Conceptual Operator/Task Model

- The commanded control is perturbed by motor noise and filtered by a first-order lag to generate the control input applied to the system.

Differences between the conceptual model described here and the operational model actually implemented in the SSOCM computer program are identified in the following section.

THE SSOCM SOFTWARE SYSTEM

This section of the paper is intended to give the reader a feel for the organization and operation of the computer software that implements the foregoing conceptual model. We concentrate primarily on the portion of the program concerned with problem definition.

Comments on Implementation and Program Organization

As of the writing of this paper the conceptual model described above has been implemented as the SSOCM software system for operation on a Digital Equipment corporation VAX machine using the VMS operating system and for IBM PC-, XT-, and AT-compatible personal computers using the DOS environment.

As noted above, this implementation of the "steady-state" model treats operator/vehicle tasks in which all problem variables may be considered as zero-mean Gaussian processes having stationary statistics. The steady-state model implementation takes advantage of the mathematical properties of linear systems driven by Gaussian noise to yield directly the statistics of the problem solution. Because this implementation does not yield time histories, it does not literally implement the information flow shown in Figure 4. The problem solution is, of course, consistent with the conceptual model described above.

The "inputs" to the steady-state model consist entirely of parameters that describe the task environment and the operator, as described in the preceding discussion of the conceptual model. Because no time histories are generated, there are no "input signals" directly analogous to the external forcing-function time histories that would be required in a simulation experiment. In the steady-state model implementation, the operator's internal model of the task environment must be identical to the "true" (linear) model of the task environment.

Similarly, there are no "output signals" in the form of time histories. Model outputs consist of (1) variance scores for all problem variables (and for the operator's estimates and estimation errors of problem variables), and (2) frequency-response metrics consisting of describing functions (gain and phase shift) and spectral densities of all problem variables, with each spectrum partitioned into a component that is linearly correlated with the external forcing function and a component due to the stochastic element of the operator's response ("remnant"). Under the assumption that all problem variables are zero-mean Gaussian processes , the standard deviation (SD) score and probability of excursion outside the "limits" associated with the performance index is computed from each variance score.

Figure 5 outlines the flow of information within SSOCM. The first five blocks represent interactive program modules that allow the user to specify the problem in terms of the external system and operator-related model parameters. Once the user is satisfied with the problem description, the program computes the steady-state control gains and the steady-state estimator gains.

Because the program must select observation noise covariances to meet certain noise-to-signal ratios specified by the user, computation of estimator gains is performed iteratively. Specifically, the program iteratively selects noise covariances until Equation 4 is satisfied for each perceptual variable assumed to be used by the operator.

Computation of variances for all state, output, and control variables is computed along with control and estimator gains, after which the predicted variance and standard-deviation scores (equivalent to rms performance scores for the zero-mean steady-state case) are displayed on the user's terminal. The user is then provided the option to save the model parameter values and model outputs on a log file, and predictions of frequency-domain measures may then be obtained.

Operation of the SSOCM Program

The SSOCM software has been designed to be operated interactively, with the user specifying parameters at runtime. Because the order in which parameters are supplied in a given problem is partially contingent on user responses, one cannot define a unique input stream. Accordingly, operation of the computer program is illustrated via a specific sample problem.

In general, two kinds of user/computer interactions take place during program operation: (1) a "linear" transaction, consisting of sequences of text created by the computer and by the user, and (2) "screens" of information in which a number of items are displayed concurrently, with the user given the opportunity to change items in any order desired before requesting the program to continue. Linear transactions shown below are indicated in bold type; screen-like information is enclosed in boxes. In both cases, the user's responses are underlined. (Neither the boxes nor the underlining would appear on the screen during actual running of the program.)

The user must specify the following sets of task- and operator- related model parameters during the problem-definition phase: system dynamics, output-related parameters, control-related parameters, parameters related to the external forcing function, vehicle and operator time delays, and noise/signal ratios. In addition, the user specifies certain parameters that control program operation.

System Dynamics. Once the program has been loaded and the program title information is displayed, the user presses the ENTER key to initiate the following transaction:

```
Dynamics on file: SAMPLE_PROB_1.DYN
Test problem for SSOCM
Filtered rate control driven by first order disturbance
OK?: Y
```

In response to the prompt, the user specifies the name of the "dynamics file" containing the linear coefficients that describe the external task environment. If the file is found, the program responds by displaying the title information contained in the target file. By responding Y to the query "OK?", the user indicates that the file read is the file that was intended, and the program continues. If the user answers N, or if the file is not found, the user is given the opportunity to specify another file.

The remaining user/computer interactions occurring during the problem-definition phase pertain largely to the selection of independent "operator-related" model parameters. Except for noise/signal ratios that are expressed in dB, all parameters must be non-negative. The reader is referred to the

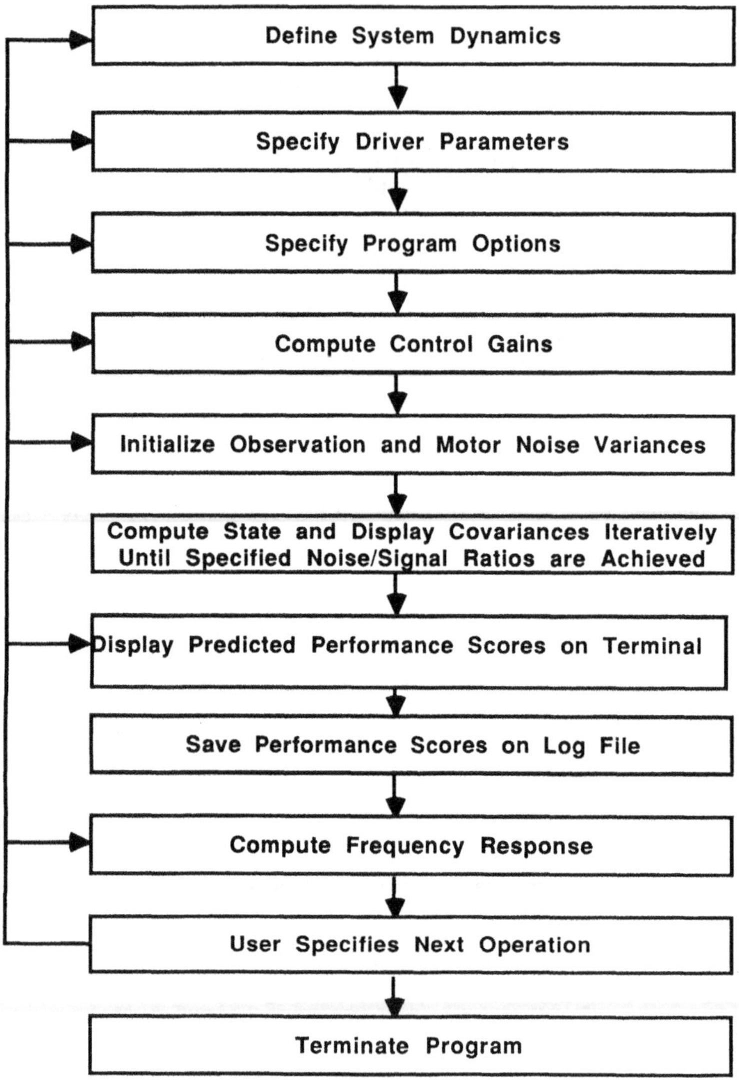

Figure 5. Information Flow in the Steady-State Optimal Control Model

preceding section for ranges of typical values. The default values appearing in the following figures are representative of idealized single-variable tracking tasks using high-bandwidth "vehicles", force sticks, and optimal display and control gains.

Output-Related Parameters. Once the desired dynamics file has been read in, the program presents the screen of information shown in Figure 6. Except for the "residual noise" entries, the numerical values shown in this figure are the default entries that appear when this screen is first presented. They are appropriate for a simple single-variable tracking task in which the error variable to be minimized is the first item in the output vector, and in which the operator is assumed to pay "full attention" to the first two output variables (assumed to be error and error-rate).

OUTPUT PARAMETERS

	Variable	Units	Value for Unit Cost	Rms residual Noise	Fractional Attention
1	error	deg	1.000E+00	5.000E-02	1.000E+00
2	error-rate	deg/sec	0.0	2.000E-01	1.000E+00
3	error-acc	deg/sec2	0.0	0.0	0.0
4	plant	volts	0.0	0.0	0.0
5	plant-rate	volts/sec	0.0	0.0	0.0

Figure 6. Specification of Output Parameters

The user has the capability to modify any of the numeric entires shown in Figure 6. Names and units of problem variables, however, are read from the dynamics file and may not be changed within SSOCM.

The first column of numeric information indicates the relative importance of maintaining problem variables within desired limits. Specifically, the entries indicate the rms error excursions that contribute one unit of "cost" to the quadratic performance index. For the default problem, the operator is penalized one unit for one degree of tracking error; no specific penalty is assigned to any other output variable.[1] For many operational tasks, the user will assign penalties to more than a single variable. Note that an entry of "0" signifies that this variable does not contribute to the scalar performance index, not that a unit cost is assigned to zero error.

Philosophies for selecting values for "residual noise" and "fractional attention" have been discussed above. The nonzero values assigned to residual noise in this example are included simply for illustrative purposes. In the SSOCM program, a value of "0" assigned to the fractional attention signifies that the operator obtains no information from the associated display, and this output variable is temporarily removed from the "output vector" when the estimator gains are computed.

Control-Related Parameters. Once editing of this information screen is terminated, the screen shown in Figure 7 is presented. The user specifies up to three parameters for each of the operator's control inputs. The first parameter shown -- the control "limit" -- represents the control excursion associated with one unit of cost in the performance index. For this problem, no specific penalty is associated with control force, and a value of 0 is assigned.

CONTROL PARAMETERS

	Variable	Units	Control "Limit"	Ctrl-rate "Limit"	Motor Time Const
1	control	pounds	0.0	1.000E+01	1.000E-01

Figure 7. Specification of Control Parameters

[1]Even though no penalty is explicitly assigned to excursions of other output variables, the operator will generally need to keep all variables under control in order to achieve system stability and minimize the objective performance index.

The second column indicates the rms control-rate associated with one unit of cost. If a motor time constant of 0.0 is shown in the third column, the value for the control-rate limit remains unchanged during the gain computation that takes place once the problem-definition phase has been completed.

A non-zero value for the motor time constant (3rd numerical column) is interpreted as a desired value. During the computational phase, the control-rate "limit" will be adjusted in an iterative manner until the resulting motor time constant is within 1% of the desired motor time constant. Once a problem solution has been obtained, subsequent entry to this point in the program will show the adjusted control-rate limit.

The algorithm for computing control gains requires that the control-rate limit be non-zero positive. The user must therefore specify a "best guess" for the control-rate limit to initialize the solution when a specific motor time constant is desired.

External Noise Parameters. The user then specifies two values for each of the external white-noise forcing-functions: the value associated with the "real" noise process, and the value associated with the operator's internal model of the noise process (Figure 8). Expect for unusual situations, these two values should be identical.[2]

EXTERNAL NOISE SD

	Variable	Units	External Noise SD	Model of Noise SD
1	noise-in	w.n.	1.000E+00	1.000E+00

Figure 8. Specification of External Noise Parameters

Delay and Operator Noise Parameters. The final set of operator parameters to be selected is shown in Figure 9. Two time delays (pure transport lags) are specified: one associated with the vehicle, and one associated with the operator. Closed-loop performance in terms of rms error, etc., depends only on the sum of the two delays. The relative allocation of delay between the vehicle and the operator will, however, influence the phase shift associated with the prediction of the operator's describing function. To maintain good correspondence between model prediction and experimental results, delays known to be associated with the vehicle (or the simulation of the vehicle) should be properly designated as "vehicle delay".

MISCELLANEOUS PARAMETERS

1	Vehicle delay (sec):	0.000
2	Operator delay (sec):	0.200
3	Motor noise ratio (dB):	-90.0
4	Internal mot nse ratio (dB):	-50.0
5	Observation noise ratio(dB):	-20.0

Figure 9. Specification of Remaining Operator Parameters

The next two parameters specify the motor noise parameters in terms of noise/signal ratios. The value of -90 dB indicates that the physical noise process associated specifically with control execution is negligible. This value would be made non-negligible except for cases where the control sensitivity is extremely high or where the user wishes to represent significant sources of neuromuscular noise (e.g., tremor). As explained in the preceding section, a small but non-negligible value is assigned to the "internal" motor noise to reflect limitations on the operator's ability to predict the effects of his control actions. The last parameter is the observation noise/signal ratio, which is the parameter that accounts in the model for most of the effects of the stochastic ("remnant") portion of the operator's control response.

[2]For cases where the user wants to model the effects of imperfect knowledge of the forcing function due to lack of exposure to the task, the value associated with the internal model may be smaller than the value associated with the "external" noise.

Before starting the problem-solution phase, SSOCM allows the user to review certain "program options" having mainly to do with convergence criteria for the observation and motor noises.

The following transaction shows the user terminating the problem-definition phase of the SSOCM program by requesting the program to begin computations:

```
 0: Cancel
 1: Specify dynamics file
 2: Review all parameters
 3: Review output parameters
 4: Review control parameters
 5: Review external noise parameters
 6: Review miscellaneous model parameters
 7: Review program options
 8: Begin computations
 9: Initiate printouts
10: Compute frequency response
To part: 8
```

Parts 1-7 allow the user to revise all or part of the independent parameters entered to this point. Once the problem solution has been obtained through execution of Part 8, the user may then execute Part 9 to store the independent model parameters and the primary model outputs (predicted SD scores for all state, control, and output variables) on a disk file for subsequent printout. Execution of Part 10 will yield displays of frequency-response metrics (describing functions and spectra), computed at frequencies determined by the user, which can be included in the output file.

CONCLUDING REMARKS

Application of the model to a new control situation requires the user to (1) develop or acquire a linearized math model of the control task, (2) quantify performance objectives in the form of a scalar performance index, and (3) specify values for the independent model parameters related most directly to operator limitations. The amount of effort involved in these tasks will depend strongly on the details of the specific situation. For example, if the model is to be used as an aid in quantifying operator performance in a laboratory-type tracking task, the user will most likely be using a linear simulation, the equations for which directly lend themselves to the state-space formulation required by the model. On the other hand, if the task involves control of an actual or simulated vehicle for which only a nonlinear set of equations is known, the user will have to derive a linearized set of equations, either from the nonlinear equations, or by performing a small calibration experiment with the physical device and using an appropriate identification technique to determine the linear-system coefficients from the data.

As we have noted above, analysis of operational control situations (as opposed to idealized laboratory-type tracking tasks) will require some effort to determine the values for unit cost ("limits") associated with variables contained in the performance index. As demonstrated in the references, a relatively straightforward analysis of task requirements and operator limitations will yield useful values for these parameters.

Use of "typical" values for operator time delay and for observation and motor noise/signal ratios found in past model studies will generally suffice for the prediction of reliable performance trends. As noted above, fractional attention can often be handled by simply assuming no attention-sharing penalties (e.g., unit attention to all relevant output variables), or by sharing attention equally among the variables most needed for adequate control. (The model may be used in an iterative fashion to determine which cues are needed.)

Analysis of the display environment may be required to determine "residual noise" levels for control situations in which limitations on perceptual resolution are likely to be important. Such limitations are generally relevant where the human operator relies upon "real-world" visual scene cues. In this case, analysis is performed first to determine the set of cues most likely to be used by the operator, and second to define the linearized relation between perceptual cues and system state variables. The user then refers to the manual control literature (particularly as it relates to previous application of the OCM) or to the psychological literature for data on which to base values for residual noise. In the unlikely case that the perceptual environment contains important cues for which no relevant quantitative data exist, the user may wish to perform a calibration experiment to obtain data on perceptual resolution limitations.

As is clear from the foregoing discussion, a new user of the SSOCM program will have a substantial amount of detail to master before he or she can use the model with confidence; one cannot expect to become a master practitioner within an hour of opening the user's manual. Furthermore, intelligent application of the model to a specific control situation requires that the user obtain a solid understanding of the control task and its objectives. These considerations should not be taken as criticisms of the model, however, as considerable expertise will generally be required to master the use of other predictive operator models, or to perform a relatively costly manned simulation in place of model analysis.

REFERENCES

1. Baron, S. and Levison, W.H., "An Optimal Control Methodology for Analyzing the Effects of Display Parameters on Performance and Workload in Manual Flight Control," IEEE Trans. on Systems, Man and Cybernetics, Vol. SMC-5, No. 4, pp. 423-430, July 1975.
2. Baron, S. and Levison, W.H., "Display Analysis with the Optimal Control Model of the Human Operator," Human Factors, Vol. 19, No.5, pp. 437-457, October 1977.
3. Hess, R.A. and Wheat, I.W., "A Model-Based Analysis of a Display for Helicopter Landing Approach," IEEE Trans. on Systems, Man and Cybernetics, pp. 505-511, Vol. SMC-6, No. 7, July 1976.
4. Kleinman, D.L., S. Baron and W.H. Levison, "An Optimal-Control Model of Human Response, Part I: Theory and Validation", Automatica, Vol. 6, pp. 357-369, 1970.
5. Kleinman, D.L., S. Baron and W.H. Levison, "A Control Theoretic Approach to Manned-Vehicle Systems Analysis", IEEE Trans. on Auto. Control, Vol. AC-16, pp. 824-833, No. 6, December 1971.
6. Levison, W.H., "Analysis and In-Simulator Evaluation of Display and Control Concepts for a Terminal Configured Vehicle in Final Approach in a Windshear Environment," NASA CR-3034, August 1978.
7. Levison, W.H., "A Model for Mental Workload in Tasks Requiring Continuous Information Processing," in Moray, N. (ed): Mental Workload Its Theory and Measurement, Plenum Press, New York and London, 1979.
8. Levison, W.H., "The Optimal Control Model for the Human Operator: Theory, Validation, and Application," in Frazier, M.L., and Cromble, R.B., (eds): Proceedings of the Workshop on Flight Testing to Identify Pilot Workload and Pilot Dynamics, AAFTC-TR-82-5, Air Force Flight Test Center, Edwards Air Force Base, CA, May 1982.
9. Levison, W.H., "Application of the Optimal Control Model to the Design of Flight Simulation Experiments," Society of Automotive Engineers Paper No. 851903, presented at the Aerospace Technology Conference and Exposition, CA, October 1985.
10. Levison, W.H., Baron, S., Kleinman, D.L., "A Model for Human Controller Remnant", IEEE Trans. on Man-Machine Systems, Vol. MMS-10, No. 4, December 1969, pp. 101-108.
11. Levison, W.H., Elkind J.I., and Ward J.L., "Studies of Multivariable Manual Control Systems: A Model for Task Interference," NASA CR- 1746, May 1971.
12. Levison, W.H. and Rickard, W.W., "Closed-Loop Criteria for Assessing Longitudinal-Axis Handling Qualities of Transports in Final Approach," Proc. of the AIAA Guidance and Control Conference, Albuquerque, NM, August 19-21, 1981.

MODELS OF MULTI-TASK SITUATIONS

INTRODUCTION: MODELS OF MULTI-TASK SITUATIONS

L. van Breda

TNO Institute for Perception
P.O. Box 23
3769 ZG Soesterberg, The Netherlands

OVERVIEW OF THE PAPERS

In this session, modelling techniques for analyzing complex technological man-machine systems in multi-task situations are discussed. These modelling techniques describe human monitoring, control, and supervisory tasks including the dynamic characteristics of the environment in which the systems operate.

In the first paper, Baron and Corker discuss engineering-based models that grew out of control theory, and incorporate developments in estimation theory and statistical decision theory. For multi-task situations, the Procedure-Oriented Crew Model (PROCRU) is described and recent modifications and extensions are discussed.

An extensive description of the MICRO-SAINT simulation language is provided by Laughery. This network modelling language can be used to model a large variety of tasks. An application to helicopter workload analysis is described.

The next two papers are focused on simulation studies in which highly automated man-machine systems play an important role. In his paper, van Breda describes the analysis of a multiple operator procedure on board future submarines using network modelling techniques. The model describes the operator's activities and the dynamic characteristics of the submarine. The study is meant to investigate the procedure with respect to information flow and operator task performance. The SAINT simulation language is used for model implementation.

Döring describes the analysis of a pilot's behaviour during an automated landing approach. The pilots' tasks are mainly based on predetermined procedures. The goals of the study are to develop a simulation methodology which can be used to determine the required knowledge of the pilot in rule-based situations, and for human engineering design requirements for cockpits. In this study, the SLAM simulation language is used.

Wickens gives an overview of models that focus on the sequential or serial aspects of multi-task performance and describes their limitations. He presents the components of multiple resource theory as an approach that can address these limitations and discusses the extent to which different performance models have employed features of multiple resource theory.

In the last paper of this session, Harris, Iavecchia, and Dick give a general introduction to the Human Operator Simulator (HOS) computer model which was demonstrated at the workshop. HOS is designed to be used in the evaluation of complex crew stations. HOS is a deterministic model based on experimental data.

DISCUSSION

The models presented during this session are mainly combined discrete/continuous event models based on normative task situations. The applicability of these models is good, but a common weakness is the lack of validation data.

Simulation languages such as SLAM, SAINT and MICRO-SAINT are general purpose modelling tools that can be applied effectively at a rather early stage of man-machine system design. Models based on these simulation languages are stochastic networks that rely on samples from data distributions for parameter specification. During the discussion, users indicated a preference for this type of modelling technique because it is suitable for many task situations and supports the simulation of system behaviour. There is no model of human behaviour embedded in these tools, however the human factors and system engineer can define the task characteristics and the aspects of human behaviour that are relevant to his problem using the tools that are provided. The costs of these models can be relatively low. However, the use of simulation languages requires specialists and may be too complex for some potential users.

Models like PROCRU, HOS, etc. have components (embedded data or models) that describe human behaviour and are used to get more insight into human information processing and workload. In the audience, there was some concern about the single channel assumption of these tools. This was seen as a restriction in their applicability. However, modifications for parallel information processing are foreseen in the near future.

ENGINEERING-BASED APPROACHES TO HUMAN PERFORMANCE MODELING

Sheldon Baron and Kevin Corker

BBN Laboratories Incorporated
70 Fawcett Street
Cambridge, Massachusetts 02238

INTRODUCTION

Complex technological systems, such as aircraft, power plants, and weapons systems, almost always require humans to monitor, control and/or supervise their operations. These "engineered" systems are designed and developed using a variety of methods for analysis, testing and evaluation that rely heavily on mathematical models that describe the inanimate systems involved and the environment in which the system operates. The desirability, if not the outright need, to account for the human component in the design/development process for these combined person-machine systems, in a manner similar to that in which the rest of the system is treated, has given rise to a large number of models of relevant human performance. (We refer to these models as human performance models or, more concisely, as HPM's.) The class of models that tend to consider the operator more from an engineering perspective than a psychological one can be viewed as engineering-based HPM's.

Although engineering based models of the human operator treat many psychological phenomena, e.g., psychomotor performance, perception, human information processing and even cognitive behavior, the aims of the models and the methods employed to describe and/or predict human performance are quite distinct from those of mathematical and/or experimental psychology. In particular, the focus of these models on prediction of total, i.e., combined, person-machine performance leads inevitably to engineering characterizations of human operation and performance that are significantly different from, but relate to, those of traditional psychology.

By far, the most significant developments in engineering-based models are associated with the class of models that are based on control theory and related systems developments in estimation theory and statistical decision theory. This paper will concentrate on this class of engineering-based models. It begins with a discussion of the applications focus of the models. Then a brief history of their development from simple servo-mechanism type models through multi-variable closed-loop control to advanced multi-task models involving discrete and/or continuous control, decision-making and procedural and higher level activities is discussed. This history will be accompanied by a discussion of the changes and advances in model structure that allowed for the expanded scope of such models. The main emphasis of the paper is on a state of the art engineering-based model with respect to modeling multi-task situations, PROCRU. The original PROCRU model is described and, then, recent modifications and extensions of it are discussed. The paper concludes with a consideration of future research needs and anticipated directions in engineering-based model development for large, complex systems.

CONTROL THEORY MODELS

The theory and techniques of control systems design, analysis and evaluation have served as a basis for developing a class of HPM's commonly referred to as control theory models. The technological systems that motivate development of these HPM's are dynamic in nature and are described by differential (or difference) equations. In continuous manual control problems an alternative representation for the system that is often used is a frequency-response or transfer-function (i.e., a frequency-domain representation). A central assumption of most models based on this approach is that trained operators will approximate the characteristics and performance of "good", or even optimal, inanimate systems performing the same functions but that their performance, and therefore that of the overall system, will be constrained by certain inherent human sensory, cognitive and response limitations. Control theory models require that these human limitations be described in terms that are commensurate with other elements of the dynamic system description. This imposes a need for human performance data appropriate to limitations in dynamic processing and response rather than those appropriate to discrete task completion.

The performance issues of interest in control theory models are associated primarily with overall person-machine performance and tend to relate to measures such as accuracy of control and information processing, system stability and responsiveness, and ability to compensate for disturbances. A major focus is the interaction between system characteristics and human limitations and the consequences that flow therefrom. Thus, the models are intended to help system designers determine whether or not the information provided and/or the control ("handling") characteristics of the system are adequate to allow a trained operator to perform the task with a reasonable amount of physical and mental effort.

Control theory models view the human operator as an information-processing and/or control/decision element in a closed loop system, (sometimes referred to as the cybernetic view of the human), as illustrated in Figure 1. In the figure the upper portion represents the system and the lower part the human. The human operator observes the system via its displayed outputs (which herein, is meant to include all relevant sensory input as well as system displays), assesses the situation, decides on a course of action and implements that action. The representations used in this straightforward "see, think, act" description of the operator have evolved as more complex problems have been considered and as the state-of-the-art in human performance modeling has advanced. This evolution has mirrored that in control theory itself, and reflects the changing nature of the systems being considered and the concomitant development of appropriate techniques for analyzing them.

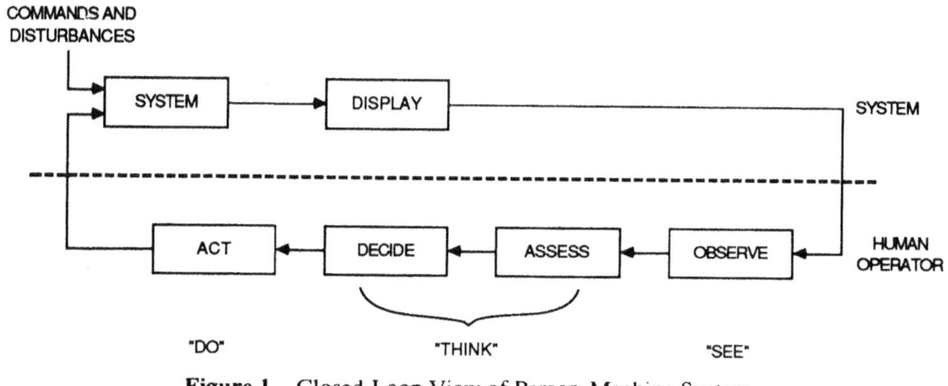

Figure 1. Closed-Loop View of Person-Machine System

The earliest engineering models of note dealt principally with problems in which human manually controlled a dynamic system (e.g., motion of a gun or an aircraft) to track a single input (e.g., a target) so as to maintain small tracking errors. These tracking problems often required very rapid and precise responses. To perform them well, human operators had to be highly practiced and had to virtually "automate" their response to given situations (e.g., specific system dynamics). In such cases, the models for the operator could be relatively simple input-output models with very little need for detail in the observe, assess, decide portions of the model. The HPM's that emerged grew out of a servo-mechanism paradigm and the operator was represented by a transfer function or, later, by a describing function and a remnant. Model parameters were those appropriate to such a description (i.e., gain, phase, delay and time constants). By far, the most successful approaches involved frequency domain representations of the human. Verbal rules were used to transform empirical results into mechanisms for selecting the forms and parameters of transfer functions as a function of task parameters.

A major increase in complexity occurred when interest turned to multi-variable problems (ones in which there are multiple outputs to be controlled by one or more controls). Because the human operator of such systems cannot observe or process simultaneously all of the outputs to be controlled, he/she must attend selectively to individual variables. This leads to a need for more complex models for the observation process as well as a requirement for modeling the decision processes associated with goal-oriented selective attention. Furthermore, more complex forms of control structure were needed to model multivariable control. The challenges to control theory modeling posed by the need to address multivariable control problems, were met in two ways, first by modifying the existing methods and, then, by the introduction of a new approach.

The single-variable control models were modified by elaborating the observation process. In particular, visual sampling models were added to the basic models which were also augmented to account for the "reconstruction" of the continuous signals needed for continuous control from the visually sampled data. Finally, multi-loop control structures were developed to deal with control complexity. (The reader is referred to McRuer and Krendel, 1974 for an excellent overview of the single-variable control models, the multi-variable extensions and the applications of the technology).

The second approach to the problem of modeling human performance in multivariable control tasks was essentially new, though it drew on the insights and data of the previous work. The approach grew out of, and built upon, contemporary (late 1950's, 1960's) developments in control theory - at the time, called modern control theory. It emphasized time-domain representations, quantitative performance criteria, state-space formulations and stochastic processes. The HPM that emerged from this development was called the Optimal Control Model (OCM) (Baron and Kleinman, 1969; Kleinman, Baron and Levison, 1971). The OCM has proven to be a successful tool for predicting human and system performance in complex manual control tasks; it has also provided a springboard for engineering-based modeling of other tasks and for multi-task situations, as discussed below.

The OCM has been well documented (see Baron and Levison, 1980 for a review and an extensive bibliography) so we will only focus on the factors that are relevant to the present discussion. The model structure for the OCM is illustrated in Figure 2. This structure can be related to that of Figure 1 as follows.

The "OBSERVE" portion of the OCM is comprised of the perceptual processing limitations of observation noise and time-delay.[1] Note that the observation noises are affected by the selective attention decision logic, incorporated in the OCM via a model for task-interference (Levison, Elkind and Ward, 1971).

The Kalman estimator/predictor of the OCM, the function of which is to produce an estimate, $\hat{X}$, of the current state of the system on the basis of the perceived outputs, corresponds to the "ASSESS" function in Figure 1. This limited form of "situation assessment" is all that is required to perform the control task in an "optimal" fashion under the assumptions normally given in deriving the OCM. Moreover, it models the (cognitive) process whereby the operator constructs a set of expectancies concerning the state of the system on the basis of his understanding of the system and his incomplete knowledge of the moment-by-moment state as accessible to him from limited and noisy observations. Though it was somewhat controversial at first, the inclusion of an estimator/predictor in a manual control model proved to be an extension of significant consequence. Not only did it support a model of improved predictive capability, it also provided a means for applying the same basic approach to a variety of problems in monitoring and decision-making.

Given the best estimate of the current system state, as provided by the Kalman estimator/predictor, the next block in the OCM (labeled L*) assigns a set of control gains or weighting factors to the elements of the estimated state, in order to produce control actions that will minimize a defined performance criterion. As might be expected, the particular choice of the performance criterion determines the weighting factors and thus the effective control law gains. This choice of control weightings or gains is a relatively simple decision process that results in a "commanded control", u_c.

---

[1]Other sensory limitations, if appropriate, would be included here.

The production of a control action from the commanded control is a process affected by certain response limitations. Just as an observation noise is postulated to account for the input processing inadequacies, a motor noise is introduced to account for an inability to generate noise-free output control actions. In many applications this noise level is insignificant in comparison to the observation noise, but where very precise control is important to the conditions being analyzed, motor noise can assume greater significance in the model. Finally, the noisy output is assumed to be filtered or smoothed by a filter that accounts for an operator bandwidth constraint. In the OCM, this constraint arises directly as a result of a penalty on excessive control rates introduced into the performance criterion. The constraint may mimic actual physiological constraints of the neuromotor system or it may reflect subjective limitations imposed by the operator.

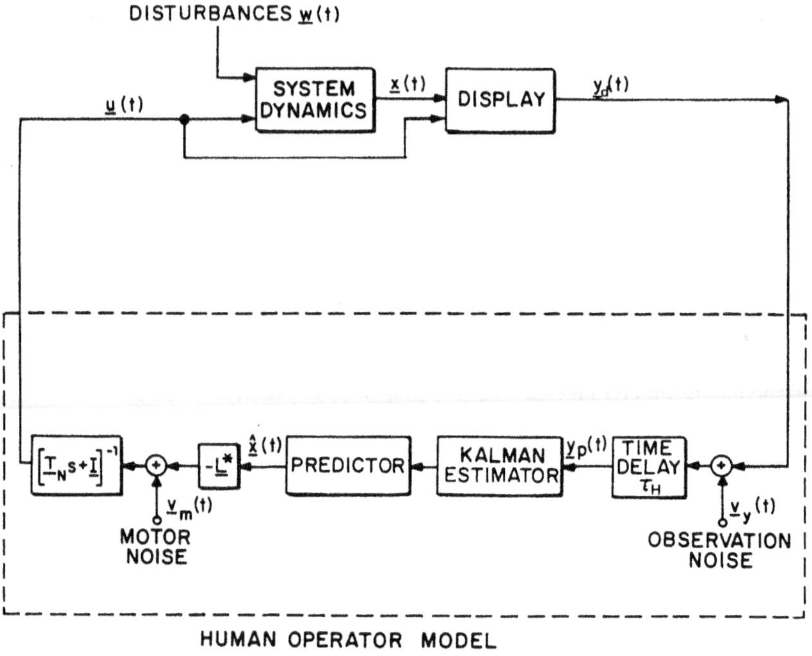

Figure 2. Model Structure of Optimal Control Model (OCM)

This, then, provides a conceptual description of the elements of the OCM that model the human operator. The parameter values that must be provided by the user of the model correspond to the human limitations that constrain behavior. With these limitations as the constraints within which system performance is produced, the model predicts the best that the operator can do (i.e., it is normative). A large backlog of empirical research provides the data necessary to make realistic estimates of the appropriate parameter settings in the manual control context. This research has shown that these parameters are relatively invariant with respect to changes in task environment, thus enhancing the model's predictive capacity.

Because of the pace of continuous manual control it seems unlikely that human operators could execute the kind of sequential processing described above for the OCM at a conscious level in such tasks. Nonetheless, the input-output behavior of the OCM predicts that of highly skilled human subjects with remarkable fidelity. This suggests that the subjects are behaving "as though" they did such processing or that, with practice, it is possible to condition or automate these fairly complex processes. Regardless, the model has proven useful for manual control analyses and as a basic for developing models for other tasks.

MODELS FOR MONITORING AND DECISION-MAKING

Several different monitoring and decision-making models have been derived using the optimal estimator and the models for human perceptual limitations that are incorporated in the OCM. The models thus have a structure that is identical to that of Figure 2 except that the control portion of the model is replaced by a decision algorithm that operates on the outputs of the state estimator to produce a discrete decision. These models have been focused largely on the detection of failures or "abnormal" events, which are characterized either by system variables exceeding prescribed tolerances or by the systematic deviation of system outputs from their expected behavior.[2] The optimal estimator generates information that can be used for detecting either type of abnormal situation.

Levison and Tanner (1971) studied the problem of how well subjects could determine whether a signal embedded in added noise was within specified tolerances (a continuous, visual analog of classical signal detection experiments). From a practical standpoint, the problem is analogous to that of deciding whether conditions warrant a given action; e.g., deciding whether an aircraft is sufficiently within approach tolerances to proceed with the landing, or deciding whether target tracking errors are small enough for weapons release. They modeled this situation by assuming that the operator is an optimal decision maker in the sense of maximizing expected utility. For equal penalties on missed detections and false alarms, this rule reduces to one of minimizing expected decision error. The decision rule is simple a likelihood ratio test that, effectively, uses the densities generated by the optimal estimator.

In a theoretical paper, Phatak and Kleinman (1972) suggested a model for failure detection based on monitoring of the estimator residuals (the deviations of observed outputs from predicted ones). Gai and Curry (1976) and Wewerinke (1981) each used such a residual monitoring scheme to model detection of instrument failures in monitoring of complex, multivariable systems and compared the predictions of these models with corresponding experimental data. In each of these models, the perceptual and information-processing portions of the OCM, including the attention-sharing model, were used. In both cases "optimal" sequential decision algorithms were used for decision-making. The models were shown to be capable of predicting experimental results collected in the empirical validation efforts.

The above studies provide further, independent validation of the display processing and state estimation models developed for the OCM. They also extended the control-theoretic models into the realm of simple decision-making by replacing control laws with decision algorithms. Because of the nature of the decision in these studies, it is probably more helpful to think of them as extending the notion of situation assessment in this framework to go beyond that of pure state estimation and to include detection of state-related, or state-defined, events (or situations).

Two other models go still further in extending the control-theoretic approach to modeling human decision-making. These are the dynamic decision model (Pattipati, Ephrath and Kleinman, 1979, 1980) and the DEMON model (Muralidharan and Baron, 1980).

The dynamic decision-making model (DDM) is a model for human task sequencing, i.e., a model for human decisions concerning what task to work on in a dynamic multi-task environment. It uses essentially the same information processing sub-model as the OCM and the other models discussed herein but its decision-making portion differs from them in several respects. First, Pattipati et al. introduced the notion of a decision state as distinguished from a task state. Decision-states are variables that are oriented towards the decision action; they are a memoryless, nonlinear transformation of the task state. In the particular paradigm modeled with the DDM, the decision states for each task are the time required and time available for completing that task. Then, they assumed the decision strategy was completely myopic; which is to say, the decision policy of the human disregards possible future actions and rewards. This is implemented by defining an instantaneous attractiveness measure for each task which balances the expected reward for performing the task against the potential losses for not working on the remaining tasks. The main distinction of the DDM's decision-making model is that human randomness is introduced into the decision-making algorithm itself. This is accomplished by assuming a distribution for the payoff values (attractiveness measures) and then incorporating Luce's stochastic choice algorithm (Luce and

---

[2]Here, we do not discuss control-theoretic models for visual scanning such as may be found in Baron and Kleinman (1969) or Kleinman and Curry (1977).

Galanter, 1963) for decision making. Though the practical value of including this randomness in performing system design and analysis may be argued, it is clear that such randomness in human decision-making does exist. Moreover, the introduction of the stochastic choice axiom allows the DDM to be used to compute performance statistics analytically, rather than by Monte Carlo simulation, at least in cases where the simplifying assumptions used in the DDM apply.

DEMON is a decision-monitoring and control model for analyzing the en route control of multiple remotely piloted vehicles (RPVs). Very briefly, the en route operator's task is to monitor the trajectories and estimated times of arrival (ETAs) of *N* vehicles, to decide if the lateral deviation from the desired preprogrammed flight path or the ETA error of any of them exceeds some tolerance threshold and to correct the paths of those that deviate excessively by issuing appropriate control commands ("patches"). In addition, the operator must decide when to "pop-up" or "hand-off" RPVs under his control.

DEMON models both monitoring and control decisions. The basic monitoring decision is to select which of 2N+1 displays[3] to observe. The control decision involves choosing among the alternatives of patching (correction), pop-up, hand-off or no-control action. In both cases, DEMON assumes that the decisions are rational and are governed by the operator's knowledge of the situation, his goals and priorities and his instructions. These factors are incorporated in an expected net gain (ENG) criterion for each display. It is assumed that the operator selects the display with the highest net gain.

DEMON extended previous control-based approaches to a multi-task environment that involved discrete monitoring and control decisions. Control of each RPV represented a separate task with a payoff for maintaining errors within tolerance and for timely pop-ups and hand-offs. Inasmuch as only one RPV (and only one RPV-state) could be observed at any time, the DEMON operator had to rely on memory and prediction (based on the state-estimates) to assess the relative priorities for "servicing" of the RPVs under his control.

In addition to providing useful models of selected monitoring and decision-making tasks, the developments and studies discussed in this section provided further, independent, validation of the display processing and state estimation models developed for the OCM. When these results are added to the implicit validation provided by the tracking data, one has a strong case for this approach to modeling human information processing. The studies also extend the notion of situation assessment, in the control theoretic modeling context, to include the detection of state-related events. These events may be defined in terms of "regions" in the multidimensional state-space or in a suitable output-space. Or, they may relate to events that are characterized by a signal deviating from expectancy which can be detected by monitoring and testing the residuals. The detection of these state-related events will, in general, trigger subsequent actions (such as selecting what to do next as in the DDM).

However, by and large, the models discussed thus far do not address total, multi-task system problems of interest today. For example, the single task[4] or single decision models do not consider multiple tasks with different, perhaps conflicting, objectives, multiple operators as is often the case for large systems, or events (e.g., failures) not directly related to dynamic state variables. Perhaps the major shortcoming of these models, with respect to modeling realistic tasks, is that they do not include the procedural activities of the operators, or the discrete tasks that are often part of such procedures.[5] DEMON began to address some of these issues and needs; we now turn to a model and approach that goes significantly further in this direction.

---

[3]ETA and lateral deviation errors for each RPV are displayed separately (yielding 2N displays for N RPVs). An additional display option is included to allow for attention to other, non-modeled, activities.

[4]The DDM is considered here as modeling the single task, or decision, of what to do next.

[5]Interestingly, these neglected features are often the prime concern of psychologically-oriented HPMs.

MODELING MULTI-TASK SYSTEMS PROBLEMS

More recent developments in advanced control-theoretic modeling have been directed at the development of comprehensive models that are suited to the analysis of large, person-machine systems. These models are comprehensive in that they cover a range of operator activities including monitoring, continuous and discrete control, situation assessment, decision-making and communication. The models were developed in a variety of applications contexts as discussed in Baron (1984), where an outline of a general model for supervisory control based on the control theory approach is also presented.

The development of the comprehensive, engineering-based, multi-task models has been driven by the following fundamental requirements. First, a system model is required, one that represents the interaction among human operators, automation, system equipment, procedures, vehicle or plant, and environment. Second, the cognitive and decision-making aspects of human performance and control must be explicitly represented. This includes the decision criteria, rules of application, attributes of selection alternatives, decision time, and cognitive or processing effort that characterize human-machine interaction in intensive information processing environments. Third, communication among operators and crew-member and between human and autonomous agents must be considered and the content and load associated with that communication must be modeled. Fourth, each crew member's knowledge about the state of the world, the condition of the system, the state of his/her goals, and the constraints on procedures to meet those goals should be represented. Finally, for purposes of analysis, a range of performance metrics tied to the mission or operation under study should be provided.

Of the control theoretic models developed thus far, PROCRU (Procedure-Oriented Crew Model) (Baron, Zacharias, Muralidharan and Lancraft, 1980) is the most comprehensive in its range of coverage of human activities in monitoring and control of a large system. This model was developed with the goal of providing a tool that would allow for the systematic investigation of questions concerning the impact of procedural and system design changes on performance and safety of commercial aircraft operations in the approach-to-landing phase of flight.

PROCRU is a closed-loop system model incorporating sub-models for the aircraft, the approach and landing aids provided by the air traffic control system, (three) aircraft crew members, and the air traffic controller (ATC). For convenience in development, only two crew members (the Pilot-Flying (PF)) and the Pilot-Not-Flying (PNF)) were represented by detailed HPM's. The models for PF and PNF had the same basic structure; differences in behavior resulted from specifying separate task assignments, task priorities and information sources for the two models. The models for PF and PNF are comprehensive in that they account for the wide range of crew activities associated with conducting a "typical" commercial ILS approach to landing, namely: display monitoring, information-processing, decision-making, flight control and management, execution of standard procedures and communication with other crew members and with the ATC.

Briefly, PF and PNF are each assumed to have a set of "procedures" or tasks to perform. The procedures include both routines established "by the book" (such as checklists) and tasks to be performed in some "optimizing" fashion (such as flying the airplane). The particular task chosen at a given instant in time is the one perceived to have the highest expected gain for execution at that time. The gain is a function of mission priorities and of the perceived estimate of the state-of-the-world ("situation") at that instant. This estimate is based on monitoring of the instrument displays and/or the external visual scene and on auditory inputs from other crew members and air traffic control (ATC).

The basic structure of the PROCRU model for either PF or PNF is shown in Figure 3. The system and system displays of the generic model are broken out laterally to illustrate the system states relevant to the problem, and the display cues available to the crew. The monitor (or display processor) of the model is partitioned to separate visual and auditory cues. The information processor is expanded to include both discrete (event detection) and continuous (state estimation) information processing. Finally, the procedure processor is separated into a procedure selector (which accounts for major decision-making in terms of procedure or task sequencing) and an effector block for implementing the actions called for by the procedures.

The monitoring and information processing portions of PROCRU are not unlike those of the OCM or other models discussed above, though they have some novel features and extensions (the reader is referred to Baron, et al. (1980) for a more detailed discussion). One such extension was the inclusion, and treatment of, auditory input information. Auditory information, whether from the system (e.g., as an alarm) or from another human is assumed to be "acquired" instantly and correctly by the operator and stored in a short-term memory buffer. Upon acquisition of this information, the

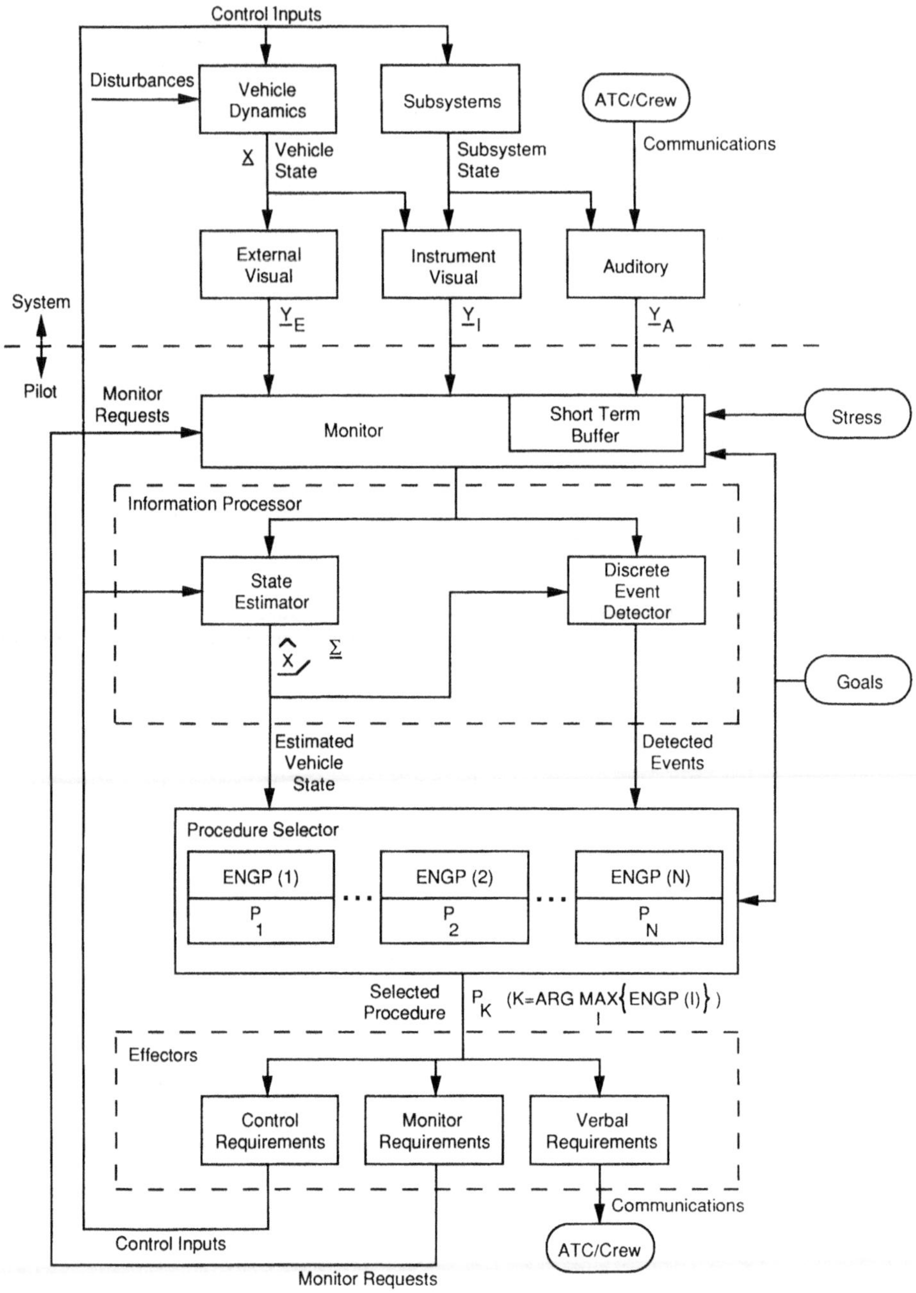

Figure 3. PROCRU Model Structure

event detector is "notified" that a message is waiting to be processed. The subsequent processing of this message depends on the nature of the message (alarm or communication) and on the time elapsed since its occurrence. If the message has a sufficiently high priority, the operator's activities will be interrupted and the auditory display will be selected for processing. The information in the buffer may also disappear or degrade in reliability with time, depending on the nature of the message. This treatment of auditory input has the advantage that alarms are priority interrupts, but that they may be missed or "unattended to" during times of high activity or workload stress (e.g., when there are many alarms).

As noted previously, the operator is assumed to have a number of procedures or tasks that may be performed at each instant. The definition of these procedures is an essential step in the formulation of PROCRU. All crew actions, except for the decision as to which procedure to execute, are determined by the procedures. We emphasize that we use the term procedure here to apply to tasks in general; a procedure in these terms could have considerably more cognitive content than might normally be considered to be the case. Table 1 illustrates the scope of the flight procedures for the PF and the PNF in approach to landing that are incorporated in PROCRU. For each crewman, six categories are shown, and for each category, specific types of procedures are itemized. These categories and types are discussed in detail in Baron, et al. (1980).

Table 1. Procedures for PF (Pilot Flying) and PNF (Pilot Not Flying) in PROCRU

PF	PNF
VEHICLE CONTROL/MONITOR	
MANEUVER	VEHICLE STATUS DETERMINATION
REGULATE	FAILURE DETECTION AND IDENTIFICATION
RETRIM	
REQUEST/CALLOUT	
FLAP REQUEST	VEHICLE POSITION CALLOUT
GEAR REQUEST	ALTITUDE CALLOUT
CHECKLIST INITIATE REQUEST	APPROACH STABILITY CALLOUT
SUBSYSTEM	
ALTITUDE ALERT MONITOR/CONTROL	FLAP MONITOR/CONTROL
MISC. SYBSYSTEM MONITOR/CONTROL	GEAR MONITOR/CONTROL
	MISC. SUBSYSTEM MONITOR/CONTROL
ACKNOWLEDGMENT	
CHECKLIST ITEM	CHECKLIST ITEM
	ATC REQUEST
SAP/MAP	SAP/MAP
MISCELLANEOUS	
GENERAL MESSAGE PROCESSING	GENERAL MESSAGE PROCESSING
LANDING PARAMETER SELECTION	

It is assumed that the PROCRU operator knows what is to be done and, essentially, how to accomplish the objective. However, he must decide what procedure to do next. This is a decision among alternatives and the procedure selected is assumed to be the one with the highest expected gain for execution at that time. The expected gain for executing a procedure is a function that is selected to reflect the urgency or priority of that procedure as well as its intrinsic "value". For procedures that are triggered by the operator's internal assessment of a condition related to the vehicle state-vector, the expected gain functions are appropriate subjective probabilities, as determined by the state estimation and event detection portions of the model. Procedures that are triggered by events external to the operator, such as ATC commands or communications from the crew, are characterized by expected gains that are explicit functions of time. For either type of function, the gain for performing a procedure will increase, subsequent to the perception of the triggering event, until the procedure is performed or until a time such that the procedure is assumed to be "missed" or no longer appropriate for execution. The rate of increase of the expected gain functions depends on the relative urgency of executing the procedure subsequent to the triggering event. This allows for distinguishing between procedures requiring immediate or fast action and those for which there is more latitude in the time of execution.

The selection and execution of a procedure will result in an action or a sequence of actions. Three types of actions are considered in PROCRU: control actions, monitoring requests and communications. The control actions include continuous manual flight control inputs to the aircraft and discrete control settings (switches, flap settings, etc.). Monitoring requests result from procedural requirements for specific information and, therefore, raise the attention allocated to the particular information source. (Note that verifying that a variable is within limits may not require an actual instrument check, if the operator already has a "confident" internal estimate of that variable.) Communications are verbal requests or responses as demanded by a procedure. They include callouts, requests or commands, and communications to the ATC. Verbal communication is modeled directly as the transfer of either state, command or event information.

Associated with each discrete procedural action is a time to complete the required action. (It is possible to modify PROCRU to allow for a probabilistic distribution of action time). When the operator decides to execute a specific procedure, it is assumed that he is "locked in" to the appropriate mode for a specified time (the "lock-up" time). For example, if the procedure requires checking a particular instrument and it is assumed that it takes T seconds to accomplish the check, then the "monitor" will not attend to other information for that period, nor will another procedure be executed.

In PROCRU, procedural implementation is modeled as essentially error free. However, errors in execution of procedures can occur because of improper decisions that result from a lack of information (quantity or quality) due to perceptual, procedural and workload limitations. If the effects of action errors are also to be analyzed, this is accomplished by deliberately inserting such errors directly into the model.

PROCRU generates a number of outputs that are useful for analyzing crew performance and workload. First, one can obtain full state trajectory information. In addition to this information, one can obtain each crew member's estimate of the state and the standard deviation at that time and PF's control inputs. PROCRU also produces activity time lines. It should be emphasized that the time lines generated by PROCRU are closed-loop time lines; unlike those normally used in human factors analyses. That is, actions are not completely preprogrammed but depend on previous responses, disturbances, etc. Thus, one can change a system or human parameter in PROCRU and automatically generate a new, different time line.

MODIFICATIONS AND EXTENSIONS TO PROCRU MODELING METHODOLOGY

In developing PROCRU, the goal was more than just a model for commercial aircraft in approach-to-landing. Rather, the major aim was to provide an engineering-based modeling methodology that was general enough and sufficiently extensible to be applicable to a wide range of problems. In early attempts to apply or extend PROCRU, it became fairly clear that conceptually the model was robust enough to be useful widely (Baron, 1984). However, it also was evident that there were major drawbacks in the computer implementation of the PROCRU model, as well as shortfalls in modeling some aspects of human performance.

In the evaluation of prototypical and evolving technologies, flexibility in the modeling of the system and its environment is critical. The analyst and designer must have the ability to efficiently make changes in equipment definition, modeling assumptions, task environment and procedures that are under investigation. The original computer implementation of PROCRU, which was developed to demonstrate the modeling concept rather than as a general analysis tool, did not satisfy this need. Indeed, the original program imposed severe constraints on the problem formulation and was even limited with respect to analyzing relatively modest variations in the approach-to-landing problem. After some modest attempts to modify this software, it became clear that to more fully exploit PROCRU modeling concepts it would be necessary to develop a generic computer implementation of the ideas.

The approach we have taken to development of a more general model/simulation draws heavily on concepts and software methodologies that have emerged largely from artificial intelligence; in it, object-oriented simulation techniques, artificial intelligence planning and problem solving methods, and human performance and information processing models are combined in a system modeling structure.

The basic model structure in the new implementation follows that of PROCRU. Process and procedural control is assumed to move from external and system-state inputs through perceptual processes (displays and human operator perceptual models) to a monitoring and decision making module. Decisions and monitoring are guided by consideration of goals, rules, constraints, and task knowledge. After a decision is reached, procedures are implemented through human operator effector models (psycho-motor or communication output) interacting with system control mechanisms. The effect of action on the system is then calculated and the loop is closed through feedback to the perceptual and display processes.

This processing flow has been modularized in object-oriented, LISP software. Extensions to the basic PROCRU model for the operator have also been made. In particular, the procedure selection process and the modeling of operator workload have been modified to be more reflective of current psychological models of human decision-making and human resource constraints. These implementation features and model extensions are discussed in more detail below.

Object-oriented Representation

In this paradigm, objects are used to represent functions of operator and equipment at varied levels of detail. The fundamental structure of this representational scheme is a set of objects that are linked to each other by a procedure for passing messages. An object can represent a function, an equipment, or (at a finer level of analytic detail) an individual performing that function, within a framework that supports all simulation objects. That framework consists of objects composed of a local memory, a set of operations, and procedures for passing messages to their neighbors.

An object's memory corresponds to the data structures and rules associated with the particular function or feature that it represents, e.g., a store of geometric relations for spatial planning. The operations of an object depend on its "type", i.e., which function it represents. Elaborating the example above, an object/model of spatial planning will have operations such as distance calculation, continuity analysis, and flight path planning. In all cases, the object/models will have operations that correspond to an "input-processing-output" formulation for information processing.

Several benefits accrue from the implementation of a PROCRU-Like system model in object representation:

- Modularity: The predominant aspect of an object representation is its modularity. Each domain function can be represented as an individual object. Within each domain, particular models can be implemented as representative of processing served by either human or automatic function. This modularity provides more than just programming benefits in that it enables the refinement of particular model elements independent of the state of development of others. In short, the most important models can be developed first and more fully than those that are merely supportive. Functional modularity allows the analyst to specify the model content at a convenient level of detail.
- Levels of representation: The object-oriented paradigm supports the representation of compound objects. Processes and subprocesses can be modeled so that the designer/analyst can choose among several levels of detail in his investigation of system function.
- Communication Protocols: The use of a message-passing function for information transmission among models allows for easy identification auditing of the assumed state of the human and automatic operators.
- Clock-Based and Event-Based Simulation: Object activity and message passing is implemented as an event-response process. This captures the flavor of systems in which signals are discrete drivers for the process of planning response tactics. This approach allows the simulation to proceed in either a clock-based or event-based manner or with some combination of the two.

In this new representation, procedures will be composed through the use of active objects. An active object is an object that has a list of activities that it is carrying out. When an active object receives a clock-tick message, it sends a tick to each of its activities before sending a tick to its component objects. Activities themselves can be organized in procedural hierarchies. An activity has a parent that spawned it, a list of children that it has spawned, and a tick procedure, (i.e, LISP code that it executes when it receives a tick), and procedures to carry out when its created, terminated, and/or aborted. In addition, an activity has an agent (i.e., the active object that has the activity on its activity list).

There are several types of activities which can be employed in the model and which, together, provide for a wide range of procedural definitions. They allow for procedures, or procedural steps, to be executed in a variety of ways: sequentially; in parallel; repetitively; for a fixed time; intermittently; by choosing among options; or, by combinations of these basic ways. This approach provides both flexibility in modeling and, relative ease in software implementation and modification.

Procedural Selection

We are implementing a procedure selection model that captures human performance characteristics to serve as the procedural selection mechanism in the evolving PROCRU model. In complex systems the operator must decide what it is he wants to do (his goals) and then must select the methods or procedures to meet those goals. In the original PROCRU, the instantaneous goals were encoded in the Expected Net Gain (ENG) functions and procedure selection simply involved choosing for execution the procedure with the maximum ENG.

The new implementation draws on current theories from the decision-making and process-control literatures which suggest that people exploit a variety of selection strategies when choosing between alternative courses of action (Rasmussen, 1983), and that the implemented strategies will be chosen so as to minimize the effort involved in making the selection while also preserving the quality of the result (Payne, 1982).

The evolving procedure selection model applies a staged selection mechanism that is capable of invoking any one of three decision strategies. We describe these strategies using the terminology of Rasmussen (1983), as either skill-based, rule-based, or knowledge-based. Rasmussen has popularized this tripartite classification scheme within the supervisory control literature.

The most efficient method for identifying an activity to execute is the skill-based strategy, which in the extreme, can be represented as involving no conscious decision-making activity at all, and might even be likened to an automatic reaction to an environmental condition. In our early system design, it will be invoked without the consumption of any appreciable time or cognitive resource. However, the structure of the object-simulation system is flexible enough to accept any fixed time and workload cost, following the invocation of the skill-based selection procedure. A skill-based decision is expected to take place whenever the operator becomes aware that a condition holds for which an established action or contingency plan is available.

When compared to skill-based decisions, those that we call rule-based are more involved. They are represented as taking a greater amount of time to complete, and therefore tie up the cognitive resources of the operator for a longer time. The information applied to these types of decisions, like that applied to skill-based decisions, takes the form of a production system. They differ from skill-based decisions in terms of the number of contingencies that must be in place before an alternative action is chosen. Unlike skill-based decisions, where a given action is triggered regardless of the context in which the precipitating environmental condition occurs, rule-based decisions are considered to be more difficult because they not only require that a primary enabling condition be in place, but also that one or more additional conditions hold before an action can be selected.

In our current system, this distinction is reflected by the fact that rule-based decisions are triggered by rules associated with specific script states, and therefore require an awareness of at least one more contingency (What state is currently active?) before the rule's dictates can be enacted. It is assumed that a fixed time cost will be incurred for each enabling condition that was required to be in place before the decisions of this type are implemented. Furthermore, it is assumed that a substantial amount of cognitive resources were applied to this process during that time.

The knowledge-based decision process is more complex and resource-demanding than the other two, in terms of the time requirements and the nature of its inputs. It is a two-stage, sequential process. The first stage applies a simplified lexicographic heuristic based on the Elimination by Aspect (EBA) decision heuristic (Tversky, 1972). The last stage applies a simplified compensatory multi-attribute utility model (similar to the ENG calculation). While the skill-based and rule-based processes require only the satisfaction of logical arguments for the determination of what procedure will be done next, the knowledge-based process requires that each alternative course of action being considered is represented and compared in terms of the values of a set of attribute variables. In the current version of the knowledge-based process, the EBA stage requires that the set of attribute variables are rank-ordered in terms of their importance, and for the compensatory process they must be weighted in a manner consistent with their ranking. We have implemented such ranking for attributes that are particular for each domain to which the object simulation has been applied.

The EBA stage serves to eliminate activity alternatives that are unsatisfactory based on their values with respect to the highest priority attributes. It does this by first comparing the value associated with the most important attribute to a situationally defined criterion. Alternatives that do not meet this criterion are dropped from consideration with respect to the second most important attribute. The alternatives meeting the criteria for the highest priority attribute are judged against the criteria set for the next most important attribute's criteria. Those attributes that do not meet the contextually sensitive criteria set for this attribute are dropped from further consideration. This seemingly crude yet efficient evaluation process is repeated until only two alternatives remain.

Next, the compensatory multi-attribute utility method is invoked. This process involves the assignment of a single value to each of the two remaining alternatives, calculated as a weighted sum of its particular set of attribute values. The alternative with the highest net utility (defined as the sum of the weighted attribute values) is then executed by the model after a period of time, which can be determined from the cognitive decision processing model developed by Johnson and Payne (1985) that provides estimates of the time required for an individual to have made such a decision.

The three decision processes are automatically implemented in the order in which they have been described. They are instituted as activity choice filters that are applied to the choice-form activity alternatives. When the conditions are in place for the implementation of a doctrinal prescription, the human's decision process will be skill-based. If alternatives are still available after the skill-based procedure has been applied, the rule-based procedure will be enacted to further prune the set of remaining alternatives. Finally, should alternatives remain after both of these strategies are applied, the (most costly) knowledge-based procedure is used.

Workload Calculation

In PROCRU there was no direct calculation of workload. Examination of various timelines generated by a PROCRU simulation allowed one to make inferences about workload or, more precisely, about the effects of excessive task demands. Implicit measures of workload based on PROCRU calculations were also possible. For example, the sum of the ENG's at a given time for all procedures (or tasks) is a measure of the instantaneous task demand and the integration of this sum over time could be a surrogate for cumulative load. We are now incorporating a more direct indication of workload. This indicator is reflective of modern multiple resource models of psychology. In particular, we are proceeding with the following general approach. For each task or activity an assessment is made of the load incurred in performing. The granularity of task definition to be used in load assessment is open to selection by the analyst. These load assessments are provided by an expert's subjective opinion and are represented on a scale from 0-7 with 7 indicating complete or full load. The task/activity loads are further refined by being divided according to resources that a human can bring to bear to perform the task. We have partitioned those resources to be visual, auditory (which includes both speech and hearing), cognitive and psychomotor. At this stage of development we are simply computing the load on each resource as a function of time but future work will be directed at comparing load with resource availability and developing appropriate moderators of performance.

CONCLUDING REMARKS

In various guises, and at different levels of detail, the engineering-based models for multi-task, complex systems described in this paper have been, or are being, applied in a wide range of application areas: commercial aircraft operations, anti-aircraft weapons systems, helicopter attack missions, command and control of tactical air missions and spacecraft telerobotic control. The manual control models, which are their ancestors, have a long history of application. It is clear that this approach to human performance modeling has both power and generality beyond what might have been expected from its somewhat narrow engineering origins. On the other hand, the models and methodology are evolving and are certainly not completely mature. There remains research areas to be addressed both near term and long term, that relate to overall validation of the models, further improvement and elaboration of them, and the manner in which they can be incorporated in an integrated approach to person-machine system design and development. Some of the more significant needs and directions are discussed briefly below.

A major issue concerning comprehensive engineering-based models, such as PROCRU or its extensions, is the lack of experimental validation for the overall integrated models. The core, continuous information processing model, has been validated many times and in different contexts as have some of the single task models (such as control and detection models) that would be used in such models. However, it must be noted that even if all sub-models in an integrated model had been validated, it would not guarantee that the integration of them yields a valid comprehensive model.

Thus, there is a need for appropriate model validation studies. However, the collection of such validation data will almost always present severe technical and economic problems. Indeed, the proper definition and scoping of model validation is an open question which, itself needs to be addressed. Another issue, related to validation, concerns the parameterization of comprehensive models. It is to be expected that as the complexity of the models increase and as they attempt to account for more aspects of human behavior, the number of parameters that must be specified to model that behavior will increase significantly. How to arrive at an appropriate number of parameters and what techniques to use in estimating them from data are important, unresolved issues.

There are a number of developments at the component, or sub-model, level that are needed or suggest themselves. The information processing portion of the model, though quite advanced, could be extended to cover nonlinear systems or imperfect mental models. The modeling of situation assessment by the operator(s) is still relatively simplistic and would benefit from an expanded view of this process and the development and testing of algorithms for representing it. There are similar needs in the area of modeling the diagnosis and problem-solving processes. The definition of goals and sub-goals in a way that supports the modeling process and implementation is a very significant aspect of this approach to modeling human performance but procedures for accomplishing this for complex systems are just being developed and need further work. Though means are being added to the model to calculate operator workload, there is need for a mechanism for accounting for the performance impact of that load, particularly with respect to discrete tasks. Finally, in virtually all aspects of the model, development has focused on performance of a highly skilled operator. Though this is perhaps the best single assumption for design purposes, there needs to be more attention devoted to individual differences and to the effects of training.

We see the models described here as becoming one part, albeit a very significant one, of an integrated approach to the human-centered design of advanced person-machine systems. The other parts involve formal methods for: mission decomposition; scenario generation; goal decomposition and task analysis; models for interaction with physical control and display devices that allow one to explore basic human engineering design considerations; and software tools. We envision these elements integrated within a workstation environment that will support a range of person-machine design, analysis and evaluation of problems, such as the feasibility of mission success, what (and what not) to automate, the determination of information requirements, and cockpit (workplace) layout analysis.

REFERENCES

Baron, S. 1984. "A Control Theoretic Approach to Modeling Human Supervisory Control of Dynamic Systems." In Rouse, W.B., (ed.), Advances in Man-Machine Systems Research 1:1-47. Greenwich, CT: JAI Press Inc.

Baron, S., and Kleinman, D.L. 1969. "The Human As An Optional Controller and Information Processor." IEEE Transactions on Man-Machine Systems., Vol. MMS-10, pp. 9-16. (Also, NASA CR-1151, September 1968).

Baron, S., and Levison, W.H. 1980. "The Optimal Control Model: Status and Future Directions." Proceedings of the IEEE Conference on Cybernetics and Society, Cambridge, MA, pp. 90-100.

Baron, S., Zacharias, G., Muralidharan, R., and Lancraft, R. 1980. "PROCRU: A Model for Analyzing Flight Crew Procedures in Approach to Landing." Proceedings of Eight IFAC Work Congress., Tokyo, Japan, Vol. XV, pp. 71-76 (Also, NASA Report No. CR152397).

Gai, E.G., and Curry, R.E. 1976. "A Model of the Human Observer in Failure Detection Tasks." IEEE Transactions on Systems, Man and Cybernetics., SMC-6, pp. 85-94.

Johnson, E.J. and Payne, J.W. 1985. "Effort and Accuracy in Choice," Management Science, 31, pp. 395-141.

Kleinman, D.L., and Curry, R.E. 1977. "Some New Control Theoretic Models for Human Display Monitoring." IEEE Transactions on Systems, Man and Cybernetics., Vol. SMC-7, No. 11, pp. 778-784.

Levison, W.H., and Tanner, R.B. 1971. "A Control Theory Model for Human Decision Making." NASA CR-1953.

Luce, R.D.. and Galanter, E. 1963. "Discrimination" in R.D. Luce, R.R. Bush and E. Galanter (Eds.). Handbook of Mathematical Psychology., Vol. 1, New York: John Wiley and Sons, Inc.

McRuer, D.T., and Krendel, E.S. 1974. "Mathematical Models of Human Pilot Behavior." (AGARDograph No. 188) North Atlantic Treaty Organization: Advisory Group for Aerospace Research and Development.

Muralidharan, R., and Baron, S. 1980. "DEMON: A Human Operator Model for Decision Making, Monitoring and Control." Journal of Cybernetics and Information Science., Vol. 3, pp. 97-122.

Pattipati, K.R., Ephrath, A.R., and Kleinman, D.L. 1979. "Analysis of Human Decision Making in Multi-Task Environments." Technical Report EECS TR-79-15, University of Connecticut (Also, Proceedings of International Conference on Cybernetics and Society, 1980).

Payne, J.W., 1982. "Contingent Decision Behavior," Psychological Bulletin, 92, 383-402.

Phatak, A. and Kleinman, D.L. 1972. "Current Statues of Models for the Human Operator as a Controller and Decision Maker in Manned Aerospace Systems." Automation in Manned Aerospace Systems., AGARD Proceedings No. 114, Dayton, Ohio.

Rasmussen, J. 1983. "Skills, Rules and Knowledge; Signals, Signs and Symbols, and Other Distinctions in Human Performance Models," IEEE Transactions on Systems, Man and Cybernetics SMC-13, 257-266.

Tversky, A. 1972. "Elimination by Aspects, a Theory of Choice," Psychology Review, 79, pp. 281-299.

Wewerinke, P. 1981. "A Model of the Human Observer and Decision Maker." Proceedings of the Seventeenth Annual Conference on Manual Control., Pasadena, California, 1981.

MICRO SAINT - A TOOL FOR MODELING HUMAN PERFORMANCE IN SYSTEMS

K. Ronald Laughery

Micro Analysis and Design
Boulder, Colorado

BACKGROUND

Over the past few decades, there has been an increasing call for the development of quantitative tools for the analysis of human engineering problems. One of the emerging engineering technologies for quantitative analysis is computer simulation. In the mid 1970s, the Air Force Aerospace Medical Research Laboratory (AFAMRL) sponsored an effort for the development of a simulation language specifically aimed at the human engineering community. This package, entitled SAINT (Systems Analysis of Integrated Networks of Tasks), received widespread recognition as a general systems engineering tool in addition to a human performance modeling tool. SAINT kindled the interest of the human engineering community in computer simulation as a tool for human performance analysis.

With all of SAINT's power, however, came several disadvantages. First, SAINT is essentially a programming language with a set of subroutines providing the specific constructs required for simulation. While this permits great flexibility, it also requires that users have computer programming skills, thereby reducing the base of potential users. Secondly, SAINT is hosted on a minicomputer. This also reduces the availability of the tool given the relative proliferation of microcomputers. In truth, these disadvantages of SAINT were also disadvantages of virtually every major simulation package available on the commercial market several years ago.

In 1984, Micro Analysis and Design, under sponsorship of the U.S. Army Medical Research and Development Command, began the development of a user-oriented, microcomputer-based simulation system. The target problem of this modeling system was to study human performance effects of a variety of stressors. The target audience for this modeling system was the human engineering community. From this effort emerged the modeling tool Micro SAINT.

While the original focus of Micro SAINT was rather narrow, a variety of factors led Micro SAINT to become a more general purpose modeling tool. Behind all of these factors was the simple truth that modeling human performance in a system can be every bit as complicated as modeling any other type of system component. Therefore, the requirements of a language for addressing human performance modeling problems is also very broad. In essence, to model human-machine performance, a full-featured discrete event modeling system with limited continuous modeling capabilities is required.

In the past four years, Micro SAINT has evolved into a full-featured modeling system. While its primary focus remains the human engineering community, it has also received broad acceptance in other applications of computer simulation such as system design and manufacturing. However, one thing has remained constant within Micro SAINT - the focus on the user. For example, at every decision point in Micro SAINT's development regarding the addition of a feature, we have asked ourselves these questions:

1. Will more than 5% of the users need this feature?

2. Is there another way it can be done with the current set of features?

3. Will the addition of this feature cause confusion to users who don't need it?

Before the feature would be added, the answers to these questions would have to lead us to the conclusion that the added complexity was worth the added power. In the future, we will continue this focus and future evolutions of Micro SAINT will strive to improve the tool's usability.

The remainder of this paper is organized into three sections. In the section entitled "Technology behind Micro SAINT," we will present some of the underlying constructs behind Micro SAINT modeling and accompanying descriptions. In the section entitled "A Case Study," we will provide an example of how Micro SAINT was used to address a common human engineering problem, operator workload analysis. Finally, in the section entitled "Future Developments," we will present a discussion of some of the recent and ongoing enhancements to Micro SAINT, particularly those related to the human engineering community.

TECHNOLOGY BEHIND MICRO SAINT

The basis of Micro SAINT modeling is task network modeling. Task network modeling involves the decomposition of system performance into a series of subactivities or tasks. The sequence of tasks is defined by constructing a task network. An example to illustrate this concept is shown in Figure 1 which presents a series of tasks for dialing a telephone.

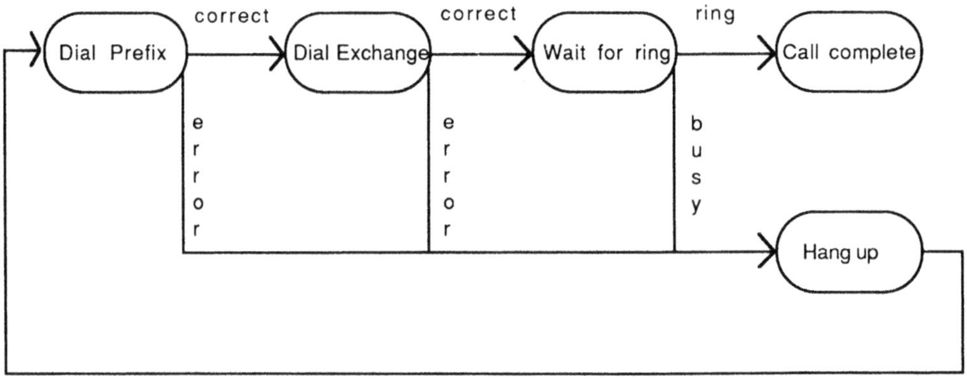

Figure 1
Sample Task Network of a Human Dialing a Telephone

The level of system decomposition (i.e., how finely we decompose the tasks) depends on the particular problem. A discussion of how this is determined is beyond the scope of this paper. However, suffice it to say that the system can be defined in as detailed or gross a level as necessary. Also, a task network may include several relatively autonomous subnetworks which, while interrelated, are also very distinctly separate. For example, in modeling a helicopter there could be relatively autonomous networks for the operators, the aircraft, and the threat environment.

While the networks may be independent, performance of the tasks can be interrelated through shared variables. Once the network is defined, the modeler must determine what variables are relevant to the modeling problem and how those variables are affected by tasks in the networks. The relationships among different components of the system (which are represented by different segments of the network) can then communicate through these shared variables.

For example, when a helicopter pilot initiates a pop-up activity, the variables associated with operator controls would be duly affected by that task. These new values would then indicate to the helicopter portion of the model that it must start executing tasks associated with increasing the aircraft's altitude which is represented by another variable. Once the altitude is above the threshold required for the threats to observe and begin firing, the threat portion of the model begins executing tasks associated with shooting at the aircraft. Representing task sequencing through a network and interrelationships among tasks through variables and changes in variables associated with tasks forms the foundation of all network models.

One explicit goal of Micro SAINT was to make the translation of task analysis information into a dynamic computer simulation model much faster and more intuitive. In Micro SAINT, we have taken three primary approaches to facilitating model construction and use. First, all model construction and execution is done through menus. Second, we provide modeling power to the user through a parser which allows the user to create application-specific logic and subprograms without ever needing to access computer code or execute a compiler. Third, we have provided a series of interactive execution modes to make model debugging much easier. Let us present these concepts in more detail.

```
(1)   Current Model Name in Memory:   dialing
(2)   Model Development
(3)   Model Execution
(4)   Analysis of Results
(5)   Utilities
(6)   Show Models
(7)   Animation


Select an item from the menu (1-7):
```

Figure 2
Micro SAINT Main Menu

One of Micro SAINT's main advantages is that it permits the modeler to construct and run all segments of a task network model by using menus. This "user friendly" interface dramatically increases the ease of learning, using, and understanding task network models. To briefly illustrate the simple process of using Micro SAINT, let us present several menus associated with constructing a network model. Figure 2 is the main menu for Micro SAINT. In using Micro SAINT, the user has three activities as shown by menu items 2, 3, and 4 in Figure 2. First, the user must develop the model. Then, the user must execute the model. Initially, the user will execute models to debug them. Once debugged, the user will execute the models to gather data on disk. Finally, the user can analyze the data with a series of built in statistical and graphical tools. Menu option 5 in Figure 2 presents the user with a series of utilities for model printing, merging, etc. Option 6 simply presents a list of the models on the user's disk. Finally, option 7 presents an animated view of the task network. Let us present several of the menus and user interfaces associated with model development, execution, analysis, and animation.

```
(1)  Current Model Name in Memory:  dialing
(2)  Task Network
(3)  Job Queues
(4)  Variable Catalog
(5)  Function Library
(6)  Continuous Variable Changes
(7)  Simulation Scenario
(8)  Snapshots of Execution
Memory Available = 181322 bytes


(C) Copyright 1987  Micro Analysis and Design, Inc.
        Serial number:  MS00001  All Rights Reserved.


Aspect of model development to work on (enter a number 1-8) ?
```

Figure 3
Model Development Main Menu

Figure 3 presents the menu associated with model development. At this level, the modeler would select an aspect of model development on which to work. The various options deal with the network model, any continuous portions of the model (e.g., vehicle location in space), the simulation scenario, and data collection (snapshots). If the modeler selected option 2, then he or she would be presented with a screen which presents the list of tasks which have been defined as depicted in Figure 4. The user may then ADD a new task, MODIFY an existing task, DELETE an existing task, COPY the information from one task into another task, or SAVE the updated task network (NEXT and PREVIOUS are commands for paging when the list of tasks exceeds one screen). If the user were to add or modify a task, he or she would be presented with a screen similar to Figure 5.

```
               TASK  NETWORK   Model Name:  dialing

     Number:         Name:                     Type:
     1               dial pre-fix              Task
     2               dial last 4 digits        Task
     3               wait for ring             Task
     4               call complete             Task
     5               hang-up                   Task


Add  Modify  Delete  Copy  Save  Next  Previous  View
Command (a, m, d, c, s, n, p, v) ?
```

Figure 4
Model Development Task Selection Menu

```
                 MODIFY  TASK
Task Number:  1
        (1)  Name:  dial pre-fix                          (2)  Type:  Task
        (3)  Upper Network:  0  dialing
        (4)  Release Condition:  1;
        (5)  Time Distribution Type:  Normal
        (6)  Mean Time:  .9;
        (7)  Standard Deviation:  .5;
        (8)  Task's Beginning Effect:
        (9)  Task's Ending Effect:
        (10) Decision Type:  Probabilistic
             Following Task/Network:      Probability of Taking
              Number:        Name:         This Path:
        (11)  2              dial 1  (12)  1 - errorprob;
        (13)  5              hang-u  (14)  errorprob;
        (15)                         (16)
        (17)                         (18)
        (19)                         (20)
        (21)                         (22)
        (23)                         (24)

Enter number of the field to change or m to modify another job:
```

Figure 5
Modify Task Menu

At this point, the modeler would add or modify information in each of the fields associated with that task. All of the other model development options shown in Figure 3 involve a similar sets of steps. At any point in model development, the user can view and/or print a graphical presentation of the model as shown in Figure 6.

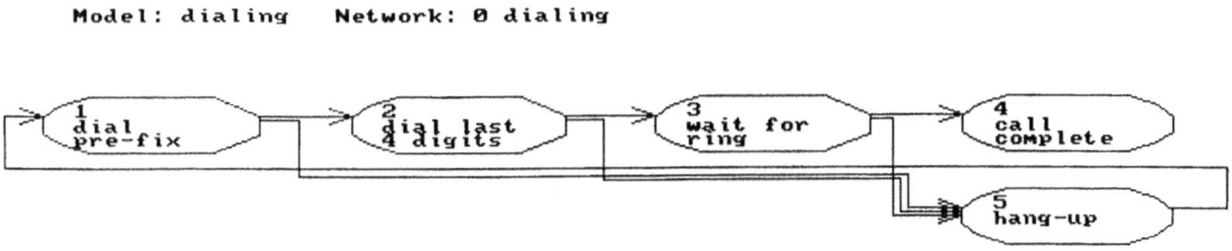

Figure 6
Graphical Presentation of a Model

Given the fairly rigid menu structure within Micro SAINT, one might ask how users can design custom algorithms within Micro SAINT models without having to access and modify the actual computer code. This is accomplished via a parser. The parser is, essentially, a run-time interpreter. A parser will take a series of algorithmic expressions (e.g., mathematical, logical, control of flow expressions) that are stored in a database in "source code" form and then, as the program runs, these statements will be "parsed" and executed in a manner identical to that which would occur if the algorithm were embedded within the object code of the program itself. Parsers are quite common in microcomputer software such as Lotus 123 where it would be clearly unacceptable to recompile the Lotus 123 program every time a formula in a cell was changed.

We have developed and incorporated a parser within Micro SAINT. At every menu item in Figure 5 followed by a semi-colon, a parsed expression or series of expressions can be included. The syntax for these expressions resembles that of the C programming language. Through the use of the parser, some of the things we can do within Micro SAINT are as follows:

- Create complex decision and branching logic within a model
- Control the interactions of tasks with other tasks in the system through interactions among variables
- Constrain task execution until certain conditions have been satisfied

In essence, the parser allows the user to build subroutines associated with task execution without program recompilation. It is this feature which gives Micro SAINT the power of other simulation languages without having to resort to computer programming.

When executing the model, the user is normally presented with an interface as shown in Figure 7. From this interface, the user can review the current simulation time, the task just completed, all tasks and other events scheduled (i.e., the event queue), and the values for up to 15 selected variables. At any point, the user can stop the simulation and manipulate variable values.

```
RUN 1        OF 1                                        System Time: 73.37
               Completed Task:  4              pack
----------------------------------------------------------------------------
VARIABLES:                            | Tag:          EVENT QUEUE:
radios                     3          | 7 At 73.37 Queue: 3q test queue
                                      | 7 At 74.09 Task end: 1 assemble parts
                                      | 6 At 77.90 Task end: 3 test
discards                   2          |
                                      |
                                      |
testq                      1          |
                                      |
                                      |
                                      |
                                      |
                                      |
                                      |-----------------------------------
                                      |      Error Count:  0
----------------------------------------------------------------------------
    paused>
```

Figure 7
Normal Micro SAINT Model Execution Screen

Recently, Micro Analysis and Design has added the ability to review model execution through animation. An example of the screen displayed during model animation is presented in Figure 8. During animation, tasks that are active are highlighted, queues building up in front of tasks are illustrated, and the flow of entities (e.g., information, components) between tasks can be followed with icons. Model animation requires no additional model development effort.

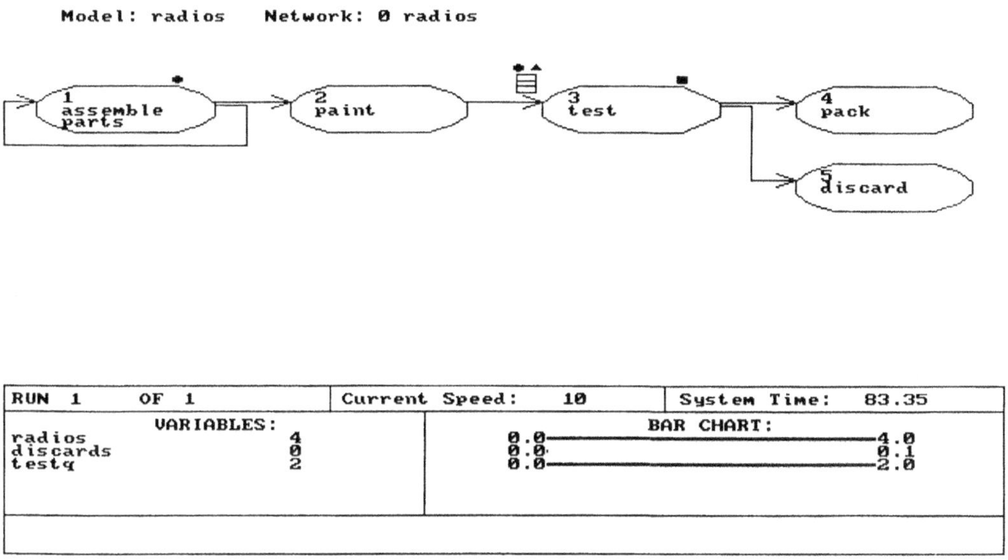

Figure 8
Animated Micro SAINT Model Execution Screen

Finally, Figure 9 presents an example of one of the graphical data analyses created using the Micro SAINT analysis software (reached through option 4 from the menu depicted in Figure 2). There are, of course, a host of other analyses included as well. If other analyses beyond those provided within Micro SAINT are required, the model data can be readily loaded into virtually any analysis package such as Lotus 123, SAS, SPSS, or one created by the user.

The above discussion presents a brief tour of the Micro SAINT menu structure. We have strived to make these menus as intuitive and straightforward as possible to facilitate ease of learning and use. The feedback we receive from the user community indicates that these goals are being met.

A CASE STUDY - OPERATOR WORKLOAD MODELING ON THE LHX HELICOPTER

This section will provide a case study of the use of Micro SAINT in addressing a common human engineering concern, operator workload analysis. It is by no means intended to represent the only kind of problem to which Micro SAINT can be successfully applied. Micro SAINT is a tool that can be applied to virtually any human performance analysis problem where quantitative predictions of human performance and/or the interaction of the human with his environment is of interest. Rather than trying to restrict the problem domain of Micro SAINT, this section is intended to show one of the successful applications of Micro SAINT and task network modeling.

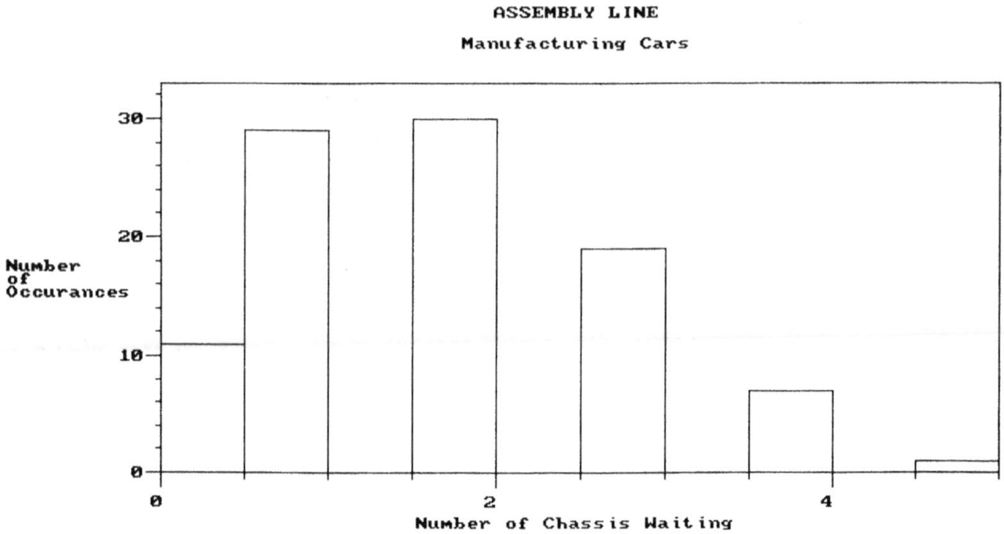

Figure 9
Example of Micro SAINT Data Analysis

The immediate issue in the LHX crew modeling effort was to predict operator workload in the early stages of design so that operator overload situations could be avoided. Micro SAINT was used to predict relative reductions in operator workload for a variety of cockpit configurations and, therefore, to evaluate the effectiveness of cockpit automation.

The fundamental approach to evaluating operator workload in the future attack helicopter was to develop a computer simulation of the helicopter system. In this model, the focus was on the human operator. However, since the helicopter operated in a highly closed-loop manner (i.e., operator activities depended on other aspects of the environment and vice versa), we believed that the operator's performance and workload had to be evaluated with a truly closed-loop model. Therefore, the following three distinct components of the model were developed:

1. The human operator
2. The helicopter
3. The threat environment

By developing detailed submodels of the human operator supported by simplified, albeit realistic, submodels of the helicopter and threat environment, a far more realistic representation of the human's workload demands could be obtained. Furthermore, by developing a model of the overall system, we could obtain some system-oriented performance measures such as survivability, kill-ratio, etc. Based on this analysis, computer simulation submodels were developed for each of the above three components and integrated into an overall model for evaluating operator workload and helicopter system effectiveness for four alternative cockpit designs. Three of these integrated models represented cockpit configurations that were being considered for the LHX, and one model represented a two-man Apache. The basic LHX was an enhanced current state-of-the-art aircraft. The other two aircraft incorporated segments of the virtual cockpit that is being developed at the Armstrong Aerospace Medical Research Laboratory. The Furness wide model simulated a wide field of view virtual display whereas the Furness medium model simulated a more limited field of view.

We focused on what was believed to be a high workload portion of the mission, the anti-armor engagement. A prior analysis using the Siegel-Wolf method of crew system modeling was used to make this determination. Each of the four models (representing the different cockpit configurations) began with the aircraft moving to a hover zone, popping up to acquire the targets, processing the targets and selecting priorities, a series of pop-up engagements and relocations to new hover zones, and finally a battle damage assessment.

How Workload was Considered

To consider workload, we employed a variation on a technique which had been used by McCrackin and Aldrich (1984).[1] Using this technique, each operator activity was characterized by the workload demand required in each of four attention channels: the auditory channel, the visual channel, the cognitive processing channel, and the psychomotor output channel. McCrackin and Aldrich present benchmark scales for determining demand for each channel. Using this approach, each task in the operator models could be characterized as requiring some amount of each of the four kinds of attentional demand. All tasks in the operator models were analyzed with respect to these demands and values were assigned accordingly.

However, the pilot was usually not performing simply one task at a time. For example, he may have been required to monitor his hover position while he received a communication. Given this, the workload literature indicated that the operator may either accept the increased workload or begin dumping less important tasks.

To factor these two issues into the computer simulations, two approaches were incorporated which provided ways of evaluating 1) combined operator workload demands for tasks which were being performed concurrently and 2) when the operator would begin ignoring tasks due to overload.

In the computer simulations, we were able to assign values for the attention required in each of the four channels for each task. Then using "task beginning effects" and "task ending effects" to adjust variables in the simulation associated with each task, we could at any point in the simulation obtain an estimate of the total attentional demands across all tasks. By taking "snapshots" every two simulated seconds during a simulation run, we were able to characterize the operator's attentional requirements graphically. By examining the points in the mission at which these attentional demands were dangerously/unacceptably high, we could assess mission segments for which automation would be helpful.

We examined operator task dumping behavior in one segment of the operator's activities associated with his "updating situational awareness." There were a series of tasks during which the operator's objective was to look out the cockpit window to maintain an awareness of his situation outside the cockpit. However, we hypothesized that at some point his visual attention demands would become excessive and he would

---

[1]McCracken, J.H. and Aldrich, T.B., Analysis of Selected LHX Mission Functions Implications for Operator Workload and System Automation Goals. Technical Note MDA903-81-C-0504, June 1984.

stop attempting to maintain situational awareness until the visual load lightened. In the models, we stopped the operator's maintaining of situational awareness whenever his combined visual attention demands exceeded a value of 5. We should note that some may argue our choices of specific values beyond which attention overload is achieved and/or the specific tasks that the operator would eliminate. These arguments are valid and, again the models are such that these assumptions are transparent and easily changed.

Study Results

We collected data for each of the four aircraft under two different threat conditions, one where six threats had to be engaged and one where two threats had to be engaged thereby creating eight experimental conditions. Under each of these eight conditions, we collected the following data:

1. Percentage time that visual attention demands exceeded the threshold to permit the update of situation awareness

2. Visual attention demands throughout a mission

3. Auditory attention demands throughout a mission

4. Cognitive attention demands throughout a mission

5. Psychomotor attention demands throughout a mission

6. The vulnerability of the aircraft as measured by the number of times it was "killed" by the threats in the simulation

Attentional demands were analyzed through the review of graphs such as the one presented in Figure 10. From this, the following conclusions were reached:

- The basic LHX was overly demanding with respect to visual attention

- The Furness cockpit design appeared to reduce visual attention demands dramatically

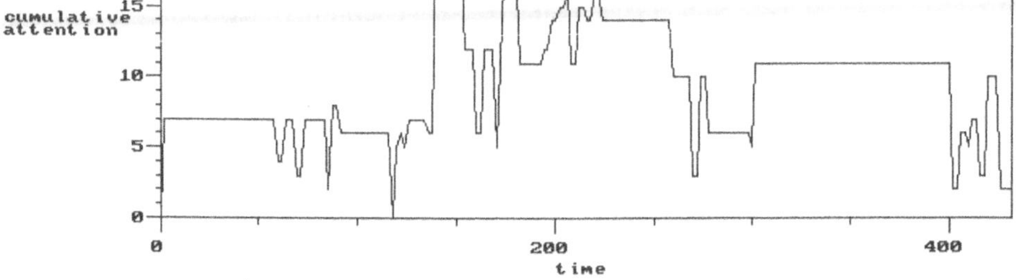

Figure 10
Predicted Workload Profile Throughout a Mission

With respect to the measures of aircraft vulnerability, we were able to identify aspects of the operator tasks which led to more vulnerable aircraft. Therefore, we were able to directly link the tasks performed by the human (as dictated by the system design) to the probability that the aircraft would survive in combat.

What did we learn in building this model?

Perhaps that most significant finding in this study relates to the methodology employed. Task network modeling with Micro SAINT was an efficient method for obtaining insight into some of the workload issues associated with the LHX. To those who have constructed and used computer models of operator behavior, the above effort may seem Herculean. However, the total time elapsed during the entire study was 10 weeks during which approximately 20 man weeks of labor were expended. Furthermore, it was estimated that a significant change in the model could be performed in roughly four man-days. Clearly, this type of effort is well within the scope of most human operator analysis efforts.

Again, we want to emphasize that this is simply an example of a successful application of Micro SAINT to a human engineering problem. We at Micro Analysis and Design have applied it to other human factors issues including manpower analysis, training system design, and human error analysis. Others have applied it to many different types of human performance analyses, from simple task analysis to neural network analysis. Micro SAINT is a modeling tool and, as such, can be applied to many different types of human performance modeling problems. Micro SAINT is limited more by the creativity of the human engineering community than the power of the tool itself.

FUTURE DEVELOPMENTS

Any software product must continually evolve and improve. Changing needs of the marketplace and ever growing computer capabilities dictate this. Currently, there are several projects underway and recently completed at Micro Analysis and Design to support the evolution of Micro SAINT. We have grouped these changes into two general categories, 1) enhancement of the basic modeling tool and 2) customization of Micro SAINT to specific users and/or problems.

Enhancements to Micro SAINT

Two projects have recently been completed at Micro Analysis and Design which have expanded the capabilities of Micro SAINT. First, under the sponsorship of the Armstrong Aerospace Medical Research Laboratory, we have developed a version of Micro SAINT for the DEC VAX line of computers. This product, named Saint Plus is virtually identical to Micro SAINT from the user perspective. Aside from the obvious difference in the hardware environment, Saint Plus also allows the development of significantly larger models. While few users have identified the need for larger models, we feel that this enhancement will expand the base of problems to which task network modeling can be applied.

The second recently completed enhancement we have already mentioned - the development of an animation package for Micro SAINT. Animation allows users to view their graphical task networks as a model executes. During animation, tasks which are being executed are highlighted, queues that are building up are presented, and the flow of entities through the system can be viewed with icons. Animation is primarily designed for the presentation of models. Additionally, it has been found to be a powerful debugging tool. A picture is, in this case, worth a thousand graphs.

A project that we are now involved in is the development of a Macintosh version of Micro SAINT named Paradigm. Paradigm will initially provide all of the basic modeling features of Micro SAINT coupled with the powerful graphics capabilities of the Macintosh. In the future, we will use the increased processing power of the Macintosh to expand the features and power of task network modeling. Our current expected completion date for the first version of Paradigm is mid 1989.

Customization of Micro SAINT

While Micro SAINT is a relatively easy to use tool, it still requires creativity in designing the underlying task network model to address specific problem types. Recently, we have seen an increasing interest in developing custom software packages for specific classes of simulation problems and/or for specific users.

Currently, we have three projects under development which use Micro SAINT as the "kernel" of the modeling application package. Each of these projects have three elements in common.

1. Customized user interfaces are developed that require the user to provide only the model information that is unique to a specific application of the model. Therefore, if segments of the model are always the same across all applications of that model, these segments will be embedded within the customized application. Also, the customized interfaces are able to "speak the language" of the specific user population. The analysis package is also customized so that the simulation data are provided to the user in terms that are particularly relevant.

2. Integrated data bases are included which provide, essentially, libraries of data relevant to the particular area of application. This reduces the need for users to collect data from external sources during model development.

3. Micro SAINT simulation software "engines" are used to provide the simulation capabilities. The simulation management and data base manipulation are supported from existing subsets of the Micro SAINT code.

Our new applications include models which support the Army's MANPRINT initiative. These models are used to examine new system performance, both in terms of setting criteria for minimally acceptable system performance and in evaluating the "likely" performance given a specific weapon system design. These applications are well suited to our customization scheme, because the users have very task-specific knowledge and terminology. They recognize and understand terms like "probability of hit" and "target acquisition time." Also, the questions which these users are trying to answer are inherently dynamic. Simulation enables them to play "what if" with mission conditions and scenarios. Since most of the required data consist of function and task performance times, fielded weapon system data will be supplied to the user in the libraries. The user will then be able to "mix and match" and modify data from other systems in order to build an estimate of what the new system's mission will look like and how the new system must perform.

We are also developing a maintenance modeling tool for the Army. This tool includes a structured model for which the user will supply maintenance failure rates, usage rates, personnel requirements and other logistical variables. This application differs from the products in the preceding paragraph because the user will not actually change the structure of the model. Rather, the user will only be changing the parameters of the tasks in the network model. This enables the user to analyze a highly variable and volatile system without requiring simulation development expertise.

During these projects, we have maintained close contact with the users. This is extremely important during the development of customized applications, and we have been very encouraged by the enthusiasm the tools have generated. We believe that tools like these can broaden the horizons of simulation applications and users.

SUMMARY

Micro SAINT has proven to be a useful tool for modeling human performance in a wide variety of system types to address many different kinds of problems. The key concept, though, is that Micro SAINT is a tool. Micro SAINT is an attempt at doing to computer simulation what Visicalc did to computerized spreadsheet analysis. It provides an easy to learn and easy to use environment in which the human factors engineer can construct a variety of types of models depending upon the problem to be addressed and, more importantly, the creative mind of the user.

ANALYSIS OF A PROCEDURE FOR MULTIPLE OPERATOR TASK PERFORMANCE ON SUBMARINES USING A NETWORK MODEL

L. van Breda

TNO Institute for Perception
P.O. Box 23
3769 ZG Soesterberg, The Netherlands

OVERVIEW

The use of network models describing complex man-machine systems is generally accepted in the field of human engineering research as a tool for predicting system performance. To analyse multiple operator procedures on board submarines of the Royal Netherlands Navy such a network model was developed. The analysis was meant to verify the appropriateness of prescribed procedures consisting of a number of tasks partly performed by operators and partly by automatons. The network model describes the operators' activities by algebraic equations and by production systems whereas the submarine's manoeuvring behaviour as well as the dynamics of relevant platform systems are mathematically described by differential equations.

This paper emphasizes the use of the network model in the analysis of the snorkeling procedure. The network model is based on a normative task performance as defined by the Navy and meant to investigate the procedure with respect to information flow and operator task performance under different conditions.

Results show that partly manual control does not slow down the course of the procedure. It also appears that the busy time of the operators never exceeds 50% of the available time. To prevent excessive overshoot of the snort top valve when appearing above sea level practising with special ballast condition is required. This is in contradiction to the assumption that the procedure has to be executed with a pitch angle of zero degrees.

INTRODUCTION

Background

On board Walrus-class submarines which are under construction now an advanced system for monitoring and control of platform systems is developed. The equipment for information presentation and control setting is implemented in the control centre of the submarine and consists of a console equipped with pushbuttons, indicators and graphic displays. The system design is based on a considerable manning reduction compared to existing Zwaardvis-class submarines, increasing complexity of the platform systems and resulting consequently in a system in which many functions are partly or completely automatized.

Monitoring and control tasks on board submarines have a procedural nature. In order to quantify the effect of workstation design and man-machine interaction on total system performance in the designing stage of the submarine, the TNO Institute for Perception conducted a study under contract to the Royal Netherlands Navy in which several procedures were analysed.

The study is based on mathematical modelling techniques using network models. The network model is implemented in a personal computer by means of the SAINT (System Analysis of Integrated Networks of Tasks) simulation language and used to analyse procedures with different initial conditions (scenarios).

This paper describes the use of the network model in the analysis of the so-called snorkeling procedure. During this procedure the submarine rises up to periscope depth and starts to recharge as soon as possible its batteries by means of diesel engines. The snorkeling procedure is critical because various tasks have to be performed by several operators simultaneously. As the operator task definition is different for each scenario, it can be expected that this will influence the course of the procedure. By using the network model the effect of operator normative task execution on the course of the procedure and on the manoeuvring behaviour of the submarine can be determined.

Snorkeling with a Submarine

Under normal operational circumstances a submarine is manoeuvring deep below sea level in a state of non-stable balance. This means that a disturbance without any compensation will cause a continuous change of rate. The vertical speed of the submarine is controlled by means of propulsion and diving rudders on stern and sail.

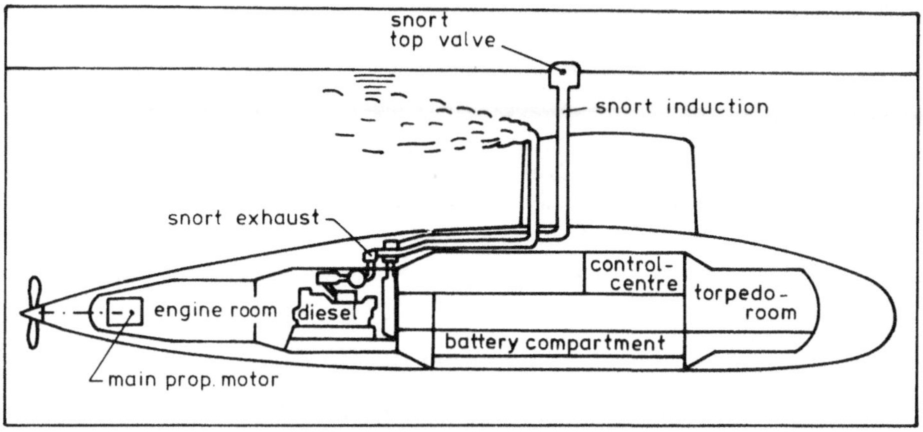

Figure 1. Snorkeling with a submarine.

Changes in ballast are compensated by vertical and horizontal trim. The propulsion is affected by the battery driven main propulsion motor which is directly coupled to the propellor (Figure 1).

Being non-detectable is of course essential for a submarine. In fact this can only be achieved by manoeuvring at great depth during long periods of time and rising only for special missions. As Walrus-class submarines are conventionally propelled submarines it is necessary after a certain amount of time to rise to periscope depth and recharge the batteries and the pneumatic system by diesel engine operated generators and compressors. The air inlet for the diesel engines is effected by the snort induction mast with the snort top valve. During the snorkeling procedure the submarine is manoeuvring just below sea level and can easily be seen. The high noise level produced by the engines also increases the detectibility. Under operational circumstances this is a dangerous situation for a submarine. By optimizing the design and use of the on board installations the Navy aims for reduction of the detection risk by minimizing the duration of the snorkeling procedure.

There will be a disturbance of the vertical balance when the diesel engines are started. The snort exhaust system is filled with about two m^3 of sea water and when starting the engines this system will be emptied by the diesel exhaust gases. This is a new aspect compared with the existing Zwaardvis-class submarines. Especially when manoeuvring with a pitch angle of zero degrees this disturbance could cause upwards accelerations so the snort top valve could appear above sea level. It was questioned during the design phase whether these effects could be compensated by the automatic depth control system.

Procedure and Scenarios

In Walrus-class submarines manoeuvring procedures are carried out by three operators. The workstations of these operators are located in the Control Centre and provided with VDU's, displays and controls connected to the so-called Information Processing System. All platform systems can be activated and monitored by VDU interaction or by pushbutton control:

- the propulsion operator is controlling propulsion, diesel engines, electric supply systems and snort induction
- the trim operator is controlling trim, ballast and unballast
- the helmsman is controlling depth and heading.

During the snorkeling procedure the submarine first rises to periscope depth maintaining a pitch angle of zero degrees. All the platform systems are stand-by. When the snort top valve appears above sea level the propulsion operator will open the snort induction system and the top valve unless the top valve is tested and the snort induction is dry. The diesel engines are prepared and started once the top valve is open. Then the battery recharge system is activated by means of generators and the compressor system for pneumatic air supply must be started. At this very moment the snort procedure is started and the submarine is regaining energy storage. During the procedure the trim operator is continuously monitoring the balance of the submarine. Disturbances are compensated immediately.

Under regular operating circumstances the platform system is operated by automatic control:

- DA = Diesel engines, Automatic control
- BA = Charge Batteries, Automatic control
- TRA = TRim, Automatic control
- PA = Propulsion, Automatic control.

Under special operating circumstances the platform system is operated by semi-automatic or manual control:

- DM = Diesel engine, Manual control
- BM = Charge Batteries, Manual control
- TRS = TRim, Semi-automatic control
- TRM = TRim, Manual control (VDU interaction)
- PM = Propulsion, Manual control.

Because the installations mentioned above are independent, any combination can be used. The snorkeling procedure can be executed completely automatically (DA, BA, TRA, PA), completely manually (DM, BM, TRM, PM) or any other combination in between.

The study is focused on three critical scenarios:

- SI = snort exhaust dry
  As it is expected that the ballast disturbances caused by the snort exhaust system will influence the course of the procedure, this not-existing scenario is used to compare the situation with existing submarines (hypothetic control condition)
- SII = snort exhaust filled, no further precautions
- SIII = snort exhaust filled, pre-ballast 2 m^3.

The assumption is made that during the snorkeling procedure the forward velocity relative to the water is kept constant so the propulsion control remains automatic and will not be considered in the analysis. Since depth control is carried out automatically, the helmsman has no influence on the course of the procedure. Only the propulsion and the trim operator tasks are analysed. In total 2x2x3=12 conditions per scenario have been simulated (Table I).

Table I. Conditions to be analysed per scenario.

Diesel		DA		DM	
Battery		BA	BM	BA	BM
Trim	TRA	r	r	r	r
	TRS	r	r	r	r
	TRM	r	r	r	r

r = 25 replications

The system performance is determined with respect to three aspects:

- the quality of the procedure.

This is the overshoot of the top valve immediately after its appearance above sea level. For minimizing the detectibility, the overshoot must be limited to 0.25 m. Scenario SI is only relevant to this aspect since it deals with the manoeuvring characteristics of the submarine.

- the procedure execution time.

This is the period of time in which all the platform systems used for energy storage and pneumatic supply have been started once the top valve is open. It can be expected that he execution time of the procedure will increase during manual control. According to design specifications this may not exceed 50% of the execution time during full automatic operation.

- operator busy time.
This is the time operators spend to start the snorkeling procedure and to compensate the disturbances by trimming the submarine. The total time spent may not exceed 50% of the available time.

METHOD

Discrete Event Simulation

Operator tasks during a procedure are sequential. Therefore a top-down approach was accomplished using network techniques to analyse the procedure. The snort procedure as defined by the Navy was divided into tasks and subtasks to be performed by operators and machines. Assumptions and conditions were defined describing the relations of the activities. Operator tasks were categorized in monitoring and control tasks. Each task was described by

- the activity to be performed
- the task category
- the time duration
- the operator(s) to perform
- the resources needed
- the input/output conditions.

The task descriptions consist of production systems, i.e. condition/action rules as they are used in the development of expert systems. The estimated time to perform a task was based on literature and based on results from experiments (Vermeulen, 1987).

Continuous Simulation

The characteristics of process variables whose values change continuously over time can be modelled by means of algebraic, difference or differential equations. This technique is used to describe the dynamics of relevant platform systems like diesel engines, generators, battery charge process, trim, etc. The model of the submarine's behaviour is based on the fact that vertical forces will not affect motion in the horizontal plane. Therefore the manoeuvring characteristics of the submarine as a continuous process are modelled by two sets of linear equations of forces and moments in the horizontal and vertical plane:

- Hydrodynamic forces and moments due to hull and sail.

- Hydrodynamic forces and moments due to hydroplane deflections on stern and sail.
- Forces and moments due to ballast, unballast, trim.
- Forces due to waves.

The model is based on a linear simulation model (Heidsieck, 1983), tested and compared with trials of a non-linear six degrees of freedom model. The model is provided with a system for automatic heading and depth control.

Implementation

The discrete event and continuous model have been implemented into an IBM PC/AT3 computer by means of the SAINT (Systems Analysis of Integrated Networks of Tasks, Wortman, 1978) simulation language. SAINT is a network and modelling technique written in FORTRAN and designed for describing complex man-machine systems in which discrete and continuous elements can be calculated simultaneously.

The discrete part of the system to be analysed is the network model. In SAINT a graphic approach is taken to model a network. There is a special graphic symbol for each element. Elements of the network are tasks, resources needed to perform the tasks and input/output relations between tasks and status variables. Additional functions can be defined by user-written subroutines.

The continuous part of the model consisting of the dynamic equations is implemented using a subroutine called STATE. This subroutine is called after periodic time intervals and/or when a discrete event occurs and calculates the process variables called state variables.

Interaction between the discrete and continuous part of the model is affected by the event related part of the model. In this part threshold functions, user functions and special attributes are used.

The values of probabilistic variables can be determined using predefined distribution sets (Monte Carlo approach). To calculate the mean values and standard deviations within reasonable confidence intervals the calculations must be replicated many times (iterations). For each condition of the snort procedure 25 iterations have been simulated.

SAINT provides a framework through which statistical information of the model can be obtained. Data collection points are user-defined and can be introduced at any point of the network. Statistical output data can be presented in both graphical or tabular form.

SIMULATION RESULTS

The Quality of the Manoeuvre

Figure 2 shows the average overshoot of the snort top valve just after its appearance above sea level. Each scenario represents 12 conditions of 25 iterations.

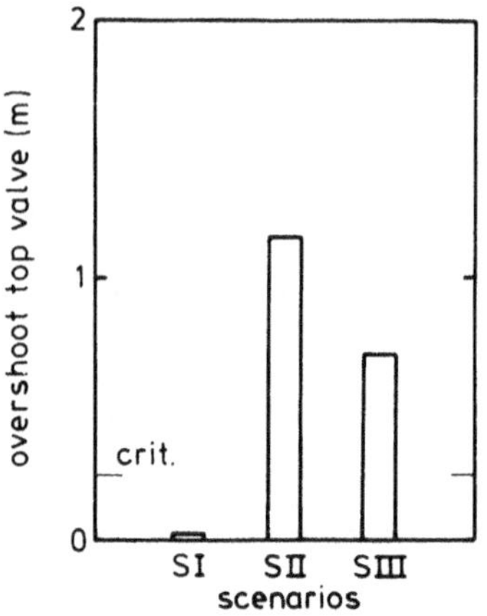

Figure 2. The overshoot of the snort top valve for each scenario, averaged over 12 conditions of 25 iterations each. On the vertical axis the criterion is presented.

It is shown that only during the hypothetical scenario SI there is no overshoot of the snort top valve. The manoeuvre in this scenario is approximately perfect. Scenario SII and SIII show results with an average overshoot of 1.2 m resp. 0.7 m.

The Procedure Execution Time

Figure 3 shows the average procedure execution time as a function of the scenarios SII and SIII. It shows that under normal operating circumstances the snorkeling procedure is terminated after 100-200 seconds. Only during manual trim conditions is there a significant delay if no further precautionary measures are taken (scenario SII). In fact, manual trim corrections by VDU may interfere with other interactive actions. The time delay is mainly caused by trim actions resulting from sub-optimal ballast conditions.

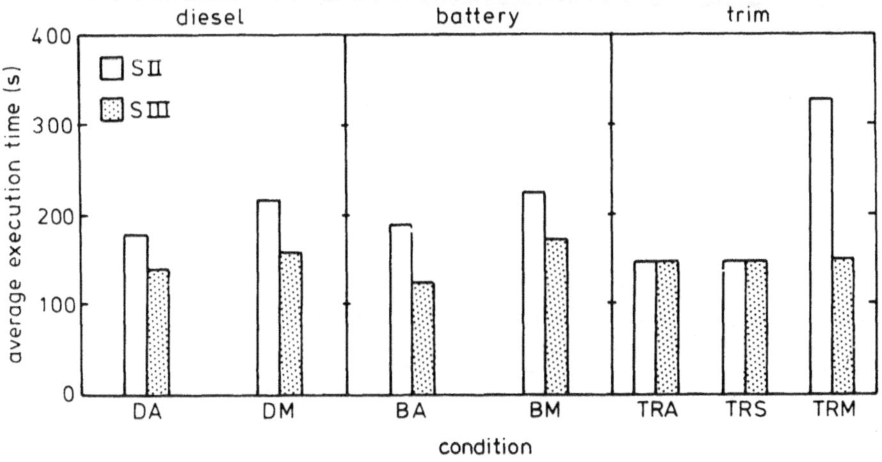

Figure 3. The procedure execution time per condition for scenario II and III, averaged over four conditions of 25 iterations each.

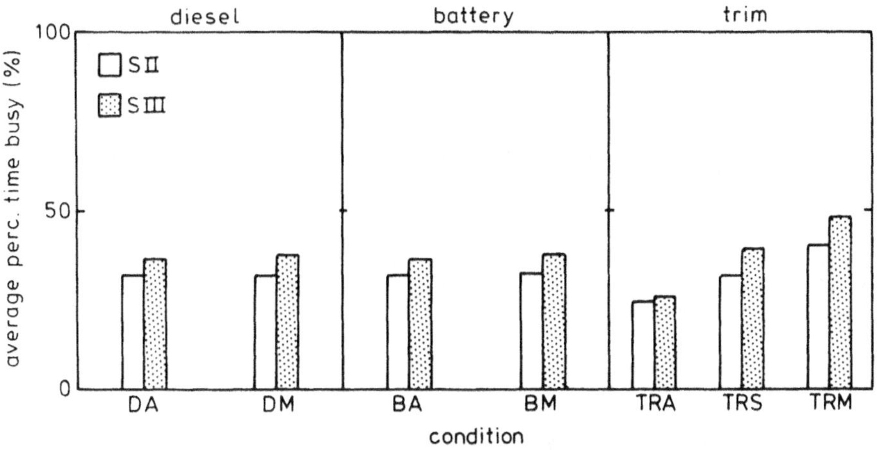

Figure 4. The percentage of time busy of the trim operator per condition for scenario II and III, averaged over four conditions of 25 iterations each.

Operator Busy Time

The model output shows that the percentage of busy time of each operator increases during manual operation. During fully automatic operation about 25% of the available time is used for monitoring and control tasks. Only a third part of it is used for control actions. In Figure 4 the busy time of the trim operator as a function of the conditions is depicted.

CONCLUSIONS

Results of the computer simulation using network models show that during the snorkeling procedure of a Walrus-class submarine, disturbances of ballast in the exhaust system cannot sufficiently be compensated. Assuming that during the procedure the pitch angle is kept at zero degrees, the top valve will have an overshoot of 1.20 m above sea level if no further precautionary measures are taken. Even by respecting pre-ballast conditions the overshoot will be about 0.70 m. This is unfavourable for the detectability of the submarine. During empirical tests performed by the Navy with existing submarines it appears that sudden vertical accelerations can be compensated by manoeuvring the submarine with a small pitch angle forward. This finding confirms the applicability of steering with a pitch angle, however, the automatic depth control system will have to be adapted for that condition. Moreover, pre-ballasting will have to be considered.

The procedure execution time varies from 100 to 200 seconds. Simulation results show that the execution time is hardly affected by the use of automatons, but mainly by the initial ballast conditions. It can be concluded that correct preventive ballasting will increase the quality of the manoeuvre and will decrease the procedure execution time.

Analysis of the operators busy time shows that even in case all platform systems are manually operated the operators are controlling and monitoring the installations during less than 50% of the available time. This suggests that there is ample capacity left for correcting unexpected errors.

REFERENCES

Döring, B., 1983, "Analyse des Arbeitsprozesses bei der Fahrzeugführung am Beispiel eines Landesanflugs", Forschunginstitut für Anthropotechnik, Wachtberg-Werthhoven.

Heidsieck, R. D., 1983, "Linear Simulation Model for a Submarine Equipped with a Depth Measurement System", TNO Institute for Mechanical Constructions, Delft.

Kraiss, K. F., Schubert, E., Widdel, H., 1983, "Untersuchungen zur manuellen Lenkung von U-Booten", Forschunginstitut für Anthropotechnik, Wachtberg-Werthhoven.

Meister, D., 1985, "Behavioral Analysis and Measurement Methods", John Wiley & Sons, New York.

Vermeulen, J., 1987, "Effects of Functionally or Topographically Presented Process Schemes on Operator Performance", TNO Institute for Perception, Soesterberg.

Wortman, D. B., Duket, S. D., Seifert, D.J., Hann, R. L., Chubb, G. F., 1977, "Simulation Using SAINT: A User-Oriented Instruction Manual", Aerospace Medical Research Laboratory, Wright Patterson Air Force Base, Ohio.

APPLICATION OF SLAM FOR MODELING AND ANALYSING PILOT'S RULE-BASED BEHAVIOR ON AN AUTOMATED LANDING APPROACH

Bernhard Döring

Research Institute for Human Engineering (FAT)
Wachtberg-Werthhoven, F.R.Germany

INTRODUCTION

Man plays an important role in complex man-machine systems (MMS), such as nuclear power plants, ships, and aircraft, although now many system functions are processed automatically. He has to monitor, supervise, and to control the automated machine processes; he is further responsible for detecting, diagnosing, and compensating malfunctions of those processes. Because today in modern MMS highly sophisticated technologies are applied, now man is often the limiting element in those systems if not designed properly. To enable the operator to perform tasks correctly during system operation, systems have to be designed according to human abilities and capabilities. To have an impact on system design, special attention has to be given to human contributions as early as possible during system development while it is still possible to influence major design decisions. To identify and analyse human activities and their impact on successful system operation in the development phase, the digital computer simulation has proven as a powerful tool. Highly sophisticated simulation languages like SLAM (Simulation Language for Alternative Modeling) are very helpful in that process as will be shown.

In the following a simulation study is described in which the pilot behavior and aircraft processes during an automated landing approach were analysed and modeled. This approach was chosen because required monitoring, supervisory, and control tasks of the pilot are typical for highly automated MMS (Sheridan et al., 1976). The tasks considered in this study are mainly based on predetermined procedures which the pilot has learned during training. Therefore he is familiar with such approach situations. Rasmussen (1986) defines the behavior for performing those tasks as rule-based behavior which is controlled consciously and appears in familiar work situations. Goal of the study was the development of a simulation supported methodology which can be used in early system development for determining the required knowledge of the pilot in rule-based situations and human engineering design and arrangement requirements for the cockpit-interface.

Before going into details, a short introduction to the steps of the digital computer simulation is given. According to those steps the conceptual model of pilot tasks and aircraft processes and its implementation with SLAM elements are described. After presenting simulation outputs and applied validation procedures some simulation results are explained.

STEPS OF A SIMULATION STUDY

The process of a simulation study can be divided into different phases which are described in detail in literature (e.g. Pritsker, 1986, Rouse, 1980). In order to show the application of SLAM for modeling and simulating pilot rule-based behavior it is sufficient to distinguish between five steps (Döring et.al., 1981) that will be used to structure the description of the simulation study:

- Definition of the Problem,
- Development of the Model,
- Implementation of the Model,
- Simulation of System Behavior, and
- Analysis of Simulation Outputs.

A simulation study starts with the clear definition of the problem to be solved which comprises a detailed description of the system to be studied as well as a formulation of the problem-solving objectives. Problems which we considered in our study were related to a man-machine system. Therefore the problem definition represents a description of the considered system elements, i.e. of the pilot and the aircraft, including system performance measures such as task relevant data as problem solving objectives.

Once an initial problem definition is formulated, the development of the model begins. Developing a model of the considered system means to determine relevant system elements and their behaviors excluding unnecessary details. The amount of detail which has to be considered should be based on the purpose for which the model is being built. During the development of the model, input and performance data of the considered system elements are required. The result of the model development phase is a mathematical-logical representation of the system, called conceptual model (Schlesinger et.al., 1979), which can be exercised later in an experimental fashion on a digital computer.

After developing the conceptual model the next task is the implementation of the model on a digital computer. The conceptual model is translated into a simulation program which is called the computerized model of the system under consideration (Schlesinger et.al., 1979). For implementing a model, a simulation language, e.g. SLAM, is very helpful. In addition to the savings in programming time, a simulation language also assists in model formulation by providing a set of concepts for articulating the system description. An important part of the implementation is the verification of the simulation program. The verification task consists of determining that the computerized model represents the conceptual model within specified limits of accuracy.

If the simulation program is implemented the next task is the actual simulation of the system behavior with the digital computer. The system behavior is represented by state trajectories which are generated as simulation outputs. Because in most cases it is neither useful nor possible to store all data that are generated during simulation, a simulation language mostly offers certain basic statistical functions for data reduction, aggregation, and documentation such as calculation of means and standard deviations, and certain formats presenting results such as histograms and certain sorts of plots. One important task related to simulation outputs is the validation of the model built so far. The validation task checks if the computerized model is a sufficient accurate representation of the system behavior under consideration. It is in this phase that accumulated errors are discovered and the final acceptance of the model must be achieved.

Simulation outputs describe the dynamic behavior of the system considered over time. The statistical analysis of simulation outputs is similar to the statistical analysis of the data obtained from an experiment with an actual system. The main difference with simulation is that the analyst has more control over the running of the simulation program. Thus he can design simulation experiments to obtain the specific analysis results necessary to answer the pertinent questions relating to the system under study.

SIMULATION STUDY FOR ANALYSING PILOT'S RULE-BASED BEHAVIOR

In this simulation study SLAM was applied for modeling and analysing pilot's rule-based behavior which are required during a highly automated landing approach. Performing rule-based behavior (Rasmussen, 1986), the pilot applies stored rules or procedures which he has acquired by training. During an approach he perceives information typically as signs which refer to aircraft states or approach situations in his environment. Signs are used to select actions which he can apply to act on those states and situations. For the approach bad weather conditions were assumed so that the pilot could not rely on external vision and the approach was necessarily guided by an instrument landing system (ILS).

In this presentation the distinct steps of the study are described according to previously distinguished simulation steps. The view is focused mainly on development and implementation of the model.

Definition of the Problem

The problem definition of a modeling process includes a statement of the phenomena of interest as well as a choice of performance measures (Rouse, 1980). Our phenomena of interest are highly automated flight processes of an aircraft and the corresponding tasks of the pilot during an ILS-approach. The basis of this simulation study was a real time flight simulator facility at the FAT. Because this facility simulates a HFB 320 Hansa executive jet approaching the runway 25 of the Cologne-Bonn airport this aircraft and this approach area were selected for analysis. The aircraft requires a crew comprising pilot, co-pilot, and flight engineer. Only pilot´s tasks were analyzed.

The simulation study should answer questions which are related to the specific task knowledge required by the pilot to approach correctly and to the information flow on the pilot-cockpit interface. Task knowledge is needed, e.g., to determine contents of pilot training programs. The information flow provides the basis for determining human engineering design and arrangement requirements for cockpit displays and controls (Döring, 1976a; Shackel, 1974). In detail, questions which have to be answered are:

1. Which pilot tasks in which approach segments have to be performed during an autopilot controlled landing to approach successfully?
2. Which values of which aircraft and approach state variables does the pilot need as inputs/outputs to perform his tasks successfully?
3. What is the frequency and sequence of use for those variables?
4. At which time points and during which time intervals are those variables used?

Development of the Model

To develop the conceptual model, i.e. the mathematical-logical behavior description of approach relevant systems and the pilot, an analysis was conducted. Approach relevant systems from which the aircraft system receives and/or to which it gives information are different navigation stations of the approach area, the outer and middle marker, the localizer and glide path transmitter, and the tower.

To determine system functions and especially pilot tasks, the ILS-approach was partitioned into 12 approach segments which are listed in part in Table 1. To identify required monitoring, supervisory, and control tasks, three task priorities and five tasks categories were distinguished:

First task priority:

- adjusting tasks, e.g., to set up the autopilot with new desired courses or headings;
- activating tasks, e.g., to change autopilot mode, flap position,or gear state;

Table 1. Approach segments and corresponding pilot tasks (in part)

Approach to Interception Heading 90°	**Adjust Heading Marker Position 170°** **Adjust Digital Course 150°** **Activate Lateral Mode VOR** **Adjust Heading Marker Position 90°** **Adjust Engine RPM 80 %** **Perform Cross Check** **Monitor Flight Attitude** **Monitor Heading**
On Interception Heading 90°	**Perform Cross Check** **Monitor Course Deviation** **Monitor Capture Indicator**
Interception of Radial 150° of VOR-Station	**Perform Cross Check** **Monitor Flight Attitude** **Monitor Course Deviation** **Monitor Heading**
On Radial 150° of VOR-Station	**Activate Flap Position 20°** **Adjust Vertical Speed 800 ft/min** **Activate Vertical Mode VS HOLD** **Adjust Vertical Speed 350 ft/min** **Activate Vertical Mode ALT HOLD** **Adjust Heading Marker Position 190°** . . . .

- special tasks, e.g., verbal communication when performing outer marker check or receiving landing clearance.

Second task priority:

- checking tasks, e.g., to compare current and desired values of indicated air speed, heading, altitude, etc. during a cross check.

Third task priority:

- monitoring tasks, e.g., to observe systematically the indicated altitude during descent or heading during approach to an interception heading.

The first task priority comprises adjusting, activating, and special tasks which are required for a successful approach. These tasks have the highest priority because they have to occur at special approach points. If they are not performed at those points the approach progress may become incorrect. For instance, during the approach segment "Approach to Interception Heading 90°" (Tab.1), at the beginning the pilot has to adjust the heading marker position to the value of 170°, and later to 90°. He has to adjust the digital value of the course to 150°, to activate the autopilot lateral mode VOR-APPROACH, and to adjust the engine revolutions per minute to 80 %.

Tasks of the second task priority are required to ensure the system safety. During all segments the pilot ought to check aircraft state variables to detect malfunctions of aircraft subsystems. For the checking procedure, it is assumed that he performs a cross check during which he compares desired and actual values of indicated air speed, flight attitude, heading, altitude, vertical speed, and engine revolutions per minute. The cross check is interrupted if tasks of the first priority have to be performed.

Monitoring tasks have the third priority. They are required to verify correct autopilot operations especially during approach transition phases in which the autopilot is controlling distinct aircraft state variables, e.g., heading variation, vertical speed. The pilot should monitor aircraft state variables over a certain time period, e.g., the heading during heading changes, the altitude during altitude changes. It is assumed that these tasks have the lowest priority because normally malfunctions of aircraft subsystems are detected already during the cross check.

For each approach segment, lower and upper limits of aircraft state and approach variables, which are specific to a segment were determined. Such variables are, e.g., air speed, vertical speed, heading, flight attitude, course and glide slope deviation. This information represents that knowledge which the pilot has to retrieve out of his memory or from the instrument approach chart when he performs checking and monitoring tasks.

By using task categories, 42 different pilot tasks were identified for the ILS-approach. Each task was described by that behavioral verb, which was used in the task classification above and which indicated the nature of activity being performed during that task and by that information or state variable that was being acted upon by the pilot. Additionally, the time required by the pilot to perform that task was estimated. Basic data for time estimation were taken from Miller (1976). Operating times for tasks such as adjusting heading marker position, digital course, and vertical speed were measured in our flight simulator facility. The task duration required by the pilot to perform a task was modeled by a normal distribution function characterized by mean and standard deviation.

For describing the identified tasks in detail, a production system approach (Nilsson, 1980, Barr et al., 1981) was used. The major elements of a production system are a database, a set of productions, and a control system. The production operates on the database. Rouse (1980) defines a production as a situation-action pair where the situation side is a list of things to watch for in the database and the action side is a list of things to do. Actions resulting from a production change the database. The control system determines which applicable productions should be applied and ceases computation when a termination condition on the database is satisfied. Possible control strategies are discussed, e.g.,by Barr et al.(1981).

Applying the production scheme to a pilot task, the situation side represents the actual values of those variables which the pilot has to perceive from his cockpit environment or to retrieve from his memory. The action side describes the actual values of those variables which he has to act upon when performing a task. The situation side of productions specifies signs which are the typically perceived information at the rule-based level of behavior, while the action side specifies the predetermined manipulations which are activated when the situation specific signs appear (Rasmussen, 1986). The situation and the action side represent the task input and output respectively. It is assumed that input and output are separated in time by the task duration, i.e. the time the pilot needs to perform the task. In this study, applicable productions were determined by querying expert pilots, analysing approach procedure descriptions and records obtained with our flight simulator.

To explain the structure of task specific productions in some detail, only the adjusting task 'Adjust Heading Marker Position' (AD HMP) is described. That task appears twice during the segment "Approach to Interception Heading 90°" and once during the segment "On Radial 150° of VOR-Station". It appears at distinct points of time at which distinct actions, i.e. the values 190°, 90°, 170° of the heading marker position hmp have to be supplied to the autopilot. At those points, the situation can be characterized by the approach segment and specific values of the heading hd(t), the heading marker position hmp(t), and the altitude alt(t) at time t. Denoting the duration of the task AD HMP by d(AD HMP), the following productions can be established:

IF (approach segment = 'Approach to Interception Heading 90°' .AND. $hd(t) > 198^{o}$ .AND. $hmp(t) \neq 170^{o}$)
THEN 'Adjust $hmp(t + d(AD\ HMP)) = 170^{o}$'

IF (approach segment = ' Approach to Interception Heading 90°' .AND. hd(t) < 198° .AND. hmp(t) $\neq 90^{o}$)
THEN 'Adjust hmp(t + d(AD HMP)) = 90°'
IF (approach segment = ' On Radial 150° of VOR-Station' .AND. alt(t) < 2500 ft .AND. hmp(t) $\neq 190^{o}$)
THEN 'Adjust hmp(t + d(AD HMP)) = 190°'.

By describing each pilot task with productions that are specific to a special approach segment and/or aircraft state, the large amount of knowledge required by the pilot during the landing approach could be structured and specified clearly and completely.

Flight processes were described in a simplified form. Only those state variables displayed in the cockpit and relevant to pilot tasks were modeled to determine the information flow between pilot and cockpit. To model the relevant state variables and their value changes over time, a set of difference equations was used. For example, the values of heading hd, indicated air speed ias, and vertical speed vs at time t_n were described by the equations:

$$hd(t_n) = hd(t_{n-1}) + (t_n - t_{n-1}) * hdv(t_{n-1}),$$
$$ias(t_n) = ias(t_{n-1}) + (t_n - t_{n-1}) * iasv(t_{n-1}),$$
$$vs(t_n) = vs(t_{n-1}) + (t_n - t_{n-1}) * vsv(t_{n-1}),$$

where $hd(t_{n-1})$, $ias(t_{n-1})$, and $vs(t_{n-1})$ are representing the values at the last computation time t_{n-1}, and $hdv(t_{n-1})$, $iasv(t_{n-1})$, and $vsv(t_{n-1})$ are the variation rates of hd, ias, and vs at time t_{n-1}.

The general relationships between model state variables are shown in Fig. 1. Starting with aircraft state variables the aircraft position is determined. Comparing the actual aircraft position with positions of navigation stations, actual bearings and distances to the stations can be derived. By comparing actual bearings with bearings which are desired according to the approach path, the actual course deviation and glide slope deviation are determined. For modeling the autopilot control, thresholds of aircraft state, aircraft position, and approach variables are used. When a corresponding state variable reaches such a threshold then a state event occurs causing a change to the values of other state variables, e.g., when during altitude change the required altitude is reached, vertical speed is set to zero.

The conceptual model that results from applying the previously described elements of a production system to pilot tasks and modeled aircraft

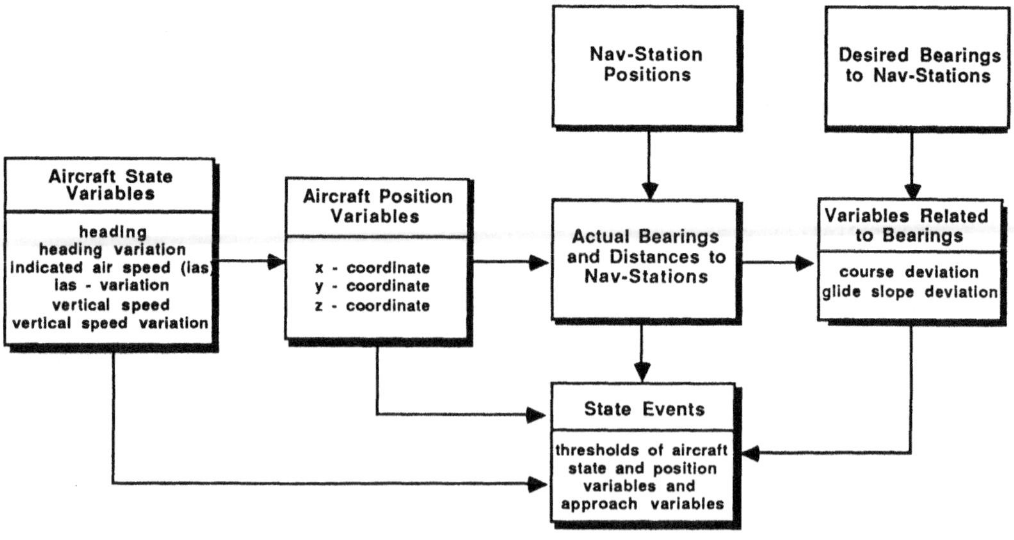

Fig. 1. Relationships between model state variables

related and approach related processes is depicted in Fig.2. It can be seen that changes of the database are not only induced by productions which characterize pilot tasks but also by aircraft and approach processes. According to the three task priorities the set of productions is partitioned in three subsets which represent the task catagories adjusting, activating, and special task, cross check task, and monitoring tasks. The control system selects a production subset according to the actual aircraft state and approach situation and its priority. Within the subset the first rule of which the condition part is matched will be activated.

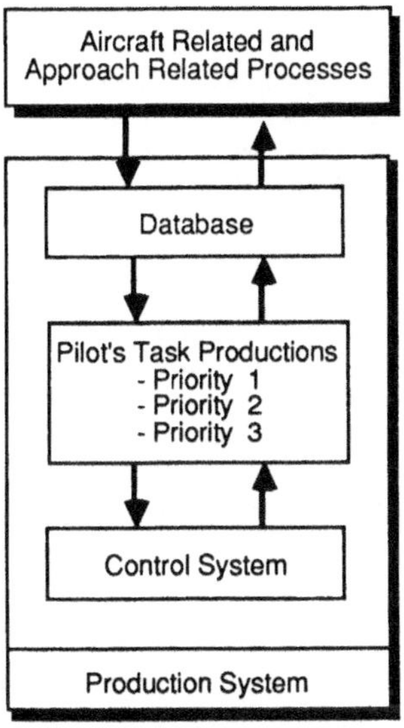

Fig. 2. Conceptual model based on a production system

Implementation of the Model

Once the conceptual model has been developed, the next step in the simulation study is the implementation of that model using SLAM, i.e. the generation of the simulation program called computerized model. SLAM has been selected mainly because it provides a conceptional framework for implementing continuous and discrete systems and combinations of them.

A continuous model is coded in SLAM by specifying differential or difference equations which describe the dynamic behavior of state variables. These equations are coded by the modeler in the SLAM subroutine STATE in FORTRAN. State variables described in the subroutine STATE are automatically updated by SLAM to calculate their values within an accuracy specified by the modeler.

For modeling discrete systems SLAM offers the possibility to apply an event oriented and a process oriented view. In the event orientation of SLAM, the modeler defines the events and the potential changes to the modeled system when that event occurs. The mathematical-logical relationships, prescribing the changes associated with each event type, are coded by the modeler in the SLAM subroutine EVENT in FORTRAN. The executive control program of SLAM controls the simulation by advancing time and initiating calls to the event subroutine at the proper points in simulated time. Hence, the modeler is completely relieved of the task sequencing events to occur chronologically.

The process orientation of SLAM employs a network structure which consists of specialized symbols called nodes and branches (see e.g. Fig.3). These symbols model elements in a process such as resources, queues for resources, activities, and entity flow decisions. The modeling task consists of combining these symbols into a network model which pictorially represents the system of interest and its processes. Entities in the system (such as people and items) flow through the network model. With special nodes, values which can be generated in a user-written function USERF are assigned to attributes of entities. To perform process simulation, each network element has to be coded by a special statement. An input file of the SLAM simulation package stores all those network statements which are examined concerning their correctness at the beginning of a simulation run.

In combined discrete-continuous models, the independent variables, e.g., the time, may change both discretely and continuously. The view of a combined model specifies that the system can be described in terms of entities, their associated attributes, and state variables. The behavior of the system model is simulated by computing the values of state variables at small time steps and by computing the values of attributes of entities at event times. A detailed description of the various modeling possibilities of SLAM is given by Pritsker (1986).

The computerized model of the ILS-approach comprises a continuous part which is described in the SLAM subroutine STATE, an event-related part realized in the subroutine EVENT, and a network part with connections to the user-written function USERF.

With the continuous part in subroutine STATE, the aircraft related and approach related processes are modeled by means of difference equations. Aircraft and approach state variables which are described previously (Fig.1) are defined by those state equations and assumed to change continuously with time. If state variables are used in subroutine STATE, the executive routine of SLAM will call STATE at periodic intervalls unless an intervening event or information request is encountered.

The combination of the network part and the function USERF models discrete changes of variables characterizing pilot task performance. Generally, the network simulates the flow of temporary entities through processes from their arrival to their departure. In our case, entities represent requests for performing a pilot task; the pilot is regarded as resource; regular activities represent his tasks. An entity that flows through the network occupies the resource and activates that activity which is selected in the function USERF. The selection occurs according to the task priority and the actual approach situation and aircraft state.

The network which models pilot tasks and their previously defined priorities is depicted in Fig.3. It consists of three part networks labeled A, B, and C. Each part represents tasks of different priority: A) Adjusting, activating, and special tasks of first priority; B) Cross check tasks of second priority; C) Monitoring tasks of third priority. The part networks are controlling the simulation by working concurrently. Each part network consists of a combination of nodes and branches. In general, an entity is moving in a cyclic manner from an ASSIGN node to a PREEMPT node, further to a FREE node and back to the ASSIGN node. The way between the PREEMPT node and the FREE node represents the performance of a single pilot task. All three part networks are similar in structure. In an ASSIGN node the characteristics of a task are determined by calling the user-written function USERF. In that function the task specific productions are selected and activated. Normally, a branching activity which needs no time leads to the PRREMPT node whose only function is to capture the resource according to the distinct task priority. The activity between PREEMPT and FREE node is used to simulate the task duration. Because in all part networks PREEMPT nodes are requiring the same resource, i.e. the pilot, the different priorities of PREEMPT nodes which correspond to task priorities are controlling which of the three task categories is performed. E.g., adjusting tasks are interrupting cross check and monitoring tasks; cross check tasks are

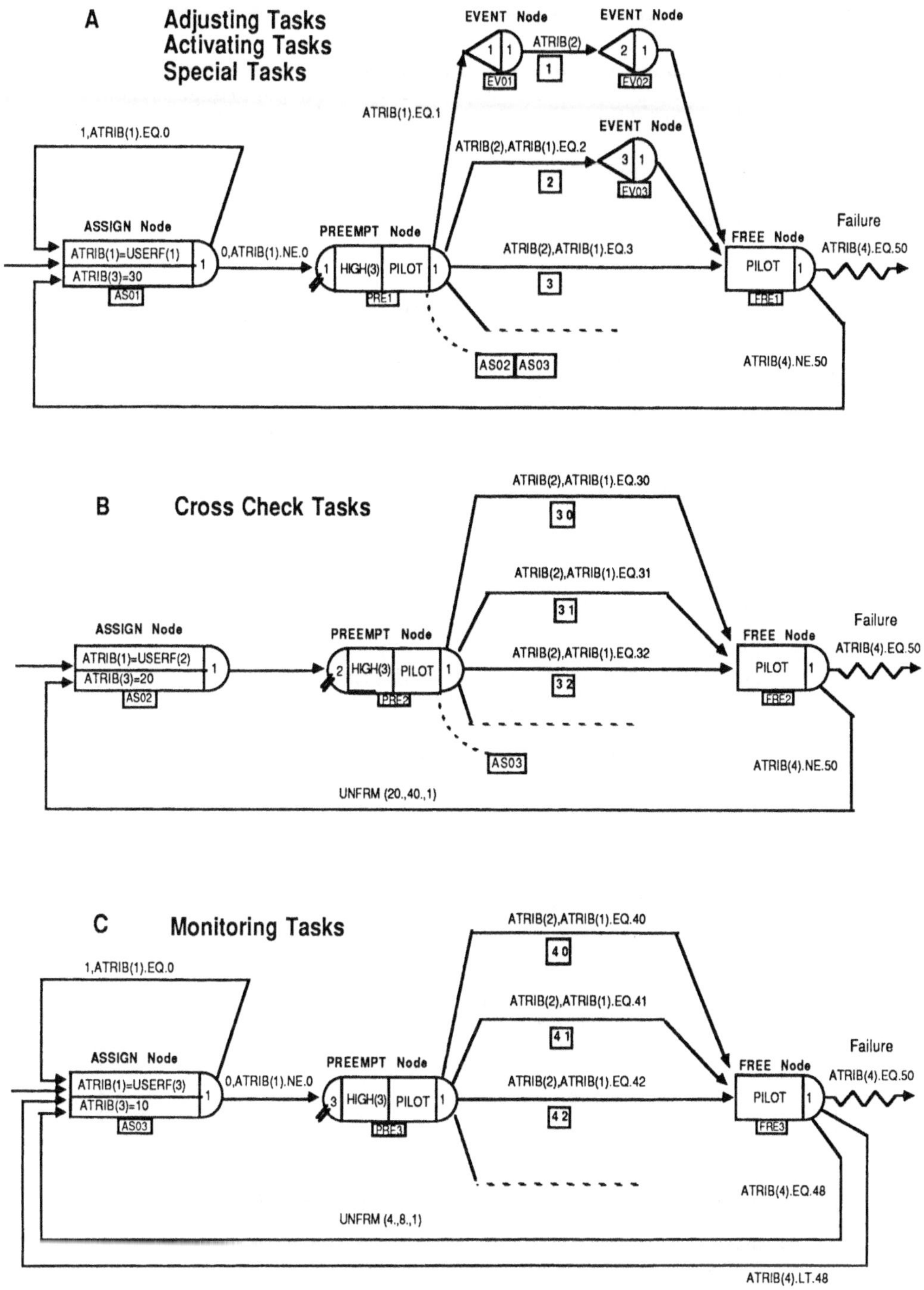

Fig. 3. SLAM network for controlling tasks selection

interrupting monitoring tasks. When a task activity between a PREEMPT node of a lower priority and a FREE node has been activated and a task is requested at a PREEMPT node of a higher priority, then the activated task activity is interrupted and the task activity with higher priority starts. The part network A includes in some branches one or two EVENT nodes which connect the network part with the EVENT oriented part of the model that will be described later. The FREE nodes release the captured resource so that it is free for performing other tasks.

As mentioned above the ASSIGN node connects the network model with the function USERF. In this function (Fig.4) the actual approach segment and the corresponding upper and lower limits of state variables, the requested

```
      FUNCTION USERF(IFN)

C***  DATABASE ***

      COMMON ...

C***  PRODUCTIONS FOR SELECTING THE APPROACH SEGMENT ***

      IF ( SITUATION 'X' )
           THEN (   APPROACHSEGEMENT = 'APPR SEGM NAME'
                    SETTING OF DESIRED VALUES OF STATE VARIABLES
                    FOR APPROACH SEGEMENT 'X')
      .

C***  SELECTION OF THE TASK TYPE ***

      GOTO (1000,2000,3000), IFN

C***  PRODUCTIONS FOR SELECTING ADJUSTING, ACTIVATING, ***
C***  AND SPECIAL TASKS                                 ***

1000  USERF = 0
      IF (  APPROACH SEGMENT 'X' .AND. SITUATION 'Y' )
           THEN (   TASK =  'ADJUST STATE VARIABLE SS'
                    USERF = 1
                    ATRIB(2) = TASK DURATION
                    ACTION  = ACTION OF TASK
                    RETURN)
      .
      RETURN

C***  PRODUCTIONS FOR SELECTING CROSS CHECK TASKS ***

2000  IF (ATRIB(4).EQ.30)
           THEN (   TASK = 'CHECK STATE VARIABLE SS'
                    ATRIB(2) = TASK DURATION
                    USERF  = 30
                    ATRIB(4) = 31
                    IF ( SS.LT.SSLOW .OR. SS.GT.SSHIGH) ATRIB(4) = 50
                    RETURN)
      .
      RETURN

C***  PRODUCTIONS FOR SELECTING MONITORING TASKS ***

3000  USERF = 0
      IF (  APPROACH SEGEMENT 'X' .AND. SITUATION 'Y' )
           THEN (   TASK = 'MONITOR STATE VARIABLE SS'
                    USERF = 40
                    ATRIB(2) = TASK DURATION
                    IF (SS.LT.SSLOW .OR. SS.GT.SSHIGH)        ATRIB(4) = 50
                    IF (ABS(SS-SSDESIRED)/DD.GT.15.)          ATRIB(4) = 48
                    RETURN)
      .
      RETURN
```

Fig. 4. Schematic structure of the user function USERF(IFN)

pilot task and its action, and the task duration are determined. The task specific production in USERF is chosen in the following way: According to the priority of a task request, the corresponding part of USERF is selected. In this part that production is selected of which the situation side is matched first.

A regular SLAM activity (denoted in Fig.3 by the activity number in a rectangle) is used to model time duration which the pilot needs to perform the task. The task duration of a specific task when performed is a sample of a normal distribution function and determined in the function USERF that is activated in the corresponding ASSIGN node.

EVENT nodes are included in the network model to interface that part of the model with discrete time events. Such event occurs, e.g., after adjusting and activating tasks are accomplished. They lead to modifications of aircraft related processes which are modeled in the subroutine STATE. The actual modification of state variables is specified by the user in the subroutine EVENT. The EVENT node causes the subroutine EVENT to be called. In contrast to time events that occur when an entity is reaching an EVENT node in the network, so called state events occur when specified state variables are crossing prescribed thresholds. To model such events, the state-event feature of SLAM which also activates subroutine EVENT is used. In the ILS-model, for instance, a state event occurs when a distinct bearing to a navigation station is reached. The user has to specify in the subroutine EVENT that the heading changes now with a rate of turn of $3^{o}/s$.

Fig.5 shows the implementation of the conceptual model based on the elements of a production system (Fig.2) with subroutines, functions, and

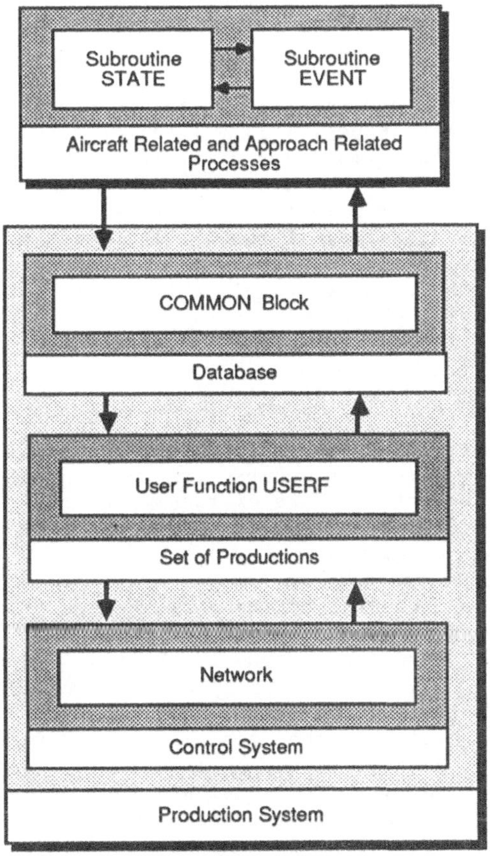

Fig. 5. Implementation of the conceptual model based on a production system with SLAM elements.

the network elements of SLAM. It can be seen that the COMMON block of SLAM represents the database of the implemented production system. Changes of database states are induced by aircraft related and approach related processes that are coded in subroutines STATE and EVENT. In the user-written function USERF, productions which describe pilot's rule-based behavior are defined by means of situation-action rules. The network part of SLAM constitutes the control system which selects a task category and its actual production according to the task priority and the actual approach and aircraft state. A task activity of the network simulates the task duration.

Simulation of the ILS-approach

After the computerized ILS-approach model has been implemented, the digital computer is used to simulate system activities. Generally, simulation implies an exercising of the computerized model to generate a chronological succession of state descriptions of, i.e. values of relevant state and task variables describing system behavior. Fig.6 shows an example of a plotted state trajectory with the state variables heading, roll angle, altitude, vertical speed, and indicated air speed. Additionally, the corresponding task time line with pilot tasks coded by numbers on the head-line is shown on the right side of the plot. The length of each line segment represents the task duration. Since the model uses probabilistic distributions for, e.g., task durations, the generation of simulation data is statistical in nature. Thus, simulation must be repeated many times to obtain a sufficient number of state and task trajectory samples in order to estimate average performance within reasonable confidence limits. 30 ILS-approaches were simulated with the computerized model in order to get corresponding trajectories and task timelines (traces).

An important task related to simulation outputs is the validation of the model. Experimental validation of a model involves using the model to predict performance and then empirically determining how close predictions are to actual occurences (Rouse, 1980). With the computerized process model the pilot's rule-based behavior, aircraft processes, and approach events were simulated. Modeled pilot behavior could be validated against performance of a real pilot in our flight simulator. Aircraft processes and event time points would be validated, if appropriate measures closely match measured values of our realistic flight simulator. Ten real time ILS-approaches were performed with one experienced pilot in the flight simulator. Pilot and simulator activities observable in the cockpit were recorded on videotape for later analysis. Data obtained during the approaches constituted the basis for validating the model.

From the various validation techniques which are applicable (Sargent, 1979), we considered face validity, traces, internal validity, and event validity. For evaluating the face validity of the model, pilots familiar with the ILS-approach were asked whether the conceptual model was reasonable. Particularly, production rules of tasks, details of flight processes described, and interrelations between tasks and flight processes were discussed and corrected in this way.

Simulation output traces were used to check the computerized model, i.e. pilot tasks as represented by the network were plotted to determine whether the simulation program correctly corresponded to the network and logic of those tasks. Deviations found in the model were modified appropriately.

Internal validity was assessed by comparing the stochastic variability of specific state variables in the model with their variability observed in the flight simulator. To test this type of validity, several stochastic simulation runs were made with the model and the variability of selected state variables, e.g., glide path interception altitude, glide path deviation, and course deviation, were determined and used for testing this validity type.

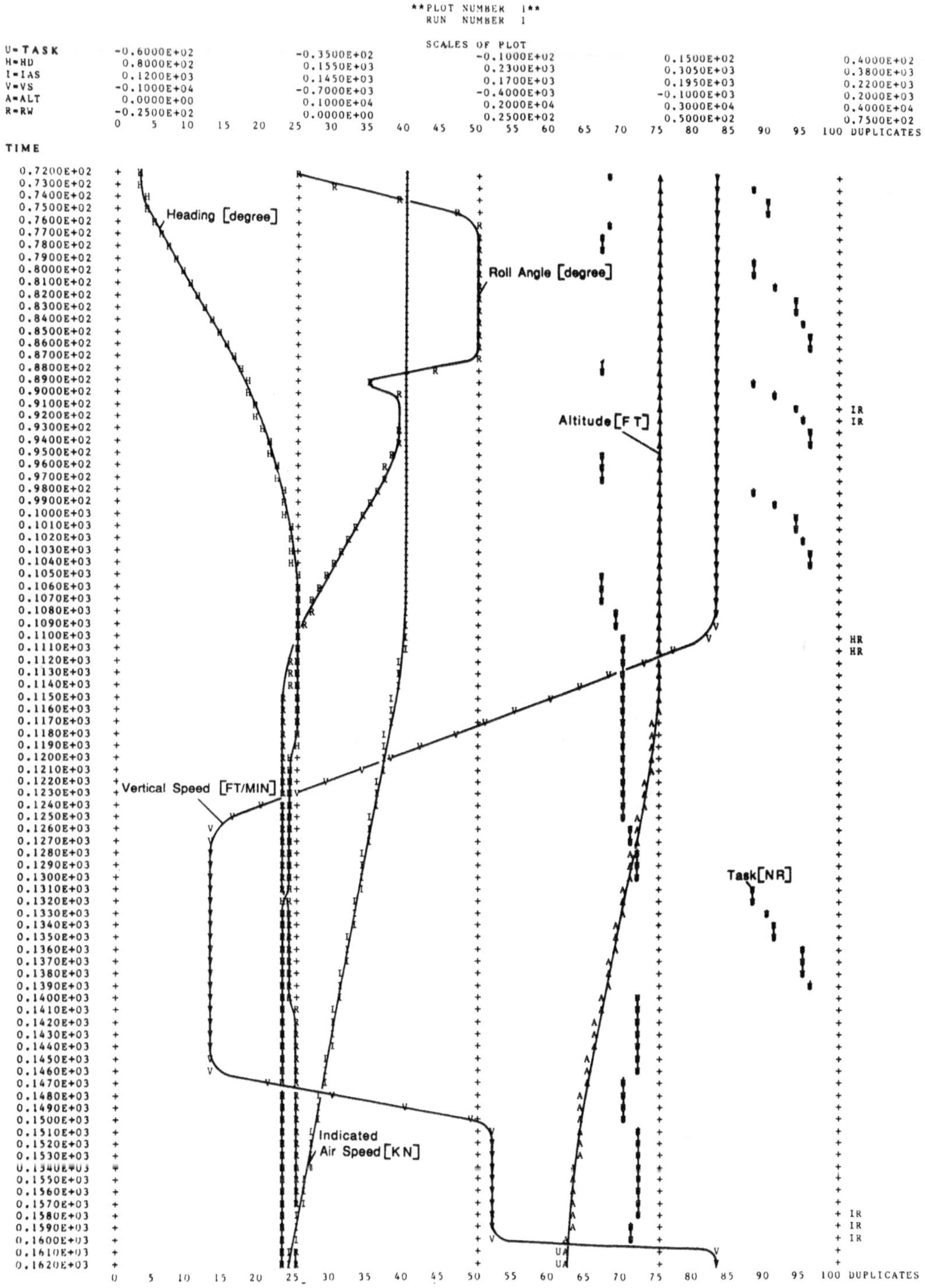

Fig. 6. State trajectory of heading (H), roll angle (R), altitude (A), vertical (V), indicated air speed (I)and the task time line (U).

To check event validity of the model, occurrence times of 17 approach relevant events were used as performance measures. Such events may be either task events which the pilot causes when, e.g., performing a task or they are system state events which occur when state variables reach specified thresholds. For all event occurrences, mean elapsed times into the approach were determined both from the 10 pilot flight simulator ILS-approaches and from 30 runs of the digital computer simulation. The times were compared statistically by means of the t-test. Existing deviations were eliminated by changing task duration distributions until no significant differences were obtained between flight simulator and computer simulation model data. Detailed descriptions of the validation process can be found in Döring (1983) and Döring et al. (1983).

Analysis of simulation outputs

After model validation, the final step in our simulation study was the experimental application of the model to generate simulation output data and analyse it. One goal of the study was to determine information flow requirements for the pilot-cockpit interface. Such requirements include all those variables, about which information is transmitted to the pilot or which are subsequently affected by control outputs of the pilot. They can be characterized by the mission required value range, time points, interval, and their use sequence, frequency, and duration. In modern MMS those specifications are necessary to design display and control units which are based on interactive electronic display concepts. To design quickly changeable electronic display formats, e.g., the following question has to be answered: Which value of which state variable at which time point or during which time interval does the pilot need to perform a specific task? Information that the pilot requires at the same time should be combined in the same format. To arrange formats correctly, the information importance and its frequence of use have to be known. Important and frequently used information/formats have to be arranged in upper priority levels. All requirement specifications can be obtained by using digital computer simulation except the importance which can be determined, e.g., by questionnaires.

In our study, output data of 30 simulation runs were analysed to get the described information flow requirements. The analysis was done by using SLAM elements for data collection and statistical calculation. Analysis results related to the state variables altitude, heading, vertical speed, roll angle, course, indicated air speed, heading marker, glide path deviation, and course deviation are listed in Table 2. For those variables the range of used values, the time interval of use, the use frequency, and the relative use duration with its standard variation are listed. As explanation only the variable altitude will be interpreted in detail. It can be seen that altitude is utilized in the value range between 385 and 3000 feet and between the 12th and 449th second during the regarded approach. It was used on the average 55 times with a relative use duration and standard deviation of 27 % and 1.2 %, respectively. Related to the average approach duration of 443 seconds, the absolute use duration and standard deviation are 120 s and 5.3 s.

Additional analysis details were obtained with SLAM by recording time points at which certain variables are used during the simulated approach. To analyse recorded data, SLAM generates, e.g., histograms using the approach time as one axis dividing it into intervals of equal duration. For distinct variables for each time interval the absolute, relative, and cumulative frequency of use were demonstrated in histograms. By using this method it could be ascertained, for instance, that heading values were used mainly in the first approach phase. Whereas during the final approach the course deviation is used instead of the heading.

The sequence of information use is another important feature to determine the arrangement of information in a format. Sequences of tasks performed (time lines) and of state variables used in their performance can be

Table 2. Analysis Results for some selected variables

Relevant Variables	Abbr.	Dimen.	Range of UsedValues		Time Interval of Use		Use Frequency	Rel. Use Duration M. Val. [%]	Rel. Use Duration St.Dev. [%]
			from	to	from [sec]	to [sec]			
Altitude	ALT	ft	364,8	3000,0	11,7	449,3	55	27,1	1,2
Heading	HD	degree	89,8	263,0	0,0	240,5	35	16,3	0,9
Vertical Speed	VS	ft/min	-983,5	0,0	13,4	449,8	27	13,9	1,1
Roll Angle	RA	degree	-25,0	25,0	9,9	448,6	29	13,9	0,7
Course	CO	degree	150,0	263,0	2,3	238,4	31	12,5	1,4
Ind.Air Speed	IAS	kn	131,4	180,0	8,4	446,5	27	9,8	0,9
Head.Marker	HDM	degree	90,0	190,0	0,0	192,9	8	5,8	0,4
Glide Slope Dev.	GSD	dots	0,6	2,2	232,5	445,5	13	5,5	0,8
Course Dev.	COD	dots	-1,4	0,4	207,2	438,0	13	4,6	0,7

obtained by plotting tasks and state variables over time in the same diagram (see, e.g., Fig. 6).

The utilization time of a format determines, among other things, its position in the display priority. To demonstrate the possibility of determining this feature with the digital computer simulation, the average utilization time in percent of total approach time of display and control components of the considered cockpit are determined by analysing output data of the simulated approaches. The most often used display components in this study are the flight director indicator, variometer, course indicator, and altimeter. Because of the highly automated ILS-approach, control components are seldom used. The control component used most often is the vertical command control with which the vertical speed is adjusted when changing altitude to the required value for intercepting the glide slope. Detailed descriptions of the simulation study and its results can be found in Döring (1983; 1986) and Döring et al. (1983).

CONCLUSIONS

The rule-based behavior of a pilot during a highly automated landing approach could be described in terms of situation-action rules of a production system. This was done by identifying tasks, their priorities, and their inputs and outputs, and randomizing task performance durations. Using a production system to describe the pilot's knowledge for a successful landing, situation oriented knowledge for monitoring, supervisory and control tasks, and procedure oriented knowledge for cross check tasks could be combined in one model. Prerequisite for determination of production rules is a comprehensive analysis of tasks that must be accomplished and the description of system processes affecting those tasks. Advantages of that method are: the model can be easily established in relatively short time; it is open to easy modification; because of its characteristics it can easily be transformed into a simulation model. This was demonstrated by using the high level simulation language SLAM to implement the model and to exercise it dynamically on a digital computer.

A general advantage of the computer simulation method is its iterative nature. As soon as the flight processes have been described in enough detail, the model and the simulation method can be used in a top-down manner for further identifying and analysing pilot tasks until the required level of detail has been reached. The modularity of production rules as well as of SLAM network elements has proven to be very useful in that modeling process.

Simulation output data are trajectories of state variables and task timelines. Task timelines are generated dynamically because tasks are not preprogrammed but depend on flight segments, approach events, system states, etc. By analysing established production rules and simulation output data, task specific knowledge which the pilot needs for a successful approach and dynamic information flow requirements which are necessary for cockpit interface design and evaluation can be determined. Although this method was applied to an existing MMS, it can be used to evaluate system concepts in early development phases, e.g. for determining the required task knowledge and information flow requirements for event and procedure oriented operator tasks.

REFERENCES

Barr, A. and E.A. Feigenbaum (Eds.) (1981): The Handbook of Artificial Intelligence, Volume I. Heuris Tech Press, Stanford, CA: William Kaufmann, Los Altos, CA.

Döring, B. (1976a): Analytical methods in man-machine system development. In K.F. Kraiss and J. Moraal (Eds.), Introduction to Human Engineering. TÜV Rheinland, Köln, pp. 293-350.

Döring, B., Berheide, W. (1981): A Review of Simulation Languages and their Application to Manned Systems Design. In: Moraal, J., Kraiss, K.F. (Eds.) Manned Systems Design - Methods, Equipment, and Applications. Plenum Press, New York, London, pp.91-120.

Döring, B. (1983): Analyse des Arbeitsprozesses bei der Fahrzeugführung am Beispiel eines Landeanflugs; Eine systemergonomische Simulationsstudie. Forschungsinstitut für Anthropotechnik, Wachtberg-Werthhoven, Bericht Nr. 59

Döring, B., Knäuper, A. (1983): A Simulation Study with a Combined Network and Production Systems Model of Pilot Behavior on an ILS-Approach. In: Automatica, Vol.19, No.6, pp.741-747.

Döring, B. (1986): Modeling Supervisory Pilot Behavior with General Systems Theory Formalisms. In: Mancini, A., Johannsen, G., Martensson, L. (Eds.) Analysis, Design, and Evaluation of Man-Machine Systems. Pergamon Press, Oxford, Frankfurt.

Miller, K.H. (1976): Timeline analysis programm (TLA-1), Final report. Boeing Commercial Airplane Company, Seattle, WA, NASA CR-14494.

Nilsson, N.J. (1980). Principles of Artificial Intelligence. Tioga, Palo Alto, CA

Pritsker, A.A.B. (1986): Introduction in Simulation and SLAM II, Third Edition. A Halsted Press Book, John Wiley & Sons, New York; System Publishing Corporation West Lafayette, Indiana.

Rasmussen, J. (1986): Information Processing and Human-Machine Interaction. North-Holland, New York, Amsterdam, London.

Rouse, W.B. (1980): Systems Engineering Models of Human-Machine Interaction. North Holland, New York, Oxford.

Sargent, R. (1979): Validation of simulation models. Proceedings of 1979 Winter Simulation Conference, San Diego, CA, pp. 496-503.

Shackel, B. (1974): Applied Ergonomics Handbook. IPC,Guildford.

Sheridan, T.B. and G. Johannsen, (Eds.) (1976): Monitoring Behavior and Supervisory Control. Plenum Press, New York.

Schlesinger, S., R.E. Crosbie, R.E. Gagne, G.S. Innis, C.S. Lahvani, J. Loch, R.J. Sylvester, R.D. Wright, N. Kheir and D. Bartos (1979): Terminology for model credibility. Simulation, March, 103.

MODELS OF MULTITASK SITUATIONS

Christopher D. Wickens

University of Illinois Institute of Aviation
Aviation Research Laboratory
#1 Airport Road
Savoy, IL 61874 U.S.A.

OVERVIEW

This tutorial discusses models that predict the loss in quality of performance of multiple tasks, which occurs as a direct result of that multiplicity. It is often assumed therefore that this multiplicity induces competition for some scarce commodity which we label "resources." At issue is whether one can predict the loss in quality given characteristics of (a) the processing on each task in isolation, and (b) the relation between tasks.

There are a number of psychological models of the resource allocation process. Unfortunately, it appears that those models which have the most precise quantitative formulation, and that have received the strongest empirical validation, have been derived in domains that may be furthest removed from complex task environments such as the flight deck, which are the focus of the current workshop. In contrast, those models that have addressed task environments of greatest complexity, are furthest removed from a quantitative formulation (or alternatively are models that have yet to achieve satisfactory validation). This disparity is unfortunate, because it is clear that the objective should be one of obtaining quantitative models, in which levels of performance in heterogeneous environments can be predicted from quantitative specification of task parameters.

Two general characteristics of the resource process have been addressed by models: the allocation of resources--the selective aspects of attention--and the sources of variance in competition between tasks--the commodity of resources that characterizes their "scarcity."

It is important to note that there are a number of elegant efforts which have discussed models of the strategic or microscopic processes by which performance is produced [e.g., whether processing is serial or parallel (Townsend, 1974; Townsend and Ashby, 1983; Kantowitz, 1986), or whether information integration and selection is early or late (Shaw, 1982; Kantowitz, 1986; Norman, 1968; Pashler, 1984)]. Therefore, a distinction must be drawn between models of how performance is produced, and models that predict how performance will vary as a function of task characteristics. The latter are clearly relevant to design environments. The former are relevant only if the modeled mechanism has robust and important implications to the level of performance obtained. By in large, this has not been the case.

In the following, we discuss first models that focus on the sequential or serial aspects of multitask performance, and describe their limitations in accounting for certain phenomena. We then present the components of multiple resource theory as an approach that can address these limitations, and discuss the extent to which different performance models have employed features of multiple resource theory. Finally, we conclude with a discussion of issues related to model validation.

SERIAL ALLOCATION: QUEUING THEORY

A number of models have dealt directly with the serial allocation of some processing resources, such as the availability of foveal vision between saccades (visual scanning), or the complete allocation of cognitive effort to one task rather than another (task selection). From the standpoint of these models, the issue of whether processing may be parallel is simply not relevant. They focus on those aspects of processing that are distinctly and unambiguously serial (i.e., aspects that require a decision of where to allocate over time).

Much of the origins of this work can be traced to the classical modeling of the distribution of visual fixations on dials, delivering information of different bandwidths, carried out by Senders (1966, 1983). His work in turn traces its origins to the analysis of pilot scanning carried out by Fitts, Jones, and Milton (1950). Subsequent developments along these lines by investigators such as Carbonnell (1966), Sheridan (1970), and Moray (1986) have incorporated costs, payoffs, and the decaying characteristics of human memory to account for when an operator will choose to fixate on a particular sample of dynamic visual information. Typically these models incorporate two components: A cost of switching from one task or channel to the next, and an expected cost of neglecting a channel or task. Following classical utility theory (Edwards, Lindman, and Phillips, 1965; Edwards, 1986), this expected cost is based upon the actual cost of negative events that will result if information in an ignored channel is not acted upon, multiplied by the probability that the negative event will occur. This probability in turn is related to factors such as event arrival rate, and the uncertainty of the state of a channel, an uncertainty that grows during the time since the channel has last been sampled. For example, if the pilot samples an instrument once, to determine that the aircraft is precisely centered on the desired flight path, his uncertainty of its location will grow with the time subsequent to that sample (Sheridan, 1970). The expected costs of that uncertainty will be linearly related to the costs of events that will occur if he is off the flight path (e.g., striking the ground short of the runway, straying into the path of a parallel approach). Moray (1986) has provided a comprehensive coverage of these models.

The more specific quantitative characteristics of visual fixation and scanning have been modeled by Harris and Spady (1985) and Allen, Clement, and Jex (1970). Harris and Spady, extending and summarizing the earlier work of Harris and Christhilf (1980), have summarized empirical data on instrument scan and dwell times to provide a data base for pilot instrument scanning. Allen, Clement, and Jex (1970) have developed a model that predicts the cost of scanning on performance of a manual control task: what changes in Crossover Model parameters of gain, time delay, and remnant can be predicted from particular requirements to fixate away from the tracking display. Both of these efforts allow for some fairly specific quantitative predictions to be made, and are of direct relevance to the aircraft environment because of the heavy visual workload and the critical role of information scanning (Simmons, 1979). Their relevance is amplified because of the ability to quantify two of the driving inputs to the models--the expected frequency with which events occur along different channels and the optimum priorities for dealing with information along these channels. However, a constraint of these models for many applications to land and air vehicle control is that they have been based upon the sampling of foveally fixated flight instruments. Their application to the processing of motion gradients distributed across space in visual contact flight has not apparently been demonstrated.

The characteristics of task selection (as opposed to visual target selection) on the basis of expected utilities and costs also lie at the core of the concurrent performance assumptions made by many of the predictive models of complex task performance that are the focus of this workshop, such as the Human Operator Simulator or HOS (Wherry, 1976; Harris et al., 1987; Strieb et al., 1981), SAINT (Wortman et al., 1978; Laughery et al., 1986)*, PROCRU (Zacharias, Baron, and Muralidharan, 1981), STALL (Saturation of Tactical Aviator Load Limits; Chubb et al., 1987), and those models developed by Siegel and Wolf (1969), Corker et al. (1986), Chu and Rouse (1979), and Tulga and Sheridan (1980). Essentially these models assume that when two (or more) tasks compete for attention (call for completion at the same time), the human operator employs an algorithm to assess the order in which the tasks are to be performed. This algorithm is based on user defined priorities (HOS; Harris et al., 1987), on computation of expected costs of ignoring those activities not immediately performed, and expected benefits for undertaking that action that is highest in the priority sequence (PROCRU; Zacharias et al., 1981), or on the application of strategy-driven decision rules (Corker et al., 1986). Network models such as SAINT and the human operator model developed by Siegel and Wolf, have been applied only to discrete tasks. Models that attempt to incorporate continuous manual control into a sequential queuing approach, assume that this task too can be characterized as a series of discrete controls, implemented when error reaches a certain critical level of urgency (Rouse, 1980). Rouse, for example, describes the validation of a queuing theory approach to predicting the interference between flying a heavy aircraft and a series of discrete fault management tasks. Recent developments of PROCRU have also incorporated continuous manual control into the human performance model (Corker et al., 1986).

Thus, the most direct manifestations of time-sharing in these models are typically accounted for by the delay induced until a task reaches the head of a queue. This delay will of course be longer as a task is more difficult, and task difficulty is modeled in terms of task completion time (either an absolute value, or a statistical distribution). In cases in which working memory is involved, Wherry's (1976) HOS model has built in quantitative assumptions regarding the rate of decay of that material, as the relevant task waits unattended. Card, Moran, and Newell (1986) have provided good algorithms for estimating this rate. Recent developments in HOS, PROCRU, and the SAINT/Siegel and Wolf approach have allowed some parallel processing assumptions. These are discussed in more detail below.

---

*SAINT is not actually a model of complex task performance but is a structured programming language that allows user-defined task sequences to be played out.

A different approach to dealing with performance deterioration under multiple task conditions has been taken in Siegel and Wolf's (1969) model. As noted above, tasks that conflict in time lead to rescheduling according to prioritization. But the Siegel and Wolf model employs a workload estimate, based upon the ratio of time necessary for task completion to time available, to determine the mean completion time for tasks and to determine the task's probability of errorless completion. Mean completion time follows a V shaped function of workload, such that this time decreases (performance improves) as workload increases to a point. Then it increases linearly to a level at which there is clearly insufficient time available, and errors as well as delays are the result.

In spite of their differences, all of these models share a common concern for the role of time. Time is the resource that is competed for in serial fashion and completion time defines the difficulty or demand of component activities. In their earlier validated versions, the models are relatively silent as to the _intensive_ aspects of task demand, and as to the differing degrees of resource _competition_ fostered by greater or lesser similarity between tasks. They do not readily allow for concurrent processing. In this regard they represent more complex elaborations of single channel models of attention that were developed in psychology (Welford, 1967).

INSUFFICIENCY OF QUEUING THEORY

Queuing theory models provide an adequate first approximation to behavior in many multitask situations. In fact, they probably provide a satisfactory _final_ approximation in task environments when demands are not high (e.g., control room or flight deck operations in routine conditions; Moray, 1986). However, during the relatively high demand conditions typical of the tactical aircraft in combat, or the control room in emergency, three important characteristics of human performance in multitask situations can be identified that call into question the adequacy of single channel queuing models as sufficient for performance prediction (although queuing assumptions for task selection are a necessary component of a good model).

Parallel processing. Ample evidence may be cited of circumstances in which the human operator is indeed "doing two things at the same time," a circumstance not accommodated by queuing theory. Research on the visual system for example suggests that operators may be extracting continuous information regarding egomotion, necessary for flight control, walking or driving, even as symbolic objects or words are recognized (Leibowitz and Dichgans, 1980; Leibowitz, 1986). Furthermore, there is little evidence that manual control performance based upon continuous visual inputs comes to a complete halt as these symbolic inputs are either operated on in working memory (i.e., rehearsed, transformed) or are responded to. This continuity of continuous response, in parallel with discrete task processing, is particularly true if the discrete task response is a vocal one (McCleod, 1977). Indeed there are many circumstances in which manual control tasks may be time-shared nearly perfectly with tasks involving comprehension and exchange of auditory verbal material, typical of the interaction between pilot and air traffic controllers (Wickens, Sandry, and Vidulich, 1983).

Structural alteration effects. Wickens (1980, 1984) has called attention to examples, often extracted from realistic aviation simulations, in which a change in the structure of a task, not affecting its difficulty (and therefore the time required for performance in a queuing model), will have a substantial effect on concurrent manual control performance. This is particularly characteristic for example of changes in response modality from manual to voice (Tsang and Wickens, 1988; Wickens, Sandry, and Vidulich, 1983; Vidulich, 1987). Assuming a queuing theory approach, unless such an alteration changes the performance time (or time away from the flight task), an interference change of this sort could not be predicted. Another example of structural alterations is provided by the substantial interference between tasks that require the use of verbal-linguistic working memory, and those that require voice responses. This interference may be reduced or eliminated by the use of manual, rather than voice responses (Wickens and Liu, 1988).

Difficulty insensitivity. This phenomenon is the converse of structural alteration effects. It describes circumstances in which the increased difficulty of a task, requiring more time to perform, and therefore according to queuing theory, producing more interference, leaves performance on a concurrently performed task unaffected (Wickens, 1980, 1984). Similar to the phenomena described in previous sections, difficulty insensitivity appears to be observed when a continuous manual control task is performed concurrently with a task involving the vocal exchange, or transformation of verbal information. Increases in the demand of either task, leaves performance of the other task relatively unaffected.

A final limitation of queuing theory models is that they do not easily deal with continuous tasks in which levels of difficulty cannot be readily expressed in terms of time demand. Two examples illustrate this limitation. One is the continuous tracking task. As control order increases, or as lags are added to the system dynamics, tracking difficulty increases, and interference with concurrent tasks is increased as well (Wickens, 1986); yet the demands of higher order tracking are not manifest in the increased frequency with which control movements are required, but rather in the increased demands for prediction and anticipation. These are effortful, but not time-based dimensions. The second example is provided by tasks that require working memory. The increased demands of retaining more information in working memory are well validated (Baddeley, 1986). Yet this increased demand cannot easily be accounted for in terms of time.

How much extra variance in dual task performance (or multitask interference) may be accounted for by augmenting queuing theory models to attempt to account for the preceding phenomena remains unclear. As described at the end of this paper, careful comparative evaluations must be carried out in order to make such a determination. However, in any case it is important to consider the resource models of human parallel (concurrent) processing that have been developed to account for these phenomena, and the manner in which different predictive models have tried to account for this complexity. We first describe models of the resource allocation process, and then consider the multiplicity of resources.

RESOURCES

Parallel Allocation Models

The emphasis of resource models has been on the loss of information processing quality that results from concurrence, and from diversions in resource allocation, rather than on the forces (such as expectancy and utility) that predict when a sequential shift will take place. Furthermore, in contrast to queuing models, resource allocation models assume that parallel processing between tasks is ongoing, and hence that interference effects result from competition for something more than time. Basically, resource allocation models have taken two generic approaches. One approach has been to model performance on two perceptual (detection or recognition) tasks given equal priority, as a function of such variables as signal strength, signal uncertainty, and signal differences (Taylor, Lindsay, and Forbes, 1967; Swets, 1984; Shaw, 1982). Several examples of this approach have been based upon the theory of signal detection. The empirical data to validate these models have been collected under fairly carefully defined laboratory conditions (near threshold stimuli in constrained display locations), and these factors may constrain their relevance to applied environments.

The second approach has focused on the differential allocation of resources to different channels or tasks, modeling this allocation from the standpoint of economic theory as a utility-based decision problem. Sperling (1984; Sperling and Dosher, 1986) provides an integrative treatment of the factors underlying this modeling approach. This approach has its origins in the assumption that resources are continuously allocatable commodities that improve performance through a function referred to as the performance resource function (Norman and Bobrow, 1975). Performance is seen to improve (or degrade) on the basis of the allocation (or withdrawal) of something other than time. Here again reported data do not extend far beyond simple detection and recognition tasks.

One important quantitative modeling approach to resource allocation however that has applicability to a more diverse set of complex tasks is found in the multitask extension of the Optimal Control Model of manual control (Levison, Elkind, and Ward, 1971; Levison, 1982; see Baron, this workshop). Fundamental to this model is a parameter of "observation noise," which is assumed to perturb the internal representation of analog signals used for tracking and monitoring. Observation noise is typically expressed as a ratio of noise to relevant signal amplitude; that is, as an "observation noise ratio." On the one hand, the causes of change in noise level are incorporated in an attention sharing model by the formula $P_I = P_o/F_i$, in which P_o is the single task observation noise ratio, P_I is the observation noise ratio under multitask conditions, and F_i is the fraction of attention allocated to the task. On the other hand, the effects of changes in this observation noise ratio on tracking error and monitoring performance may be quantitatively predicted within the model (Levison, 1982; Levison and Tanner, 1971).

The quantitative aspects of Levison's approach have been validated (e.g., Stein and Wewerwinke, 1983), and it has been applied to detection performance in discrete decision tasks (Levison and Tanner, 1971) to aviation system display design (Curry, Kleinman, and Levison, 1977; Baron and Levison, 1977), as well as incorporated into the PROCRU model of pilot performance (Zacharias et al., 1981). However, the constraints of the assumptions are clear as well. The observation noise ratio is only applicable to tasks whose inputs are linear spatial quantities (position, velocity), and not qualitative or configurational feature-defined patterns, such as those used in symbolic or verbal processing.

Multiple Resources

Early versions of resource allocation theories assumed that all tasks competed for a single "undifferentiated pool" of resources (Kahneman, 1973). As I have outlined elsewhere (Wickens, 1984, 1987) this assumption cannot easily accommodate certain phenomena described above, in particular the phenomenon of difficulty insensitivity, and the nearly perfect parallel processing of tracking tasks with verbal cognitive activity. Such phenomena instead, may be better accounted for by assuming that human processing resources are multiple (North, 1985; Navon and Gopher, 1979; Tsang and Wickens, 1988; Wickens, 1984, 1987; Wickens and Liu, 1988). Because aspects of the multiple resources model have been incorporated, to different degrees in the existing simulation models, it will be described in some detail.

According to the multiple resource model, two tasks will suffer greater interference to the extent that the component tasks are more difficult (demand more resources) and that the components compete for overlapping resources. Furthermore, the effects of difficulty and resource overlap interact. The greater the degree of resource overlap, the more pronounced will be the effect of difficulty of one task on the level of performance of another task.

These resources are described at a more general level (e.g., spatial-verbal) than are the processing mechanisms of the tasks themselves. The current version of the multiple resource model proposed by Wickens (1984, 1987; Wickens and Liu, 1988) has three dichotomous dimensions each of which defines two resources. These dimensions are processing codes (spatial/analog vs. verbal/linguistic), processing modalities (auditory/speech vs. visual/manual) and processing stages (perceptual/cognitive vs. response). However it is possible, particularly in the environment of vehicle control, that a dimension of ambient vs. focal vision described by Leibowitz and Dichgans (1980), Leibowitz (1986), and Christensen et al. (1985) might well be relevant. These two visual systems are used for the control of egomotion and the recognition of objects, respectively. Validation of the model in basic laboratory experiments has been carried out by a number of studies in our laboratory (e.g., Wickens and Liu, 1988; Wickens, 1980; Wickens and Weingartner, 1985; Tsang and Wickens, 1988). Validation in a more complex aviation simulator environment has been carried out by Wickens, Sandry, and Vidulich (1983), and the resulting model has helped to account for the phenomena of parallel processing, structural alteration effects and difficulty insensitivity described above.

Three limitations of the multiple resource model however make it difficult to move from a qualitative modeling to a quantitative modeling domain. These are all limitations that are inherent in the model's efforts to address interference between heterogeneous tasks; but they are limitations for which potential solutions exist.

(1) The amount of resource overlap between tasks depends upon the careful definition of what is a resource. Wickens' (1984) specification of resources defined by three dichotomous dimensions, allows for some quantification to be accomplished at 4 levels of resolution, according to a "shared dimensions" approach. For example, two tasks may compete for resources on 0, 1, 2, or 3 dimensions. Using this approach, Derrick and Wickens (1984) and Gopher and Braune (1984) have obtained reasonably good predictions of the degree of interference between a collection of heterogeneous tasks.

(2) There is no single metric that can quantify the demand for resources (i.e., task difficulty) which is applicable across different component tasks. Four possibilities avail themselves. First, one can use single task performance differences imposed by a change in demand, to predict dual task interference. Second, one can use some relatively generic task analytic metric such as information rate or working memory

load to quantify demands. Third, one can use subjective ratings, or estimates of single task difficulty levels. Fourth, it is possible to depend upon expert opinion ratings to code demands (i.e., 0, 1 or 2). This is the technique used by Gopher and Braune (1984), and advocated by North (1985), Laughery et al. (1986) in their applications of WINDEX and MICROSAINT models respectively, both relying upon the tabled demand values proposed by McCracken and Aldrich (1984) for coding these demand levels.

(3) There is yet no single metric for scaling the decrement or interference between tasks that may involve different performance measures (see Kantowitz and Weldon, 1985; Wickens and Yeh, 1985 for an informative debate on this point).

COMPUTER SIMULATION MODELS

The series of models described below may be ordered in terms of four levels of complexity regarding the extent to which they treat the concurrent processing or resource-like properties of the human information processing system. At the lowest level are the queuing models already discussed, and these of course make no assumptions about resource sharing.

HOS. The model of the Human Operator Simulator or HOS, developed by Wherry (1976) is currently under revision (Harris et al., 1987), and the revision, which will be available in a user-friendly microcomputer form, explicitly allows parallel processing of activities. Thus, for example, the model will allow the operator to reach while scanning, or to encode while controlling. However, parallel processing is assumed to be perfect processing. There is no mechanism for specifying interaction between tasks. In this sense, the model allows the assumption of either an infinite pool of resources, or single channel operation, depending on what activities are allowed to be parallel rather than serial. The activities that are parallel processed are user-defined, as is a pre-emption mechanism that will terminate a particular activity when one of higher priority is imposed. In addition, the software is designed to be flexible enough so that the user's own model may be substituted.

MICROSAINT-A^{3}I. At the next level of sophistication regarding resource sharing are models developed by Laughery et al. (1986) and by Corker et al. (1986). These are elaborations of earlier versions of the SAINT/Siegal and Wolf model, and the PROCRU model, respectively. Both models make some assumptions about how the interaction between tasks is modified by their structural similarity and demand level, as required by multiple resource theory.

Laughery et al. (1986) have used the programming capabilities of the MICROSAINT language (Wortman et al., 1978) to expand upon previous developments in two important respects: (1) They accommodate demand specifications of tasks (or mental operations) that are not defined only in terms of time. Rather, the model employs a set of tabled demand values for different tasks, ranging from 0 to 7. These values were generated by expert pilots and compiled by McCracken and Aldrich (1984; Aldrich, Szabo, and Bierbaum, 1988). For example, the activities "monitor, scan, survey" have a demand level of 1. "Trace, follow, track" have a demand level of 3. "Read, decipher text, decode" have a demand level of 7. (2) They acknowledge the multiplicity of processing resources by assuming that task demands on certain combinations of channels will interfere, but not on other combinations. Four "channels" are defined: visual, auditory, cognitive, and psychomotor (VACP)

(McCracken and Aldrich, 1984). Within each channel, simultaneous demands are summed, and values of greater than 5 on the visual channel are assumed to exceed a threshold which terminates monitoring to support situational awareness. The A^3I model developed by Corker et al. (1986) makes similar assumptions about the association of tasks and task demands to the 4 channels. An assumption made in this model is that demands of greater than 7 in any channel will lead to a temporary postponement of the last task to be added that caused demand to exceed the threshold.

Although the developments reported by Laughery et al. and by Corker et al. are a marked advancement over previous efforts in their treatment of dual task interaction, they still suffer from a number of limitations. First, the demand level codings of activities within a channel do not appear to acknowledge difficulty variation of tasks within a level. Thus, for example, detecting a change in size (coded demand 2 by the McCracken and Aldrich table), if it is a subtle change in a dynamic environment, could be far more difficult than reading a simple one word message (which is actually coded demand level 7!).

Secondly, the assumption of perfect parallel processing between channels (demand levels do not aggregate across the four channels) appears to be unwarranted. For example, there is clear experimental evidence that auditory and visual tasks interfere, as do perceptual (both auditory and visual) and cognitive ones (Wickens, 1984). But no assumptions are made regarding this sort of interference. The A^3I elaboration of PROCRU by Corker et al. however does model the effects of dividing resources along channels relevant to manual control, using the optimal control theory observation noise ratio described above.

WINDEX. Representing the most sophisticated level of assumptions, North's (1985) WINDEX model has incorporated many of the assumptions of the multiple resources model into a predictive workload index algorithm (see North and Riley, 1989). Applicable to cockpit design modifications, WINDEX enables specification of a task in terms of resource demand levels (rated 1-5) assigned to different channels and/or processing systems (e.g., Window, helmet-mounted-display, CRT, auditory, stick, keypress, speech and cognitive activity). Critical to the operation of WINDEX is a conflict matrix by which concurrent activities in different channels will interfere more or less, depending on their similarity in the multiple resource space. It is this feature that was absent from the Laughery et al. (1986) version of the MICROSAINT simulation, and from the A^3I application developed by Corker et al. (1986). Thus, for example in the WINDEX conflict matrix, large penalties will be assigned to tasks that impose concurrent demands on two visual channels (e.g., Window and helmet-mounted-display). Reduced, but still substantial conflicts may apply to simultaneous use of the window and auditory channel (both involving perceptual encoding); to the speech and key press channel (both involving responses); or to speech output and verbal rehearsal (both involving verbal processing). Minimum penalties would be assigned to concurrent use of the auditory and stick channel, these lying "far apart" in the multiple resource space.

MODEL EVALUATION: REQUIREMENTS AND AN EXPERIMENT

The previous section describes models lying on a continuum regarding their assumptions about human operator performance in multitask environments. A problem from the point of view of the potential model user is that an objective, competitive evaluation of the different models applied to a common data base does not exist (Meister,

1985). Model designer's have applied their own models to sets of data, and not surprisingly often reach the conclusions that their models perform satisfactorily. In order to jointly serve the interests of both science and technology such a model competition is required and is now ongoing within our laboratory (Wickens et al., 1988a, 1988b). Models may be rank ordered in terms of the amount of variance in task interference that is accounted for, and this order may then be compared with (or traded off against) the sophistication of the model assumptions. It is apparent for example that if the increased complexity of the multiple resource assumptions made by WINDEX, only gains the model user a few percentage points in variance accounted for, then simpler single resource models may be sufficient.

To be satisfactory, however, such competitive evaluation must be concerned with two issues: heterogeneity of test conditions, and the objective nature of the evaluation criterion. Regarding the first of these issues, the models should be required to predict performance across a relatively wide range of heterogeneous task conditions. Thus, for example, it would not be sufficient to merely apply the models to three tightly controlled conditions of increasing difficulty. Task combinations that are diverse, in terms of their demand levels and degrees of resource competition should be examined.

The second issue concerns the criterion measure against which models should be evaluated. In this regard, subjective measures such as SWAT or the NASA TLX rating (Vidulich and Tsang, 1987) should be avoided, in favor of performance measures. This is not because subjective measures are not useful in other contexts. But because the ultimate criterion for system design should be satisfactory performance rather than adequate subjective opinion, it is the former that should be used for validation. Of course it is true that across most systems, performance and subjective opinions do (negatively) correlate. That is, system that provide better performance are generally described as subjectively less loading. But as Yeh and Wickens (1988; Wickens and Yeh, 1986) have described in detail, there are some pervasive circumstances in which subjective workload and performance dissociate. Tasks (or task configurations) that are described as subjectively more difficult may provide better, not worse performance than tasks rated as easier. Yeh and Wickens concluded that tasks in the former category (better performance, higher workload) are likely to be those that: (a) display more precise control information (like predictor displays), (b) involve multiple task combinations (rather than difficult single task combinations), and (c) involve separate resources (rather than competition for common resources). What these conclusions suggest is that model evaluations based only on subjective ratings (Laughery et al., 1986; Bateman and Thompson, 1986), may produce systematic distortions, depending upon the sample of tasks used to carry out the validation.

The model evaluation carried out in our laboratory required subjects to fly a helicopter simulation through various phases of simulated low level and nap of the earth flight. They responded to navigational commands and periodically were required to carry out cognitive side tasks involving spatial problem solving and fuel computation, each of varying demand levels. Information for these tasks could be displayed either visually or auditorially. The flight was accomplished by reference to a visual display of the world through which the flight proceeded, presented on an IRIS 2400 display system and an IBM AT, presenting a track-up electronic map and side task information. Subjects exercised flight control with a 2 axis joystick, and responded to the side task via a keyboard (Wickens et al., 1988a).

The heterogeneous collection of task configurations, created by the time-sharing of different phases of flight with different kinds of side tasks and the navigation task, provided a set of data points against which different predicted models could be evaluated. We did not test all of the simulation models in their purest form, but rather, created models with three levels of sophistication regarding the assumed mechanism of task interference: (1) a pure task time-line prediction in which the total number of tasks/unit time was the workload driver, (2) an "undifferentiated capacity," demand model, in which the demands of component tasks, rated on a scale of 1-10, were summed, and (3) a WINDEX-like multiple resource model, which accounted for both demand and resource similarity between competing tasks (see Wickens et al., 1988b for more details). Computational algorithms for each model were implemented on an IBM AT, and the predicted model workload scores for each model, across the various task conditions, were correlated with two obtained measures: subjective workload, and RMS vertical deviation from the flight path--the most stable measure of primary task performance.

These correlations are shown in Table 1, and reveal two very salient findings: (1) subjective ratings are well predicted by the two simpler models, but not by the multiple resource implementation; (2) performance measures in contrast are predicted much better by the more sophisticated multiple resource model than by the simpler models, hence reemphasizing the dissociation between these two measures of workload (subjective ratings and performance) described by Yeh and Wickens (1988).

Table 1

	Model		
	Total Task	Undifferentiated Capacity	Multiple Resources
Correlation with:			
Subjective Ratings	0.55	0.68	0.06
Performance (RMS Vertical Error)	0.16	0.04	0.48

CONCLUSION

It is apparent that the ideally accurate model must contain both mechanisms that account for task selection, and those that concern resource sharing. This model should also be able to offer quantitative predictions of performance levels. However, there is a tradeoff between the degree of quantifiable prediction achieved (and perhaps possible) by models of task interference and the level of environmental complexity and heterogeneity in which those models are suited to operate. Three approaches are possible to extend quantitative prediction to the level of complexity existing in the complex task environment: (1) Attempt to build quantitative elements into a multiple resource model. (2) Attempt to extend the more quantitatively precise models of multichannel detection and recognition (e.g., Shaw, 1982; Sperling and Dosher, 1986) to heterogeneous task performance. (3) Establish how accurately complex performance can be accounted for by serial queuing models with assumptions of single task neglect.

Each approach has its own costs and benefits. The first approach is bound to fall short of precise prediction, because of the complexity and heterogeneity of the task environments that is its goal to predict. Yet it remains clear that the operator in real world multitask environments will often be time-sharing different tasks or mental activities that are heterogeneous in their demand. Furthermore, the preliminary evaluation reported above is encouraging. The second alternative awaits verification, to establish whether for example, the prediction of performance on a detection task when time-shared with a second simultaneous detection task will generalize to instances when the synchrony in timing is less precise, or the concurrent task is of a different qualitative sort (i.e., tracking). The attention allocation feature of the Optimal Control Model offers a good step in this direction. The third alternative already offers a great degree of promise as far as it goes but is simply not designed to handle those aspects of time-shared performance that truly are parallel (e.g., flying while communicating). All three alternatives have merit. What is needed now are more validation data.

ACKNOWLEDGMENT

This research was sponsored by a contract from NASA Ames Research Center, NASA NAG-2-308. Sandra Hart was the technical monitor.

REFERENCES

Aldrich, T.B., Szabo, S., and Bierbaum, C.R., 1988, The development and application of models to predict workload during system design, in: "Human Performance Models," G. MacMillan, ed., NATO AGARD Workshop, Plenum Press.

Allen, W., Clement, W., and Jex, H., 1970, Research on display scanning, sampling and reconstruction using separate main and secondary tasks, (NASA-CR-1569), Systems Technology, Hawthorne, CA.

Baddeley, A., 1986, "Working Memory," Claredon Press, Oxford, UK.

Baron, S., and Levison, W., 1977, Display analysis with the optimal control model of the human operator, Human Factors, 19:437-458.

Bateman, R.P., and Thompson, M.W., 1986, Correlation of predicted workload with actual workload measured using the subjective workload assessment technique, SAE Aerotech 86 Conference Society for Automotive Engineers.

Carbonnell, J.R., 1966, A queuing model for many-instrument visual sampling. IEEE Trans. on Human Factors in Electronics, HFE-7:157-164.

Card, S.K., Moran, T.P., and Newell, A., 1986, The model human processor: An engineering model of human performance, in: "Handbook of Perception and Human Performance," K.R. Boff, L. Kaufman, and J.P. Thomas, eds., John Wiley and Sons, New York.

Christensen, J.M., O'Donnell, R.D., Shingledecker, C.A., Kraft, C.L., and Williamson, G., 1985, Optimization of peripheral vision (USAFSAM Technical Report TR-85-96), Brooks AFB, Texas.

Chu, Y.Y., and Rouse, W.B., 1979, Adaptive allocation of decision making responsibility between human and computer in multitask situations, IEEE Trans. on Sys., Man, and Cyber., SMC-9:769-778.

Chubb, G.P., Stodolsky, N., Fleming, W.D., and Hassoun, J.A., 1987, STALL: A simple model for workload analysis in early system development, in: "Proceedings of the 31st Annual Meeting of the Human Factors Society," Human Factors Society, Santa Monica, CA.

Corker, K., Davis, L., Papazian, B., and Pew, R., 1986, Development of an advanced task analysis methodology and demonstration for Army-NASA aircrew/aircraft integration (Report No. 6124), Bolt Beranek and Newman, Inc., Cambridge, MA.
Curry, R., Kleinman, D., and Levison, W., 1977, A design procedure for control/display systems, Human Factors, 19:421-437.
Derrick, W.L., and Wickens, C.D., 1984, A multiple processing resource explanation of the subjective dimensions of operator workload, University of Illinois Engineering Psychology Laboratory Technical Report (EPL-84-2/ONR-84-1), Department of Psychology, Champaign, IL.
Edwards, W., 1986, Expected utility theory, in: "Handbook of Human Factors," G. Salvendy, ed., John Wiley and Sons, New York, NY.
Edwards, W., Lindman, H., and Phillips, L.D., 1965, Emerging technologies for making decision, in: "New Direction in Psychology II," T.M. Newcomb, ed., Holt, Rinehart, and Winston, New York.
Fitts, P.M., Jones, R.E., and Milton, J.L., 1950, Eye movements of aircraft pilots during instrument landing approaches. Aero. Eng. Rev., 9:1-5.
Gopher, D., and Braune, R., 1984, On the psychophysics of workload: Why bother with subjective measures? Human Factors, 26:519-532.
Harris, R.L., and Christhilf, D.M., 1980, What do pilots see in displays?, in: "Proceedings of the 24th Annual Meeting of the Human Factors Society," G. Corrick, E. Hazeltine, and R. Durst, eds., Human Factors Society, Santa Monica. CA:
Harris, R.M., Iavecchia, H.P., Ross, L.V., and Shaffer, S.C., 1987, Microcomputer human operator simulator (HOS-IV), in: "Proceedings of the 31st Annual Meeting of the Human Factors Society," Human Factor Society, Santa Monica, CA.
Harris, R., and Spady, A., 1985, Visual scanning behavior, in: "Proceedings of NAECON 85," Dayton, OH.
Kahneman, D., 1973, "Attention and effort," Prentice Hall, Englewood Cliffs, NJ.
Kantowitz, B., 1986, Channels and stages in human information processing. J. of Math. Psych., 29.
Kantowitz, B., and Weldon, M., 1985, On scaling performance operating characteristics: Caveat emptor, Human Factors, 27:531-547.
Laughery, R., Drews, C., and Archer, R., 1986, A micro SAINT simulation analyzing operator workload in a future attack helicopter, in: "Proceedings NAECON," IEEE, New York.
Leibowitz, H.W., 1986, Recent advances in our understanding of peripheral vision and some implications, in: "Proceedings of the 30th Annual Meeting of the Human Factors Society," Human Factors Society, Santa Monica, CA.
Leibowitz, H.W., and Dichgans, J., 1980, The ambient visual systems and spatial orientation, in: "Spatial Disorientation in Flight: Current Problems," in: "NATO AGARD Conference Proceedings."
Levison, W.H., 1982, The optimal control model for the human operator: Theory, validation, and applications (AFFTCOTR-82-5), in: "Proceedings, Workshop on Flight Testing to Identify Pilot Workload," Edwards Air Force Base, Air Force Flight Test Center.
Levison, W.H., Elkind, J.I., and Ward, J.L., 1971, Studies of multi-variable manual control systems: A model for task interference (NASA CR-1746), NASA, Washington, DC.
Levison, W.H., and Tanner, R.B., 1971, A control-theory model for human decision making (NASA CR-1953).
McCleod, P., 1977, A dual task response modality effect: Support for multiprocessor models of attention, Quarterly Journal of Experimental Psychology, 29:651-667.
McCracken, J., and Aldrich, T.B., 1984, Analysis of selected LHX mission functions, Technical Note ASI 479-024-84(B), Anacopa Sciences, Inc.

Meister, D., 1985, in: "Behavioral Analysis and Measurement Methods," John Wiley and Sons, New York, NY.
Moray, N., 1986, Monitoring behavior and supervisory control, in: "Handbook of Perception and Performance" (Vol. 2), K. Boff, L. Kaufman, and J. Thomas eds., John Wiley, New York.
Navon, D., and Gopher, D., 1979, On the economy of the human processing system, Psych. Rev., 86:254-255.
Norman, D., 1968, Toward a theory of memory and attention, Psych. Rev., 75:522-536.
Norman, D., and Bobrow, D.J., 1975, On data limited and resources limited processes, Cogn. Psych., 7:44-64.
North, R., 1985, WINDEX: A workload index for interactive crew station evaluation, in: "Proceedings NAECON," IEEE, New York, NY.
North, R., and Riley, V., 1989, This Volume.
Pashler, H., 1984, Evidence against late selection: Stimulus quality effects in previewed displays, J. of Exptl. Psych: Human Perc. and Perf., 10:413.
Rouse, W., 1980, "Systems Engineering Models of Human Machine Interaction," North Holland, New York.
Senders, J.W., 1966, A re-analysis of the pilot eye-movement data, IEEE Trans. on Human Factors in Electronics, HFE-7:103-106.
Senders, J.W., 1983, "Visual Scanning Processes," University of Tilburg Press, Netherlands.
Shaw, M. L., 1982, Attending to multiple sources of information: The integration of information in decision making, Cog. Psych., 14:353-409.
Sheridan, T.B., 1970, On how often the supervisor should sample. IEEE Trans. on Sys., Sci., and Cyber., SSC-6:140-145.
Siegel, A.I., and Wolf, J.J., 1969, in: "Man-machine Simulation Models," John Wiley and Sons, Inc., New York.
Simmons, R., 1979, Methodological considerations of visual workload of helicopter pilots, Human Factors, 21, 353-368.
Sperling, G., 1984, A unified theory of attention and signal detection, in: "Varieties of Attention," (pp. 103-177), R. Parasuraman and R. Davies, eds., Academic Press, Orlando, FL.
Sperling, G., and Dosher, B.A., 1986, Strategy and optimization in human information processing, in: "Handbook of Perception and Performance" (Vol. 2), K. Boff, L. Kaufman, and J. Thomas, eds., John Wiley, New York.
Stein, W., and Wewerwinke, P., 1983, Human display monitoring and failure detection: Control theoretic models and experiments. Automatica, 19:711-718.
Strieb, M., Lane, N., Glenn, F., and Wherry, R.J., Jr., 1981, The human operator simulator: An overview, in: "Manned System Design: Methods, Equipment, and Applications" (pp. 121-152), J. Moraal and K.F. Kraiss, eds., Plenum Press, New York.
Swets, J.A., 1984, Mathematical models of attention, in: "Varieties of Attention," R. Parasuraman and D.R. Davies, eds., Academic Press, New York.
Taylor, M.M., Lindsay, P.H., and Forbes, S.M., 1967, Quantification of shared capacity processing in auditory and visual discrimination, in: "Attention and Performance I," A.F. Sanders, ed., North-Holland Publishing Company, Amsterdam.
Townsend, J., 1974, Issues and models concerning the processing of a finite number of inputs, in: "Human Information Processing," B. Kantowitz, ed., Erlbaum, Hillsdale, NJ.
Townsend, J.T., and Ashby, F.G., 1983, in: "Stochastic Modeling of Elementary Psychological Processes," Cambridge University Press, New York.

Tsang, P.S., and Wickens, C.D., 1988, The structural constraints and strategic control of resource allocation, Human Performance, 1:45-72.
Tulga, M.K., and Sheridan, T.B., 1980, Dynamic decisions and workload in multitask supervisory control, IEEE Trans. on Sys., Man, and Cyber., SMC-10:217-232.
Vidulich, M., 1987, Response modalities and time-sharing, in: "Proceedings of the 30th Annual Meeting of the Human Factors Society" (pp. 337-341), Human Factors Society, Santa Monica, CA.
Vidulich, M., and Tsang, P., 1987, Absolute magnitude estimation and relative judgment approaches to subjective workload assessment, in: "Proceedings of the 31st Annual Meeting of the Human Factors Society" (pp. 1057-161), Human Factors Society, Santa Monica, CA.
Welford, A.T., 1967, Single-channel operation in the brain, in: "Attention and Performance I," A.F. Sanders, ed., North-Holland Publishing Company, Amsterdam.
Wherry, R.J., 1976, The human operator simulator-HOS, in: "Monitoring Behavior and Supervisory Control," T.B. Sheridan and G. Johannsen, ed., Plenum Press, New York and London.
Wickens, C.D., 1980, The structure of attentional resources, in: "Attention and Performance VIII" (pp. 239-257), R. Nickerson, ed., Lawrence Erlbaum, Hillsdale, NJ.
Wickens, C.D., 1984, Multiple resources in attention, in: "Varieties of Attention," R. Parasuraman and R. Davies, eds., Academic Press, New York.
Wickens, C.D., 1986, The effects of control dynamics on performance, in: "Handbook of Perception and Performance Vol. II," (pp 39-1/39-60), K. Boff, L. Kaufman, and J. Thomas, eds., Wiley and Sons, New York.
Wickens, C.D., 1987, Attention, in: "Human Factors in Psychology," P. Hancock, ed., N. Holland Pub. Co., Amsterdam.
Wickens, C.D., Flach, J., Kramer, A., Harwood, K., and Lintern, G., 1988a, TASKILLAN: A validation test bed for complex performance models, in: "Proceedings, 11th Psychology in the Department of Defense Symposium," USAFA, Colorado.
Wickens, C.D., Harwood, K., Segal, L., Tkalcevic, I., and Sherman, B., 1988b, TASKILLAN: A simulation to predict the validity of multiple resource models of aviation workload. in: "Proceedings of the 32nd Annual Meeting of the Human Factors Society," Human Factors Society, Santa Monica, CA.
Wickens, C.D., and Liu, Y., 1988, Codes and input modalities in multiple resources: A success and a qualification. Human Factors, 30.
Wickens, C.D., Sandry, D., and Vidulich, M., 1983, Compatibility and resource competition between modalities of input, output, and central processing, Human Factors, 25:227-248.
Wickens, C.D., and Weingartner, A., 1985, Process control monitoring: The effects of spatial and verbal ability and current task demand, in: "Trends in Ergonomics and Human Factors," R. Eberts and C. Eberts, eds., North Holland Pub. Co.
Wickens, C.D., and Yeh, Y-Y., 1985, POC's and performance decrements: A reply to Kantowitz and Weldon, Human Factors, 27:549-554.
Wickens, C.D., and Yeh, Y-Y, 1986, A multiple resource model of workload prediction and assessment, in: "Proceedings IEEE Conference on Systems, Man, and Cybernetics," Atlanta, GA, November.
Wortman, D.B., Duket, S.D., Seifert, D.J., Hann, R.L., and Chubb, G.P., 1978, Simulation using SAINT: A user-oriented instruction manual (AMRL-TR-77-61), Pritsker and Associates, West Lafayette, IN.
Yeh, Y.Y., and Wickens, C.D., 1988, The dissociation between subjective workload and performance, Human Factors, 30:111-120.
Zacharias, G.L., Baron, S., and Muralidharan, R., 1981, A supervisory control model of the AAA crew, (BBN Report No. 4802), Bolt Beranek and Newman, Inc., Cambridge, MA.

THE HUMAN OPERATOR SIMULATOR (HOS-IV)

Regina Harris, Helene P. Iavecchia and A. O. Dick

Analytics, Inc.
Willow Grove, PA

INTRODUCTION

The continuing evolution of complex technology carries with it a host of problems involving the relationship between humans and systems. Designing a system involves a complex set of decisions balancing a multitude of constraints and criteria. A system designer is generally supplied with a set of mission goals and criteria, major system components, the expected environment, an initial operator/system function allocation, and a set of operator and system procedures. Using this information, the system designer must determine the ability of the operator/ system to accomplish mission goals and to establish if a particular design has the correct balance of system and operator responsibilities. Ideally, this evaluation process would occur in the early states of system design while it is possible to affect the design in a cost effective manner.

For early system design process, a need existed for a simulation tool to assist in determining whether the human operator can operate the system as intended. In the past, major efforts have often been directed towards the system hardware, i.e., Will it function properly? How often will it fail? How reliable is the system? The goal was to construct a tool that would permit system developers to evaluate quickly the performance of a system in a particular mission environment and to assess the ability of the human to use the system effectively. The Human Operator Simulation (HOS) was developed to be a generic, analytical approach to simulation of human-machine system interactions. HOS utilizes a task decomposition process to simulate the cognitive, perceptual, and motor activities of a human operator while the operator is manipulating a specific system within a particular environment.

HOS was originally conceived in 1970 and since has undergone a series of major upgrades resulting in the latest version, HOS-IV. It has been applied to simulate a variety of complex sensor and weapons systems as well as commercial systems. These applications have addressed the dual purpose of providing system design guidance and demonstrating the validity of the HOS approach, models, and software. Based on experience with previous versions of HOS, HOS-IV has been specifically designed to operate in a microcomputer environment and contains numerous improvements. It has been developed using a modular approach with the initial version containing essential baseline elements. Additional elements can be added in subsequent evolutionary steps.

The required inputs to the model are descriptions of the system design, procedures for using the system, human operator characteristics, and a mission scenario. A set of operator micromodels are available to the HOS user to assist in the development of the simulation. These micromodels contain algorithms, based on human experimental literature, which can predict the timing and accuracy of basic human cognitive, perceptual, and psychomotor actions.

An important feature of HOS-IV is that it utilizes rule-based concepts, incorporating current Artificial Intelligence techniques to structure the simulation. Inputs to HOS-IV include a set of rules which activates a set of actions when conditions are appropriate. Defining these rules facilitates a "top-down" approach to simulation since it allows the user to design the simulation flow independent of the implementation of detailed actions and models. It also allows the simulation to mimic reality more closely since operators usually make decisions based on an implicit or explicit set of rules when responding to a particular situation.

Figure 1 presents the components developed for an application simulation. During a typical implementation, such as for a model of a radar operator and crewstation, the HOS analyst determines the allocation of functions between the human operator and the machine shown at the top of the figure. The analyst then describes the

- Environment (e.g., number, location, speed, and bearing of targets);
- Hardware system (e.g., sensors, signal processors, displays, and controls); and
- Operator procedures and tactics for interacting with the system and for accomplishing mission goals.

A key to the HOS approach is the descriptions of the interfaces between the operator, system, and environment, shown in the middle of the figure. In a radar system simulation for example, a hardware-environment interface routine would determine which targets were within the radar detection range at any given temporal snapshot. An operator-environment interface routine determines the effects of heat, cold, drugs, or other stresses on human performance timing and accuracy. The operator-machine interface models establish the time and accuracy of an operator performing such tasks as reading alphanumeric information from displays, manipulating controls, searching for targets in a particular field-of-view, or physically moving objects from one location to another. These interface descriptions are developed by the HOS analyst to the level of detail and scope required for a particular application.

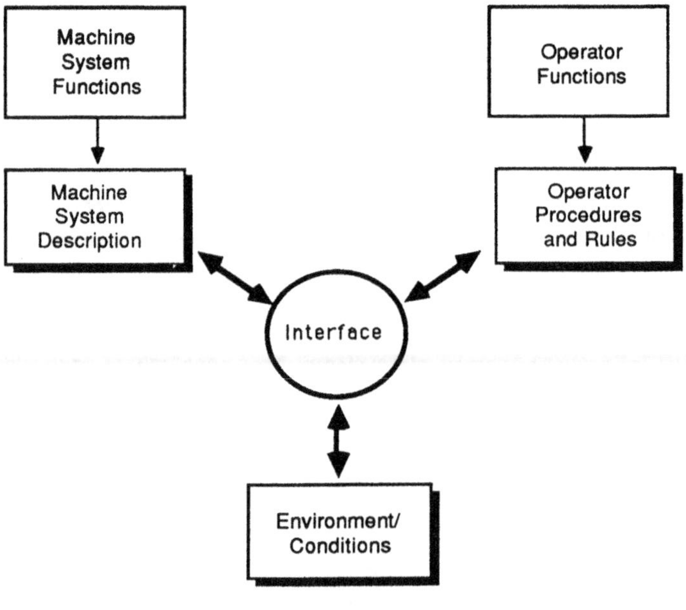

Figure 1. The three major HOS simulation components which are connected through the interface.

The execution of HOS-IV results in a sequence of operator decisions about what to do at each point in time based on moment-to-moment mission events and predefined tactics and procedures. Once the simulated operator decides what procedure to do next, HOS micromodels are invoked to carry out specific detailed actions such as reading information from a display or manipulating a control. The HOS-IV micromodels are based on data derived from the human performance literature and are available to the user through micromodel libraries. The result of a simulation is a detailed timeline of operator, hardware, and environmental events which can be summarized and analyzed to support a broad variety of purposes. HOS-IV has been designed to be a production tool to support use by system designers as well as human factors engineers to evaluate human-system performance.

SIMULATION BASICS

HOS-IV is a highly flexible simulation facility which allows the user to develop a simulation incrementally. The operator characteristics, hardware components, and environmental factors can be developed and tested independently. Furthermore, each component can be simulated at varying levels of detail. For example, the operator characteristics can be simulated simply by charging a fixed time for the operator to respond to a system alert. Alternately, more complicated perceptual and fatigue models such as those available from the set of HOS micromodels can be accessed to determine operator alert processing times and errors.

The user can build a simulation layer by layer, developing components to any level of detail desired. Thus, the user can define any or all of the following components in any order:

- The external world, or environment (aircraft, emitters);
- The system (sensors, processors);
- The operator system interface (displays, controls, procedures); and
- The operator activities (how the operator makes decisions, how the operator accesses necessary information, how the operator executes decisions, etc).

The user also defines events which can affect any of these layers during the simulation. Furthermore, a system failure can be specified to occur at some time during the simulation.

HOS Structure

To define further the characteristics of the environment, hardware, and operator, the information is specified in the following HOS terms:

- Objects,
- Events,
- Rules, and
- Actions.

The relation of these terms is shown in Figure 2. Editors are provided to create and modify the information. For example, the Object Editor is used to modify objects and the characteristics, as well as the values for those characteristics. The analyst might define an object named 'switch,' with the characteristic status, which has the initial value off. Later, during the simulation, the status of the switch might be changed to on.

Objects. The knowledge about the important entities in the simulation and their distinctive features are simulated using an object-characteristic structure. Each entity to be modeled in the simulation (e.g., displays, controls, sensors, aircraft, etc.) is described as an object. Each object has an associated list of characteristics such as size, color, status. This object-characteristic structure provides the user with the capability to define what is important to the simulation at any level of detail.

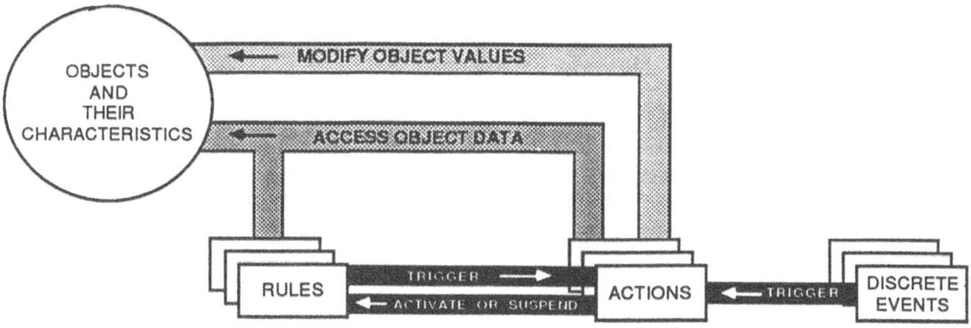

Figure 2. Illustration of the relations among Rules, Actions, Objects, and Events

The list of characteristics associated with each object describes the features of the object relevant to the simulation. Each characteristic is assigned a value which indicates its state at a particular point in the simulation. For example, the status characteristic of the generator object can have a value on or off. Furthermore, objects with the same set of characteristics can be distinguished from each other by the values of their characteristics. Thus, the characteristic, color, differentiates a dial object which is red from a dial which is blue.

Objects can be defined at various levels of detail. For example, the object emergency indicator light can have one characteristic, i.e., status, with a value of on or off. Another object may have several characteristics, such as an enemy emitter with characteristics for location, speed, bearing, frequency, pulse width, pulse repetition frequency, and pulse mode, along with assigned values. The values of the characteristics indirectly control the flow of events in the simulation by providing information to the rules and actions.

The definition of an object to represent a radar is illustrated below:

Object Name:	Radar
Characteristics	Initial Value
Status	off
Mode	automatic
X-location	46.78
Y-location	124.54

The important features of the radar are defined as characteristics and include status, mode of operation, and x-y location in the workspace. The initial values contain the state of the characteristic at the beginning of the simulation. For example, when the simulation starts, the status of the radar would be 'off' and the mode 'automatic.'

A group of objects which are of the same type (i.e., they have the same characteristics) can also be simulated. For example, a group of emitters share the same set of characteristics (location, name, frequency, etc.). If the analyst wished to simulate a group of 50 emitters, it would be very time-consuming to repeatedly enter location, name, etc. for each emitter. To alleviate this problem, HOS-IV allows the definition of groups of objects, called sets. Each object in the set has exactly the same characteristics. However, the value of the characteristics can vary among members of the set.

Events. Events represent external occurrences that affect the simulation. They simulate processes that are not originated by the operator but which have an impact on the course of the simulation. Events are defined by the event name, the time of the event, and the name of the action that is to be triggered at the event time. Typical events are special task messages communicated to the operator from a higher command, hardware system failures, or environmental changes (e.g., weather). For example:

At Time:	00:05:00.000
Event Name:	Electrical storm causes power outage.
Do:	system_failure *(action name)*

This event simulates a system failure at 5 minutes representing a power outage caused by an electrical storm.

Rules. Rules define what actions are to be taken depending on specified conditions. Rules are defined by a starting and ending condition, the name of the action to be invoked if the conditions are appropriate, and a priority assignment. Rules are separated into one of three categories (or queues) – operator, hardware, and environment. Rules can be grouped according to precedence or mission phase by the assignment of a priority. Hardware and environment can contain a maximum of 100 rules each. Rule 99 has the highest priority; 0 has the lowest. 1000 operator rules can be defined and the operator rules are pooled into 10 groups (0 - 9). Each operator group can contain 100 rules. Operator rule 999 has the highest priority; operator rule 001 has the lowest.

Each rule is defined by three clauses:

- An IF clause that specifies a rule initiation criteria,
- A DO clause that specifies the action to be taken if the initiation criteria is true, and
- An UNTIL clause that specifies the rule completion criteria.

An example of a rule is:

IF color OF alert_light EQUALS yellow

DO process_yellow_alert

UNTIL color OF alert_light NOT_EQUAL_TO yellow

This rule will be invoked whenever the color of the alert light is yellow. When the rule is invoked, the action process_yellow_alert will be performed. Thus, if the starting conditional statement is true, the action named in the DO clause will be executed if the rule is active. The action will be executed at each simulation time interval until the UNTIL condition is true or the rule is suspended.

Rules, or groups of rules, can be declared active or inactive by actions at any point during the simulation. At each simulation time unit, the condition statements of all active rules will be evaluated to determine what actions will be executed or terminated. The rules will be processed in the sequence specified by the priority assignment. This precedence relationship is particularly useful as actions invoked by a high priority rule may suspend lower priority rules. This capability permits rules to be categorized by mission critical tasks as well as mission phase. All active rules represent the processes that are occurring in parallel during a simulation time interval.

Actions. Actions describe the steps required to accomplish a process by the operator, system, or environment. An action can be invoked in one of the following ways:

- By a DO clause in a rule,
- By another action, or
- By an event.

Actions update the values of object characteristics, invoke other actions, and activate or suspend rules. The values of the characteristics of objects can only be changed by an action.

Actions are defined using a small set of standard verbs (e.g., DO, SET, SUSPEND) known as the HOS Action Language (HAL). Verbs are the 'primitive' building blocks with which the user constructs complex and application specific actions. The list includes verbs for Information modification, object, rule, and action manipulation, computation, conditional evaluation, and termination.

HOS-IV also contains provision for including models written using the C language (Microsoft C Version 4.0). These models can be directly integrated into the HOS actions.

Micro-Models

HOS includes operator micromodels which generate performance time and an indication of success/failure when appropriate. The current version of HOS-IV contains cognitive (decision time, memory), perceptual (visual and auditory), and psychomotor (anatomy movement) models which are based on experimental data from the human performance literature. The HOS analyst is able to tailor the models to the needs of a particular application by modifying key parameters of the micromodel or incorporating a new model. All models included in HOS-IV are written utilizing the same HAL language as used to define simulation actions.

HOS Outputs

The HOS-IV outputs include:

- A timeline of events for the operator, system, and environment,
- User-defined measures of effectiveness, and
- Standard analysis, such as:
  - Mean time to complete an action,
  - Number of times an action is performed,
  - Proportion of the operator's time spent on each action.

HOS-IV allows for user-controlled detail of simulation outputs. When building the object library, the user can identify objects whose values are to be tracked throughout the simulation. User-defined measures of effectiveness, such as the number of contacts processed per hour or a history of error rates, can also be defined and stored in a separate simulation file at simulation end. With this data, the user can identify operator and system bottlenecks and determine periods of operator overload and the circumstances surrounding those overload periods.

Development of HOS-IV was supported by the Army Research Institute and the USAF Human Resources Laboratory, Wright-Patterson AFB, Contract F33615-86-C-0019

The views, opinions, and/or findings contained in this report are those of the authors and should not be construed as an Official Department of the Army position, policy or decision.

CREW PERFORMANCE MODELS

INTRODUCTION: CREW PERFORMANCE MODELS

Michael H. Strub

U.S. Army Research Institute Field Unit
Fort Bliss, TX
USA

OVERVIEW OF THE PAPERS

When one considers the complexity of some of the single task models and the increased complexity of multiple task models, one might wonder how it would be possible to capture all of the richness of the behavioural complexity displayed by a team of operators. It seems clear that as we move from a single task to a multi-operator level of description, we sacrifice behavioural purity, in terms of a perfect human performance description, in favor of capturing the criteria of interest to the system designer. Thus, the how of performance is secondary to productivity issues involving quantity, quality or accuracy, and speed. The concern centers around the ability of the model to capture key input variables, conduct a valid processing function, and deliver a result that is meaningful to the system designer.

The three reports which follow represent three very different approaches toward predicting crew performance. The first paper, "Use of Crew Performance Models and the Development of Complementary Mathematical Techniques to Promote Efficient Use of Analytic Resources", by W. Peter Cherry and Susan M. Evans, describes the application of Jackson Network and Markov Renewal Theories to develop an analytic rather than simulation tool for conducting crew workload analyses. The mathematical techniques are described, examples of task scenarios are provided, and comparisons made to simulation results. The second paper, "Systematic Modeling of Multioperator Systems to Evaluate Design Concepts", by Gerald P. Chubb and Constance M. Hoyland, points out that success in the use of modeling tools is a function both of the quality of the tool and the skill of the user. The paper presents a concept called IDEAL (Integrated Design Evaluation and Analysis Language). IDEAL combines two previously developed tools. The first is SADT (Structured Analysis and Design Technique), and is proprietary. The second tool is SAINT (Systems Analysis of Integrated Networks of Tasks). The paper describes how IDEAL was used to develop a surface to air missile model. The third paper is entitled, "METACREW: Simulating Operator Teams in a Sensor Data Processing System", by Thomas Plocher. It describes a computer simulation of operator performance and work flow in a specific system, the Joint Army-Air Force Surveillance Target Attack Radar System (Joint STARS). Team and system performance are described in terms of a variety of information throughout measures.

DISCUSSION

The discussion period which followed presentation of the above papers highlighted two areas for further development. The first might be called the addition of a cohesion factor. It was pointed out that very often teams will show very marked performance differences based on their ability to work together as a team. The present state of the art in crew models does not address team cohesiveness at a level adequate to describe demonstrated effects due to experience in working together as a team. The second area identified as needing further development is the ability to represent the team in terms of the relevant command and control hierarchy. In reality, many crew and team operations are conducted in an environment which either requires coordination with higher levels of command or interruptions occasioned by sudden requirements from higher headquarters. For the most part, the current crew models tend to focus on a team or crew responding only to the requirements which flow from the mission they have been assigned. It is possible that a merger of recent expert system capabilities with the crew performance models themselves might be useful in addressing the supervisory and management aspects of crew performance.

USE OF CREW PERFORMANCE MODELS AND THE DEVELOPMENT OF COMPLEMENTARY MATHEMATICAL TECHNIQUES TO PROMOTE EFFICIENT USE OF ANALYTIC RESOURCES

W. Peter Cherry and Susan M. Evans

Vector Research, Incorporated
P.O. Box 1506
Ann Arbor, Michigan 48106

INTRODUCTION

Operator workload analyses relative to integrated system performance rely, in many cases, on detailed, high-resolution, Monte Carlo simulations of task networks. This paper describes the application of Jackson Network and Markov Renewal Theories to develop an analytic, rather than simulation, tool for such analyses. Key advantages of such an approach are faster predictions and greater control over causal factors influencing performance and workload. The mathematical techniques are described, examples of task scenarios are provided, and comparisons made to simulation results. Use in a design environment, specifically, to evaluate the impact of system and task parameters on critical crew workload factors (e.g., task completion time, average and maximum operator loading, etc.), is discussed.

Considerable attention is currently being focused by the US Army on the need to design systems which accommodate the soldiers that will operate and maintain them. As a consequence, increased emphasis has been placed on modeling the human operator during the earliest stages of the weapon system design process and on making design tradeoffs which specifically address soldier attributes, hardware attributes, and software attributes in the context of contribution to force effectiveness. This paper describes such an effort carried out during the functional system design of the All Source Analysis System (ASAS), an intelligence system designed to support the collection management, processing, and dissemination of intelligence at division, corps, and echelons above corps.

Background

During the mid 1970's, the US Army undertook the development of a system to support the production of intelligence at division, corps, and echelons above corps. The development responded to a requirement for faster and better intelligence in the context of modern warfare and to the introduction of sensor systems that would produce volumes of data orders of magnitude greater than those hitherto available. The system, ASAS, was to exploit automated data processing and artificial intelligence to improve product timeliness and quality, to accommodate increased input, and to reduce manpower requirements. The system was to support five major functions: Intelligence Development, Target Development, Electronic Warfare (EW) Support, Operations Security (OPSEC) Support, and Collection Management and Sensor Command and Control. As originally configured, the system had 15 workstations, each with multiple operators. At the division level, input was anticipated to be 6000 reports per hour of various types. To illustrate the complexity, the situation analysis workstation had five types of input and seven types of output. It utilized (read from and/or wrote to) 19 different data files and performed 52 different tasks depending upon the data or information input to the workstation.

The overall approach adopted to analyze the impact of human operator performance was to focus on the accuracy and timeliness of critical products by tracking inputs to those products to the ASAS through a network of operators. Input messages, derived from a large-scale combat simulation, originated from collection assets, taskers or users, and were characterized by type, content, and timing. Operator task performance was described via distributions of time and accuracy for tasks that fell into five categories: interpretation, processing, updating, querying, and routing. The critical products, characterized by distributions of timeliness, completeness, and accuracy, included estimates, queries, taskings, targets, and warnings. The alternative system designs were differentiated primarily by the degree of automated support provided to the operators, ranging from none to a data base management system to comprehensive and extensive use of Artificial Intelligence and other types of decision support.

Human Operator Performance in ASAS

At the earliest stages of the design process, attention was focused principally on task performance time and accuracy, and consideration of other dimensions of cognitive workload was minimal. Operators were presumed to respond sequentially to the stimuli/data arriving at their respective duty positions. At the time, great hopes were held out for the use of artificial intelligence and decision support techniques as tools which would reduce cognitive workload. As work on the design progressed, in particular in the decision support area, it became clear that the key issues were pattern analysis, numerical computation and long-term memory. These issues were added to models of human performance and are discussed below.

METHODOLOGY

To represent human operator performance in the ASAS, a methodology later referred to as the ASAS Performance Assessment Model (ASAS PAM) was developed (IBM and VRI, 1980). It consisted of two distinct components. The first of these, the Operator Workstation Model, was a data driven Monte Carlo Simulation of the functions and processes carried out within operator work stations. It had resolution to individual messages, tasks, and products. The second component, the ASAS System Model, was an analytical model of the development and contents of the ASAS data base, specifically a probabilistic description of the current perception of entities, events, and activities on the battlefield and within the system itself derived as a function of input, system performance, and operator performance.

Before discussing the two components we first describe an element that is common to both, namely the representation of pattern recognition or templating.

Templating

Central to much of what ASAS was to accomplish was the concept of templating. In essence, a template is nothing more than a pattern or frame which specifies the different attributes of an entity or activity. Given that an appropriate combination of those attributes have been associated, the template can be said to be filled and the appropriate inference (detection, classification, location, identification, etc.) drawn. Figure 1 illustrates the use of a template in a broad context which includes selecting and rejecting different templates, associating new

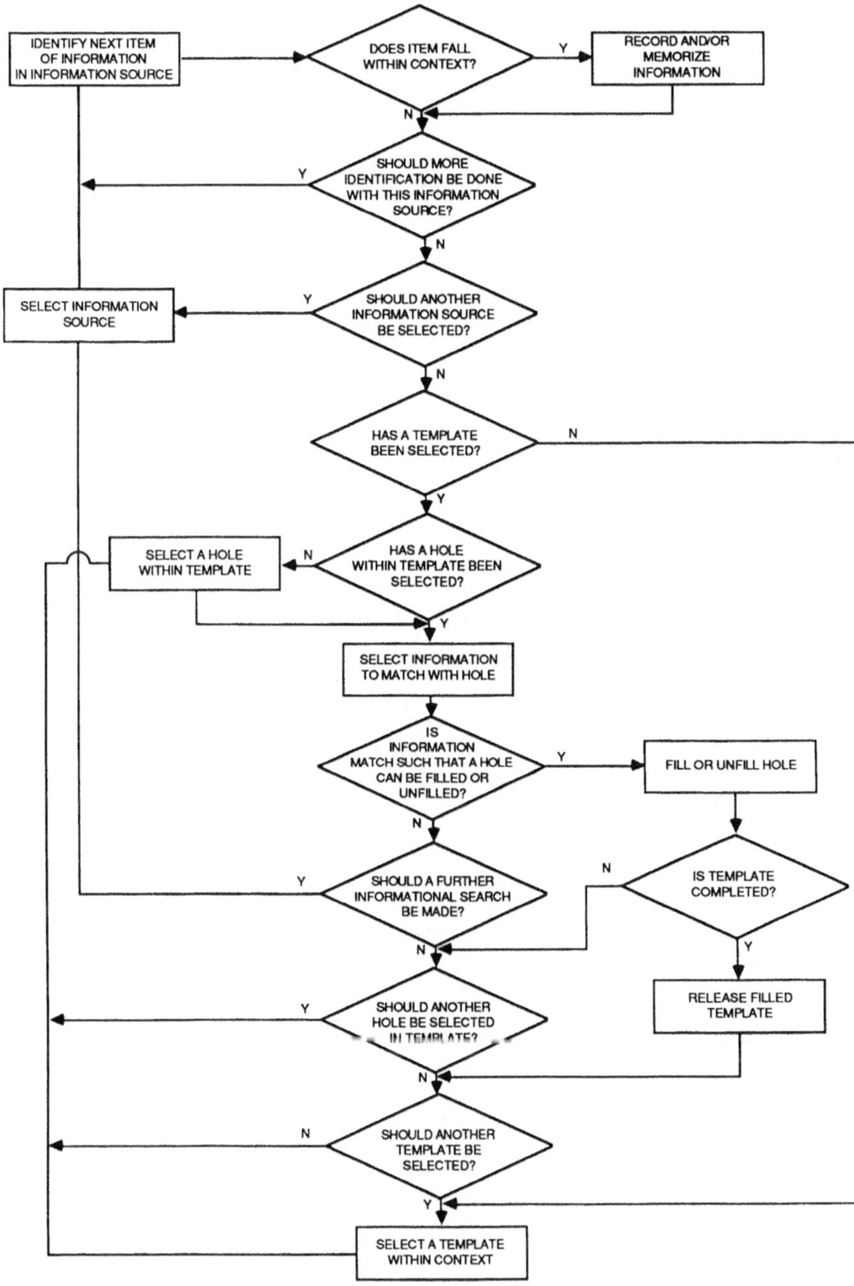

Fig. 1. The Templating Process

information with and/or disassociating old information from templates which are partially filled. Templates were represented by Bayesian likelihood functions. A set of alternatives $A_1,...A_n$ were postulated for each pattern recognition task, together with a set of prior probabilities $Pr[A_1],...Pr[A_n]$, that the alternative in question was producing the data forming the pattern. As a set $\underline{S}$ of pertinent data was associated with the pattern, a posterior probability:

$$Pr\,[A_i|_{\underline{S}}] = \frac{Pr\,[\underline{S}|_{A_i}] \cdot Pr[A_i]}{\sum_k Pr[\underline{S}|_{A_k}] \cdot Pr[A_k]}$$

was calculated as a measure of the degree to which pattern recognition was complete or the template was "filled." In a theoretical sense, this model of the process is not unreasonable (and it proved to be a useful analysis and design tool). (However, in an applied sense, particularly for complex systems, it is quickly overcome by the combinatorics of the possible elements of the set $\underline{S}$, the applicable data.) In reality, this set must include not only what has been reported and its reliability, but also what has not been reported and the associated conditions under which data were developed.

The Operator Workstation Model

After considering the commercially available Monte Carlo simulation packages, the decision was made to develop a simulation specifically for the ASAS application. Developing and debugging the simulation was not difficult. However, as was expected, its initial use was not straightforward. First, generating task performance data was difficult. Although the simulation was designed to accommodate a wide range of distributions, it was hard to find expectations and variances, let alone full distributional specifications. Second, also not unexpected, the system description was vague. Third, the specification of frames and templates was speculative. Nonetheless, the first runs of the simulation on the preliminary design concepts proved to be of high value. In particular, they demonstrated that the system design had not considered the load placed on the human operators or the capacity of those operators to respond. For example, early designs called for less than 30 operators to meet system timing, quality, and quantity requirements. Runs of the simulation indicated that at least 60 would be needed. Moreover, the distributions used to characterize task performance were optimistic. As a result of these early operator simulations, design emphasis focused on decision support applications and data base management to maintain performance under the system requirement of reduced manpower.

While the Monte Carlo simulation proved to be useful in the design process, it was not as responsive as desired. The simulation was executed on an AMDAHL 470/V8 mainframe and typically required approximately 4 CPU seconds to execute sufficient replications to provide a statistically valid description of the performance of a design alternative for a workstation. It consisted of 3500 lines of FORTRAN code and required approximately 55,000 words of memory. As a tool to identify sets of tasks or operators which were the source of performance deficiencies, it was difficult to interpret and not as timely as was desired.

Partly as a preprocessor to filter designs prior to executing the simulation, the authors utilized Jackson's theorem to develop an approximate model of the system. Jackson's theorem applies to a network of N nodes. At each node, there are M_i servers whose service times are independent and identically negative exponentially distributed random variables with service rate μ_i. Customers from outside the network arrive at node i according to a Poisson process with parameter γ_i. A customer completing service at node i goes to node j with probability r_{ij}. Defining

$$\lambda_i = \gamma_i + \sum_{j=1}^{N} \lambda_j r_{ji} \quad ,$$

k_i = number of customers at node i,

and requiring that

$$\lambda_i < M_i \mu_i \quad ,$$

Jackson's theorem (Jackson, 1957) states that

$$\Pr\,[k_1, k_2, \ldots k_N] = \prod_{\ell=1}^{N} P_\ell\,(k_\ell) \quad ,$$

where $P_\ell\,(k_\ell)$ is the steady state probability for an $M|M|M_i|\infty$ queuing system.

The model as implemented treated operators as servers and focused on the workload presented to them by data, queries, etc., arriving from outside or inside the system. It proved, perhaps because of the numbers of servers and the relatively high input intensity, to be a reasonable model in the sense that it approximated closely the equilibrium results of the high resolution simulation. It proved particularly useful in balancing the load throughout the system by allocating operators to different workstations or by decision support concepts to the work stations.

Application of Jackson's theorem to systems such as that considered in the ASAS design process is not unusual and frequently subject to criticism. Assumptions of independence and Markovian processes are required and these most certainly do not accurately reflect human operator performance. Furthermore, the theorem presents its results in terms of a steady state or equilibrium and as noted by one of the authors, (Cherry, 1977), a military system must be designed to meet the requirements of transient situations. Nonetheless, the Markovian/Jackson approach, implemented on a handheld calculator, led to system designs which were changed after evaluation in the high resolution Monte Carlo simulation in only approximately five percent of the cases evaluated. The authors have similar experience in the case of other weapon systems in which the price of meeting timeliness requirements is the provision of extra and, over the long run, under utilized system capacity. It appears that designers frequently overlook this fact, particularly in the case of human task performance, and that a simple model, such as a Jackson network, is sufficient to identify such problems and correct the situation in the early stages of design.

The ASAS System Model

The ASAS System Model was designed to provide a probabilistic, time dependent description of the contents of the ASAS data base files and records. The system model is analytic in nature and focuses on the status of the perception available in the system of events, activities and entities on the battlefield. Figure 2 illustrates the basic concept underlying the model which is combined with the Bayesian model of the templating process to provide the system based description.

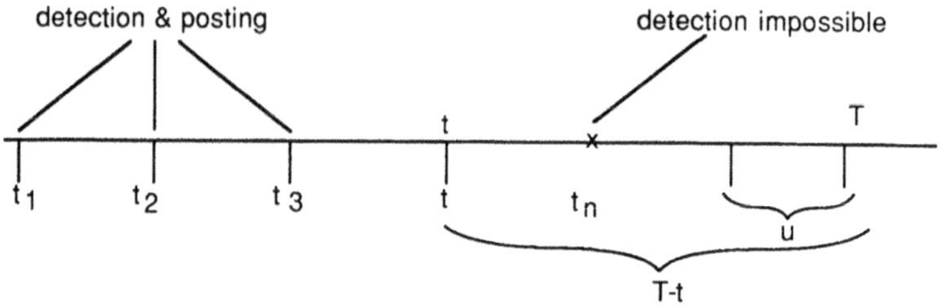

P_c = Probability of Coverage

$F(T\text{-}t)$ = Probability that communications delay plus processing delay exceeds T-t

$f(u)du$ = Probability that communications delay plus processing delay is between u and u+du

$\lambda(t)$ = Rate at which signatures occur at time t

$P_{d|c}$ = Probability of detection given coverage

Probability that currently exceeds T-t

$$= (1-P_c) + P_c \left\{ F(T-t) + \int_0^{T-t} \exp\left(-\int_t^{T-u} \lambda(x) P_{d|c}\, dx\right) f(u)\, du \right\}$$

Fig. 2. File Contents Methodology

The methodology used to represent the changes in the contents of a file or record corresponding to an element, event, or activity on the battlefield is based upon viewing the updates to the file as a random point process in time. The methodology will be described for a single sensor; the extension to multiple sensors is made by utilizing the results produced for each sensor capable of detecting the element, event, or activity. Consider the time line contained in Figure 2. Suppose that at times t_1, t_2, ... t_n, signatures have been produced which could have been detected by the sensor under consideration. Suppose that at time T the record is examined and the following questions are asked:

(1) What is the age of the last report from this sensor that has been posted to the record? and

(2) How many reports were posted prior to the posting of the last?

Addressing the first question, suppose it is rephrased as:

(1) What is the probability that the age of the most current report is greater than T-t?

The age will be greater than T-t if:

(1) the sensor does not have coverage;
(2) the sensor does have coverage, but:

(a) the delay comprised of communications and ASAS processing for this sensor and this report is such that the age of information that could be posted at time T exceeds T-t (this usage could be infinite or reflect errors or lost information); or

(b) the age of information that could be posted at time T is u, less than T-t, but in the interval t to T-u either no signatures occurred or if signatures occurred they were not detected.

Under the assumption that:

(a) reports from a single sensor do not "leapfrog" in communications and processing; and
(b) signature occurrence can be described by a non-time-homogeneous Poisson process.

The formula shown at the bottom of the figure is an analytic solution for the probability that currency, (i.e., the age of the most recent information), exceeds T-t. This formula can be used to calculate the probability that currency lies in any specified interval. The leapfrog assumptions apply to the currency of postings to the file. All reports prior to the most recently posted are considered relative to that posting event. Conditional on that event, the probability that a specified number of reports were posted prior to the most current is calculated via standard formula for a non-time-homogenous Poisson process.

The structure described above possess a number of advantages. First, it clearly separates the occurrence of signatures, sensor performance, and communications performance from the performance of the ASAS. Second, ASAS performance is represented in terms of a probability distribution that is specific to individual messages, which can be derived from the Operator Workstation Model. The level of resolution in terms of signatures and report characteristics is variable according to user requirements. Finally, the use of the "age" of information at time T provides a natural means of properly representing the impact of communications and ASAS processing in a combat model by looking back, for example, with knowledge of Electronic Counter Measures (ECM) or combat damage, rather than predicting forward.

Strictly speaking, the system model does not explicitly model human operator performance; instead it captures that performance through the distribution of processing time and probabilistic treatment of accuracy. The system model has been used, however to investigate the impact of operator performance on the system products, including performance parameters related to long-term memory, pattern analysis, and numerical computation. Figure 3 illustrates the impact of operator errors and imperfect memory for an example in which there are two templates under consideration, one of which (Template 2) is incorrect. For one component of Template 1 there is a probability of 0.2 that the analyst will reject an element of data that in fact should be associated with the template. For another component there is a probability of 0.10 that an element of data that should be rejected will be accepted. (These specific errors are typically associated with numerical computation in the real situation from which the example templates were derived.)

Memory for one component of the templates is imperfect with recall dependent upon the amount of data present at last consideration and the time elapsed since that consideration. That component serves as an exclusion threshold for Template 2, i.e., once the component is filled the fact that Template 2 is incorrect becomes apparent. Memory is perfect once the exclusion threshold is crossed. The input stream is a non-homogenous Poisson process corresponding to sensor reports. As illustrated, early in the process it is more likely that an operator, if called upon, would incorrectly perceive that the second template was filled. As time progresses, the likelihood that the exclusion threshold is reached increases, the probability that Template 1 is filled increases and the probability of incorrectly filling Template 2 decreases. However, it should be noted that correction of the inappropriate use of Template 2 is delayed because of memory constraints relative to the exclusion threshold.

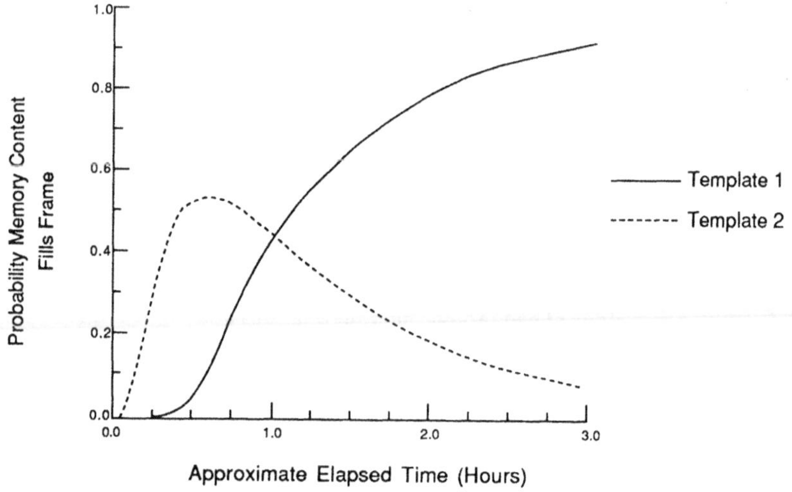

Fig. 3. Interaction of Memory Constraints and Exclusion Procedures With Performance Errors

In addition to investigating the impact of operator performance on the quality and quantity of system output, the system model was used to evaluate a series of measures of system performance including, for example, time to respond to a query, time to locate and identify second echelon regiments and divisions, time to classify a target, target location accuracy, etc. These measures of system performance, input to and used in a force-on-force combat model permit analysis of the impact of operator performance on force effectiveness and the evaluation of different measures to enhance and/or support human task performance in the context of force effectiveness.

DISCUSSION

The approach taken by the authors to modeling human operator performance corresponds closely to that of classical industrial engineering and methods-time measurement (MTM). Monte Carlo simulation is extremely compatible with this approach but does not in general lend itself to responsive analysis or extensive parametric variation. As a substitute for simulation the authors have used stochastic processes; the Jackson Network approach outlined in this paper is an example. We have also made extensive use of Markov Renewal processes defined on finite state space. In modeling operator performance, states correspond to activities or tasks which an operator can be observed to be performing and occupancy times are random variables with distributions depending upon the state occupied and the state to be entered next. Transitions among states also reflect accuracy and completeness of task performance. To our knowledge, this approach was first used by Bonder and Farrell, (1970), to develop weapon and situation based attrition rates for use in differential models of force on force combat. These rates, widely used in deterministic models of combat, explicitly represent parameters of human operator performance for tank crews, anti-tank guided missile teams, machine gunners, riflemen, etc. The approach has also been applied to sensor operators as described in Meerschaert and Cherry (1988). In supporting a variety of system design processes, we have found these analytic models of human operator performance to be very useful, particularly during the earliest stages of the design processes.

The state-task approach using Markov Renewal processes has not been as useful in supporting the design of systems such as ASAS where multiple functions and large numbers of operators are involved. However, as described in this paper, focusing on product quality is analogous to the rate based approach, and the use of the Operator Workstation Model to develop distributions related to operator performance yields a technique by which that performance can be represented in large scale force-on-force combat models. This approach centers on relating the quality of soldier/machine system output or product to operator performance and then implicitly representing operator performance and its dynamics in the force-on-force model. In a balanced corps level model where the number of "operators" on both sides exceeds 100,000 this is clearly necessary.

Representing the impact of conditions and environment on the human operator requires that functions or tables be used within the force-on-force simulation. For example, in the case of ASAS, sensor rates and coverage depend upon deployment and attrition and communications distributions depend upon ECM and load as well as the physical status of communications hardware. These parameters are maintained (and change) in the force-on-force models. Distributions of processing delays, corresponding to events in the templating process described previously, depend upon and can be represented as functions of these parameters with the representations derived via preprocessing using models such as the Operator Workstation Model. In the case of the rate based submodels, less preprocessing is required but relationships between conditions and operator performance must be available as data within the larger model.

A note of caution should be sounded with regard to the impact of including models of human operator performance in the large-scale combat simulations. The extent to which system operator performance influences the dynamics and outcomes of combat depends to a large extent on how the system is deployed and how small units (platoons, teams) are led. If the systems are not in the right place at the right time, human operator performance will have little if any impact on overall results. It has been our experience in simulating performance that this issue is rarely considered. It has been our experience in analyzing contribution to combat that it is a dominant factor.

REFERENCES

Bonder, S., and Miller, G., 1982, "Human Factors Representations for Combat Models," Vector Research, Incorporated, Ann Arbor, Michigan.

Bonder, S., and Farrell, R. (editors), 1970, "Development of Analytical Models of Battalion Task Force Activities," Systems Research Laboratory, The University of Michigan, Ann Arbor, Michigan.

Cherry, W.P., Alden, J., and Witus, G., 1981, "Parametric Analysis Human Performance in Templating," Vector Research, Incorporated, Ann Arbor, Michigan.

Cherry, W.P., 1977, "Analytic Methods for Studies of Command and Control in Anti-Ship Cruise Missile Defense," Vector Research, Incorporated, Ann Arbor, Michigan.

Jackson, J.R., 1957, "Networks of Waiting Lines," Operations Research, Volume 5, pp. 518-521

Meerschaert, Mark M., and Cherry, W. Peter, 1988, "Modeling the Performance of a Scanning Radio Communications Sensor," Naval Research Logistics, Volume 35, John Wiley & Sons, New York, pp 307-314.

IBM and VRI, 1980, "Users Manual ASAS PAM, All Source Analysis System Functional System Description (ASAS FSD)," International Business Machines Corporation and Vector Research, Incorporated, Ann Arbor, Michigan.

SYSTEMATIC MODELING OF MULTIOPERATOR SYSTEMS TO EVALUATE DESIGN CONCEPTS

Gerald P. Chubb and
Constance M. Hoyland

SofTech, Incorporated

INTRODUCTION

Model development requires use of modeling tools. The use of any tool depends on skill in application. The product therefore depends on at least two factors, the quality of the tool and the skill of its user. When multiple users are needed to put the tools to work producing a final model, there is a further need to organize the model development and implementation process to facilitate communication, assure proper coordination of effort, and provide a basis for controlling and documenting what is done. This is especially important for large, multidisciplinary efforts.

The purpose of this paper is to illustrate an approach to model development, implementation, and validation that has proven effective in a series of studies. The concept itself has been termed IDEAL (Integrated Design Evaluation and Analysis Language) by its sponsors, the Armstrong Aerospace Medical Research Laboratory. IDEAL is based on the combined use of two previously developed tools. The first is SofTech's Structured Analysis and Design Technique (SADT(TM)) also known as $IDEF_0$ (ICAM Definition language). $IDEF_0$ is a non-proprietary version of SADT provided to the Air Force as part of the Integrated Computer Aided Manufacturing (ICAM) program. SADT and $IDEF_0$ are therefore virtual equivalents (though not strictly identical). The second tool used in IDEAL is the Air Force developed Systems Analysis of Integrated Networks of Tasks (SAINT). IDEAL provides a systematic method for developing a static description of system functions before building a dynamic model of system behavior. This approach is an integral part of what Wallace, Stockenberg, and Charette refer to as Unified System Development Methodology.[1]

This introduction will review the historical evolution of the IDEAL concept. It will also provide additional references to each of the component parts (SADT and SAINT). The application of the concept will be discussed in the context of an Air Defense missile system,

(TM) SADT is a Trademark of SofTech, Inc.

although the entirety of that model will not be presented because of its scope and detail. The progressive implementation, validation, and use of that model will be discussed.

AN INTEGRATED APPROACH FOR MODEL DEVELOPMENT

The IDEAL methodology uses SADT as a front-end analysis tool for deciding what makes up the system and how that might itself be modeled. Then the SAINT model can be built on a firm foundation.

SADT or $IDEF_0$: Functional Decomposition

SADT began as a software requirements engineering tool[2]. It was subsequently extended to other applications[2] where a top-down, hierarchical decomposition was needed in order to describe a system, particularly when the intent was to examine and describe how that system does or should function. The principal goal of SADT or $IDEF_0$ is to provide a structured approach for breaking a complex system down into more elemental components that are simpler to deal with. This is the basis for contemporary system engineering practices where a large, multidisciplinary team must be employed to design, develop, and produce the product.

The original Greek root for our word analysis is λύω. The meaning of that Greek word is enlightening. While it means to loose or free, it also may mean to break or destroy. The Greeks apparently recognized that breaking something down can destroy the nature of the whole. This is well recognized in biology where studies of living organisms done in vivo may provide different results than laboratory studies of specific organs done in vitro. A similar problem was noted by the Gestalt psychologists studying perception when they recognized that the whole is something more than the aggregate of its parts. One must also account for structure and relationships. SADT is a disciplined analysis method that keeps track of functional structure and data relationships. SADT has proven especially useful when a team of analysts must be used to get a complete, accurate description of a large, complex system. It has been used to describe hardware, software, and human functions in the context of system operation.

SAINT: Dynamic Behavioral Modeling

A static description of the system's functional structure does not convey the evolution of performance over time. That takes a model capable of representing the dynamic aspects of real-time system behavior. Those behavioral dynamics capture the sequential dependencies among groups of tasks, the uncertainties of concurrent activity sequences, variations in activity duration, conflicts in resources demands, situation specific rule implementation, and a variety of other interactions between operators, tasks, the equipment, and the environment. SAINT was developed to provide a tool for analyzing the behavior of large, complex systems.[3] The general character of SAINT and some simple models were presented at a prior NATO symposium.[4] Early applications of SAINT were reviewed by Seifert and Chubb,[3] and more recent applications were subsequently reviewed by Chubb.[5] Additional examples will be briefly discussed later.

SAINT provides a general-purpose, FORTRAN-based, simulation language within a network-oriented framework. It can be used to exercise Petri-net representations quite easily, but is not limited to them. It has often been used to represent and assess the implications of a task analysis, but not every use of SAINT has to begin with task analysis. Nor does every use have to begin with SADT. Chubb, Stodolski, Fleming and Hassoun recently used SAINT to perform sensitivity analyses of a closed-form analytic model of pilot workload.[6]

This flexibility in the use of SAINT is both a strength and a weakness. As a strength, SAINT's flexibility makes it generally useful to a very broad class of systems modeling problems. The wide variety of SAINT applications to date aptly attests to this utility. As a weakness, SAINT's flexibility does not guide the modeler into any systematic definition of the problem or its representation. Often, the biggest difficulty one faces in solving system design problems is to define the nature of the problem itself and to decide how that problem can be resolved by studying a suitably constructed representation of the system's dynamic behavior. Typically, this requires an identification of the performance requirements the system must meet. Then an analysis of the behavioral model is performed to discern whether a particular design can be expected to meet or exceed those requirements. If not, the design may need to be changed, the requirements relaxed, or both.

IDEAL: Combining SADT and SAINT

Backert, Evers, and Santucci first described the need for a front-end analysis tool as an aid for SAINT model development.[7] They identified the close correspondence between concepts used in SADT and similar constructs in SAINT. They also noted where there were gaps in the transition from SADT to SAINT. These gaps are bridged by building a performance data base that captures information needed for implementing and executing a SAINT model. The performance data base contents are based upon the functions and data relationships identified in the SADT static, structural model of the system, but augment these data with descriptions of how activity duration will be specified, how activity sequencing will be controlled, and what may affect the values assigned to various attributes incorporated into the model. These attributes typically describe characteristics of the system, resources, tasks, and information. Activity completion often affects such variables, changing their value.

Bachert, Evers, and Santucci also describe the modeling and simulation process in an SADT diagram. Figure 1 is a slightly modified version of that description. The activities are described by the verbs in the boxes. The results of an activity are shown as labels on the lines leaving a box on its right side. The input data required to perform that activity are shown by lines entering the box from the left. Arrows at the top of a box signify control data that will influence how an activity is executed. Controls may describe conditional dependencies that affect the implementation of a particular function. Finally, arrows entering a box from below are termed mechanisms. They are the means by which the function or activity is performed. Mechanisms are usually synonymous with resources. Resources may be hardware, software, people, or anatomical (or cognitive) components one wants to treat as resources. In some cases (for example, workload analyses), it may be desirable to represent anatomical features of the individual human operator, such as eyes, right hand, left hand, etc. or mental resources.

As the diagram implies, there are important aspects of modeling that must be addressed besides describing the system and representing its dynamic behavior. One must also identify the objective of the studies to be done with the model. This will define what outputs the

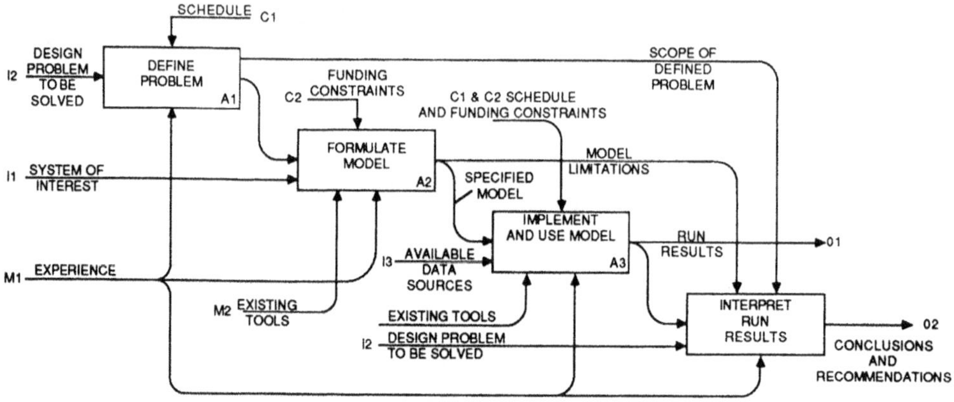

Figure 1. The Simulation Process

model should produce, what functions require more or less detailed treatment, and the quality and quantity of input data that will be needed to support model development and validation. Figure 1 also implies that the quality of the behavioral model is critically dependent on the quality of the system description available to the modeler. The utility of SADT for SAINT model development is that it forces a careful examination and integration of the available system technical information. IDEAL further supports the SADT to SAINT transition by suggesting a standardized Performance Data Base (PDB) format for organizing source document information that will be needed during model implementation.

IDEAL DEVELOPMENT OF A SURFACE TO AIR MISSILE MODEL

The IDEAL methodology was first applied to the description of a surface to air missile (SAM) battery.[8] In this particular case, the SAM battery was actually a generic simulator rather than a specific system. However, it was instrumented for studies of human operator performance in the context of air defense. It therefore provided an excellent test bed for applying the IDEAL concept and validating the resultant SAINT model. The simulator was designed to use three trained operators to acquire, track, and engage simulated penetrating aircraft. Simulating missile launch, the impact of tracking errors and operator strategy shifts then could be evaluated in terms of miss distance, probability of hit, and probability of kill. Consequently both proximal and distal criteria of operator performance were identified and measurable.

Figure 2 presents a schematic diagram of this system. Three sensor subsystems were included: 1) an acquisition radar, 2) a tracking radar, and 3) a tracking television. Three responsibilities were allocated among the operators: a) elevation tracking, b) azimuth tracking, and c) launch control. The last of these three responsibilities was given to an operator called the Fire Control Operator (FCO). The FCO was also responsible for aircraft acquisition and overall system control, as well as missile launch decisions. Two tracking operators shared the responsibility for keeping track of where the aircraft was. This was to be done in the azimuth-elevation plane. Thus, the two-axis problem was reduced to a single-axis tracking problem for each of two operators. The FCO monitored the coordinated

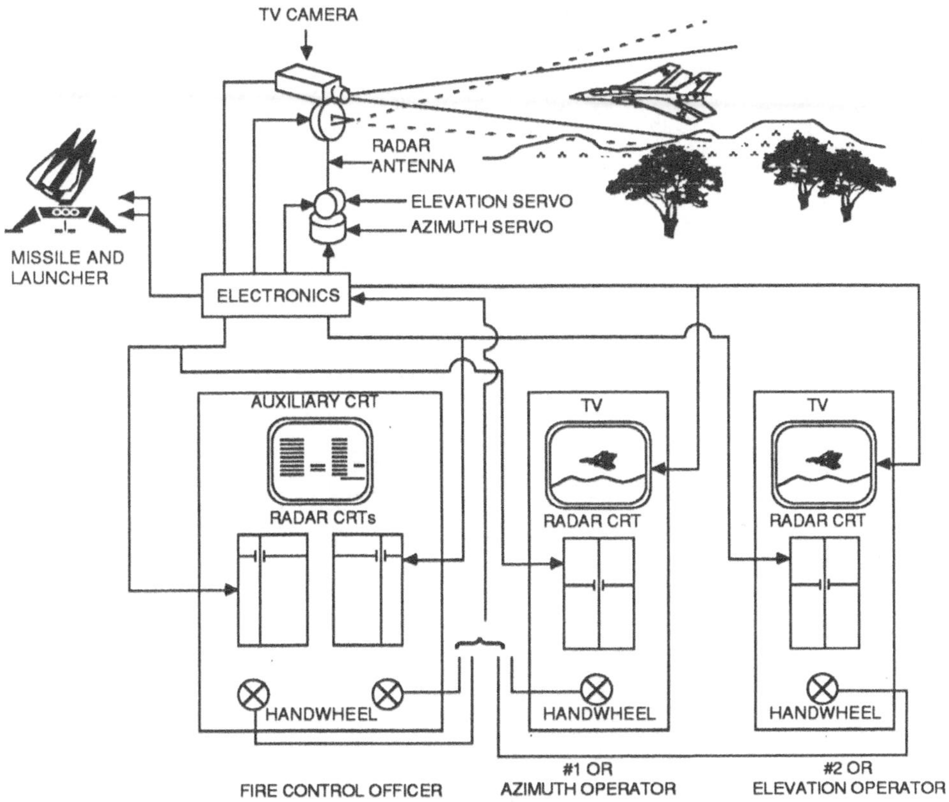

Figure 2. Schematic of the Generic SAM Simulator

activity of the two tracking operators and decided when their joint error was small enough to permit successful launch, once the aircraft was within range.

As a detected aircraft flies inbound toward the SAM site, the crew must evaluate any Command, Control, and Communication (C^3) data available from an external area defense surveillance team. This forms a perceptual set influencing the acquisition phase of their mission. The next major decision after acquisition is determining when to enter the tracking mode and selecting whether tracking will be done by radar or by television. The terminal decision is whether to launch a missile or not. Figure 3 identifies the overall decision sequence the crew must execute.

Static Model Development

Table 1 presents a ten-step model development process that has proven useful for guiding the evolution, test, and use of a model.

What one chooses to do in modeling a system depends on a number of factors. These include the training and experience of the modelers, the people they are working for, and whoever might review the results or conclusions drawn from the model. These are not irrelevant to problem formulation, but they are issues independent of IDEAL.

The stages in table 1 are listed sequentially, but in practice there may be a reordering or even repetition and looping among these

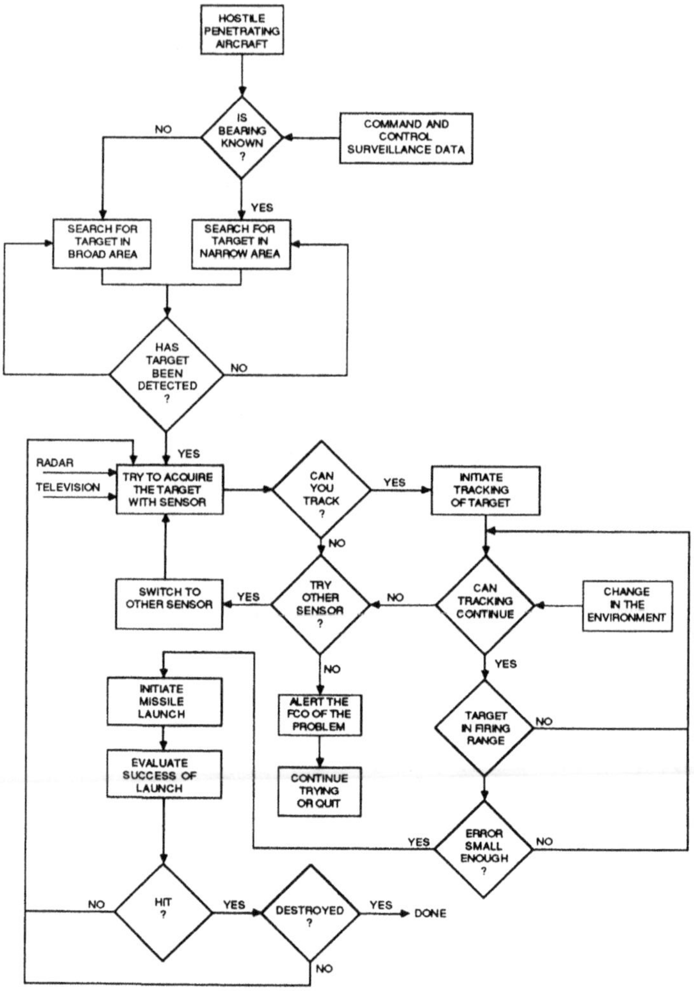

Figure 3. Overall Decision Sequence

various stages. This is an idealized description of a much more dynamic process, especially when it involves more than one person.

A seven person team was involved in SADT model construction and included subject matter experts familiar with the system, human factors experts, and project managers. The SADT model was developed in a set of IDEF$_0$ diagrams to produce a static model of the system's functional structure. An IDEF$_0$ diagram results in a top-down hierarchical decomposition of system functions. The process of developing IDEF$_0$ diagrams is more completely described in Marca and McGowan.[9] Details of the system's architecture and operation are progressively refined in greater detail as the decomposition continues to break apart a parent function into its children. The relationships among these children (siblings) are also described at each level in the decomposition. Emphasis is placed on identifying logical dependencies and data relationships among the various functions at any particular level. At this stage of modeling, the sequencing and duration of those functions is intentionally ignored. Sequencing and duration of activities is later treated as the performance data base is constructed to implement the dynamic behavioral model of system performance.

Table 1. Stages of Model Development

(1) Problem Formulation

The definition of the problem-solving objective. What design uncertainties need to be resolved? How will system performance be evaluated so alternatives can be compared? What model outputs are needed as dependent variables?

(2) Model Building

The abstraction of the human/machine system into mathematical/logical relationships in accordance with the problem formulation. What is critical? What can be ignored, left out, or treated later? This is where IDEAL is applied, specifically SADT.

(3) Data Acquisition

The identification, specification, and collection of data. For example, estimates of subtask durations, the frequency/probability of taking optional pathways, etc. This is where the Performance Data Base (PDB) is generated.

(4) Model Translation

Preparing the model for computer processing as prescribed by the selected simulation language (e.g., applying the contents of the SAINT User's manual, as recently updated by the C-SAINT User's Manual).

(5) Verification

The process of ensuring that the computer executes as intended. In this context, this step is done when no error codes occur and the run results appear reasonable.

(6) Validation

The process of establishing that the desired accuracy or correspondence exists between the simulation model behavior (reflected in run results) and what is known about, has been observed, or has been measured with respect to the real system's behavior. This requires subsequent study using results from empirical testing (e.g., using results from a prior experiment's data, from other studies, or results from new experiments.)

(7) Strategic and Tactical Planning

The process of establishing the experimental conditions for using the model, including statistical experimental design. Identify the independent variables, range of variation, and number of levels; specify what will be held constant; and establish what factors will be randomized across treatment conditions.

(8) Experimentation

The execution of the simulation model to obtain output values that achieve the desired precision in estimating dependent variable statistics.

(9) Analysis of Results

The process of analyzing the simulation outputs to draw inferences, fit regression equations or response surfaces to run results, and make recommendations for problem resolution.

(10) Utilization

The process of implementing decisions resulting from the analysis of simulation results. This step is crucial, but not really technical in nature. It is important at the beginning to remember that this is the payoff. Consider this step in all of the foregoing steps, especially problem formulation.

The Author of the IDEF$_0$ diagrams is responsible for identifying the purpose and perspective for the static description. The first step is to define the context within which decomposition will occur. This consists of identifying system outputs, inputs, mechanisms, and controls at a global level. The Readers of the diagram have the responsibility of making explicit any objections or questions they may have about the stated purpose, perspective, or context as presented by the Author. All disagreements over terminology are resolved before proceeding to the next level of decomposition. This assures that all participants agree on the objectives, definitions, and descriptions as they are initially stated, later modified (if necessary), and ultimately validated as a correct representation of known facts about the system. An IDEF$_0$ diagram is not unique. If the purpose or perspective is changed, the nature of the decomposition of system functions may change as well. Each team member may have access to different information about various aspects of the system's operation, based on their training, experience, and understanding of available documentation. Development of the IDEF$_0$ model as a precursor to developing the SAINT model of system behavior serves an important role: it makes sure the known facts about the system are gathered together, interpreted in a well-integrated description, and certified as acceptably correct by all members of the project. In that role, it serves as a knowledge engineering tool.

This process incorporates information such as the system's specifications, various operating procedures, and mission performance parameters. It focuses attention on where source data are available to define various aspects of system function, and where there are data voids that need to be filled by analysis, measurement, or speculation. Thus the development of this static model can eliminate building a non-supportable behavioral model.

This methodical, systematic evolution may seem frustrating to those who want quick results, but in large-scale modeling, this process precludes disaster. It is worth the investment because it reduces the risk of developing a good model for the wrong purpose as well as assuring you get a good model. Premature entry into behavioral modeling may inadvertently leave important details undiscovered until late in development when it becomes more difficult to implement changes. Moreover, the IDEF$_0$ description provides a foundation for building the roadmap that will guide the behavioral model development. It forces the team to agree on what level of detail is sufficient, how to quantify various aspects of the problem, and what factors will be intentionally ignored or suppressed. Bachert et al. claim the use of SADT produces a 60% savings compared to building SAINT models less systematically.[10]

Performance Data Base Construction

The Performance Data Base (PDB) is the bridge between SADT and SAINT. It serves to map the IDEF$_0$ model into a corresponding SAINT model. This transition is essentially a two step process. First, the modeler examines the IDEF$_0$ diagram and fills in the slots in a PDB frame. Second, these data are then used to construct the SAINT representation of behavioral dynamics that corresponds with the IDEF$_0$ decomposition of system functions. Table 2 summarizes the lineage between IDEF$_0$ constructs and SAINT modeling concepts. Table 3 briefly describes the SAINT concepts identified as column headings in table 2. The letters in the cells of the mapping matrix (table 2) refer to the following major categories of information:

Table 2. $IDEF_o$ SAINT MAPPING MATRIX

IDEF$_o$ \ SAINT	LABL	TIME	PREDECESSOR	RESOURCE	PRIORITY	INCM	DMOD	UTCH	SWIT	MODRF	BRANCHING	STATE VAR.	MONITOR	CLEARING	NETWORK
INPUT			d			b, d	b				b		b		
FUNCTION	d	d						e		d		b		d	
CONTROL					d	b, d	d		d		d				a, b
MECHANISM		c		c				e		c, e	c			c, e	
OUTPUT											a, b, c d			d	
NETWORK		b							b	b, c, e	a, b, c d	b	b,d	b,c	a, b

a. Global System Characteristics - variables and their parameters that relate to all functions and system constraints.

b. Senario Specific State Conditions - variables and conditions relating functions and information flow through the network for various operational scenarios.

c. Resource Attributes - A list of machine or operator attributes, the values of which describe such things as physical characteristics, stress levels, skill levels, etc.

d. Function or Task Characteristics - detail of activity, may include label, time statistics, priority level, probability of successful completion, precedence relationships, resource requirements, etc.

e. Environmental Factors - variables which influence an activity, e.g., lighting conditions, noise, etc., should also be noted.

Table 4 presents an example of a representative PDB frame. The upper portion is a header for record keeping purposes. This is useful when several efforts may be on-going or when a project may be of a size or duration where it becomes important to keep track of who filled in the information and when it was last reviewed.

There will be one PDB frame for each pair of an $IDEF_o$ function with a corresponding SAINT node. While this frame is the key data collection record, it is not the only record a SAINT modeler should be keeping. Several other tabular lists need to be generated in conjunction with the PDBs. However, these secondary lists simply keep track of what has already been done on prior PDBs. The modeler will therefore want to keep the following lists updated as new PDB frames are filled in:

- o Distribution Sets: number and parameter value
- o Moderator Functions: number and definition
- o System Attributes: number and meaning
- o Information Attributes: number and use
- o Resources: number, nature, and how may attributes
- o Resource Attributes: resource number, nature of resource attribute, and assigned resource attribute number

Table 3. SAINT MODELING CONCEPTS (ABRIDGED)

LABL:	An eight character mnemonic used as a task or activity name (label).
TIME:	The duration of a task or activity when it is defined by a Monte Carlo sampling from a specified distribution or is a fixed, constant value.
PREDECESSOR:	A prior event which must have occurred before the present activity is permitted to begin.
RESOURCE:	If required (and some activities may not require resources), the activity is not to begin until one (or all) of the specified resources is available.
PRIORITY:	Controls which activity will be started first if more than one could begin at the same time.
INCM:	Information choice mechanism for determining which arriving SAINT information packet will be saved and examined (the first arriver, last, or some other option); information packets flow through a SAINT network and packet contents may be used many different ways in modeling system dynamics.
DMOD:	Distribution modification allows TIME samples to be drawn from an alternate distribution when some particular event occurs.
UTCH:	User task characteristics are scalar values the modeler may use to describe the attributes of a task (e.g., level of difficulty, task complexity, etc.).
SWIT:	Switches are binary state indicators that may be used to represent status changes of any sort, usually some equipment event (mode change, indicator light, etc.).
MODRF:	An alternate way of computing activity duration, where the user defines an equation that determines how long an activity takes and codes this in a FORTRAN subroutine so SAINT can call it.
BRANCHING:	Control of node exits; determines what comes next after the current activity is completed.
STATE VARIABLE:	A continuously changing value (like aircraft azimuth and elevation as it flies by the SAM site).
MONITOR:	A mathematical function that looks at state variables to determine when they reach certain values of interest to the modeler.
CLEARING:	The process of pre-emptively interrupting either on-going tasks or presently busy resources.
NETWORK:	The interconnection of $IDEF_0$ functions or SAINT nodes; IDEAL attempts to retain a high degree of commonality between SAINT networks and their corresponding $IDEF_0$ counterparts.

Table 4. A REPRESENTATIVE PDB FRAME

FORM	CONSTRUCT AIRPLANE		
TITLE:	PERFORMANCE DATA BASE		
IDEF:	AO.2	AUTHOR:	SOFTECH
NODE:	4	DATE:	AUG. 84

FUNCTION DESCRIPTIONS
THIS ACTIVITY REPRESENTS THE CONSTRUCTION OF THE MODEL AIRPLANE BUT NONE OF THE DECORATING. THIS ACTIVITY IS DECOMPOSED TO A LOWER LEVEL OF DETAIL. BASED ON THE TYPE OF INSTRUCTIONS READ, THE BRANCHING WILL GO TO ONE OF THE THREE LOWER ACTIVITIES.

PERFORMANCE TIME:		MECHANISMS	MECHANISM COND.
DISTRIBUTION	- - -		
MEAN			
MINIMUM			
MAXIMUM			
STD DEV.:			
PREVIOUS COMPLETIONS REQS.			
PRIOR TASKS			
AO.1			

SUBSEQUENT BRANCHING		
TASK NUMBER	BRANCHING LOGIC	MULT. BRANCHING COND
AO.1	IF MORE INSTRUCTIONS ARE TO BE READ.	
AO.3	IF CONSTRUCTION IS COMPLETE	

NOTES

- o State Variables: number and name
- o Monitors: number and definition
- o Task Characteristics: number and meaning
- o User Functions: number and definition

Using the PDB as a template, the modeler has a guide for collecting the information that will be needed to represent dynamic task execution. One of the first questions addressed is what governs activity duration. A particular activity might be of fixed or variable duration. If it is variable, it may be determined as a function of some set of factors or may vary randomly according to some particular stochastic distribution. If randomly varying, then the modeler will specify the distribution type and its parameter values. If these are the same as for some other activity, then the modeler can simply refer to the existing distribution set number that was used before and annotate his PDB accordingly. If this task duration is a unique distribution type or uses a unique set of parameter values, then a new distribution set needs to be defined and added to the list of distribution sets that are being separately recorded.

Alternately, task timing may be a function of task characteristics, the present value of system attributes, the value of one or more information attributes, and/or the present value of one or more attributes of the resource that is required for task execution. If task duration is to be specified as being a function of such factors, then the timing is described differently. The modeler needs to specify the mathematical formula to be used to calculate the time value and the argument list (set of independent variables for that equation). This will then become a statement (or set of statements) in MODRF. The modeler will then assign a number to this particular function and add it to his list, or he will simply reference some function on the list if a suitable function was already defined for some previous PDB frame.

Before or after timing considerations have been defined, it is also necessary to ask what governs when the activity will start. At least two factors control initiation. One is precedents and the other is resources. Precedents are simply activities which logically must have been completed before the present one is allowed to occur. A distinction may also be needed between the first time something is done versus all subsequent repetitions of that activity. For example, start your automobile engine, you must first insert the key then turn it, but if the engine fails to start, you will need to turn the key again, but it doesn't need to be reinserted because that precedent was satisfied on the first attempt to start the engine. The second factor governing task initiation was availability of the requisite resources. Two situations routinely arise. The first is where several different resources are all required. The second case is where any of several resources may be suitable substitutes for one another. The PDB will specify both precedence and resource requirements for this particular node in the network, and if some new resource is implicated, it needs to be added to the list being kept.

It may happen that an existing resource is to be used, but in this case, the nature of that resource influences how long the task takes. Perhaps the modeler did not note this situation previously. Now the modeler may want to add a new attribute to the list of previously defined resource attributes and also go back and modify the definition of activity timing. In this fashion, the PDB allows for continuing review and updating as the modeling process evolves.

As a part of task accomplishment, various attributes may take on new values, reflecting the impact of starting or completing this particular activity. For example, a switch may be set to a new position, an error may occur, or some state variable may be regulated (changing its value). Or alternately, the performance of that activity may change a resource attribute (e.g., skill level) or some task characteristic (e.g., perceived difficulty). These changes are also noted on the PDB along with annotations about how the new attribute values will be assigned. They can be fixed or variable just as time values were. Attribute values that are to be changed according to a functional relationship require that the modeler specify the nature of the equation to be used and the argument list for that equation.

Finally, attention must given to what happens after activity completion. Branching on exit from a node is along one or more paths to successor nodes and can be controlled three basic ways. So the PDB has slots for identifying what the successor (next) nodes will be and which set of rules will control exit path selection (deterministic, probabilistic, or conditional). If branching is conditional, the conditions must be stated and the form of evaluation must be identified. SAINT recognizes two forms of conditional branching evaluation: 1) take the first branch for which the stated condition is met, or 2) take all branches for which the conditions are met.

While these are the basic elements of a PDB frame, there are a variety of special cases that can be represented in a SAINT network. The PDB has a slot for notes to permit annotations that call attention to the need to exercise any of these specialized modeling features (e.g., task or resource clearing).

In the transition from $IDEF_0$ to SAINT, the dynamic behavioral modeling can be done at more than one level of detail. The $IDEF_0$ decomposition of static functions will provide a hierarchical partitioning. The SAINT model could easily be implemented at any particular level or strata in the hierarchy. However, it is also possible to build a model which operates at one level of detail for some functions and at another level of detail for other functions. Thus a single activity at one (more molar) level of detail may be replaced by a set of functions representing that activity at a more detailed (molecular) level.

As a standard rule, when a function at one level of detail is to be replaced by an activity network representing a more detailed level of breakdown, two dummy tasks are added to initiate and terminate the detailed SAINT network. These consume no time but control the branching and tunneling of information when going from one level of $IDEF_0$ to another. This procedure also simplifies control of looping and interactions among various modules. This convention of using dummy nodes assures a higher degree of modularity, simplifying the number of changes one must make if the model is changed later. The dummy tasks tend to insulate and isolate changes to localized areas while maintaining a tight coupling between the static $IDEF_0$ model of functions and the dynamic SAINT model of behavior.

Construction of the SAINT Performance Assessment Model of a SAM System (SPAMSS)

The next stage of model development re-examines the problem being addressed and lays out an architecture for various submodels that need to be constructed. From the $IDEF_0$ decomposition, it became apparent that there were interfaces where inputs or controls existed in the static $IDEF_0$ model that needed to be provided but were not incorporated in the PDB describing operator function execution.[10] For example, the threat aircraft state needs to be represented since it drives all other activities in the model. This was achieved by writing appropriate equations and incorporating them in subroutine STATE, which is linked with the SAINT code after it is successfully compiled. Correspondingly, there is a need for a tracking model and for a scoring model, the latter being incorporated in USERF which is also linked to SAINT after compilation.

Figure 4 shows the overall SPAMSS architecture as presented by Bachert et al.[10] Figure 5 portrays a sample threat aircraft flight

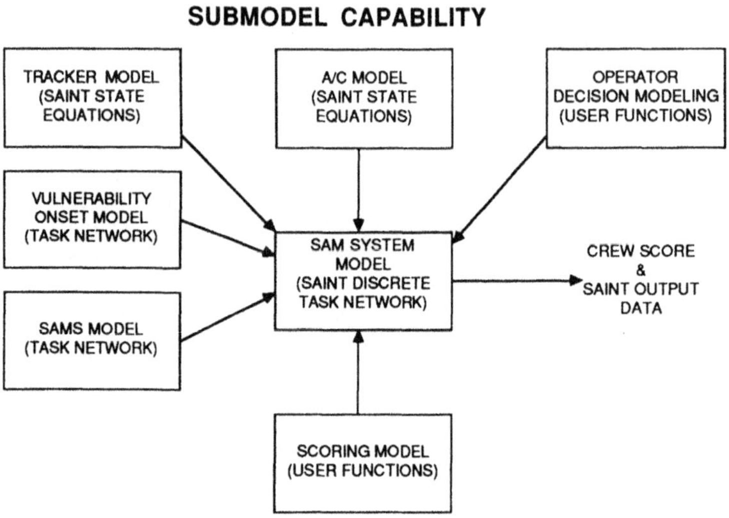

Figure 4. Overall SPAMSS Submodel Architecture

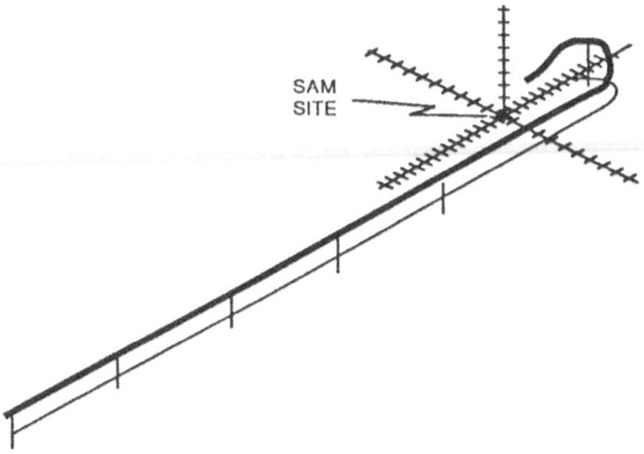

Figure 5. Aircraft Flight Path (Typical)

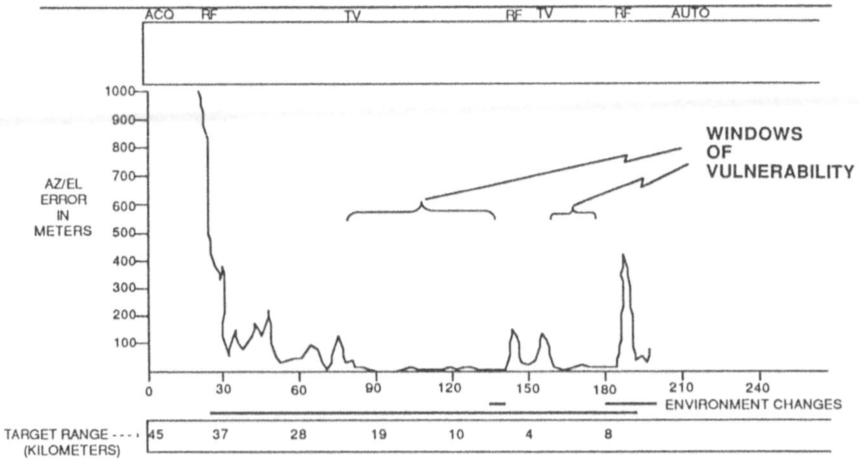

Figure 6. Model Results

path. Different flight paths can be represented by altering the module that contains the A/C Model SAINT STATE Equations. This can be done without affecting the contents of the other modules in the model. Four primary paths were incorporated with two to five variants of each. Path representation was accomplished by using least squares regression to fit a polynomial to actual flight path measurements taken from the generic SAM simulator. The three factor polynomial equation was then implemented in SAINT subroutine STATE. The simulated flight paths were then correlated with the simulator by comparing the velocity and acceleration data with respect to the SAM site. This assured that the polynomial was a valid representation of the aircraft flight path.

The operator tracking model considered multi-task time sharing, as the tracking operators continually rotate the firing pedestal and also perform other associated duties. The pedestal was modeled as moving at a constant rate, using rate equations implemented in subroutine STATE. The operators regulated that rate (up or down), and this regulation task was treated as a discrete activity in the SAINT task network. When operators had no competing responsibilities, they could dedicate themselves to evaluating tracking error and adjusting pedestal rate accordingly, which produces a smaller tracking error. By contrast, as diversions increase, the rates are not changed as often and must be of greater magnitude, so tracking errors increase.

The scoring model provides a record of how well the operators performed under the various conditions studied. It keeps track of the elevation and azimuth tracking errors and the range from the SAM site to the aircraft. These scores are then used to determine when the aircraft is vulnerable to a missile strike. The Fire Control Operator can then be scored on how long it takes to make a firing decision relative to the first possible choice, the optimal (maximum hit probability), and last feasible firing solution. If a decision was made to launch, the outputs from the scoring model are used as inputs to a missile fly-out model. That model will then evaluate the estimated miss distance, probability of hit, and probability of kill for this particular engagement. Consequently, both behavioral and system measures are available.

VALIDATION AND UTILIZATION

Because the SAINT model was based on an instrumented simulator, there was a considerable body of data available from prior experiments that could be used to estimate parameter values for the SAINT model. Moreover, the simulator was available to validate model predictions subsequent to model implementation.

It was possible then to validate the SAINT predictions in terms of several measures and for more than one set of operating conditions. Figure 6 shows some sample results from SPAMSS.[11] Along the top of this graph are indications when particular mode changes occurred. SAINT predictions were not significantly different from those found in the empirical studies. The continuous tracking error is also plotted over time, and windows of vulnerability are shown were the errors were small enough to permit successful missile launch. Again, SAINT provided valid predictions of error and start/stop times on the windows of vulnerability.

Subsequent tests looked at changes in the use of countermeasures that impact the use of automatic tracking and the use of various sensors for manual tracking. In this case, the SAINT model was first used to estimate what should be observed when the empirical tests were run. Following the empirical studies, actual operator and system performance was compared with the SAINT predictions. Again, no statistically significant differences were found.

Having developed a valid representation of the existing simulator based on data from prior studies, it was then possible to use the model two ways. First, one could try out new studies before running human subjects. Second, one could examine conditions that could not be tested in the simulator. Both uses were pursued.

The first set of studies is especially usefully in systems analysis and empirical research planning. One can examine the hypothetical implications of changes in system design or in operating procedures strategies. Alternatives can be quickly examined to see which are worth validating by empirical test. Some concepts may perform so poorly there is little reason to spend money on real-time, human-operator studies. In other cases, the model may predict results that appear more promising and clearly seem to warrant empirical test to validate those predictions.

The second set of studies is more controversial. Here we speculate on conditions that cannot (or at least will not) be empirically tested. For example, it is of some interest to estimate how well operators might perform under actual combat conditions where they may be exposed to various weapons effects. Because those weapons effects cannot be safely administered at large exposure levels, modeling provides a means of estimating what could occur under conditions that would be unsafe for empirical study. A general methodology was developed for treating this problem, one that could be used for a variety of environmental effects.[12]

CONCLUSION

The SPAMSS use of the IDEAL concept was done manually. Using Business Filevision, SofTech has mechanized portions of the IDEAL concept to provide a Computer Aided Engineering (CAE) tool for our own use. DESIGN IDEF by Metasoft is another recently released tool for automating SADT or IDEF diagram construction. Under the Cockpit Automation Technology (CAT) Program, SofTech has also recently developed a customized version of SAINT called C-SAINT (Hoyland, Ganote, and Chubb, 1988). C-SAINT is a customized version of SAINT to help build workload and information analysis kinds of models more easily. C-SAINT also includes several improvements and enhancements to SAINT. A function set facility has been added to allow attribute assignment and moderator function specification without FORTRAN programming in USERF or MODRF routines. Conditional branching tests now allow comparisons of any pair of attributes. Resource allocation can be dynamically specified by an information attribute. For its own use, SofTech has developed several versions of SAINT for use on IBM/PC and Macintosh computers. While these are not sold as commercial products, they are made available to use, with some restricted rights.

REFERENCES

1. Wallace, Stockenberg, and Charette. A Unified Methodology for Systems Development, McGraw-Hill Book Company. 1987.

2. Ross, Douglas T., "Applications and Extensions of SADT." IEEE Computer. April, 1985. pp 25-34.

3. D.J. Seifert and G.P. Chubb. SAINT: A Combined Simulation Language for Modeling Large, Complex Systems. AMRL-TR-78-48. Aerospace Medical Research Laboratory, Wright-Patterson AFB, OH. October, 1978.

4. G.P. Chubb. "SAINT, A Digital Simulation Language for the Study of Manned Systems," in J. Moraal and K.F. Kraiss (eds.), Manned System Design: Methods, Equipment, and Applications. Plenum Press, NY. 1981.

5. G.P. Chubb. "Human Factors in Systems Engineering: New Analysis Requirements," Proceedings. International Topical Meeting on Advances in Human Factors in Nuclear Power Systems. Oak Ridge National Laboratories, TN. 20-23 April, 1986.

6. G.P. Chubb, N. Stodolski, W.C. Fleming, and J.A. Hassoun. "STALL: A Simple Model for Workload Analysis in Early System Development." Proceedings of the 31st Annual Human Factors Society Meeting. NY. 1987. pp 363-367.

7. R.F. Bachert, K.H. Evers, and P.R. Santucci. "SADT/SAINT Simulation Technique. "Proceeding IEEE National Aerospace and Electronics Conference (NAECON). Dayton, OH. 1981. pp 4-9.

8. R.F. Bachert, K.H. Evers, C.M. Hoyland, and E.P. Rolek. "Static and Dynamic Model Integration for a SAM C^3 Simulation via $IDEF_0$/SAINT." Proceedings of 1982 IEEE International Conference on Cybernetics and Society. Seattle, WA. 1982. pp 144-148.

9. Marca, David A. and Clement L. McGowan. SADT: Structured Analysis and Design Technique. McGraw-Hill, NY. 1987.

10. R.F. Bachert, K.H. Evers, C.M. Hoyland, and E.P. Rolek. "IDEFo/SAINT SAM Simulation: Hardware/Human Submodels." Proceedings of the IEEE 1983 National Aerospace and Electronics Conference (NAECON). Dayton, OH. May, 1983. pp 1066-1070.

11. C.M. Hoyland, K.H. Evers, and D.E. Snyder. "Incorporating Human Operator Considerations into Existing Weapon System Analysis and Quantification Capabilities." Proceedings of the IEEE 1985 National Aerospace and Electronics Conference (NAECON). Dayton, OH. May, 1985. pp 911-916.

12. K.H. Evers, C.M. Hoyland, and R.L. Hann. "Measuring and Integrating Environmental Effects into Preliminary Crew Station Design." Proceedings of the IEEE 1986 National Aerospace and Electronics Conference (NAECON). Dayton, OH. May, 1986. pp 1350-1354.

13. C.M. Hoyland, D. Ganote, and G.P. Chubb. "C-SAINT: A Simulation Modeling Tool Customized for Workload Analysis." Proceedings of the IEEE 1988 National Aerospace and Electronics Conference (NAECON). Dayton, OH. May, 1988. (in press).

METACREW: SIMULATING OPERATOR TEAMS IN A SENSOR DATA PROCESSING SYSTEM

Thomas Plocher

Honeywell Inc.
Systems and Research Center
Minneapolis, Minnesota

OVERVIEW

A fundamental concern during the development and evolution of a tactical data system is that of defining appropriate roles for human operators and teams. Roles must be defined such that crew workload is kept to a tolerable level, while still achieving the overall mission performance goals of the system.

Under the sponsorship of the U.S. Army CECOM, Honeywell has been exploring human operator concepts for a new sensor system, the Joint Army-Air Force Surveillance Target Attack Radar System (Joint STARS). To aid these investigations, Honeywell has developed METACREW, a computer simulation of operator performance and work flow in the system.

The METACREW simulation treats the Joint STARS crew as a queue-servicing operation. Simulated targets are acquired by the Joint STARS sensor and presented to the operator team for processing and interpretation. Each detection event waits in queue until a qualified operator becomes available to service it. The operator then performs a task or series of tasks to process the event. The time required by the operator to service an event is determined by his nominal task performance capabilities. Work timelines are recorded throughout the simulation exercise for individual operators. Team and system performance are described in terms of a variety of information throughput measures.

The METACREW simulation has been validated against the performance of experienced Joint STARS data handling teams during exercises at USAICS, Ft. Huachuca, Arizona. In these validation trials, the simulation was shown to account for 76-96% of the variance in the performance of the real operators. Further, the simulation was shown to respond to workload challenges in a manner similar to actual operators.

Since its validation, METACREW has been used to explore a variety of system development issues on the Joint STARS program. These have included identification of optimal crew configurations, estimation of the impact of new missions on crew performance, and assessment of the system's contribution at a force level.

INTRODUCTION

A fundamental concern during the development and evolution of a tactical data system is that of defining appropriate roles for human operators. As Meister (1976) has suggested, operator roles should be viewed at three levels: individual, team, and mission or force level.

Individual operator roles in the system must be defined such that no single operator is overburdened. However, definition of individual roles must be done in a manner that is cognizant of how the individuals will operate as a team. Smooth work flow between team members, minimal processing delays, and optimal utilization of manpower are the design objectives. Finally, at the highest level, these individuals and teams must be configured and deployed in a manner that achieves the overall mission performance goals of the system at a unit or force level.

Within the world of tactical data system design, our attention typically has focused on the individual operator and his displays and controls. Performance of the team is often assumed to be the sum of the performance of the individual operators composing the team. Roles are usually evolved to the extent that the team must have a supervisor and one or more operators to supervise. Design changes that affect individual operator performance are usually assessed in terms of measures of individual performance (i.e., reaction time, time to complete a series of menu entries, etc.). The effect of the design modification on the flow of work through the team, and on the overall accomplishment of the system's mission, commonly is ignored.

Our work with the Joint STARS and other programs has cautioned us about focusing too heavily on the role of the individual in tactical data system design efforts. Some general observations from our work are worth mentioning here. First of all, the performance of a team is rarely the simple sum of the performance of its constituent operators. In terms of time, the work flow rarely takes a linear path through the system. Quite often, when the designer creates a true team of operators, with an associated division of labor, and exercises that team in an environment of multiple mission priorities and scarce resources, something other than an additive, linear performance model applies. At times during the mission exercise, tasks will sit and wait to be executed by some key member of the team, who is preoccupied with something of higher priority. At the same time, other team members might be sitting idle and waiting for tasks to execute. The goal of system design should be to balance these two extremes. Tasks should be distributed among the team members in a manner that minimizes delays in task processing, yet maximizes the utilization of available operator personnel.

The above considerations apply as well to the insertion of automation or decision aids into a system. Increasing the processing speed or capacity of any single operator in the system does not necessarily ensure improved team or mission performance. Rather, such technologies will result in a significant enhancement in mission performance only if they are inserted strategically into the team operation, i.e., where the work flow is typically impeded or delayed. Thus, understanding the dynamics of work flow in the team is essential to making sound design and technology insertion decisions.

The Joint STARS Program

Under sponsorship of the U.S. Army Communications and Electronics Command (CECOM), Honeywell has been exploring human operator concepts for a new airborne sensor system, the Joint Army-Air Force Surveillance Target Attack Radar System. Honeywell's work has focused on the Ground Station Module (GSM) component of the system. The GSM crew is responsible for the identification and tracking of moving tactical ground targets on a radar imagery display. The crew interacts with intelligence and fire support staff members at the Division and Corps Tactical Operations Centers to receive Essential Elements of Information and mission taskings, and disseminate target information they develop. Honeywell's work has addressed numerous system design, manpower, and system utilization issues including: 1) crew configuration required in each GSM, 2) the number of GSMs required to support a U.S. Corps, and 3) the impact on crew performance of supporting new, additional missions such as targeting for the Army Tactical Missile System (ATACMS).

To aid these investigations, Honeywell has developed the METACREW, a fast-time computer simulation of operator and team performance in the Joint STARS GSM. The

METACREW simulation, its validation, and some recent force-level applications, are discussed on the remainder of this paper.

SIMULATION DESCRIPTION

Overview

Figure 1 shows, in very schematic form, the basic models in METACREW. To summarize briefly, the simulation maintains a set of taskings and priorities (Mission/Command Rules) that describe 1) battlefield events that are of interest to the commander and 2) what kind of processing is required of the GSM crew to describe and report them. As the Battlefield Scenario unfolds over time, the simulation selects tasking-relevant scenario events and places them in a queue to await processing by the GSM crew. Events wait in the queue until a GSM crew member with the appropriate task and skill qualifications (as stated in the Resource Model) has time to process them. Order of event processing depends upon the priorities set in the Command Rules. The crew's processing of the events can involve any one of the dozen operational task sequences or "processes" that comprise the current Behavioral Model. These range from basic target tracking and updating to aircraft vectoring and fire direction sequences. The time required by the crew members to process a scenario event is determined by their nominal performance capabilities and the number of events competing for attention at any one time.

The simulation describes the resulting system performance both in process and product terms. For instance, a complete work history can be obtained showing what tasks each operator performed on a minute-by-minute basis during the simulation exercise. Various summary measures also are available that describe information throughput during the simulation exercise and the team's performance in terms of their assigned mission taskings.

Each of the models shown in Figure 1 is described in more detail below.

Behavioral Model

Tasks, Decisions, and Processes. The foundation of the METACREW simulation is a model of GSM operator behavior. The model is built upon three basic constructs: tasks, decisions, and processes.

Tasks are the most fundamental units of behavior used in the model. They are the basic blocks from which the mission-oriented team processes are constructed. Further, they are the level at which operator skills, responsibilities, and performance are described in the model. In concept, tasks could be defined at any level of detail, depending upon the application. For the Joint STARS application, they were defined with the following philosophy in mind. First, a task is defined here as a sufficiently small unit of behavior that it is always performed by a single operator, rather than shared between two or more. This is simply to avoid ambiguity in accounting for the time expenditures of each operator during an exercise of the model. Second, a task is a large enough unit of behavior that it has a clear event-related goal or purpose. Thus, individual keypresses, menu selections, and cognitive behaviors are, by themselves, not tasks. Rather, tasks are defined here as groups of these elemental behaviors combined to perform a function specifically related to a mission event. A final concern in defining tasks was the level at which man-machine interface changes could be anticipated to have an effect. Thus, all other things considered, tasks were defined at the highest level at which the effects of major design modifications still could be evaluated.

Decisions mark the points of departure for alternative behavioral paths within each operational task sequence or process. They are included in the Behavioral Model to add the element of flexibility and dynamism. For example, various alternatives for target analysis and reporting are provided in the Behavioral Model. Different alternatives will be selected at different times in the exercise depending upon their nominal probabilities of occurrence or scenario event characteristics. These decision points, together with tasks, form processes.

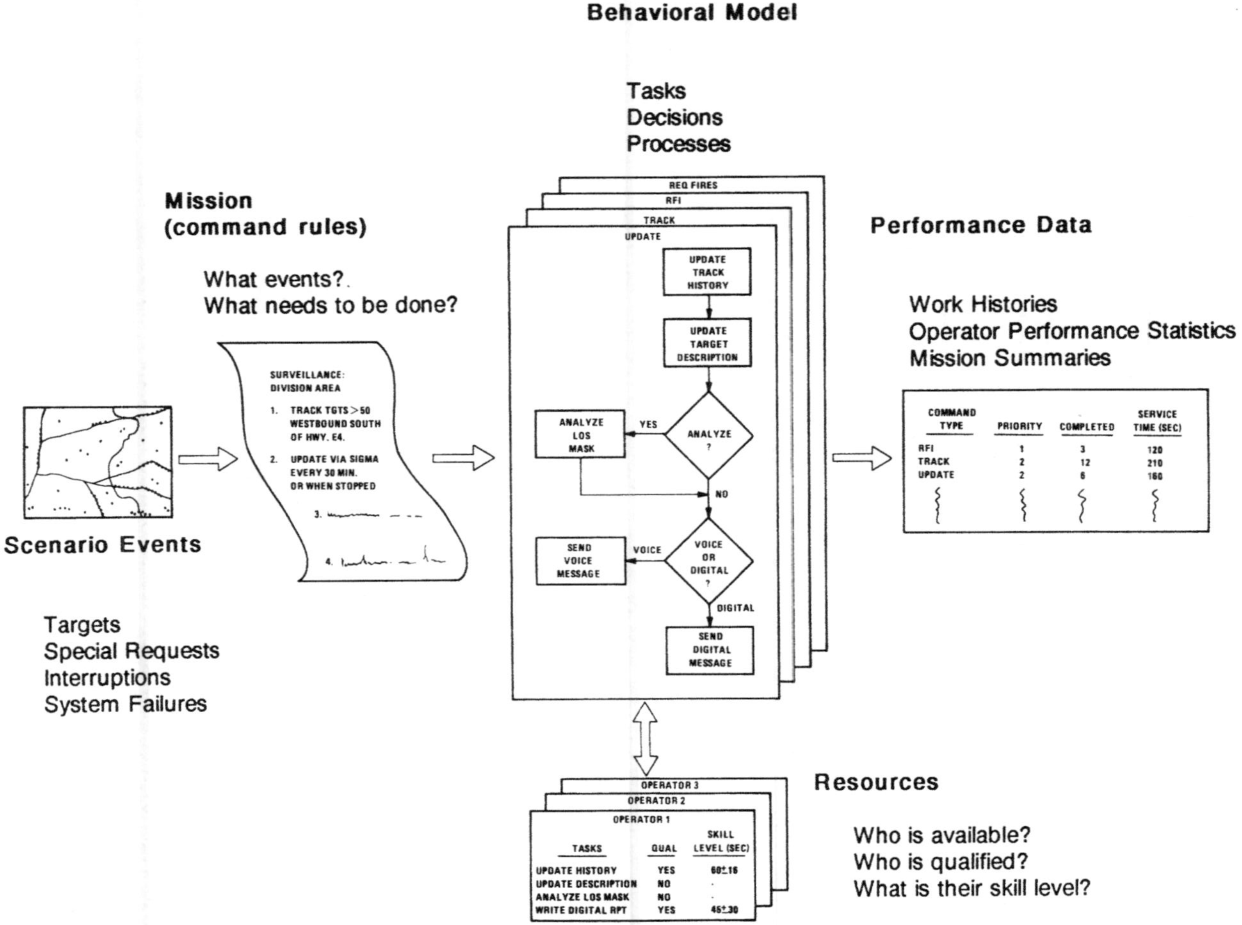

Figure 1. Overview of the METACREW Simulation.

Processes are sequences of individual tasks and decisions linked together to perform the missions stated in the Command Rules. They are the means by which the interworkings of the GSM crew are described and related to tactical missions. The emphasis here is on the GSM crew as a *team*—communicating, making decisions, and managing their work according to the commander's priorities.

Stimulus Events. The principal stimuli or "initiators" of crew processes are discrete events occurring in the Battlefield Scenario. These include: 1) targets appearing, stopping, deploying, crossing boundaries, 2) special commander's requests for information, and 3) nontactical interruptions (malfunctions, etc.). However, crew processes also can be initiated according to a time schedule, using the process of "spawning". A number of processes in Joint STARS are periodic, re-occurring at specific time intervals following the completion of, and relative to, some initial activity. For example, crews often are tasked to update their observations on a given target periodically, after an initial observation has been made and reported. These "spawned" processes are different from strictly scheduled processes in that their time of occurrence cannot be specified, a priori, before an exercise. Rather, they are dynamically scheduled, depending upon the ongoing activity of the crew.

Work Interruptions. In the simplest case, a process, once started, is played out to completion without interruption. However, in the real world, and in our simulation of it, interruptions in the continuous flow of process task activity are quite common. These interruptions in work flow derive from two sources.

First, processes are usually team efforts—one operator performs several tasks on a target and then hands it off to another operator for additional processing or for reporting. If that second operator happens to be involved in another process at the desired hand-off time, the target generally will have to wait until that second operator is available to continue its processing. An interruption in the process flow is the result.

Secondly, any process in the Behavioral Model, due to its priority, may be allowed the privilege of interrupting other processes being executed. For example, special requests for information from the commander may be of such importance that they are permitted to preempt other ongoing operator activity. The analyst can specify and vary such interruption privileges through the priorities he prescribes in the Command Rules. Further, the analyst can specify and vary where, within each process, interruptions can occur.

Battlefield Scenario

As shown in Figure 1 (Simulation Overview), the Battlefield Scenario provides the principal source of stimuli for the simulation. METACREW is designed to use lists of discrete battlefield events as simulation drivers. In our Joint STARS work, the events are, of course, Threat movement events detected by the Joint STARS airborne radar.

Typical movement events used in a METACREW scenario for Joint STARS are: 1) target begins movement, 2) target enters sector, 3) target leaves sector, 4) target changes formation, and 5) target stops. The sectors referred to above are geographical areas of interest or tactical boundaries. They are defined as desired by the analyst to reflect different intelligence or targeting requirements.

Each scenario event also has a set of attributes associated with it. These include: 1) time of occurrence, 2) location of occurrence, 3) number of vehicles in the target group, and 4) a target identification number. These attributes, together with the type of movement event, are the basis upon which the Command Rules are specified by the analyst and the crew processing priorities are ordered. As the simulation is exercised, target events with attributes desired by the commander are selected from the scenario listing and placed in the queue for processing. Events without the desired attributes are ignored by the simulation and remain in the list.

The use of a discrete event scenario driver in METACREW allows us to stimulate our crew/simulations with explicit, rather than abstract representations of the Threat. The standard scenarios for Army combat developments (i.e., SCORES) can be used as basic data sources. A discrete event scenario for METACREW constructed from these standard scenarios, has great tactical fidelity in terms of Order of Battle, force density, and movement patterns and rates. A doctrinally realistic level of stress on the simulated crew therefore is ensured. The use of standard combat scenarios also facilitates the exchange of scenario data and simulation results with other members of the combat developments community. For example, we recently completed a joint study with the U.S. Army TRADOC Analysis Center (TRAC-White Sands) in which all participants exercised their models and simulations against TRAC's Europe 6.5 Threat movement scenario.

Command Rules (Mission)

The mission taskings and processing priorities stated in the Command Rules act as a filter between the Battlefield Scenario and the Behavioral Model. As shown in Figure 1, the Command Rules literally tie the scenario events to behavioral processes. The rules determine:

1. Which target events and special requests are selected from the scenario for processing by the stimulated GSM crew.
2. Which processes in the Behavioral Model are performed on each selected target event.
3. The order in which target events are processed.
4. Which decision alternatives for target analysis and reporting are followed under various conditions.

Thus, the rules laid down by the analyst prior to the exercise quite literally determine the mission emphasis and crew task loadings for the exercise.

Resource Model

The Resource Model in METACREW provides a vehicle for describing and manipulating certain characteristics of the GSM crew. Currently, the crew can be described in terms of three variables: 1) number of operators, 2) operator task assignments/responsibilities, and 3) operator performance level.

Number of Operators. The number of operators exercised in the simulation can be varied from one to N. However, any manipulation of crew size needs to be accompanied by a corresponding change in underlying crew concept. This is done through the task assignment, teaming arrangement, and operator performance level parameters discussed below.

Operator Task Assignments. Operator task assignments are a variable input to the simulation. Thus, tasks in the Behavioral Model can be assigned to operators in any manner. During an exercise, then, these "task qualifications" act as a resource constraint. When a task must be performed during the exercise, the simulation considers only those crew members who are qualified for the task assignment. If all qualified operators are busy, the task waits.

Operator Performance Level. Each task in the Behavioral Model has a distribution of performance times associated with it. During an exercise of the simulation, task performance times for each operator are selected for play from these distributions. Since these operator performance times are a variable input to the simulation, they can be changed quite easily as new data become available or to reflect new assumptions about crew performance level.

Performance Data

The METACREW simulation describes performance at three levels: the individual operator, the team, and the system or mission level. Performance can be summarized at the conclusion of the exercise, or in "snapshots" collected at specified time intervals during the exercise.

Individual Performance. METACREW describes individual operator performance in terms of two reports: an Activity Log and Processing Queue Statistics.

The Activity Log provides an exhaustive chronological record of the task activities of each operator in the team during the exercise, i.e., a mission-task timeline. For each task performed by the operator, the Log reports the scenario event that was being acted upon, the start time of the task, the time of task completion, and the service time. Periods of time in which the operator was idle also are highlighted. Finally, descriptive statistics on individual operator task performance are summarized at the end of the log.

METACREW also collects and reports statistics on the work processing queue of each operator in the team. These reports show the number of events waiting in queue and the average age of queued events as a function of exercise time. The queue statistics are particularly useful in assessing the effects on operator performance of various mission disruptors such as special requests for information, malfunctions, etc.

Team Performance. The overall performance of operators working as a Joint STARS team is summarized by METACREW in two ways: a Mission Event Throughput Summary and a Mission Event History Summary.

The crew's information throughput performance for an exercise is summarized and compared to their assigned mission taskings in the Throughput Summary. The number of battlefield events processed by the crew in support of each tasking is recorded, as well as the speed of event processing. These end-to-end event processing times incorporate the contributions of all operators in the team, plus any delays in the event processing. Deviations of throughput performance from that predicted by a strictly additive task model are easily highlighted.

Event Histories are recorded by METACREW for each scenario event serviced by the crew during an exercise. Each event history includes a complete description of the scenario event and a timeline describing the tasks performed on it by crew members. The timelines highlight points of hand-off between operators in the processing sequence, and any associated processing delays. The Histories thus are a convenient way to diagnose suspected bottlenecks in the team operation.

Mission/Force-Level Performance. METACREW records a list of discretely-identified scenario events processed and "reported" by each crew exercised in the simulation. This event list provides the basis for a variety of mission or force-level measures of performance. All of the measures are based on a comparison of the Threat events reported by the crew(s) with "ground truth", i.e., the events that actually happened in the scenario. Of interest are both the proportion of actual events perceived and reported, and the time lag between actual event occurrence and the crew's report. Performance can be summarized across all crews in the system to determine a force-level contribution. The value of this contribution (or figure of merit of the system) can be estimated by comparing it to the relevant doctrinal values for Commander's Information Needs. For example, a Division Commander must know the location of X percent of the Threat battalions opposing him in order to identify the direction of the main attack and effectively commit his reserve brigade. The proportion of battalions reported by the combined Joint STARS crews can be compared to this doctrinal value to describe the system's contribution to that element of the Division's mission.

METACREW MODEL VALIDATION

The METACREW validation exercises were conducted at Ft. Huachuca, Arizona, using the Joint STARS Ground Station Simulator (GSS). The GSS is a valid, high-fidelity, human-in-the-loop simulation of a Joint STARS Ground Station Module. The GSS is used at Huachuca to train operators for Joint STARS field test and demonstrations.

Three full-mission tactical exercises were conducted using the GSS. Each represented a variant in terms of scenario content, mission taskings, and crew behavioral requirements and demands. Two experienced Joint STARS instructor cadre personnel served as exercise participants. Individual exercise performance data were collected, as well as data describing crew and overall system performance.

Figure 2 shows the paradigm by which these human-in-the-loop exercise data were used to validate METACREW. First of all, the data base of individual task performance times, collected during the cadre exercises, were combined into statistical distributions and inserted into METACREW's Resource Model. Next, METACREW's Scenario and Command Rules were set up to reflect the same exercise parameters (in terms of scenario and mission) that had been used in the cadre exercises. Programmed in this manner, any differences in performance between the real crews and those simulated in METACREW could be attributable to deviations in the fidelity of the operator and team behavior modeled in METACREW. Conversely, the degree of similarity in their performance could be taken as a measure of the validity of the simulation.

Results

Figure 3 compares METACREW with cadre exercise performance in one of the three validation exercises. Cadre performance is shown in the single heavy line in the figure. Each of 10 random replicate trials of the METACREW simulation is shown as a lighter line. Specifically, the figure shows the cumulative number of TRACK and UPDATE processes completed over the time period of the exercises. Thus, they show the "rate" at which both the real and simulated crews performed their target-processing work.

Qualitative Analysis. Figure 3 shows that both the cadre and simulated crew performed in the manner expected, given the scenario and mission. The first hour of the exercise was devoted primarily to the process of initiating new target tracks. During the second hour, as directed by the mission tasking, their attention turned to the process of updating these target tracks. Commander's Requests for Information (RFIs) usually resulted in a momentary pause in this baseline tracking and updating activity. This was quite pronounced in the exercise illustrated in Figure 3, where the crew had to process six Requests for Information during the second hour. Baseline processes of tracking and updating were, for all practical purposes, suspended during this period. Significantly, the METACREW responded to this challenge in a manner similar to the actual cadre. Further, diagnostic analysis showed that, throughout the exercises, the METACREW executed tasks, decisions, and processes in a manner that was highly faithful to the behavior of the actual crew.

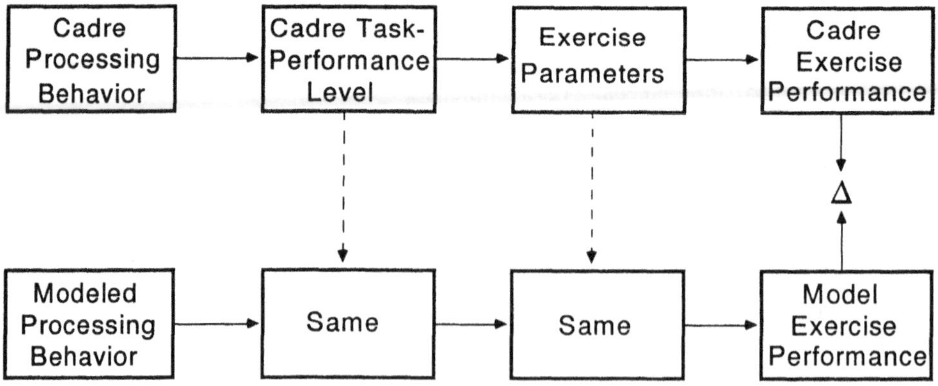

Figure 2. Paradigm used for METACREW Validation.

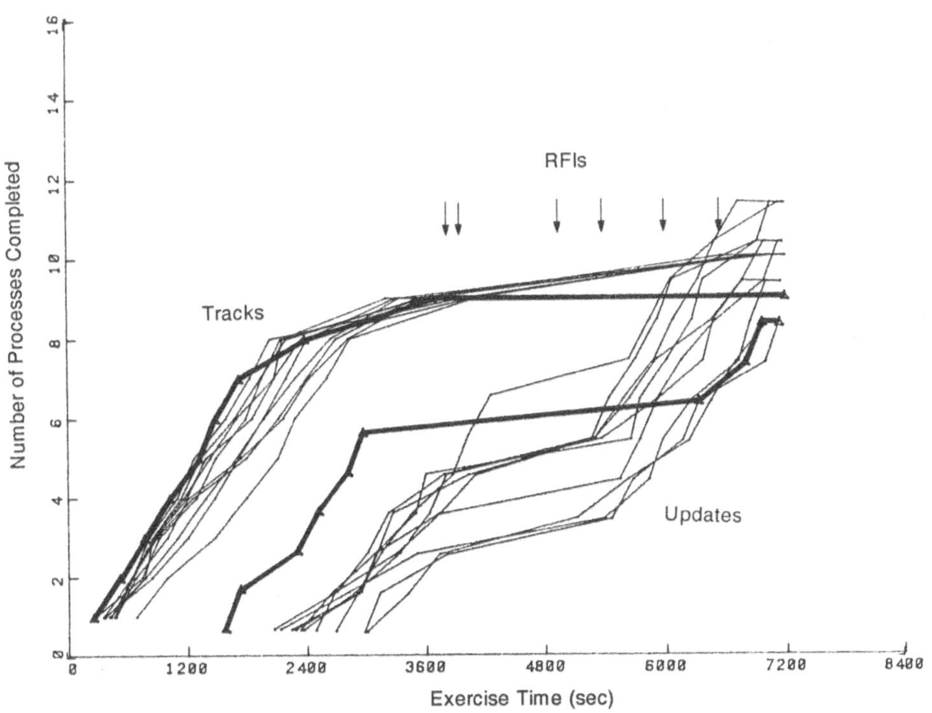

Figure 3. METACREW versus Actual Cadre Performance in Validation Exercise 3

Correlational Analysis. A correlational analysis was performed on the data collected in the three validation exercises to measure the degree of relationship between METACREW and cadre performance. For each exercise, correlations were computed between the cadre performance curve and each of the ten METACREW random replicate performance curves. The correlations were based on pairs of process completion times. That is, the time at which the cadre completed its first process was paired with the time at which the METACREW completed its first process in each of the ten replicate trials. Such pairs of data points were constructed for all N processes completed during each exercise. TRACK and UPDATE processes were analyzed separately. The ten correlation coefficients that resulted from these computations then were averaged together to yield Rc, an estimate of the average degree of relationship between METACREW and cadre performance. The square of this average correlation value, Rc^2, indicates the proportion of the cadre's performance variance that the model takes into account. Rc and Rc^2 values are shown in Table 1 for TRACK and for UPDATE processes completed in each of the three validation exercises. The values show that the METACREW simulation accounted for 75% or more of the cadre's performance variance in each case. The METACREW thus was determined to be a statistically robust and highly valid simulation of the cadre's performance in these exercises.

FORCE-LEVEL APPLICATION OF METACREW

The Army and Air Force are conducting, jointly, an Operational Utility Evaluation (OUE) of the Joint STARS. The purpose of the evaluation is to provide results and recommendations in support of the Decision Advisory Board at the Milestone IIb decision review. Honeywell is one of several contractors and government agencies that has been supporting the OUE study effort. Honeywell's role has been that of simulating the contribution of the Army GSM, including its human operators to the tactical missions of Division and Corps-level forces.

TABLE 1. RESULTS OF CORRELATIONAL ANALYSIS OF METACREW VERSUS CADRE PERFORMANCE

	Track Processes Exercise			Update Processes Exercise		
	1	2	3	1	2	3
Rc	.97	.92	.98	.96	.95	.84
Rc^2	.95	.84	.96	.93	.91	.76

The basic METACREW simulation was modified during the OUE effort to simulate the simultaneous target-processing activities of an entire Corps complement of Joint STARS GSMs. Figure 4 shows that, as in the actual Joint STARS, all GSMs in the Force-Level METACREW process radar-detected Threat movements from a common database. However, each GSM has its own peculiar mission tasking, designed to support the Essential Elements of Information required by the force element it supports. Thus, each simulated GSM extracts and processes a unique portion of the Threat events detected by the radar and reports them to its supported element. The overall force-level contribution of the system is summarized across GSMs. Any GSM deployment concept can be exercised with Force-Level METACREW, and force-level contributions assessed. For example, the number of GSMs in the force, supported element relationships, and mission taskings all can be varied systematically.

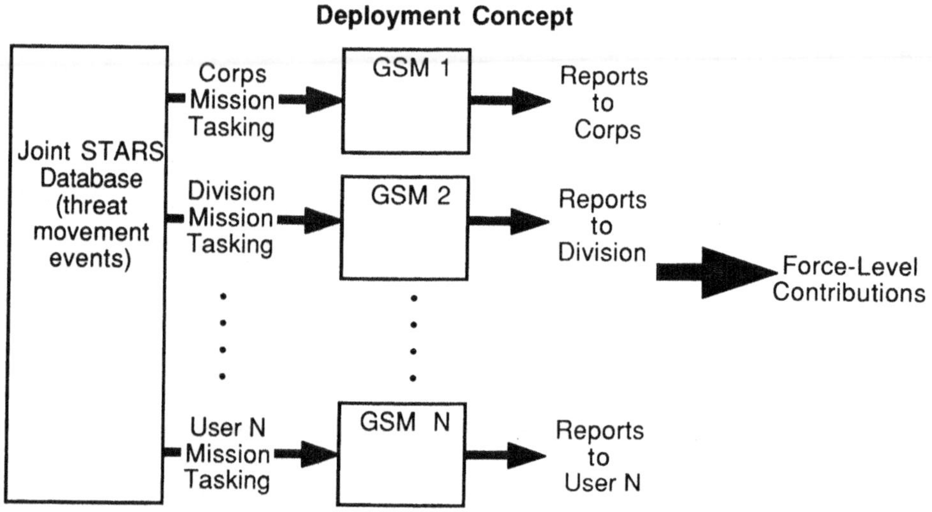

Figure 4. Illustration of Force-Level Version of METACREW.

Figure 5 shows the force-level evaluation methodology used in the OUE. A Joint STARS deployment concept, supported element relationships, and mission taskings all were specified as input to METACREW. A scenario of the Threat movement events, developed from the SCORES Europe 6.5 scenario by the U.S. Army TRADOC Analysis Center, provided the stimulation for the METACREW exercises.

Exercising Force-Level METACREW against this discrete Threat event scenario produced a time-sequenced list of specific targets detected and reported by each GSM simulated in the evaluation. The numbers and types of targets reported by each, and any time lags in reporting, reflect the realistic human operator parameters built into the basic METACREW model.

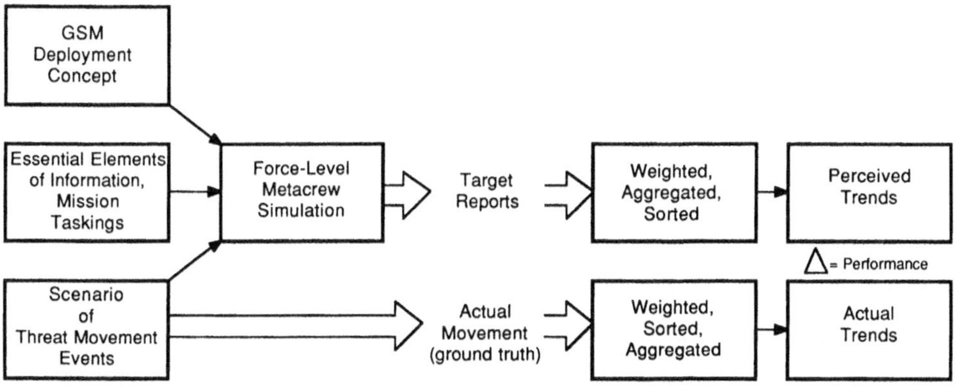

Figure 5. Methodology Flow for Assessment of Joint STARS Force-Level Contribution.

The data produced by exercising force-level METACREW (i.e., target reports) were subjected to a series of statistical manipulations that essentially "mimicked" the intelligence analysis process. First, in order to achieve a common metric of "target value", each report was weighted according to the size of the target it reported on. Target reports then were sorted into geographical areas of interest. Geographical patterns of Threat movement and build-up thus were identified. Thirdly, the individual reports of Threat activity were aggregated across intervals of time to highlight temporal patterns in the reported activity. Reports were processed and analyzed in this manner for individual GSMs, and then aggregated across the entire Joint STARS system to estimate overall contribution.

Various measures of system performance were used in the OUE analysis, all based on the "delta" or difference between Threat movement trends detected and perceived by Joint STARS versus the Threat movements that actually occurred in the TRAC scenario. Specific measures used included 1) quantity of Threat movement activity over time, 2) geographical concentration of Threat forces, 3) route and assembly area usage, and 4) latency to detect Threat second echelon commitments.

REFERENCE

Meister, D. *Behavioral foundations of system development.* New York: Wiley, 1976.

WORKSPACE DESIGN-ANTHROPOMETRICAL AND BIOMECHANICAL APPROACHES

INTRODUCTION: WORKSPACE DESIGN - ANTHROPOMETRICAL AND BIOMECHANICAL APPROACHES

David Beevis

Defence and Civil Institute of Environmental Medicine
Downsview, Ontario
Canada

OVERVIEW OF THE PAPERS

Anthropometrical and biomechanical models of operator performance are probably the closest to current engineering design techniques of any reviewed in the Workshop. In their use of two and three-dimensional geometry they are compatible with the Computer Aided Design (CAD) techniques which are increasingly being used in system design. In fact some currently available models are offered as modules of CAD systems. The number of available models appears to be expanding rapidly as model developers take advantage of the capabilities of mini and personal computers. Models are known to have been developed in Australia, Canada, France, Germany, The Netherlands, UK, and USA.

This is a welcome development which has been recommended by several studies of human factors applications problems, including the NATO Workshop on Applications of Systems Ergonomics (Merriman 1984), and a recent US DoD study which lead to the publication of MIL-HDBK-763. It is a development which will continue, however, only if the models continue to meet the expectations and requirements of users. As one of the papers suggests, user experience indicates that some current models fall short in these respects.

The various anthropometrical and biomechanical models can be classified in several ways. One classification, employed by the US National Academy of Sciences, is reported by Kroemer. The three classes are: anthropometric(al), biomechanical, and human-machine interface models. Kroemer points out that research is needed in all three classes, particularly in the collection of anthropometry data, the development of dynamic biomechanical models, and the development of a fully integrated model of man and machine. His paper also highlights the need, common to nearly all types of model, for validation.

In several cases one reason for the lack of validation has been the developmental history of the models themselves. When funding has been made available to produce a working model it has seldom been provided to support the lengthy process needed to validate a general purpose design tool. The developmental history of one model, SAMMIE, is reviewed by Bonney et al. They also survey applications of what is a typical "human-machine interface" model.

Models such as SAMMIE are sometimes classed as automated design aids rather than models (Meister 1985, BAE 1987). Their use is associated with the "detailed design" phase of man-machine system development; however their capabilities permit consideration of such design issues earlier in the design process. These models also impact the design process by permitting design issues to be re-iterated in a way which is not possible with traditional methods of analysis.

The impact of adopting CAD man-modelling is touched on by Rothwell in a review of the issues related to the selection and use of a model, and the limitations of available "human-machine interface" models. Rothwell cautions the potential user not to expect such models to import expertise: the user must have relevant skills if they are to exploit current models.

Anthropometrical and human-machine interface models address performance at its most basic level, in the operator's ability to see and reach displays and controls, perform maintenance tasks, and fit into, and get out of, the workspace. Biomechanical models deal with human performance in terms of the operator's ability to pull, push, lift, and carry. Although most of the other models reviewed in this Workshop apply to advanced technology systems, the importance of having reliable tools for the study of such tasks should not be underestimated. Despite the impressive developments in mechanical materials handling in recent years, such tasks remain important in military, as well as civilian, occupations (Celentano and Nottrodt 1984).

Evans reviews several types of biomechanical model, and illustrates their application. Again it appears to be the advent of personal computers which has improved the availability of this class of models. As noted by Kroemer, most of these models are not "proactive", in the sense of predicting the operator's capabilities; they are "reactive" in that they calculate internal stresses imposed by specified external loads.

By far the majority of available models are static; the manikins must be positioned by the user and lack rules governing the sequences of movements required for task performance. Badler demonstrated an animated model, TEMPUS, which integrates anthropometrical and biomechanical features, and which offers several approaches to manikin positioning and movement. SAFEWORK, another model which integrates anthropometric and biomechanical features, was demonstrated by Gilbert et al.

Several of the papers suggest directions for model development. Badler reviews the approaches used to develop TEMPUS and indicates the potential for producing improved models by exploiting developments in computer science. Badler, Bonney and Evans outline attempts to build some expertise into biomechanical and man-machine integration models. Some of those developments are compatible with the concept of the "supermodel" mentioned in Kroemer's paper.

Future developments may also facilitate a change in the approach taken to dealing with the underlying data (the anthropometrical models, in Kroemer's terms). Most current man-models can use either specific body size data for an individual, or general "percentile" data drawn from descriptive statistics of the population being represented. As Rothwell notes, the concept of an "Nth percentile" manikin is not realistic, however, because the intercorrelations between individual body dimensions exclude a growing percentage of the population as more dimensions are considered. Thus a manikin based on 50th percentile body segment dimensions would represent much less than the 50th percentile of the population being modelled.

Computer-based man-models permit the representation and manipulation of multi-variate distributions of body segment lengths through appropriate sampling strategies, but few have exploited this capability. CAR and its predecessor CAPE (see Kroemer's paper), and TEMPUS are exceptions. The work of Bittner (1976) in the identification of the body segment combinations which pose most "challenge" to the workspace, and earlier work on the mathematics of the multi-variate body dimensions (Bittner and Moroney 1974) appear to have been ignored. It is to be hoped that future developments take a second look at this problem area.

DISCUSSION

Four main issues emerged during the discussions, which apply in principal to all the types of model reviewed. These were the relative advantages of model complexity vs simplicity of use, the need for control of posture, the need for dynamic models, and the cost-effectiveness of such models.

In terms of complexity, the models which were demonstrated ranged from the comparative simplicity of the Werner personal computer-based manikin (Werner 1988) or the NIOSH guide for lifting (NIOSH 1981), to the animated display of operator activities simulated in a zero-G environment, demonstrated by Badler. One argument put forward was that simple models can be used early in the design process, and can therefore have more impact on the design than a complex model which must be used later when more design details are known. User experience is a factor, however, because complex models can be used early in the design process if "default" values based on previous experience are available.

Complex models require higher levels of skill from their users because they incorporate more degrees of freedom than simple models. The need for relevant skills in anthropometry and biomechanics was turned around by the suggestion that such models could be modified to make potential users more sensitive to some of the design issues involved. Although the idea has merit, and is an extension of current work aimed at providing on-line aiding and "intelligence", there was not a lot of interest in this suggestion. This was possibly because the teaching community were not strongly represented at the Workshop.

Related to the question of model degrees of freedom and user skill is the need to control the posture of the manikins. In the paper presentations and in the discussion it became clear that the manikins do not have internal rules governing posture. This is unfortunate given the demonstrated effect of discomfort on performance (Corlett and Manenica 1980). The suggestion that posture checks should be built into models was discussed, but it was concluded that such checks would not reduce the user's skill requirement significantly.

Most current models are static, and must be worked through a sequence of postures in order to study tasks involving significant limb movements. No strong requirement was expressed for dynamic models. Understandably there was general agreement that dynamic capabilities, while desirable, are difficult to implement. Dynamic strength models need much more research before they can be implemented. In fact existing models require more development; current models can produce widely differing estimates of the acceptable levels of physical workload, as unpublished studies at DCIEM have shown.

The cost-effectiveness of the models was questioned by a potential user with a limited budget. His argument was that he had only enough money for either a mockup or a computer system; he could not afford both. He asked if he should have chosen computer modelling over the mockup. The response was that the two techniques are complementary, and that CAD should be used to evaluate a range of possible solutions (a large solution space), and select one solution for refinement with a mockup, which deals best with a small solution space. It was noted that the Public Relations aspect of a full-size mockup is important and that the models which had been reviewed were no substitute for a full scale mockup.

The need for model accuracy was also discussed. The requirement for manikins accurate to within 1 cm. was questioned. It was argued that a design which requires such accuracy in the manikin is undesirable because it implies that there is little tolerance in the workspace: a workspace with so little tolerance is likely to inconvenience a significant proportion of the user population. This question was not resolved. The answer is related to the impact of model accuracy on the percentage of population accommodated by a design: it appears dependent on the particular application and use being made of the model.

An attendee summarized the session by arguing that the technology does not yet seem to be mature; the necessary data are not available to run either anthropometrical or biomechanical models. There was general agreement on that point, particularly in the context of data for three-dimensional models. It is probably a fair generalisation that those from academic backgrounds expressed the need for more research, and those from industry the need to develop models to the point where they can be more readily used. Bonney et al's point that the technology currently being exploited is at least ten years old was reiterated. There have been no significant improvements in that time, despite the potential for development indicated in several of the papers.

In summary, the discussions were seldom conclusive, particularly when dealing with the needs for future development. It was shown that, in the hands of skilled users, existing models do have considerable potential for assisting the design process, and for integrating human engineering considerations with other engineering design issues. There are limitations to their application, however. As with all tools, users must have a clear understanding of the capabilities and limitations of the models. The general recommendations are that more research is required into the human capabilities which are modelled, particularly in the areas of anthropometry and dynamic biomechanics, and that more development is required to integrate the models which are available into a suite of user-friendly tools, and to develop more general purpose models based on advances in human factors knowledge.

REFERENCES

BAE, 1987, Human Factors for Designers of Equipment (Systems) - The Preparation of Part 12 of Defence Standard 00-25, British Aerospace, Sowerby Research Centre, Draft Final Report JS 10752.

Bittner, A.C. Jr., and Moroney, W.F., 1974, The accommodated proportion of a potential user population: compilation and comparisons of methods for estimation, in: "Proceedings, Human Factors Society 18th Annual Meeting", E.L. Saenger, M. Kirkpatrick, eds, Human Factors Society, Santa Monica, 376-381.

Bittner, A.C. Jr., 1976, Computerized accommodated percentage evaluation: review and prospectus, in: "Proceedings, 6th International Ergonomics Association Congress", University of Maryland, Human Factors Society, Santa Monica, 157-164.

Celentano, E.J., and Nottrodt, J., 1984, Analyzing physically demanding jobs: the Canadian Forces approach, in: Proceedings of the 1984 International Conference on Ergonomics, D. Attwood, C. McCann, R. Sugerman, eds, Human Factors Association of Canada, 421-424.

Corlett, E.N., and Manenica, I., 1980, The effects and measurement of working postures, Applied Ergonomics, 11: 1, 7-16.

Meister, D., 1985, "Behavioral Analysis and Measurement Methods", John Wiley & Sons, New York.

Merriman, S., (Ed), 1984, Workshop on Applications of Systems Ergonomics to Weapon Systems Development, Vols 1 & 2, NATO Defence Research Group, Brussels, DS/A/DR(84)408.

NIOSH, 1981, Work Practices Guide for Manual Lifting, National Institute of Occupational Safety and Health (NIOSH), USDHEW Publication No. 81-122.

Werner, B.K., 1988, Werner - a small computer based tool for ergonomic design, Poster presentation to this Workshop.

A SURVEY OF ERGONOMIC MODELS OF ANTHROPOMETRY, HUMAN BIOMECHANICS, AND OPERATOR-EQUIPMENT INTERFACES

Karl H.E. Kroemer

Human Factors Engineering Center, IEOR Department
Virginia Tech (VPI), Blacksburg, VA 24061

INTRODUCTION

The Committee on Human Factors of the National Academy of Sciences convened a two-day workshop in June of 1985 to assess the feasibility of developing an integrated ergonomic model of the "human at work." The specific objectives of the Workshop were to: (1) assess the usefulness of current anthropometric, biomechanical, and interface models; (2) identify critical points of compatibility and disparity among such models; (3) review the feasibility of using these existing models for the development of an integrated ergonomic model; and, if feasible, (4) recommend research approaches to the development of an integrated ergonomic model. Fifteen experts in anthropometry, biomechanics, bioengineering, work physiology, human factors engineering, psychomotor performance, computer modeling, and system design and operation participated in the Workshop. The Workshop was co-chaired by Stoover H. Snook and the author. The following text relies extensively on the Proceedings of this Workshop (Kroemer, Snook, Meadows, and Deutsch, 1988).

The efficient and safe operation of civilian and military systems requires their "ergonomic design," i.e., that the work task, the work equipment as well as the overall work environment be compatible with the user's capabilities. If the equipment is, instead, designed as if it worked on its own, and if the task is required as if it were independent of human characteristics, then the human-technology system is bound to be inefficient, overly demanding for the human operator, often unsafe, and occasionally destined for a breakdown or destruction. Apparently, equipment and system failure, if it occurs, is often believed to be caused by human error while in fact the equipment or system may have been developed with too little consideration of the capabilities and limitations of the person who operates it. Even if the maintaining or operating people are in fact considered during the design phase of the system, too often that consideration is incomplete, inaccurate, or otherwise insufficient. This may be due to a lack of knowledge or of thoroughness on the side of the designer.

Given the complexity of the interactions among the user, the equipment, the task, and the environment, many different models of the human operating equipment, and of these human-equipment interfaces have been attempted. To be successful, the underlying models describing the interactions between human, equipment, and task should be complete and realistic. In the past, only incomplete models of "person systems" have been achieved, which is due in some part to the inability to describe the versatility and mobility of the human body and mind. Even if one limits modelling goals to represent, as far as the human is concerned, such physical characteristics as body size, visual field, reach capabilities, loading of muscles and bones, strength capabilities, etc., existing models are incomplete and inaccurate.

In the prospectus for the Workshop, Kroemer limited its scope to three major classes of models:

- anthropometric, i.e., representations of static body geometry such as body dimensions, reach, position of the body and/or its parts, posture;

- biomechanical, i.e., representing physical activities of the body in motion, using primarily anthropometric data as inputs; and

- interface, specific combinations of anthropometric and biomechanical models with regard to their interfacing with the technological machine, i.e., representing human-machine interactions.

The integration of anthropometric, biomechanical, and interface models into a comprehensive ergonomic model of the human operator could provide a valuable tool for researchers, program planners and designers.

The major question discussed at the Workshop was which research approach appears, or which approaches appear commendable for the development of such an integrated ergonomic model. Of course, before this overall goal can be addressed, it needed to be assessed whether current anthropometric, biomechanical, and interface models are useful and how they should be developed for use as parts of an integrated ergonomic model.

ANTHROPOMETRIC MODELS

In the past, human body models have been mostly physical in forms of templates, manikins, and dummys. The following discussion will concentrate on computer analogs of the human body. Such models need exact anthropometric information in order to be accurate representation of body size, shape, and proportions.[1] In the U.S.A., anthropometric information is most often drawn from the "anthropometric data bank" at the U.S. Air Force's Armstrong Aerospace Medical Research Laboratory. The AAMRL anthropometric data bank includes the data from 50 anthropometric surveys, most of them of U.S. military populations, but about 20 foreign surveys are included as well. In all, more than 300 different measured variables are covered in the data bank. Given all this information, probably the largest repository of such data in the world, it needs to be stated that the information on civilian populations, including the U.S., is comparatively weak. No comprehensive anthropometric study of the civilian population has ever been undertaken in any western country. Hence, information for civilian populations must be deducted from the more abundant military data.

Most anthropometric computer models rely, even today, on the concept of interconnected body lengths ("links"), as originally developed by Braune and Fischer in their classical biomechanical analysis published in 1889. This approach was refined and expended by Dempster in 1955. Von Meyer (1873) reduced the body form to a series of elipsoids and spheres to simulate the

---

[1] The discussion of anthropometric models utilizes extensively the contributions of Dr. John T. McConville during the 1985 Workshop.

shapes and masses of body segements. This elementary work is still, in many respects, the approach taken today. The number of body segments, and of their connecting lengths, has been increased in recent decades to describe particularly the dynamic response of the body to external forces and impacts. To represent varying body sizes and proportions, the current approach no longer relies on preconceived single percentile models, but allows the introduction and manipulation of relative link sizes through the use of regression equations or of direct survey result inputs. Modern powerful computational capabilities offer a large variety and much variability in modeling approaches.

The currently existing anthropometric data base does not contain three-dimensional body data, but only univariate descriptors. Furthermore, many of these univariate dimensions lack a common point of origin to which the individual measurements are related. This fact causes much conceptual and practical difficulty in the development of computer models of the human body size. Hence, various techniques for three-dimensional anthropometric data acquisition have been proposed. Early techniques relied mostly on mechanical measurements in two or three dimensions. Newer photographic techniques use the principle of projecting a regular geometric grid onto the irregularly shaped human body. The projected grid remains regular when viewed along its axis of projection, but appears distorted if viewed at an angle. While such stereophoto techniques are promising in theory, they have numerous practical problems, among them subject alignment, coordinate origin, data acquisition, manipulation, summarization, and display. The laser as a distance measuring device can be used, in theory, for the determination of the shape of the human body. Current techniques either rotate the body to be measured, or the sending and receiving units of the laser device move around the body. However, similar problems of data acquisition as in the stereophoto technique exist. Furthermore, mathematical-statistical techniques need to be developed that collect, organize, and summarize as well as display the huge number of collected data. Such advanced mathematical procedures no longer rely on presumed Gaussian distributions but instead use robust techniques such as the M-estimation. Surface definition has been much improved by "facet algorithms" which allow a complete topographic description of the body surface.

Of course, the current use of landmarks and reference points on the body often palpated below the skin, will need to be strongly modified by the use of photographic or laser measuring techniques. This poses the question of whether traditionally measured dimensions can be compared with body dimensions gathered by newer technologies.

In summary, the status of anthropometric models may be described briefly as:

- even today, most existing anthropometric data are univariate which severely limits their application in the development of accurate three-dimensional body models.

- At present, no standardized reliable procedure exists for determining the three-dimensional body shape based on classical anthropometric data.

- Three-dimensional techniques for data collection are being developed. Data acquired by these means may not be easily compatible with classical anthropometric information.

- Theoretical understanding and computational capability exist, or will be developed quickly, to describe the true size and shape of the static human body, and further of the moving body.

BIOMECHANICAL MODELS

Interest in the biomechanical properties of the human body is basic, as evidenced by the early attempts of da Vinci, Gallileo, and Borelli in the fourteenth, fifteenth, and sixteenth centuries demonstrated. In the 1960's, models still simulated the body as a series of rigid lengths, in two or three dimensions, reacting to external impulses, forces and torques.[2] Many of the early models were built to describe body displacements as a result of externally applied vibrations and impacts. Other models were used to study body segment positions in work or in motion, such as gait. Another series of models was developed to predict the external static forces and torques applicable to outside objects.

A separate set of models describes the stresses in human bones resulting from external loads. These are often combined with models of body articulations, a difficult task particularly because of the involvement of many muscles and ligaments, and the consideration of elastic or plastic properties of human tissues. In fact, quite often simplifying assumptions must be made to reduce the number of unknowns in the model equations to the number of available equations. Partly stipulated by the interest in artificial joints, the knee, ankle, and intervertebral joints have been the objects of many modeling efforts.

Models of the whole body, or of large portions thereof, have been of particular interest for the design engineer. Most current models provide an analog of the human body manipulable in size and motion envelope, together with limited information about the static forces and torques that can be generated within the body and applied through hands and feet to outside objects, such as control devices.

However, very little is achieved yet with respect to the true internal activation of muscles, and the loading of joints, bones, and connective tissue. For example, the simultaneous use of agonistic muscles and their antagonistic counterparts is neither well understood, nor modeled. Hence, the loads on joints are calculated simply from the resultant force, and therefore may depict the internal loading incorrectly, i.e., often too small.

A large variety of models, different in inputs, outputs, model structure, optimization, etc., exists. An extensive table (prepared by Marras and King) contained in the Proceedings of the 1985 Workshop, lists the model types, their input and output variables, and particularly their underlying assumptions. This list shows not only the successes made in modeling, but also indicates the often severe restrictions in model coverage, usually making the applicability and validity of models very limited to given cases and conditions.

One desirable goal of biomechanical modeling is to create a universal model that represents the great variety of use situations. This model should accurately reflect the loading on the body caused by both internal and external forces. It should also be capable of evaluating the "wear and tear" of body components (such as the vertebral column) under realistic, i.e., static as well as dynamic, three-dimensional conditions. Finally, it should predict motions, forces, power, and other actions (of the proactive as opposed to reactive type) of the human.

---

[2] This text relies much on the contributions of Drs. W. S. Marras and A. King to the 1985 Workshop.

To achieve such a goal, advances are needed in several areas of model improvement. For example, more data are needed to describe the material and functional properties of body tissues. The properties of bone must be more realistically incorporated. Elastic tissues and noncontractile elements, such as found in and around body joints, need to be investigated and modeled. The control of muscular activities, particularly of groups of muscles around the same joint, must be investigated and incorporated in models. Deterioration by age, illness, or injury (including repetitive trauma) should be considered. Such models could serve for a variety of applications beyond work design, such as for prosthesis design and for diagnosis and treatment of musculo-skeletal diseases, including rehabilitation. However, at present even the aspect of three-dimensonality is still only incompletely considered, and active motion is virtually missing.

An area untouched by biomechanical modeling is that of incorporating cognitive characteristics. People are information processors who can modify the interaction with their musculo-skeletal system. Under circumstances of great stress, such as life-threatening danger, one can short-circuit internal protective mechanisms and is capable of exhibiting usually "impossible" actions. Such cognitive control processes are virtually nonexistent in biomechanical models at present.

It appears that progress in biomechanical modeling is currently more hindered by our limited basic understanding of the body rather than by computational abilities.

In summary, the current biomechanical models might be characterized as

- mostly relying on a rigid skeleton of links joined in articulations of defined degrees of freedom, embellished with volumes and mass properties.

- These body models usually serve to describe simplified static (isometric) capabilities for exerting forces or torques to the outside, and/or to describe some motion characteristics.

- Most of the whole body models are by design static, with some having passive kinematic properties.

- Nearly all whole body models are reactive (instead of pro-active) in nature.

- Optimization algorithms and objective functions are usually of fairly simple nature and not thoroughly validated.

- Another major class of biomechanical models describes the properties of bones, and of joints, with regard to their load bearing capabilities. The bone models usually use finite-element analyses.

- Major disadvantages of practically all models is their lack of realism, lack of pro-active capabilities, and for whole body models, of their inability to represent internal forces. The control of motions and of muscular actions is not sufficiently represented.

- The existing models are built on diverse principles, use varying computational procedures and techniques, require different inputs and yield different outputs. Thus, the models are mutually incompatible.

HUMAN-MACHINE INTERFACE MODELS

Anthropometric and biomechanical models combine to build the next higher model in the hierarchical structure, that is, the interface models. Interface models describe the interactions between the modeled person with the equipment in a human-technology system.

While the origin of such models is difficult to trace, the first published models in today's sense of the term appeared in the late 1960s and in the 70s. Further developments are usually known by their acronyms, such as ATB, BOEMAN, CAPE, CAR, COMBIMAN, PLAID-TEMPUS, SAMMIE, and, currently being developed, CREW CHIEF. These models were discussed in the 1985 Workshop.

The ATB (Articulated Total Body) model is a modified version of a crash victim simulation program to study human responses during automobile crashes. This model was modified to reflect the human body dynamics during ejection from high-performance aircraft. The model relies on rigid body equations of motions, where the body segments do not deform, but deformation occurs only in the joints. The standard model consists of fifteen segments, but that number can be modified. The output of the model is a time history of linear and angular displacement, velocity and acceleration, for each segment, depending on restraint harness forces, wind forces, etc.

BOEMAN, developed at the Boeing Company, is a computerized model for the design and evaluation of cockpit and other crewstations. The operator model consists of thirteen links constrained by hard angular limits at each body joint. For joint displacements, a time-cost function is used. Mathematical programming minimizes the total time for operator hand reaches from one point to another. The links are enfleshed by truncated cones. Cockpit boundary surfaces are defined. The output of the model is a description of the effort and time required to reach to hand controls. BOEMAN was the first major interface model in the U.S.A. It provided both conceptual basis and motivation for other workplace assessment models such as CAPE, CAR, and COMBIMAN.

The CAPE (Computerized Accommodated Percentage Evaluation) model assessed the accommodation of an aircrew population in a given cockpit design. This program relied on a Monte Carlo simulation to generate the user sample, with thirteen anthropometric variables represented. Major features of this model were implemented in the subsequent CAR model.

The CAR (Crew Assessment of Reach) model is a design evaluation tool for determining the percentage of a user population that can be accommodated properly by a particular crew station design. The model allows to define the geometry of the workstation and to select an operator sample for which the relative geometric fit is evaluated. The body model is located in space within the cockpit, either sitting or standing. Major evaluation points are the design eye point, the line of sight, seat location, head clearance, and hand or foot operated controls. The user population can either be generated by a Monte Carlo process or by direct inputs reflecting the actual measurements of test individuals. The model is built on nineteen links. The output of the model indicates the percentage of the population that can achieve visual accommodation and by the percentage capable of reaching each control.

COMBIMAN (COMputerized BIomechanical MAN) model is an interactive graphic technique developed for U.S. Air Force aircraft workplace design and evaluation. It is also used for selecting persons who fit a given workplace, and for formatting visibility plots. The human model consists of 33 links, most reflecting the major long bones of the human skeletal system. The model is seated at the Seat Reference Point at all times. The link dimensions reflect anthropometric data that may be entered directly or taken from the AAMRL anthropometric data bank. Each link is limited in its angular deviations from its adjacent link to reflect the true range of mobility of a body joint, and to permit the repositioning of a distal link by moving a proximal link. The body surface is represented by an enfleshment technique, with only essential surface lines shown on the screen.

The workplace consists of predetermined panel dimensions, restrictions and contraints which are entered by light pen or keyboard, or taken from a data storage. The user has the option of displaying all or but a few of the characteristics of the workplace at any time.

The workplace is evaluated by interaction with the three-dimensional human body model. The display is, of course, two-dimensional, but the model can be rotated and spaced to any angle, and any details can be magnified.

The output of the program indicates reach capabilities to specified items of the workstation, considering clothing and restraints of the operator. For points that can be reached, the model indicates the amount of static force that can be exerted in that location. For mapping the visual field, head and eye positions are defined. Visual restrictions, such as window size, can be varied. The output can be viewed on the screen or plotted.

PLAID-TEMPUS (not acronyms) are related modeling programs used by NASA in crewstation design and evaluation. PLAID is a system for analyzing the crew interaction with the workstation and its components. It is based on a three-dimensional solid geometry computer software model created interactively by the user. Elements called primitives are assembled in the computer and viewed on the monitor. The compatibility of human body size, and human reach capabilities, can be viewed on the screen, or plotted. TEMPUS is the complimentary software package used to create the human model that interacts with the PLAID-generated workstation. The user selects a specific crewmember from the existing data base, or a "random body." The underlying body model relies on the CAR approach. It can be in shirt sleeves or in a space suit. A dictionary of units of motions is incorporated to reflect extravehicular activities in space. The dictionary entries are isolated motions that can be combined to describe complex tasks. The output is information regarding time, forces, and restraints as well as physical aids required to perform extravehicular tasks. TEMPUS has also an animation capability in which the movements of subjects and objects can be coordinated.

SAMMIE (System for Aiding Man-Machine Interaction Evaluation) is a model developed at the University of Nottingham, England. It was originally meant to evaluate the design of simple workstation layouts. A three-dimensional model of equipment and environments can be built by specifying and assembling geometrical shapes. The anthropometric model was pre-programmed to represent a male of average height and weight, but can be modified to represent other anthropometric data. SAMMIE consisted of two independent modules: one builds models of equipment or workplaces by assembling primitives, i.e., geometric shapes. The other is the human model that consisted of 19 connected lengths representing a schematic skeleton around which three-dimensional solids such as boxes, cones, and cylinders were placed to show the outer contours of the human body. The idealized flesh contours could be varied to simulate different body builds. All body segments were connected by pin joints. SAMMIE also had the capability to create concave, convex, or plain mirrors superimposed on any workplace surface, and could examine the reflections found in these mirror surfaces. Another module is used to assess visibility around the head. The output of SAMMIE was information about reach ability, fit of a person in a confined workspace and visibility including mirrors views.

In summary, the interface models discussed in the Workshop represent the state-of-the-art in the mid-1980s:

- The interface models are specific to given designs, purposes, and characteristics.

- Their usefulness is basically limited by their anthropometric and biomechanical components. Predictive models of the effects of the dynamics of either their workstations, their tasks, or of the modeled human, are not available.

- Effects of stress and motivation are not adequately quantified, hence not modeled. The same is true for the effects of fatigue, trauma, or injuries. Furthermore, the effects of environmental factors on human performance are not included.

- Validity of the models is largely unknown.

DISCUSSION AND CONCLUSIONS

Aspects of existing models, generating a hierarchy in which anthropometric and biomechanical models are basic inputs to the interface models, have been briefly discussed in the preceding text. At the end of the discussion of each model type, summary statements indicate existing limitations. These restrictions also indicate, by inference, research and development needs. These are spelled out in the Proceedings of the Workshop (Kroemer, Snook, Meadows, and Deutsch, 1988) and will therefore not be repeated here in detail.

Major discussions among the participants in the Workshop resulted in the following three conclusions:

1. There is a need for an integrated model of the human body, of its performance characteristics and limitations, and of its interactions with technological systems. Such an integrated ergonomic model would be a valuable tool for the development of specifications for designing the physical parameters of the work site.

2. The development of an integrated model of the human body is feasible. Advances in research methods and instrumentation make research feasible on the many details in anthropometry, biomechanics, and human-equipment interface. For this, the establishment of a standard nomenclature is essential.

3. An integrated ergonomic model would guide future research as well as improve engineering applications.

A basic requirement for an integrated ergonomic model is standardization of model structure, model inputs, and model outputs, and model language so that the model is generally available and not limited to the specifics of a given situation, or to an expert user only. Other requirements (the list position does not imply importance) include that the model

- simulate the "real world,"

- have three-dimensional structure,

- be dynamic,

- be predictive,

- be validated,
- be time- and cost-effective,
- permit rapid analysis,
- permit on-line documentation,
- have graphical display capability,
- be user-friendly.

Two approaches to the development of an integrated model were discussed. The first relies on the development of one "supermodel" which integrates the best qualities of all other models. The other approach is to develop "modules" which can be linked together as needed.

Current interface models such as COMBIMAN, PLAID-TEMPUS, and SAMMIE appear to represent the "supermodel" approach. But these models are not compatible with each other, due to the different data formats, different modeling complexity, different model theories and techniques, and the use of different computers. Whether one of the existing models should be further developed to become "the" supermodel, or whether a new approach should be taken needs to be determined.

The modular approach is a building block process of joining compatible modules. This also requires a standard structure, i.e., a sort of a "supermodel," but for the user the modular approach means that only the module of interest needs to be operated.

The Workshop members concluded that an integrated ergonomic model is needed, feasible, and useful, whether it be of the supermodel or of the modular type. For this, a number of overall Research Recommendations were formulated. Some are:

RR1: Establish the objectives, procedures, and outline for the development of a general integrated ergonomic model.

RR2: Review and integrate existing anthropometric and biomechanical data bases.

RR3: Develop submodels and modular groups.

RR4: Develop generic interfaces between human models and workstation models.

RR5: Develop methods and criteria for the validation of ergonomic models.

Given recent advances made in understanding the human body and mind, and how the human interacts with the equipment, and also considering the advances made in computer modeling and simulation, one should expect that "better" models of the human-machine system can be achieved: models which are realistic, predictive, dynamic, accurate, and easy to use.

REFERENCE

Kroemer, K. H. E., Snook, S. H., Meadows, S. K., and Deutsch, S. (Eds.) 1988, Ergonomic Models of Anthropometry, Human Biomechanics, and Operator-Equipment Interfaces. Washington, D.C.: National Academy Press.

APPLICATIONS OF SAMMIE AND THE DEVELOPMENT OF MAN MODELLING

* Maurice Bonney, **Keith Case and Mark Porter

* Department of Production Engineering
University of Nottingham, Nottingham, UK
**University of Technology, Loughborough, UK

INTRODUCTION

SAMMIE is a computer aided design system which assists the designer of workplaces suitable for the human operator. The original development of SAMMIE started over 20 years ago, and this paper traces its development and shows the links with other associated software. The paper also describes the current facilities within SAMMIE and some applications for which it has been used. It then goes on to set this work in the context of the needs for a workplace design system which includes man-modelling.

SAMMIE - A BRIEF HISTORY

SAMMIE, System for Aiding Man-Machine Interaction Evaluation, was developed at the University of Nottingham. The need for a tool such as SAMMIE came to light when a computer model was built to evaluate some engineering aspects of an equipment. It was soon apparent that by examining the equipment alone only one part of the man-machine system was being investigated. It was immediately identified that there was a need for a system which functionally included a man-model, a workplace modelling system and methods to help the users communicate with the system and also to help designers evaluate the suitability of work places and work tasks.

The SAMMIE name was an obvious corollary to that specification. To achieve the functionality, four separate developments were undertaken. The first was on the computerised man-model which began in 1967. This started with modelling a single arm. This was later extended to represent two separate arms joined to a single rigid link representing the spine. A separate representation of the lower part of the body was also undertaken. The posture control was by means of an approximate method called the 'natural planes algorithm' designed for computational simplicity but suffering from data deficiency. This was later replaced by other methods using the end position to be reached and 'comfort' algorithms.

From the beginning it was recognised that it was impossible to automate the design process and that it was necessary to involve the designer intimately in the design procedures. This was achieved by having pictorial displays and a good communication mechanism. In the late 1960's computers were slower had less memory and were more expensive. In order that user involvement and interest would be maintained a requirement was placed on the system that it should respond within seconds. The consequences of this decision were an emphasis on good data structuring and on model simplicity. These features later proved invaluable.

The other strands in the development were the construction of a workplace modelling system which consisted of a 3D solid modeller with associated data structure, methods to help the user communicate with the system and finally work study evaluations to help designers evaluate the suitability of work places and work tasks.

The philosophy behind the inclusion of a work study system was that it might allow an assessment of the work content to be evaluated simultaneously with the dynamic representation of man. In 1971 a short film was produced showing a stick man carrying out assembly tasks while at the same time a work study analysis based on an MTM-2 assessment of the task appeared on the screen. This computerised work study system was reported by Bonney and Schofield (1971). From 1971 the work task evaluator, already distinct from SAMMIE, continued as a separately funded development. At that stage limited MTM assessments also existed in SAMMIE. Some of the ideas were also included in the computer aided control and panel layout work reported by Bonney and Williams (1977).

The end of the first phase of the SAMMIE development was reported in Bonney et al., (1974). The second phase of the development which completely restructured SAMMIE began in 1974. For this, unlike the early developments which were produced by research students using external computers, the team was properly resourced and had the primary objective of producing software which could solve practical problems. In order to achieve this many of the interesting but not fully evaluated features in the 1974 SAMMIE though potentially useful, were not included in the new version. By 1977 the reconfigured software, now FORTRAN based, was being used regularly to carry out industrial design studies and had met the objective of being useful for solving practical problems.

The next few years proved to be both exciting and very trying to the University research group. SAMMIE was launched on to the market by Compeda Ltd in 1980. As a result of a take-over this was relaunched by PRIME computers in 1984. SAMMIE has proved itself to be a valuable computer aided design product and is being used by an increasing number of companies. However associated with the technology transfer, research funding for SAMMIE at the University was greatly reduced and the implementation of planned important functional man-model developments was delayed and so during the period from 1979 to 1986, although important restructuring of the software took place, few facilities were added to the man-model. However limited software development and design studies continued at the University of Nottingham and Loughborough University of Technology. The design studies are discussed later.

Because of the high level of interest from industry, academia and public bodies a new company, SAMMIE CAD Ltd, was launched in 1987 to develop the functionality of SAMMIE, carry out industrial design studies and license software to be used on a range of computers and workstations.

THE SAMMIE FACILITIES

SAMMIE is based on a CAD solid modelling system which is used to represent equipment and workplaces. The human operator is represented by an anthropometric and biomechanical model of man using the same general solid modelling technique. This man model provides the ergonomic evaluative facilities such as reach, vision and fit which are the principal objectives of the design system. The complete system is accessed via a user interface which assists in the interactive and iterative nature of the design and evaluation process.

A relatively simple Boundary Representation form of solid modelling is used for geometric modelling as it is sufficiently precise for the needs of the application and also responds adequately to interactive changes in design. Solids are defined in a variety of ways including simple parametrically defined shapes such as cuboids and prisms. Relatively complex models can be built and swiftly manipulated to change the geometry or to change the view as seen on the screen.

The SAMMIE modeller uses a hierarchical data structure to represent the logical functional and spatial relationships between geometric items in the model. Thus the opening of a door of a car model is a meaningful operation, and the use of hierarchical data structures in this way is strongly developed within SAMMIE as an essential evaluative tool. Interactive modification of the geometry in ways relevant to the design situation is also an important part of the system. Hence, for example, if a table were modelled as a table top and four legs, then increasing the length of the top would automatically re-position the legs to maintain a valid model.

A wide variety of viewing options is available including orthographic projections, perspective, viewing point and centre of interest control, scaling, etc. In addition to being part of the user interface enabling better comprehension of the model, viewing is also available as a model evaluation facility in its own right. Hence the view as would be seen by the human operator model from within the geometric model can be presented, as can special views such as those seen in mirrors. The production of two-dimensional visibility plots and three-dimensional visibility charts is described in Porter et al (1980). Figure 1 shows a three-dimensional visibility chart.

The man-model provides much of the evaluative power of SAMMIE through its anthropometric and biomechanical modelling capabilities. The pin-jointed, rigid linked model represents the major points of articulation at the ankles, knees, hips, lumbar and thoracic spine, shoulder, elbows, wrist, head and neck. Additionally, left and right eyes can be used as part of the vision evaluation package. A three-dimensional flesh shape is arranged about this link structure using the modelling methods described above. The user can control this flesh shape by combinations of stature and weight related to Sheldon's (1940) somatotype classification methods. The seven-point scales of endomorphy, ectomorphy and mesomorphy may be used to describe a body shape, or alternatively the most commonly found combinations of these factors may be used to specify general body shapes. Body segment parameters (mass, centre of gravity etc) are also available from the database for use in evaluative procedures involving the assessment of balance and static strength.

The anthropometry can be varied by changing the overall body percentile, an individual link percentile, or an explicit link dimension.

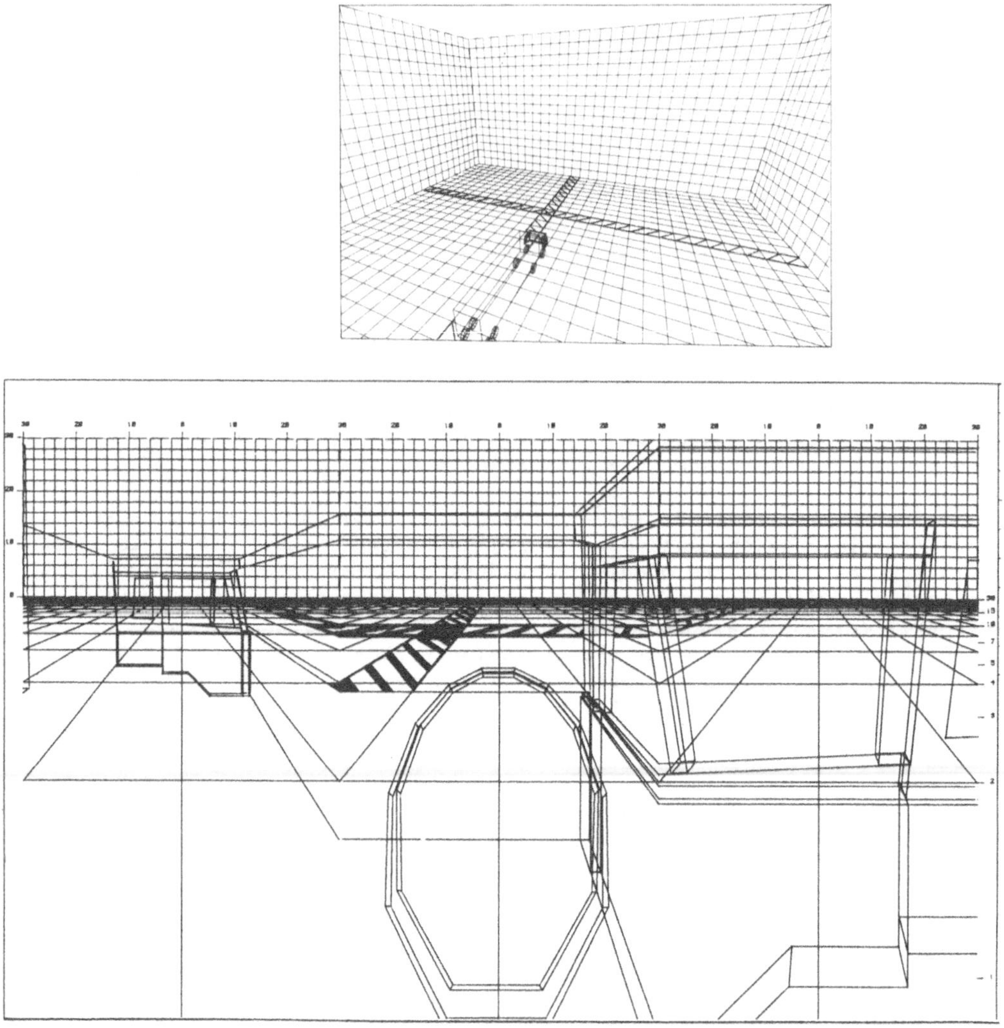

FIGURE 1. THREE DIMENSIONAL VISIBILITY CHART SHOWING 180° VIEW FROM THE DRIVING POSITION WITHIN A COMMERCIAL VEHICLE

This data could originate from measurements taken on an individual, or be a sample from a user population held within a database. Correlation equations are used to relate externally measurable dimensions to the internal link dimensions. Clearly the selection of an adequate database and the subsequent manipulation of the information requires the user to have a thorough understanding of the anthropometric implications.

Several methods are available for the manipulation of the man model within the workplace model. Postures can be created and stored within the database to be subsequently recalled. These postures can simply be a set of potentially useful starting points for the investigation of actual working postures or they can be a precise set against which designs must be evaluated. Each body segment can be articulated about its proximal joint. Movement is permitted in the flexion-extension, abduction-adduction and medial-lateral rotation senses and the resulting joint posture is compared with joint constraints in the database where some of these degrees of freedom (e.g. abduction-adduction of the elbow) will be constrained not to occur. The system reports whether the joint is within the 'normal' range of movement, within the maximum, or infeasible. The 'normal' constraint data is intended to allow the user to define design criteria in terms of body posture. Hence, for example, if high gravity forces precluded any attempt to raise the arms above shoulder height, then this could be accommodated in the database. In more usual conditions this facility is used to define preferred working volumes related to a joint postural comfort criterion.

Reach algorithms are available which predict a feasible posture for a sequential set of links such as the arm or leg. In the evaluation situation, the ability to test reach to the specific points where, for example, controls are located is a useful facility. However in a design situation it may be necessary to determine suitable areas or volumes within which controls could be placed and for this application 'reach contours' have been developed see Figure 2. These enable envelopes or areas within reach to be overlaid on any surface of the model as an aid to assessing suitable positions for control locations. A major study involving this facility to determine reach zones for the drivers of agricultural tractors and machinery is described in the next section.

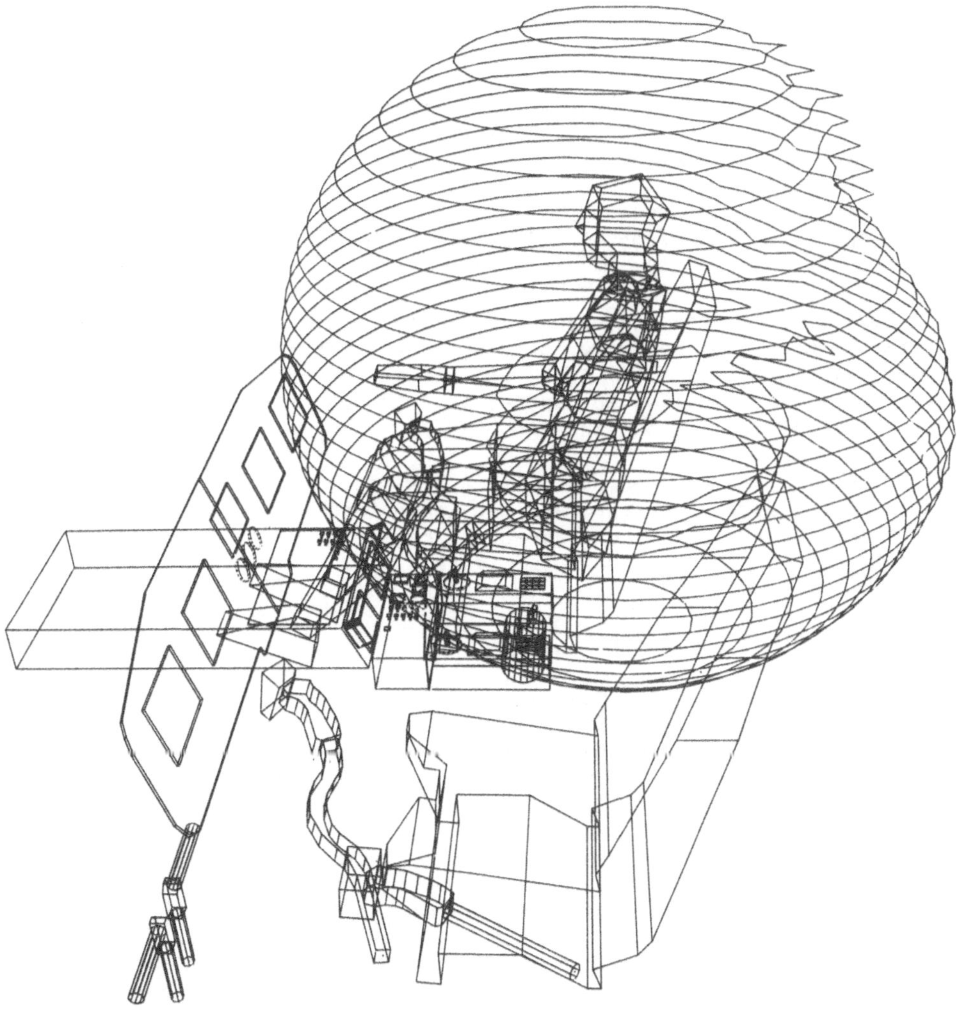

FIGURE 2. VOLUMETRIC REACH CONTOURS WITHIN A COCKPIT

APPLICATIONS

Simulation reduces the necessity to build physical mock-ups for user trials and improves the quality and usefulness of any prototypes that are built. In this way the increasing pressures for good ergonomic design can be more easily accommodated within the functional design process.

An indication of the range of applications for which SAMMIE has already been used will be found in SAMMIE (1987). Two important broad application areas are apparent. The first is for the design of workstations for transport such as cockpits for aircraft and helicopters, cabins of trucks, ships bridges, trains etc. The second broad application area is the design of computer based workstations such as CAD terminals, shop check-out stations, individual bank workstations, complete bank layouts and work stations for financial dealers. There is no essential difference in these applications from the modelling point of view. However it is interesting to note that each raise problems which are difficult and time consuming to solve using conventional methods and each requires a considerable investment in time and money. There is a need to produce a good design first time.

Much of the detailed work of the design studies performed is company confidential but among the reported studies have been work on shop check-out design reported in Gibson et al., (1985) and studies to develop reach volumes for users of agricultural machinery reported in Reid et al., (1985), The work on reach volumes was eventually published as a British Standard (BS 6735, 1987). This provides information on volumetric hand reach space and volumetric foot reach space. The main variables used in the investigation were sex (M/F), posture (upright, leaning sidewards, twisted to the rear) reach type (fingertip, toe, heel), limbs used (both, left, right) and the direction of reach (upwards, downwards, forwards, backwards and sidewards).

Some applications including visibility from a fork lift truck, layout of tractor controls and design of rear view mirrors for buses were briefly reported in Bonney et al., (1979) and the design of mirror systems was more extensively reported in Case et al., (1980). An example of the mirror facilities is shown in Figure 3.

Some work was done on modelling a jig into which the components of the front fuselage of an aircraft are loaded and subsequently assembled. This showed considerable cost advantage over conventional mock-up methods. One part of the study is shown in Figure 4.

SYSTEM DESIGN NEEDS - GENERAL

A large number of man-models now exist in various software systems created around the world. Their attributes, the way that they interface with the workplace models and how they, in turn, interface with the other parts of the total system determine their usefulness. A good user interface is also essential.

With SAMMIE a system requirement has always been that a fast response was needed so that truly interactive working was possible. This, together with the important limitation arising from the lack of appropriate data to build complex models, meant that a relatively simple linkage arrangement was chosen and efficient methods of dealing with these were developed. At the same time it was necessary to model moderately complex workplace models and to handle these in an efficient manner.

At one level then the system design was a natural reaction to knowledge and technology. Since 1974 there has been some growth of knowledge and enormous growth in computing power so that more complex man-models and workplace models can be easily handled. However it must be remembered that all people are different and do things in different ways and hence accuracy and realism in a man-model, above a certain level, may be potentially misleading. At the end of the day the modelling system designer still needs to ask what is the man-model to be used for and what attributes are required to achieve this?

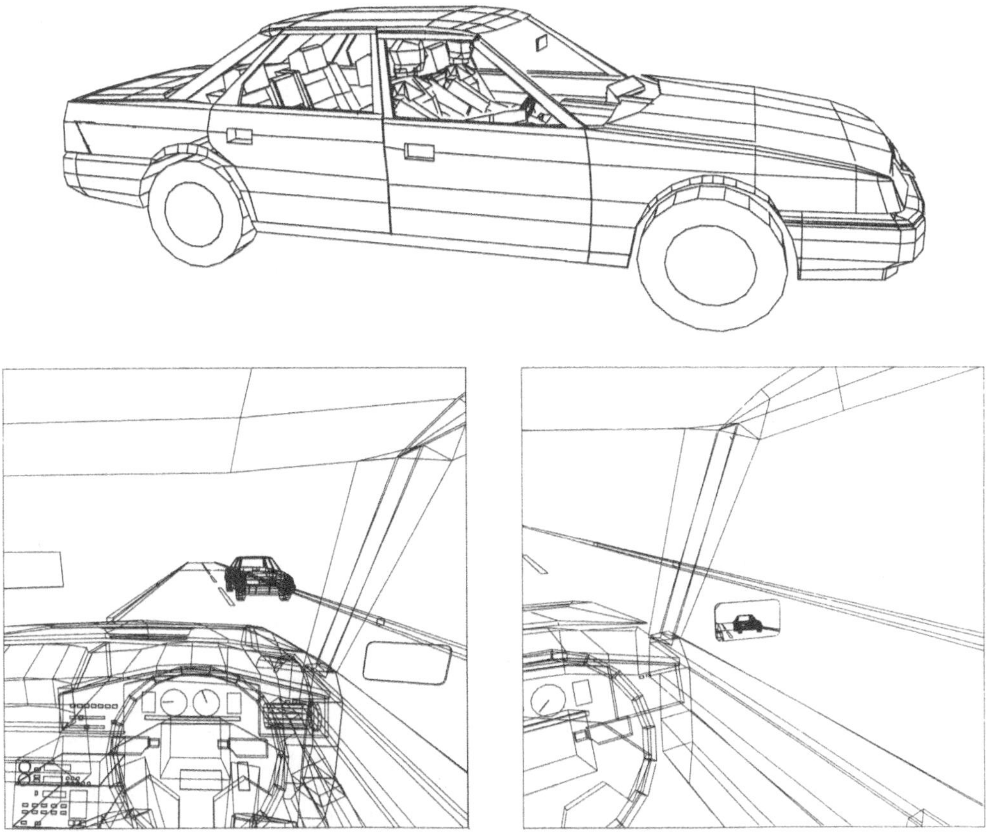

FIGURE 3. FORWARD AND REAR VIEWS FROM A PASSENGER VEHICLE

In Bonney et al., (1982) some of the general points made were that

'Human factors information can be computerised either as a predictive model or as data - raw or summarised. An obvious approach is to use a predictive man-model. The reasons for this include:

- the necessity to represent the man-model geometrically because of the interactions with the workplace model

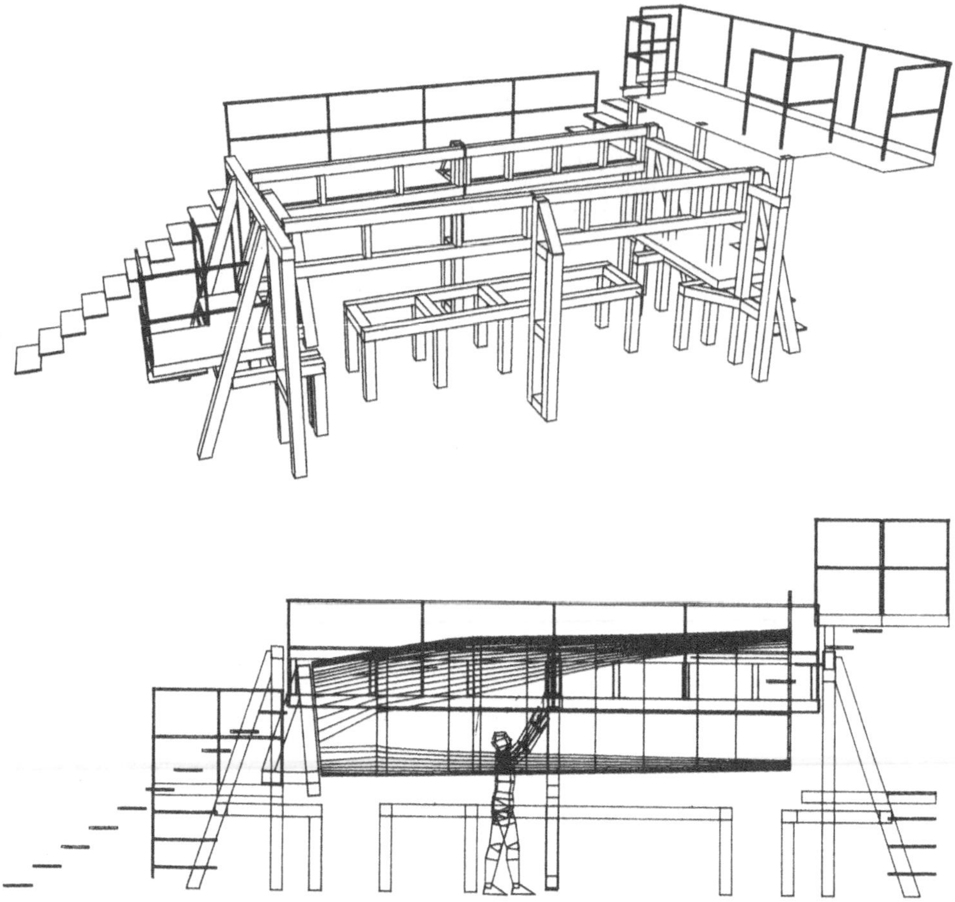

FIGURE 4. MANUAL ASSEMBLY OF THE FRONT FUSELAGE SECTION OF A FIGHTER AIRCRAFT

- predictive models can be combined with flesh algorithms, strength algorithms etc. in order to represent the man-model in working situations rather than the artificial situations of data collection. Evaluative algorithms can be used for example to test reach, joint constraints, interference with the workplace etc.

However, it is necessary to compromise the natural wish to model all attributes of man because:

- most applications do not need a detailed model.

- a detailed man-model would take so long to manipulate and display that it could not be used as an interactive tool.

- except for some specific populations (mostly military) comprehensive data are not available.

- even with good data there are problems with generalising the recorded actions into predictive rules.'

That long quotation was the lead into a plea for an agreed man-model to guide and encourage data collection on a unified basis. Eight years later the need is probably just as strong in that the data that are and will be collected will meet the needs of man modellers by chance if at all. A suggestion is that it may be possible to construct an idealised man-model or man-models specifically as a framework for data collection, processing and presentation. This presupposes that the data to be collected is to aid a parametric representation.

So far two classes of man-model, predictive and data models, have been mentioned. It is probably helpful to look briefly at some examples of how man-models might be used as the use will determine the facilities that should be included in a particular model.

The first use of man-models is, as in SAMMIE, for workplace and worktask design purposes. These workplace design systems will add progressively more facilities, attributes and functions to the man-model and so enable them to be used for a wider range of design problems. Sometimes it may be more efficient to generate a sub-set of the attributes for specific-design problems. The fundamental question is what model attributes are needed to solve specific problems.

The second use of man-models could be for medical diagnostic purposes e.g. identifying the consequences that certain work places and work tasks will have on back problems. The aim would be to use the man-model to determine the consequences of performing proposed or actual work. This may be looked upon as an extension of the previous model but needs more extensive work task evaluation methods together with highly detailed and realistic modelling.

Other uses of man-models are for specific purposes such as crash simulations or as needed by legislation. Although there may be technical difficulties the clear objectives clarify the system design problems.

A final example of man-models is to link with expert system design approaches. This was discussed in greater detail in Bonney et al., (1986). The broad argument is that there would appear to be merit in using models, probably geometric and algorithmic, which can be combined with rule based models. Some of the expert system work carried out at Nottingham is discussed in the next section.

An important point is that each additional attribute added to the man-model for realism not only adds to the complexity but could change the number of degrees of freedom. For example the number of degrees of freedom will increase if further linkages are added. On the other hand the number of degrees of freedom may be reduced by adding muscular constraints. Complexity makes the man-model control problem progressively more difficult. A corollary to this is the need to develop posture recording and controlling devices or transducers which correspond in complexity and which enable one to communicate easily the desired human 3D movement characteristics to the computer.

DEVELOPMENT OF MAN-MODELS FOR WORKPLACE DESIGN

As far as can be gathered from the literature surprisingly few major developments appear to have taken place over the last 10 years either with SAMMIE or other man-models. Yet it is clear that there is considerable potential for adding extra ergonomic modelling and evaluation facilities to man-models.

A range of possibilities for the development of the man-model can be listed by systematically taking each characteristic in turn such as flesh, man structure, strength models, posture prediction models etc., and adding the current knowledge. For example 'better' modelling of the spine could allow forces and moments to be calculated in relation to each vertebra and muscle grouping. This could aid posture prediction and allow identification and assessment of many of the problems associated with planning work tasks. These problems can thus be eliminated at the design stage.

Unpublished research along these lines by Friedrich and Corlett has produced a CAD model with static force and torque calculations between each vertebra. This provides considerable potential for development. In other research Tracy (1988) has used improved biomechanic models for the analysis of muscle and joint forces, particularly in relation to a wide range of awkward working postures. Together these two research projects clearly show that there is room for considerable advances to be made in the computer aided design of human work.

Another kind of facility which exists in some systems is to represent items of 'clothing' such as helmets and backpacks so that cockpit type evaluations can become more realistic. This is straightforward and by constraining the model may make it easier to represent than a full linkage representation e.g. by reducing the flexibility of back movement the modelling of the spine does not need to be so exact. At the end of the day, however, judgements still will be needed on which of the attributes to include or exclude. Examples of such attributes are how many links should be included in the man-model, should the hand be modelled, should the flesh be modelled etc. and if so to what level of precision?

In recent years an expert system called ALFIE (Auxiliary Logistics for Industrial Engineers) has been designed (Taylor and Corlett, 1987). Although the shell is widely applicable it has been implemented in the areas of heat stress, thermal comfort, work load, inspection lighting and general lighting. Other areas of ergonomics knowledge engineering are being actively pursued. Part of the design philosophy is that ALFIE will link with SAMMIE in order to provide the capabilities for wider ergonomic design assessments by combining the capabilities of evaluating geometric, anthropometric and environmental effects within the same system.

CONCLUSIONS

It has been argued that the knowledge base has and is increasing, computer power has increased and this has taken place in an environment where many more people are interested in solving complex human workplace design problems. CAD has become commonplace and many more experimental man-models exist. The time appears right to bring these developments together and for there to be a further major surge forward in this exciting and promising field.

REFERENCES

M. C. Bonney and N. A. Schofield, 1971, Computerised Work Study using the SAMMIE/AUTOMAT system. International Journal of Production Research. 19, 321:336.

M. C. Bonney, K. Case, B. J. Hughes, D. N. Kennedy and R. W. Williams, 1974, Using SAMMIE for Computer Aided Work Place and Work Task Design.

Paper presented at the Automotive Engineering Congress and Exposition of the Society of Automotive Engineers, Detroit, February 1974. Paper 740270.

M. C. Bonney and R. W. Williams, 1977, CAPABLE A computer program to Layout Controls and Panels. Ergonomics Vol 20 No 3 297:316.

M. C. Bonney, C. A. Blunsden, K Case and J. M. Porter, 1979, Man-Machine Interaction in Work Systems, International Journal of Production Research, Vol 17 No 6 619:629.

M. C. Bonney, K. Case and J. M. Porter, 1982 User needs in Computerised Man Models In: 'Anthropometry and Biomechanics, Theory and Application'. R. Easterby, K. H. E. Kroemer and D. Chaffin eds 97:101, Plenum Press. Previously presented at the NATO Symposium, Cambridge, July 1980.

M. C. Bonney, N. K. Taylor and K. Case, 1986, Using CAD and Expert Systems for Human Workplace Design, Proceedings of the IBM UK Scientific Centre Geometric Reasoning Conference, Winchester, December 1986.

BS 6735, 1987, Reach Volumes for location of controls on agricultural tractors and machinery. British Standards Institution.

K. Case, J. M. Porter, M. C. Bonney and J. Levis, 1980, Design of Mirror Systems for Commercial Vehicles. Applied Ergonomics Vol II, No 4.

S. A. Gibson, C. J. Osborne and M. C. Bonney, 1985, Computer Aided Ergonomic Design of a Cash Payment Unit (CPU) for a large retailer. in: 'Ergonomics International '85'. Proceedings of the 9th International Congress of the International Ergonomics Association. I. D. Brown, R. Goldsmith, K. Coombes and M. A. Sinclair eds Taylor and Francis.

J. M. Porter, K. Case, M. C. Bonney, 1980, Computer Generated Three-Dimensional Visibility Chart, in Human Factors in Transport Research D. J. Oborne and J. A. Levis. eds Academic Press London Vol 1 365:373.

C. J. Reid, S. A. Gibson, M. C. Bonney, D. Bottoms, 1985, Computer Simulation of Reach Zones for the Agricultural Driver. Proceedings of the 9th International Congress of the International Ergonomics Association. I. D. Brown, R. Goldsmith, K. Coombes and M. A. Sinclair eds Taylor and Francis.

SAMMIE, 1987, SAMMIE Systems Information Booklet Edition 5. SAMMIE CAD Ltd

W.H.Sheldon, 1940, The Varieties of Human Physique. Harper Bros. New York.

N. K. Taylor and E. N. Corlett, 1987, ALFIE - Auxiliary Logistics for Industrial Engineers, International Journal of Industrial Ergonomics, 2, 15:25.

M. Tracy, 1988, Strength and Posture Guidelines: A biomechanical approach. Unpublished PhD thesis. University of Nottingham

ACKNOWLEDGEMENTS

The authors would like to thank the Science and Engineering Research Council and the British Technology Group who between them funded much of the work described. Thanks also are due to SAMMIE CAD Ltd for permission to reproduce the figures which appear in SAMMIE (1987).

USE OF BIOMECHANICAL STATIC STRENGTH MODELS IN WORKSPACE DESIGN*

Susan M. Evans

Vector Research, Inc.
P.O. Box 1506
Ann Arbor, Michigan 48106

INTRODUCTION

Human anthropometric models have traditionally focused on the factors of human performance related to size, fit, clearance, or range of movement. An underlying structure in most of these models is a link system, similar to a skeleton, but concerned with functional, rather than anatomic joint centers, and used to position the human in space. Biomechanical models represent a logical extension of anthropometric models, adding segmental mass properties, load moments, and often muscle strength characteristics to assess operator performance over a wider range of static and dynamic conditions.

Both anthropometric and biomechanical models are appropriate for representing and modeling the physical stresses present in human-operated systems. Biomechanical models of operator strength or low-back stress are particularly relevant for tasks involving manual material handling, where the operator is required to move loads from one location to another, such as by lifting, lowering, pushing, or pulling. Quite often, these exertions are performed in confined spaces, at a high frequency or over extended periods of time, and with extreme or awkward postures. Palletizing tasks immediately come to mind. Equally stressful are tasks involving munitions handling, or routine maintenance and repair, where several components must be removed and replaced in the process of accessing and repairing the failed item.

This chapter describes a class of models used to evaluate physical stresses and operator performance. It begins with a brief overview of the biomechanical strength prediction models which have been developed at The University of Michigan. Programs currently applicable to system design and operating on a range of host hardware are then discussed. This includes PC-based programs which focus solely on static strength and low back biomechanics. It also includes a discussion of an ergonomic design system which integrates several measures of operator performance within a single designer interface and operates on a MicroVAX II engineering workstation. General issues for selecting such human perfor-

* The models described here have been developed at the University of Michigan's Center for Ergonomics. Don B. Chaffin, Center Director, leads the development of strength and low back biomechanical models. Jay Elkerton and the author lead the development and implementation of the current ergonomic design system (EDGE) on a MicroVAX II workstation at the Center. Inquiries regarding the single factor biomechanical or low back models should be directed to Don B. Chaffin, at the Center for Ergonomics, The University of Michigan, Ann Arbor, Michigan, 48109. All other inquiries should be sent to the author at Vector Research.

mance models are presented as is a wish list for future design-oriented biomechanical models. Enhanced graphical and decision support interfaces, as well as more robust models, are key concerns for the future.

BACKGROUND: STATIC STRENGTH AND LOW BACK BIOMECHANICAL MODELS

Computer-based biomechanical models have been present since the 1960s. Over the years biomechanical models of static strength have been applied to the design of hand tools, safe lifting limits, and control panel layout and activation force guidelines. Dynamic models, which consider human motion, have been applied to crash impact and human vibration studies as well as to the design of pushing, pulling and lifting tasks. Biomechanical strength prediction models have been a research topic at The University of Michigan's (U of M) Center for Ergonomics for nearly 20 years. Historically, the research has concentrated on static strength models, strength databases, and the use of these models in assessing the strength demands of industrial tasks. More recently, research projects have also considered dynamic strength modeling, specifically with respect to cart pushing, dynamic lifting, ladder climbing, and slip and fall prevention.

The static strength models consider the resultant moments acting on the joints, and the reactive moment produced to sustain the force and maintain equilibrium. While resultant moments can be measured directly from postural (i.e., body geometry), segmental mass, and force data, reactive forces are dependent on the strength-producing capability at the joint in question. For these models, the strength producing capability is based on an underlying database of population strength data, derived from standardized strength tests administered to thousands of industrial workers, both male and female. Strength capability of a specific exertion direction, force magnitude, posture, and anthropometry is directly related to the ratio of forces acting on the joint over the strength capability to resist the force.

Low back biomechanical models address the stress in the lumbar region, where a significant number of overexertion injuries occur. These models depict individual muscles and abdominal pressure forces acting to stabilize the torso in a specific task and posture.

The early U of M static strength model was based on a six-link co-planar parallel static force system (Chaffin and Baker, 1970). The model relied on mechanics to determine the resultant torques at a given articulation due to external and inertial forces. Resultant torques, based on body segment weight, load at the hands, and posture, were compared with experimentally derived reactive torque data for the same articulation. Several vital assumptions applied to the original biomechanical strength model and to those that followed:

(1) The total strength of the body is a function of the weakest muscle group's action;

(2) The ability to produce maximum torques at each articulation is independent of the activity level of adjoining articulations; and

(3) The coefficient of variation of the strengths of a group of people within the same age group and sex is independent of sex, age, and the hand and arm positions.

The coefficient of variation was used to adjust experimentally obtained muscle strength values for a given population to determine the population strength norms for performing a specified task.

Muscle strength moments for the trunk and upper extremity were developed from tests of 18 subjects in 270 positions involving 20 muscle groups (Shanne, 1972). Because muscles often span two joints, the prediction of strength at one joint required consideration of the angle of joints adjacent to the primary joint. Thus, muscle strength prediction equations for 19 muscle group strengths were developed as a function of body configuration. Burggraaf (1972) extended the strength equations for hip, knee and ankle extension as a function of leg position.

In 1973 the torso and upper extremity strength prediction model was extended to three dimensions, based in Shanne's three-dimensional torso model (1972). At that time, the model was validated for seated postures, using data from Shanne and from the Air Force. Jointly, this validation included 18 civilian and 71 military subjects in over 3000 test positions. The logic for the biomechanical strength prediction models are described in Chaffin and Anderson (1984).

Research conducted by Stobbe (1982) produced a static strength library, which related major muscle moment strengths to standardized postures, and obtained means and standard deviations for adjusting laboratory-obtained range-of-motion strength distribution data for industrial population strengths. Joint moment strength prediction equations derived in the laboratory from a limited number of subjects were adjusted for gender and industrial population strengths at standardized postures. While the contour of the strength curve for a particular range of motion remains relatively constant across populations, the location of the curve, i.e., the specific strength moment value at a given joint angle, varies widely among populations (Stobbe, 1982).

A basic assumption of these static exertion models is that load movement occurs slowly and smoothly, so that the effects of acceleration are negligible. Current biomechanical modeling research includes development of two and three dimensional static strength models on PCs, enhancement of the low back model through application of linear programming techniques to modeling the complex muscle forces acting in the torso, and development of dynamic biomechanical strength models for cart pushing and pulling, ladder climbing, and lower extremity gait analysis.

In summary, these biomechanical models have focused on representations of static muscle strength and low back (L5/S1) spinal compressive forces. Models of static strength depict the external forces acting on the body. Internal loads, due to muscle contractions, are implicitly modeled through the strength equations available at each of the model's joints. Both external and internal loading are considered in models of spinal compression, however. Dynamic or environmental forces are being addressed in current research. Despite the restriction to slow, controlled (i.e., static) exertions, the models have considerable relevance and validity to warrant their application to system design, particularly when the human's tasks involve controlled exertions.

SINGLE FACTOR BIOMECHANICAL MODELS USED IN DESIGN

Unlike the high performance crew stations of jet aircraft or even the interior of contemporary automobiles, most material handling workstations are designed without the benefit of a computer. Yet, the stresses imposed on the human operator and the impact on performance can be equally severe. In a study of 40 workspace designers in six industrial firms (Evans, 1985), the author found that computers, particularly mainframes or CAD workstations, generally were not available to the engineers and designers involved with workspace design. Only 20 percent of the designers cited computer graphics as a tool in design, and no more than 10 percent indicated that the computer provided human performance, work methods, or design related information. At that time, few of the respondents had micro-computers available to them. That has changed considerably since the 1985 study. Clearly, though, the mainframe-based models, such as the original biomechanical strength models discussed above, are unsuitable for design tools. The remainder of this section discusses single-factor, PC-based models developed to better address design requirements.

The algorithms and strength equations for the two- and three-dimensional biomechanical models have been programmed to operate on IBM-PCs. The models are interactive and use graphics to present the operator's posture and the resulting muscle strength requirements and low back compressive forces.

2D Static Strength Prediction Program™

The 2D Static Strength Prediction Program™ is appropriate for symmetric postures and exertions in the sagittal plane. The program is useful in the design and evaluation of workstations and tasks to ensure safe muscular exertion levels (Center for Ergonomics, 1986). Posture, exertion, and operator input for the 2D model consists of:

(1) body posture, specified as joint angles at the elbow, shoulder, hip, knee, and ankle;

(2) magnitude and direction of force acting at the hands; and

(3) male and female anthropometry, as percentiles or numeric values for height and weight.

Joint angles are provided either through the keyboard, or a mouse and cursor to position each link. A stick figure representation of the operator's posture shown on the screen provides immediate posture verification. The primary input and output screen from the 2D model is shown in Figure 1.

Hand coordinates are displayed as well. Coordinates are cakulated as the horizontal and vertical distance of the hand grip center from the midpoint between the ankles for the specified posture and anthropometry. The exertion at the hands may be in any direction in the sagittal plane; thus the model is not restricted to purely vertical lifts or horizontal pushes. Model output indicates the muscle strength requirements, relative to population norms, at each of the six joints, for the specified exertion. Strength percent capable at each joint is provided for males and females in a bar chart. Calculated back compression force is also displayed relative to the NIOSH standards for acceptable limits (NIOSH, 1981). Body link lengths, masses, resultant forces and joint moments, abdominal pressure predictions, and torso muscle force predictions are available on additional information screens.

3D Static Strength Prediction Program

While the biomechanics algorithms for the 3D static strength prediction program are more complex, the program does share a similar interface with the 2D program. Posture angles can be supplied via either the keyboard or mouse and cursor control. Muscle strength requirements are collapsed over both right and left sides and applicable axes of rotation (at the shoulder and trunk) and displayed in a bar graph similar to that shown in Figure 1. The key differences between the two models are:

- Three dimensional postures are possible, allowing limbs to be positioned asymmetrically and permitting torso twisting and bending. While the right and

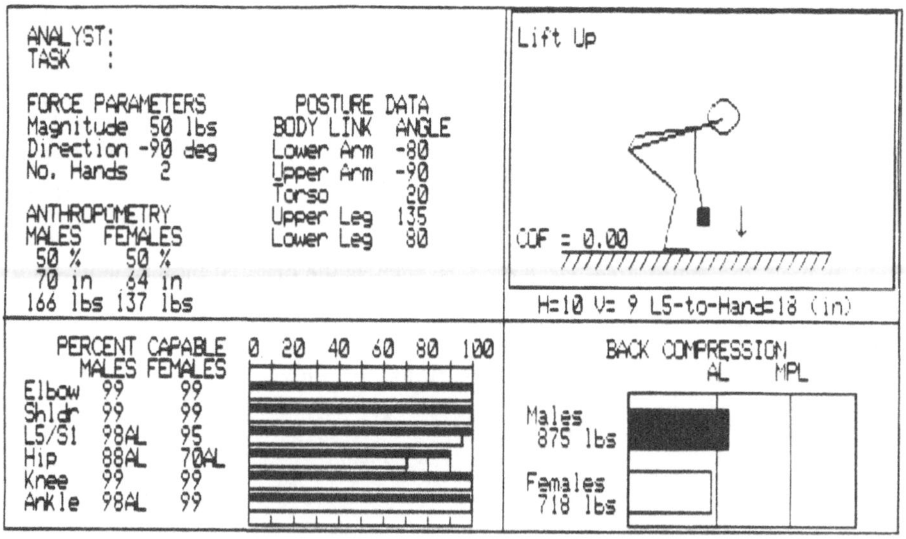

Fig. 1. Main input and output screen from 2D Static Strength Prediction Program™ (Center for Ergonomics, 1986), reprinted with permission, (software and screen copyrighted, 1986, The Regents of The University of Michigan, Center for Ergonomics).

left leg may be positioned asymmetrically, their movement is restricted to the sagittal plane. Fifteen angles are used to define the operator posture. Postures are displayed in three orthogonal views on the screen.

- Right and left hand exertions (both direction and magnitude) are independent, allowing more accurate representation of complex tasks. For each hand, two direction vectors (angle from horizontal and vertical) and a magnitude uniquely specify the force and direction.
- Muscle strength requirements, or percent capable, displayed by body region, represent the most limiting strength at that region. In the case of the trunk, this would be based on the values for flexion/extension, lateral bending, and rotation. Elbow, shoulder, hip, knee, and ankle strengths are compared over the right and left side values.

Both the 2D and 3D programs are structured to separate the biomechanical algorithms from the population strength data (e.g., the means and standard deviations for muscle strengths at specific joints). The standard population strength database is derived from strength tests on over 2000 industrial workers. These civilian strength profiles may be significantly different from either the Army's enlisted population or the Air Force's population of crew chiefs, however. As data on different populations are available they can be applied to the 2D and 3D programs to increase the accuracy of the muscle strength predictions for the population in question. With either program, results aid in identifying stressful task conditions and postures, and provide a means of directly identifying posture/weight/strength tradeoffs in early design.

AN INTEGRATED WORKSPACE DESIGN MODEL

The biomechanical focus of the models just discussed represents only one of several aspects of human performance relevant to workspace designs. The highly repetitive nature of the operator's tasks or the inefficiencies due to poor work methods influence the effectiveness of the operator-equipment interface but cannot be studied with static biomechanical strength models alone. A need exists for an integrated design system which considers the interactions and interdependencies among several measures of operator performance (Evans, 1985; Kroemer et al., 1988).

The EDGE System: Ergonomic Design using Graphic Evaluation

One approach to developing an integrated ergonomic design model capitalizes on the availability of relevant single factor performance models, and achieves the integration through a common designer interface which accesses the models. The EDGE (Ergonomic Design using Graphic Evaluation) system employs this modular approach and uses existing models of strength, reach, metabolic energy expenditure, and elemental time prediction to aid in the design and analysis of manual tasks. Muscle strength requirements and low back compression force estimates are obtained from a variation of the 3D strength model described earlier. The variation includes a posture prediction feature which simplifies task input considerably. The posture prediction feature also serves as a mechanism for performing reach analyses. Strength and energy expenditure are combined for sagittal plane lifting tasks in a prediction of lifting limits provided by the NIOSH Work Practices Guide (NIOSH, 1981). Predictions of metabolic energy expenditure are based on research by Garg (1976); elemental time predictions are obtained from MTM-2 tables.

The prototype EDGE system was developed on an HP-1000 mini-computer. The current system is being implemented on a MicroVAX II engineering workstation. EDGE developers have been concerned with two key issues:

- design of a user interface which represents the design tasks and goals common to workspace design; and
- development of a design tool which is of use to engineers and designers who have limited background in ergonomics.

The basic EDGE system framework is shown in Figure 2. A designer interface serves as the bridge between the operator performance models and the designer. The interface is also the means for providing ergonomic information in formats of use to designers, and providing design guidance to aid engineers who are not trained in ergonomics. The framework is sufficiently flexible to allow the addition, deletion, or modification of performance models with only minor modifications to the EDGE system itself. Special-purpose subroutines handle the input and output to the individual models, and to related operator and workspace graphic routines, human performance databases, and design criteria. Operator performance models within EDGE share information pertaining to the operator, workspace or environment, and task. The information categories, and their overlap among the current set of performance models, are shown in Table 1.

System Components

The EDGE system contains a number of components which support the workspace design process and facilitate the interface between models and designer. Among these components are 3D representations for the human operator, workspace locations, and objects, a methods table for defining sequences of task elements, and operator performance criteria. EDGE uses the latter component, performance criteria, in evaluating model output and assessing design acceptability.

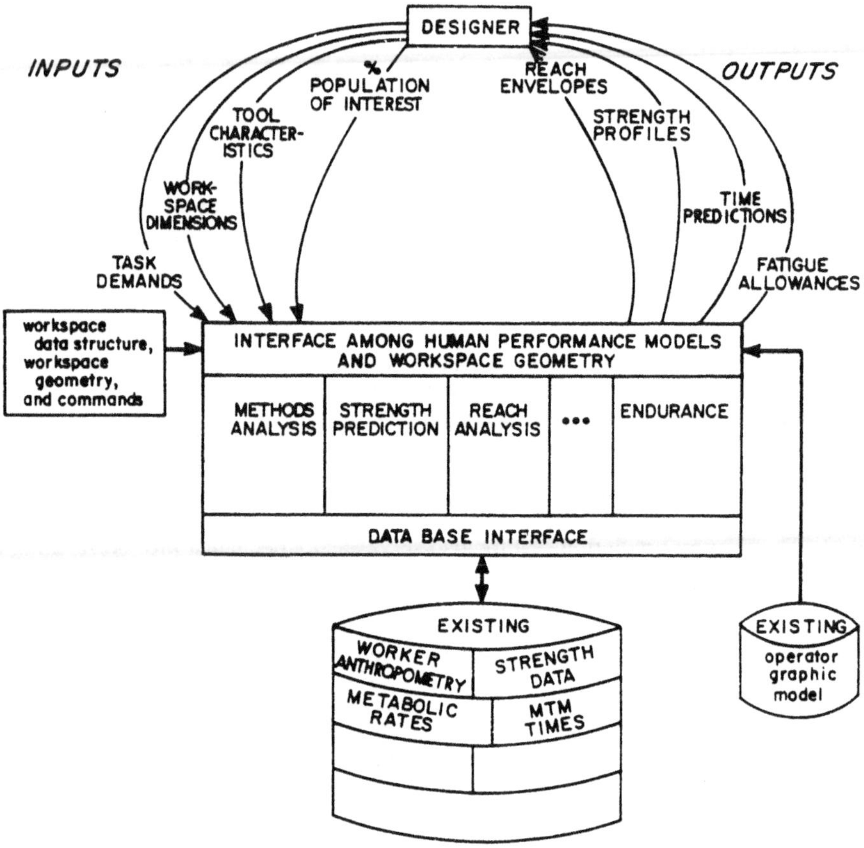

Fig. 2. EDGE system framework to support multiple models of human performance in ergonomic design (Evans et al., 1984).

Table 1. Information categories shared by selected operator performance models in the EDGE design system (adapted from Evans, 1985).

Information	Biomechanical Strength Model (Garg and Chaffin, 1975)	NIOSH Lifting Model (1981)	Energy Expenditure (Garg, 1976)	MTM-2 (Karger and Hancock, 1982)
Operator:				
Age	I[1]	-	-	-
Size (stature, weight)	I	-	I	-
Strength	I	E	-	-
Reach, range of motion	E[2]	E	-	E
Endurance	-	E	O	-
Posture	I/O[3]	E	I	-
Task element				
Action	I(Direction of Exertion)[4]	I(Lift)	I	I(Get/Put)
Frequency	-	I	I	I
Workspace				
Layout/locations	I	I	I	I
Clearances	-	-	-	I
Object				
Size	I	I	I	-
Weight	I	O(Predicted)	I	I
Number of hands	I	2 assumed	I	I
Handling characteristics	-	-	-	I
Performance Measures				
Muscle strength requirements	O	-	-	-
Balance feasibility	O	-	-	-
Back compression force	O	-	-	-
Reach feasibility	O	-	-	-
Lifting limits	-	O	-	-
Energy Expenditure	-	-	O	-
Time	-	-	-	O

NOTE:
I = Input parameter; O = Output value; E = parameter embedded within model.

1 age embedded within strength profiles of 3000 industrial workers.
2 in iterative mode, model will perform reach feasibility analysis prior to calculating biomechanical loading and strength of posture.
3 iterative model allows input of general posture orientation, but produces detailed posture as output. Predefined postures include stand, sit, squat, deep-squat, stoop, lean, and split-leg.
4 predefined exertions include lift, lower, push, pull, pull right, pull left, pull down, hold, torque-right, and torque-left. Users can also define their own exertion vector.

Operator and Posture. Operator profiles define the internal capabilities and structure of the operator model: it's mobility, strength and linkages. Profiles can be modified by an experienced user if the design population differs from the default operator description. Kinematic data is currently embedded within the strength/reach prediction model, (see Garg and Chaffin, 1975) preventing user modification. The operator consists of a 13-link model, with its origin on the floor at the midpoint between the ankles. Body segment lengths are expressed as a ratio of body height (stature) based on "average" ratios developed by Drillis and Contini (1966). Link enfleshment parameters, used to define the three-dimensional contours of the operator model during graphic display, are based on anthropometric breadth, depth, and circumference measurements, adjusted by stature.

A posture descriptor specifies the general body orientation for each task element. Posture descriptions completely define the body orientation i.e., all 17 angles required to position the legs, trunk and arm segments in three dimensions. These angles may be derived by the system's biomechanical strength prediction model, or may be explicitly defined by the designer to accommodate the restrictions or obstructions of the workspace.

Workspace Locations. At this time, workspace geometries exist primarily as lists of locations. In addition to the 3D location reference point defined relative to the workspace origin, locations also contain a location case, which specifies the expected object "fit" at the location, and is used in determining the MTM-2 **put** movement code. The code distinguishes between **Loose** and **Close** fit, with the latter requiring some correcting motion to engage, as with assembling non-symmetric parts.

Objects. Objects apply to any number of locations or tasks, in any combination. Additional properties include object dimension, weight, handling code, handhold locations, and handhold case. Handling code identifies the ease of handling for use in predicting movement time. Codes differentiate between balanced, unbalanced or awkward to handle, or requiring extra care. Handholds are point-locations, defined in the object coordinate system. Handhold class defines the **get** movement class during movement time prediction, and identifies the type of handhold provided. Classes include 1) an adequate handhold exists for a power grasp, as with a tote box or cart handle; 2) location prevents power grip, but secure hold is possible, as with an ammunition cartridge; or 3) no obvious handholds: multiple regrasps needed to gain control, as with large awkward subassembly components.

Tasks. Task elements identify the action (as a direction of exertion), the specific object, workspace locations for origin and destination of exertion, element frequency, and posture. Employing *direction of exertion* as the primary action verb is in contrast to the traditional MTM elements of reach-grasp-position. These whole body activities are more meaningful for the design applications and operator biomechanical and posture analyses considered by the system user.

Workload and Performance Criteria. Workload stress is the objective descriptor of operator physical performance under the specific combination of task actions, object weight, hand locations, frequency, and posture data supplied by the designer. Workload stress relevant to material handling tasks include biomechanical (whole body and muscle strength, body balance, and low back stress), kinematic (whole body reach with body balance), metabolic (energy expenditure), and temporal or time and motion.

Ergonomic performance criteria define specific critical values or regions of acceptable stress levels. They are compared against the predicted task-related stresses in evaluating workspace/task designs. Example criteria include the minimum population strength or reach percentile accommodated, the maximum allowable back compression force, the maximum allowable lifting limit, as a function of the NIOSH action limit, the maximum energy expenditure rate, or the maximum percent allowed for non-productive or body-assist time per task element or cycle.

System Input

Task elements, consisting of action, object, location, posture, and frequency tuples, are entered onto a spreadsheet-like work methods table. Separate screens appear for defining object dimensions or locations in the workspace. Graphical templates are provided for predefined postures or actions/exertions to guide the user during the input process. A menu-based window environment manages the various input screens and forms. The user specifies the inputs once, and the system interface processes and reformats them for each individual model.

Input screens are also provided for operator descriptions and design criteria specification. In the current system, operator anthropometry is restricted to values for height and weight . Given the expertise of the users, the design orientation of the system, and the fidelity of the models, this level of detail is appropriate. As the base of performance models supported by EDGE expands to include reach contours or visibility checks, more sophisticated methods will be required to more accurately depict operator anthropometry. The system accommodates design criteria and uses the criterion for evaluating model predictions and comparing designs. The user supplies or uses default values for male and female strength accommodation levels, maximum back compression force, energy expenditure, or target elemental time.

System Design Tasks

EDGE provides the designer with several specific functions for evaluating workspace and task configurations. The functions or design tasks are selected based on the nature of the operator's tasks and the types of physical stresses affecting performance. Specifically, the design tasks:

- evaluate operator physical stress during a single task;
- evaluate the cumulative stress during repetitive tasks;
- perform "what-if" analyses by varying task parameters; and
- compare operator performance over two or more task designs.

The designer selects the operator tasks to analyze, and then selects the performance model, or lets EDGE execute all applicable performance models. The biomechanical strength prediction and the NIOSH Work Practices Guide models operate on single task elements. The metabolic energy expenditure and the MTM-2 elemental time prediction models are appropriate for task sequences and indicate cumulative stress over the selected tasks.

EDGE provides for "what if" analyses by allowing the designer to vary task or workspace parameters along specific dimensions. Output formats aid in identifying the trends in design outcome, interactions among parameters, and overall design result in light of workload and performance criteria. As the design progresses, and parameter changes affect performance, the designer can periodically stop and review the results and compare the cause-effect relationships of the design iterations. The outcome helps to delineate further areas for investigation.

System Output

Primary ergonomic output displays employ two-dimensional graphs to show trends and three dimensional layouts to project the enfleshed operator within the workspace. The objective is to avoid overwhelming the designer with too much detail. Formats for system output correspond to the design tasks just discussed. They have been constructed to aid in detecting design deficiencies, diagnosing the possible cause, and correcting the problem.

Preliminary output screens provide performance and workload results at a very general level, across tasks if appropriate. They indicate exceptions or unacceptable cases which deserve further attention. EDGE identifies "exception" tasks by comparing performance model outcomes against stated performance criteria. Examples of the type of output information displayed at this level for the most stressful tasks are overall muscle strength percent capable, maximum back compression force, or maximum energy expenditure in kilo calories (Kcals). Data would be separated for males and females.

Subsequent displays focus on specific parameters within stressful tasks. As individual tasks are selected, a three-dimensional operator graphic depicts the posture with the object and hand location information. Muscle strength percentiles are coded based on stressfulness and superimposed over each joint, providing a direct mapping between stress and body location. The designer selects which tasks to display based on query-like commands, e.g., "Show the tasks which contain horizontal locations over 20 inches," or "Show the tasks with the three worst predicted muscle strengths." The combination of display formats and user control over what is presented leads to quick identification of task element *high drivers*, and directs the designer's attention to the critical parameters to change.

ISSUES FOR DESIGN MODEL SELECTION

While the 2D Static Strength Prediction Program is currently available for use, the EDGE system, however, is still under development. System designers should be asking questions of model developers and of themselves to identify their model needs. They must identify their design application, and the human performance issues relevant to the resulting operator tasks. They also need to identify the people who will be using the human performance models, what their skills are, and where they fit in the overall system design process. The amount of time available to study design issues, the quality and quantity of information available, the expertise of the designers, and the corporate CAD culture all influence how the model will be used, if not which one applies. These issues are discussed in more detail below.

Is it the right tool for the task?

The question addresses the relevance of the model or system to the operator's task. Static biomechanical strength models are inappropriate for assessing strength requirements in a zero-G environment. Similarly biodynamic models which assess crash impact or other high-G forces are cumbersome and data intensive when used in manual materials handling applications. The models discussed in this chapter are appropriate for estimating the effectiveness of workspace designs involving physical exertions by the operator. The current set of strength, low-back stress, metabolic, and elemental time prediction models included in the EDGE system will address the demands of both infrequent and highly repetitive tasks. The structure of the EDGE interface provides room to expand or improve the model base as models of dynamic strength, physical interference, or visual performance become available.

Who will use the model?

Decisions impacting the man-machine interface design are generally made by several diverse groups located throughout the organization and the design process. The impact of these decisions may go undetected until the pieces are assembled at system mock-up, or later. Clearly the need exists for design decision aids which answer the questions of physical stress, almost before they're asked, and within the current design structure. Often, in the interest of tight time frames and reduced human factors manpower, this means that the models and tools will be used by persons trained in industrial or mechanical engineering rather than ergonomics or human factors. The choice of system user has implications for the type of interface and the types of analyses (Askren, 1985). The human performance expertise should be provided to the designer (via expert systems or enhanced decision support systems), whenever possible, rather than expecting users to come to the system already endowed with it.

The EDGE framework assumes that the system would be available to and used by all designers, as well as any in-house ergonomic experts. The interface has been designed to address both groups, providing structure and assistance to the untrained, and permitting free-form input and analysis selection for the expert.

How easy/difficult is it to use?

Ease of use is influenced by the complexity of the model inputs and the design assistance provided by the output. Cumbersome or complicated input requirements, coupled with the time constraints of the design process, and the impatience and inexperience of the designer will preclude a model's use in all but the most severe cases. For example, automatic posture prediction, although often limited in fidelity, is adequate for rough posture estimations in the first iterations. This is especially true when the alternative requires inputting twenty angles in a range of local coordinate systems. Similarly, the availability of well documented system defaults and design templates will aid the user in the initial stages of design.

Model output, or an appropriate designer interface, should also support the design process. At a minimum the output should provide for *detection* of design problems. At the least, this involves comparing model output against available criteria and displaying exception cases or outliers. An example would be highlighting a task which yielded a muscle strength prediction of only 10 percent capable. A further step requires that the output aid in *diagnosing* the problem. With the above example, the system would locate the body region which is limiting the strength capability, in this case the shoulder. A final aid would provide *remediation*. Here the system suggests a course of action to alleviate the problem. In this example, based on the exertion at the hands (30 pounds) and the posture (standing with arms extended), the system suggested that the load location be brought closer to the body, reducing the horizontal distance. In many current systems, the first two interventions are possible. The third is considerably more difficult, particularly when multiple factors are involved, yet is worth working toward.

Is it compatible with other systems?

Human performance issues cover many dimensions which often interact. The EDGE system is an attempt at combining several measures of performance within the related domain of physical stress models. Problems arising in such integration efforts include consistent model structures (e.g., compatible link systems or angle notations), level of information detail (tasks described at the THERBLIG level (e.g., reach, grasp, position) versus aggregated actions (e.g., assemble)). Other models are appropriate within a given man-machine interface design as well. A challenge for model developers will be to investigate means by which these diverse models can be integrated together to assess overall operator performance in complex systems.

Independent of other areas of performance, the models should be developed to work within the intended system design CAD environment. The advantages for the models are immediate access to design data, such as workspace geometries, part dimensions, process or methods standards, or previous designs. The advantage for the designer is immediate access to the answers within current design activities.

CONCLUSIONS

The biomechanical static strength prediction models and the EDGE system provide valuable tools for the design of workstations where operator physical performance is an issue. There is, however, considerable room for enhancements to better represent the human operator, and to address the needs of a range of system designers.

Future research is needed to develop enhanced performance models which reflect a wider range of task conditions. The biomechanical strength prediction models presented here reflect static, or slow, controlled exertions. Predicting performance under dynamic exertions is the next step. Research is needed to develop models which reflect both dynamic responses and dynamic strength capabilities. Both static and dynamic models are in need of improved posture prediction algorithms which accurately reflect the body kinematics under loaded conditions. While "snapshot" images of operator postures may be acceptable for static exertions, dynamic activities require techniques for operator animation which depict the operator's task-oriented postures over time. Systems such as TEMPUS at the University of Pennsylvania (Badler, 1983) are moving in that direction.

The focus of these models should be on both preliminary and detailed design. The burden of supplying input information for existing models restricts their use in preliminary design, when the quantity of operator, task, and environment information is often too limited to meet the model's requirements. Catalogues of previous designs, used as templates for preliminary design may be one solution. Design integration with existing company CAD databases is also essential to make the models available throughout the design process, and to eliminate the need for redundant input of previously defined layouts, object geometries, or corporate standards.

Finally, the technical expertise of the model user should be considered in developing interfaces for a wider range of system designers. Expert design aids which assist in detection, diagnosis, and remediation should enhance the process for all designers, as well as for the operator, the ultimate recipient of the improved design.

ACKNOWLEDGEMENTS

The EDGE integrated design system was initially constructed as part of the author's doctoral dissertation, in the Department of Industrial and Operations Engineering at The University of Michigan. Dissertation funding support was provided in part by Ford Motor

Company. Funding support for the current implementation of EDGE is provided by Chrysler Corporation.

REFERENCES

Askren, W.B., 1985, "New Role for Human Factors in Equipment Design." Human Factors Society Bulletin, 28(4), 1-2.

Badler, N.I., Webber, B.L., Korein, J.U., and Korein, J., 1983, "TEMPUS: A System for the Design and Simulation of Mobile Agents in a Workstation and Task Environment." Proceedings of IEEE Trends and Applications Conference, March.

Burggraaf, J.D., 1972, "An Isometric Biomechanical Model for Sagittal Plane Leg Extension." M.S. thesis, Department of Industrial and Operations Engineering, The University of Michigan, Ann Arbor, MI.

Center for Ergonomics, 1986, "Version 4.0 User's Manual for the Two Dimensional Static Strength Prediction Program™," The Center for Ergonomics, The University of Michigan, Ann Arbor, MI.

Chaffin, D. B. and Baker, W., 1970, "A Computerized Biomechanical Model for the Study of Manual Materials Handling." AIIE Transactions, 2(1).

Chaffin D. B., and Andersson, G., 1984, Occupational Biomechanics, John Wiley and Sons, New York.

Drillis, R. and Contini, R., 1966, Body Segment Parameters (Report No. 1166-03). (Office of Vocational Rehabilitation, Department of Health, Education, and Welfare). New York University School of Engineering and Science, New York.

Evans, S.M., Chaffin, D.B., and Foulke, J., 1984, "A Method for Integrating Ergonomic Information in Workspace Design." Proceedings of the 1984 International Conference on Occupational Ergonomics, Toronto, 373-376.

Evans, S.M.R., 1985, "Ergonomics in Manual Workspace Design: Current Practices and an Alternative Computer-Assisted Approach," Ph.D. Dissertation, Department of Industrial and Operations Engineering, The University of Michigan, Ann Arbor, MI.

Garg, A., 1976, "A Metabolic Rate Prediction Model for Manual Materials Handling Jobs." Ph.D. Dissertation, Department of Industrial and Operations Engineering, The University of Michigan, Ann Arbor, MI.

Garg. A., and Chaffin, D.B., 1975, "A Biomechanical Computerized Simulation of Human Strength." AIIE Transactions, 7, 1-15.

Karger, D.W., and Hancock, W. M., 1982. Advanced Work Measurement. Industrial Press, Inc., New York.

Kroemer, K.H.E., Snook, S.H., Meadows, T.B., and Deutsch, S.(editors), 1988, Ergonomic Models of Anthropometry, Human Biomechanics, and Operator-Equipment Interfaces, (Proceedings of a Workshop), National Academy Press, Washington, D.C.

NIOSH, 1981. Work Practices Guide for Manual Lifting. USDHEW Publ. No. 81-122.

Shanne, F. A., 1972, "A Three-Dimensional Hand Force Capability Model for the Seated Operator." Ph.D. Dissertation, Department of Industrial and Operations Engineering, The University of Michigan, Ann Arbor, MI.

Stobbe, T., 1982, The Development of a Practical Strength Testing Program in Industry, Ph.D. Dissertation, Department of Industrial and Operations Engineering, The University of Michigan, Ann Arbor, MI.

REPRESENTATION OF MAN USING CAD TECHNOLOGY: USER BEWARE

Patricia L. Rothwell

Defence and Civil Institute of Environmental Medicine
Department of National Defence, Canada

INTRODUCTION

In workplace design, the ergonomist aims to optimize worker comfort, safety and performance. To do this, he must consider relationships between the operator and his work, the work environment, and the equipment used. Anthropometry, the scientific measurement of the human body, provides techniques which the ergonomist uses to estimate man's reach, vision, body clearance and body posture within the workplace. For example, two-dimensional drawing board manikins, composed of articulated scale representations of body segments, are used to describe man in side view; and stick-figure manikins are used to provide reference loci from which reach distances, visual angles and eye positions can be calculated. Partial and whole-body manikins are used to convey more realistic likenesses of man by representing anatomical landmarks, body contours and segment masses.

Each of these tools is appropriate for isolated applications but none is singularly satisfactory for general-purpose workplace design. The major reason is that they are limited in their abilities to represent the anthropometric variability between individuals. There are also problems associated with the availability of these tools, the cumbersome nature of their use, and the questionable validity of their results (Rothwell, 1985).

More and more, computer-aided design (CAD) is replacing traditional manual design functions (e.g., drafting, calculating, analyzing) with computer packages that utilize interactive computer graphics (Majchrzak et al., 1987). This has revolutionized the methods to model and simulate the physical environment. But relatively few CAD program developers have used this approach to represent the human component of the man-machine system (Bonney et al., 1979; Kingsley, Schofield and Case, 1981). A major factor influencing this deficit is the lack of mathematical models of the human body with respect to its shape, joint articulations and motions. Still, man-modelling CAD offers better ways to represent man in the man-machine system than do traditional anthropometry tools (Rothwell, 1985).

One major advantage of man-modelling CAD is the potential to model complex individual differences. The technology also offers the potential to model atypical body structures and functions (e.g., in designing for physically disabled persons). Another advantage is the ability to view the man-machine system in three-dimensions. This facilitates the consideration of issues such as cross-body reaches, asymmetrical postures and postural stability (Rothwell and Hickey, 1986). By encouraging iterative explorations of complex man-machine relationships, it can be argued that man-modelling CAD leads to more thorough workplace evaluations than do manual techniques.

The use of CAD to represent man's anthropometry also has disadvantages. Some problems stem from the limitations that anthropometric source data impose on how the man-models can be defined (e.g., using only external body measures). Other problems arise from our limited understanding of how man should be modelled (e.g., the underlying assumptions used to generate and manipulate man-models, population representation using percentile data). There are also problems that stem from the nature of CAD technology itself. For instance, demand for detailed input data and compelling graphic displays imply inherent model validity that may not be justified. These problems persist because few CAD systems are evaluated rigorously and little basic research is conducted to resolve outstanding modelling problems (e.g., how to derive internal link structures from external anthropometric data).

Man-modelling CAD is not the universal remedy to represent man's physical attributes in design. Part of the reason is that, at present, CAD systems do not provide expertise in anthropometry or workplace design. Instead, they offer the user the freedom to manipulate design elements in many different ways. It is implicit, then, that the user must understand a) how man-models can and should be used to represent man's physical characteristics, and b) how workplace elements should be manipulated to address design objectives. Furthermore, CAD man-models do not replace the use of human subjects in design fitting trials. Rather, they allow the designer to consider man-machine interactions in early design stages. Finally, as with any computing system, individual CAD packages are subject to programming errors, some of which are not immediately obvious. The user must beware of the potential for such errors, and make an effort to understand how the capabilities and limitations of a given system will affect design work.

The following sections of this paper illustrate that expertise in anthropometry and workplace design are required of those who elect to use man-modelling CAD. Some insights are provided into the availability and nature of the technology. Following this is a general discussion of man-modelling and workplace-modelling issues, and the factors that should be considered when performing an ergonomics assessment of the man-machine model. The message can be viewed as a warning insofar as the technology should not be used naively, and not all systems satisfy all workplace design objectives.

SYSTEM AVAILABILITY

A survey (Hickey, Pierrynowski and Rothwell, 1985) conducted to identify existing computer man-modelling programs found that some 33 programs employed man-models while approximately 40 more incorporated single or isolated measures of human characteristics (e.g., maximal reach, visual field angles, task time). Only a few of these programs attempted to represent the various body sizes, proportions and joint articulations of man. Also, little detailed information was available on these programs, making it difficult to classify the types of analyses they perform. Changes in the program names and gaps in reporting updates made it difficult to identify new or independent modelling efforts. The conclusion from the survey was that the written literature is currently a poor source of information regarding the characteristics and uses of man-modelling CAD systems.

Most man-modelling CAD systems have been developed for governmental or university research purposes. Although several programs are used for general military applications (Richards and Companion, 1982), the majority focus on automotive or aerospace problems. Only a few man-modelling systems are available commercially, and the costs to make these systems functional can be prohibitive. Several systems can be obtained through cooperative agreements with their developers. However, the conditions for using them can be limiting (e.g., in exchange for consulting services or cooperative developmental support). In either case, few system developers provide the user with access to the source code, or freedom to modify the software.

SYSTEM USE

The reasons for wanting to use man-modelling CAD are diverse. One may seek to animate sequences of man-machine interaction, to augment analyses of operator reach, vision and body clearance, or to obtain an effective tool for communicating design ideas to managers or customers. Although such objectives are legitimate, it is possible that no system will suit all the needs of the user. Even when useful systems are found, associated costs may overshadow their advantages.

In deciding to employ man-modelling CAD, the user may believe that the expertise of anthropometry and workplace design specialists will be provided (e.g., that the system incorporates heuristics and algorithms necessary to interpret and evaluate model specifications, and to present optimal design solutions). In fact, most man-modelling systems do not provide this expertise and are best suited to users who are already knowledgeable about anthropometry and workplace design.

Essentially, man-modelling CAD allows the user to create and manipulate entities that represent the physical characteristics of man and the workplace; the extent to which the user *can* model and manipulate these characteristics varies widely among systems. The extent to which the user *should* model and manipulate these characteristics varies according to the application.

The user may desire numerous modelling capabilities, but trade-offs exist between the conveniences and costs of acquiring those capabilities (Lane, 1982). For instance, increased modelling capabilities often imply demands for detailed data. Sometimes, these data are laborious or time-consuming to obtain, if they can be obtained at all (e.g., enfleshment-modelling, calling for breadth and depth data for each major link segment). In turn, detailed input data may reflect model detail that demands increased computational loads. As a consequence to obtaining more modelling capabilities, then, the user may require more resources to address the application. Therefore, the user must determine modelling needs on the basis of the design objectives, and not succumb to the urge to model all attributes of the man-machine interface.

Other implications of using man-modelling CAD should not be underrated. Its use can have affects that range from the way design work is structured, to how results are interpreted and communicated to others. It can impose demands for physical, operational and maintenance support of computing hardware and software, and can have significant impact on the skills and design strategies required of the user. Therefore, the technical, management and human-computer aspects of using CAD must be considered. These issues are discussed more thoroughly by Majchrzak et al. (1987).

SELECTED MODELLING ISSUES

The Man-Model

It appears that many of today's anthropometric man-models have similar origins and are based on similar principles. Yet, they offer significantly different ways to define and manipulate their respective man-models. To follow are some of the factors that should be considered by the potential user.

Individual versus Population Representation. In workplace design, anthropometry data are usually used to represent populations or individuals that must be accommodated by the design. In some cases, the user may employ anthropometric dimensions that do not represent any particular population or individual in order to explore the boundaries of body combinations that fit the geometry of the design (e.g., for the purpose of operator selection). Each of these design/evaluation approaches has implications on the way that man should be modelled using computer.

To represent the physical characteristics of a population, anthropometric data are usually expressed as means and standard deviations from which percentiles are calculated. Commonly, a percentile is interpreted as the percentage of a population having *one* body dimension of a certain size, or smaller (Damon, Stoudt and McFarland, 1966). This use of anthropometric data to assess accommodation of a population has been criticized because it fails to consider the interactions of anthropometry variables (Bittner and Moroney, 1974).

Some CAD systems deviate from the conventional use of percentiles by generating man-models that have the same percentile on *all* body dimensions. This modelling technique is misleading in that it implies that the sums of individual dimensions (expressed as percentiles) can be equated with composite dimensions (also expressed as percentiles). For example, a man-model having 95th percentile stature is mis-represented as being equivalent in height to a man-model having 95th percentile measures for all body segments that contribute to stature.

This technique also results in the generation of proportionate models. At the least, CAD systems must use percentile data appropriately. If the technology is to show superiority over traditional techniques, it must represent more than just percentiles and proportionate man-models. It must be able to represent the disproportionate body sizes that make up the population. Hence, man-modelling CAD programs should make use of sophisticated statistical methods to manipulate population data (e.g., Monte Carlo simulation, use of multi-variate statistics, etc.) and to represent individual extremes of a specific population.

To represent an individual, the CAD system must be able to accept a unique set of anthropometric data as input. These data must be interpreted in ways that convey the individual's specific physical dimensions that are important to his or her fit to the design. Therefore, the user must be satisfied that the system can generate valid man-models that are based on individual data.

Gender Representation. The standard practice in man-modelling is to generate a manikin that is derived from male data. In most cases, systems that claim to model females accept female data as input and then use data derived from studies on males to generate link lengths, represent joint mobilities, and so on. Since the skeletal and muscular features of males and females differ in numerous ways, it is not known how well the male-derived data can be generalized to females (Lane, 1982). Therefore, it is possible that no available system will satisfy the need to represent females in a design.

Body Segment Representation. Man-model link structures often dictate the number of body segments and joint articulations that are represented. Calvert, Chapman and Patla (1982) recommend that the human body can be represented using 23 links, if details of the hands and feet are ignored. Some man-modelling CAD developers count single point landmarks (e.g., an eye location reference point) as links, giving the impression that the model has a more detailed link structure than is the case. Similarly, some count different functional conditions of single links as separate, independent links (e.g., fingertip and palm reach references counted as two separate hand links). Therefore, the user must examine the link structures of candidate man-models to ensure that major body segments needed for the application are represented.

To define body segment geometries, some CAD systems require external body dimensions from which the man-model's link lengths are derived using *standard* formulae (e.g., those of Dempster (1955)). The user must be satisfied that such formulae appropriately model the individuals or populations to be represented. Other systems call for internal link length data. Because anthropometric data are not normally collected to satisfy this need, the user may have to transform external data to internal data when precise segment lengths must be modelled. In either case, the user may be faced with the task of transforming internal link lengths back to external body measures when reporting the anthropometry of the man-model.

Joint Representation. To model man's movement characteristics, most computer man-models allow rotation about body segments representing major joint centres. Usually, those rotations model movements in only one or two planes, treating all joints as either hinge or pin joints. Unless the user needs to model multi-axial joint rotations (e.g., to indicate realistic joint excursions at the spine, hip or shoulder), this level of joint modelling should be satisfactory. The onus is on the user to make this judgement.

The various movements that occur at the joints are commonly referred to as flexion, extension, abduction, adduction, medial rotation and lateral rotation. These movement notations refer to the relative change in position of body segments, with respect to their proximal joint centres. When using CAD, it is advantageous to be able to define the body segments' spatial orientations that result from these movements. Euler angle sets, consisting of three rotations about two axes (i.e., z,y,z rotations), provide one way of expressing these orientations. For example, in a movement sequence, a body segment may be rotated by 20 degrees flexion (x-axis), 5 degrees adduction (y-axis) and 5 degrees lateral rotation (z-axis). The same orientation can be achieved through one unique set of Euler rotations (e.g., in this case, by rotations of 76 degrees, 21 degrees and -82 degrees about the z, y and z axes, respectively). Flexion of an additional 10 degrees (x-axis) yields another unique Euler angle set (e.g., 79, 32, and -84 degrees about the z, y and z axes, respectively). The user may find that familiarity with this notation is necessary to interpret man-model manipulations.

To limit the freedom of a man-model's joint movements, various CAD systems impose constraints on link articulations. In man, functional joint constraints are influenced by muscle flexibilities, body enfleshment and mechanical structures at each joint. For example, most people can flex more at the hip with flexed knees than with fully extended knees. These relational constraints are hard to quantify, making them difficult to model. Perhaps for this reason, CAD man-models usually have constraints that are expressed for each joint, in isolation, without regard to body posture. Therefore, the user must incorporate an understanding of man's physical movement capabilities when manipulating the man-model about its joints.

Enfleshment Representation. Link segment enfleshment is another modelling option that requires the user's consideration. A principle decision is whether or not enfleshment modelling is necessary to satisfy the design objectives. For example, body surfaces may not be crucial for distance-based evaluations of reach and visual interference (Lane, 1982). However, they are necessary for evaluations of body clearance and operator *fit*.

Among CAD systems that model body surfaces, some do not offer the user freedom to change the enfleshment envelope of the man-model. Systems that do allow enfleshment alterations do so in different ways, and to different degrees. Some require segment breadth and depth data, or circumferences, while others require mass, volume or somatotype characteristics. In any case, the user must determine the extent to which enfleshment modelling is needed and the data that can be provided, given available data sources.

Consideration must also include how CAD systems interpret enfleshment input data. For example, if breadth and depth values are required to describe enfleshment about a link segment, where along the long axis of that link are those enfleshment values ascribed? Unfortunately, such information is usually difficult to obtain. At the least, system developers should provide the user with the source(s) of research used to make enfleshment-modelling decisions. In any case, the user must choose to rely on the validity of the man-model, or conduct an independent evaluation of its representation of human body shape.

Other Considerations. Other attributes of the man-model influence the way it can or should be used. Some systems employ reference loci to represent functional landmarks on the man-model. For example, binocular and/or monocular eye reference points are often located relative to the head link of the man-model. In this case, head link length influences assessments of what the man-model can *see*. Similarly, seated height sometimes determines the location of the man-model's shoulder joint in which case it also influences assessments of reach. The user must assess the functional implications of these model-attributes and how accurately they must represent man.

Computer-generated manikins can represent the effects of personal equipment and clothing only to a limited extent. However, if the user has access to appropriate functional anthropometry data, influences due to clothing or equipment may be considered by manipulating the man-model's characteristics. For example, movement restrictions imposed by heavy clothing may be modelled by manipulating joint constraints, or the bulk of heavy clothing may be modelled by manipulating the man-model's enfleshment envelope. Similar approaches are possible for modelling physical disabilities, and in general suggest how man-modelling CAD can facilitate non-traditional manipulations of anthropometric data.

The Workplace Model

In workplace design, there is little value in providing sophisticated models of man if there is no way of relating them to design structures. Therefore, the user who creates appropriate man-models may also be expected to create (or at least manipulate) models of workplace components.

The available CAD systems offer varying facilities to create workplace models that have different levels of complexity. The suitability of their respective modelling approaches depends largely upon three things: the requirements of the application, the availability of appropriate input data, and the user's preferred modelling strategy. Each of these issues must be considered before using CAD to model the workplace.

The great advantage of CAD is that the workplace can be described in three dimensions. The model can take several forms. At the simplest level it is defined as a series of x,y,z space coordinates. More commonly, simple primitives (e.g., boxes, cylinders spheres and cones) are used to construct desired shapes. In more sophisticated CAD programs, smooth curves that closely resemble the true shapes of the objects are modelled. Still more sophisticated systems represent the rigid or deformable characteristics of workplace items when subjected to an external force. Most commonly, edges and vertices (called wireframe models) are used to define the boundaries of the items (Majchrzak et al., 1987).

Workplace items are best modelled with respect to their relative spatial and logical (hierarchical) relationships. These relationships help to establish functions of different objects. For example, if a book lies on a desk, the hierarchy can be defined such that the book *belongs* to the desk and so moves when the desk is moved. But in order to take advantage of these relationships, the modeller must a priori specify the spatial and hierarchical structures of workplace components. Then, model data must be formatted to convey these relationships and the limits of their manipulations. These data are sometimes communicated to the system through the use of Euler operatives (as described earlier).

In representing the workplace, the onus is on the user to keep sight of the purpose of modelling, and to work only to the level of accuracy and realism that the application demands. For example, a CAD system may boast accuracy in the order of millimetres, but it may be unnecessary to define workplace models to that level of accuracy. Furthermore, the accuracy of available input data may impose limits on the accuracy that can be demanded of the models. Over-designing must also be guarded against. Complexity can cause unwanted distraction, or even counter the design objectives by encouraging modelling to inappropriate levels of detail. For example, it may be necessary only to model workplace surfaces that face the operator or influence reach, vision and body clearance. Or, items such as reach and vision targets may be best represented as reference loci rather than replicas of real objects. In all probability, schematic models of design items will suffice, minimizing other potential computer-modelling problems (e.g., storage of irrelevant data, time to update the graphics image, data error-checking, etc.). Modelling decisions such as these usually must be made by the user.

Integration of the Man- and Workplace Models

Once the representations of man and the workplace have been defined, they must be integrated into a working model that will support the analyses to be performed. Generally, this requires that the user manipulate and interpret their relative orientations. To do this, the user often relies on graphics images prepared by the CAD system. Therefore, the system should offer the flexibility to alter those images (e.g., change the perspective and orientation of the displayed model). As basic manipulation techniques, it is desirable that the system's capabilities include shift, rotate and scale functions. Facilities to store and restore working views of the model also should be accessible.

Some CAD systems check all model manipulations to assure that pre-defined modelling constraints are not violated (e.g., functional movement restrictions on workplace items). The user must define each of these constraints. If error-checking mechanisms are not provided, the user must define the constraints, plus be prepared to inspect the model visually (and perhaps quantitatively) whenever it is manipulated.

Depending on the format and purpose of the input data, the use of colour in the graphic display can contribute significantly to the interpretation of the model. For example, colour can be used to differentiate workplace components according to criteria that are relevant to the analyses (e.g., functional groupings of displays, controls, structural panels; reach targets; vision targets; etc.). If colour is employed, the user must be prepared to establish and assign the coding-conventions.

Some modelling features have associated functional costs. For example, an Aitoff projection (which gives a 360 degree flat representation of a view, superimposed on a reference grid) can present what an operator *sees* (theoretically), in a way that is fairly easy to interpret. But this is at the expense of considerable computation time. As another example, wireframe modelling is sometimes augmented by a facility that projects a display of the model with all hidden lines removed. A clear image of the model is produced, but again with a high associated computation time. Although the user may choose to use such features infrequently, they can provide effective means to communicate design solutions.

Evaluation of the Man-Machine Model

In some ways, CAD forces the user to take a standardized approach to a task that was previously more of an art-form. Where *common sense* and *good judgement* in evaluating a design were once prominent, explicit criteria must be defined to determine success or failure of design objectives. These criteria must account for the degrees of accuracy in the man and workplace models (e.g., if items in the workplace are modelled to 1 cm accuracy, assessments of reach may require an allowable *miss* distance of 1 cm). The effort required to determine these evaluation criteria must not be underestimated. Their impact on results of the man-machine system evaluation must be understood.

One example is the the placement of the computer man-model in the workplace model. Many computer systems do not account for governing variables such as gravitational forces, postural changes or tissue compression when positioning a man-model in a workplace model. Usually these are left to the discretion of the user. Hence, the user must ensure that the man-model is positioned in a stable, natural posture, and that its relative location with respect to the workplace (e.g., the relationship of bony landmarks to known reference points on a seat surface) is realistic.

Man-modelling systems currently do not produce goal-oriented reach sequences. Instead, the manikins' links are moved in the direction of a defined reach target without regard to a *comfortable* or *probable* terminal posture. This is usually done by extending connecting links in the direction of the target until it is reached or passed. Frequently, these reaches are initiated from the shoulder joint (for arm reaches) or the hip joint (for leg reaches) with no regard for movement of the links of the torso. In addition, the influences of simultaneous reaches (e.g., on posture, joint constraints, reach obstructions) usually are not considered. It must be appreciated that two reaches made at the same time are treated as single, independent reaches. For these reasons, the user's expertise is needed to interpret reaches performed by the man-model.

Some systems provide no means, other than visual inspection, to assess the physical interference of workplace components with the man-model. Systems that do evaluate body clearance generally do so by determining instances of overlap between the models of man and the workplace. They do not normally identify obstructions imposed on the man-model by the man-model (e.g., reaches made through the body). In any case, derivation of a man-model's enfleshment (e.g., surfaces, edges, or free form curves) is important for evaluations of body clearances. If enfleshment is not derived from appropriate data, or not representative of population characteristics, then body clearance assessments must be rated accordingly. Even if the enfleshment envelope correctly represents the human form, the user must consider the influence of tissue compression on the assessment.

In light of the limitations of man-modelling CAD systems to provide expertise, the user must temper evaluations of man-workplace models with knowledge of man's physical characteristics and their implications on workplace design. The user must also take into account the assumptions that were used throughout the modelling stages, and that contributed to the system evaluation criteria. These tasks are often more difficult than originally expected.

Usually the user must acquire new skills and adopt new personal strategies to interpret CAD solutions. For example, three-dimensional computer displays can pose perceptual problems. The graphic images are free-floating and ignore the influences of gravity, motion, lighting, etc., and usually provide no visual frames of reference to indicate an item's relative size, orientation or relationship to other objects. The requirement to manipulate design specifications in three-dimensions, and unfamiliar terminology add to the task's difficulty. When the results obtained contradict the user's judgement, interpretations of design solutions can be particularly challenging. This is because the computer-generated results can seem so objective and the graphics images can seem so precise. Indeed the precision and analytical capabilities perceived by the novice CAD user often exceed the system's true capabilities.

SUMMARY

Successful workplace design and evaluation call for the representation of man's physical characteristics as they relate to his environment. Man-modelling CAD is one method of obtaining this representation. The distinctions of this technology from manual anthropometric techniques are appreciable. It offers the potential to model such characteristics as disproportionate body parts, body enfleshment, joint constraints and vision parameters in ways that far exceed the capabilities of manual techniques. Furthermore, its flexibility offers great potential to explore different designs and to arrive at solutions that are derived from iterative work.

In spite of its advantages, the decision to use this technology must be weighed carefully. The objectives of the application, and the needs of the technology and the user must be considered. The capabilities and limitations of the system's software must be understood, as far as possible. This can be facilitated by asking the system developer to provide results of validation studies, data sources, precise input data requirements, examples of other system applications, demonstrations, training requirements, associated maintenance costs, and plans for future system developments.

The user of man-modelling CAD must possess expertise in anthropometry and workplace design. This is evident from the issues that must be addressed when modelling man and the workplace. First, the user must understand man's physical characteristics and how they can be represented using CAD. Second, he or she must understand the influences of system capabilities and deficiencies on design work. Finally, the results obtained must be kept in perspective; computer models are only as good as the input data and modelling assumptions used for their creation and manipulation. The design process does not end here; man-modelling CAD does not replace the use of human subjects in the design/evaluation loop.

REFERENCES

Badler, N. I., 1982, Human body models and animation, IEEE Computer Graphics and Applications, 2:6-7.

Bittner, A. C., and Moroney, W. F., 1974, The accommodated proportion of a potential user population: compilation and comparisons of methods for estimation, in: "Proceedings of the Human Factors Society 18th Annual Meeting," Santa Monica, California, 376-381.

Bonney, M. C., Blunsdon, C. A., Case, K., and Porter, J. M., 1979, Man-machine interaction in work systems, International Journal of Production Research, 17:619-629.

Calvert, T. W., Chapman, J., and Patla, A., 1982, Aspects of kinematic simulation of human movement, IEEE Computer Graphics and Applications, 2:41-50.

Damon, A., Stoudt, H. W., and McFarland, R. A., 1966, "The Human Body in Equipment Design," Harvard University Press, Cambridge.

Dempster, W. T., 1955, Space requirements of the seated operator: geometrical, kinematic and mechanical aspects of the body with special reference to the limbs, WADC TR-55-159, Wright Air Development Center, Wright Patterson AFB, Ohio.

Hickey, D. T., Pierrynowski, M. R., and Rothwell, P. L., 1985, Man-modelling CAD programs for workspace evaluations, DCIEM Contract Report, Defence and Civil Institute of Environmental Medicine, Downsview, Ontario.

Kingsley, E. C., Schofield, N. A., and Case, K., 1981, SAMMIE: a computer aid for man machine modeling, Computer Graphics, 15:163-169.

Kingsley, E. C., 1982, CAD and the human operator, unpublished document, Compeda Ltd, Stevenage, U.K.

Lane, N. E., 1982, Issues in the statistical modeling of anthropoemtric data for workplace design, in: "Anthropometry and Biomechanics: Theory and Application," R. Easterby, K. H. E. Kroemer and D. B. Chaffin, eds., Plenum Press, New York.

Majchrzak, A., Chang, T., Barfield, W., Eberts, R., and Salvendy, G., 1987, "Human Aspects of Computer-Aided Design," Taylor and Francis, London.

Richards, J. M., and Companion, M. A., 1982, Computer-aided design and evaluation techniques (CADET), AFWAL-TR-82-3096, U.S. Air Force Wright Aeronautical Laboratory, Wright Patterson AFB, Ohio.

Rothwell, P. L., 1985, Use of man-modelling CAD systems by the ergonomist, in: "People and Computers: Designing the Interface," P. Johnson and S. Cook, eds., Cambridge University Press, New York.

Rothwell, P. L., and Hickey, D. T., 1986, Three-dimensional computer models of man, in: "Proceedings of Human Factors Society 30th Annual Meeting," Santa Monica, California, 216-220.

TASK-ORIENTED COMPUTER ANIMATION OF HUMAN FIGURES

Norman I. Badler

Computer and Information Science
University of Pennsylvania
Philadelphia, PA 19104-6389

INTRODUCTION

The effective computer animation of human figures is an endeavor with a relatively short history. The earliest attempts involved simple geometries and simple animation techniques which failed to yield convincing motions. Within the last decade, both modeling and animation tools have evolved more realistic figures and motions. A large software project has been under development in the University of Pennsylvania Computer Graphics Research Facility since 1982 to create an interactive system which assists an animator or human factors engineer to graphically simulate the task-oriented activities of several human agents. An interactive system called TEMPUS and its high performance successor is outlined which is intended to graphically simulate the task-oriented activities of several human agents. Besides an anthropometric database, TEMPUS offers multiple constraint-based joint positioning, dynamic simulation, real-time motion playback, a flexible three-dimensional user interface, and hooks for artificial intelligence motion control methods including hierarchical simulation, and natural language specification of movements. The overall organization of this project and some specific components will be discussed.

HUMAN TASK ANIMATION

With the widespread acceptance of three-dimensional modeling techniques, high-speed hardware, and relatively low-cost computation, modeling and animating one or more human figures for the purposes of design assessment, human factors, task simulation, and human movement understanding has become quite feasible. Though not recent, the demand for creating, modeling, and controlling one or more human figures in a 3-D world is expanding and the application base is growing. Human figure models have long been used in cockpit and automobile occupant studies (Dooley, 1982); now they are finding application in vehicle and space station design, maintainence assessment, product safety studies, and computer animation for its own sake (Badler, 1987). When motion information is measured directly off human subjects the result is natural motion but little theory of how such motion can be synthesized.

The scope of the task animation process is much broader than usually realized: to produce convincing animation without an expert animator requires a computational understanding of motion and its "semantics"; in other words, a synthetic "expert." Our intention is to extend the capabilities of the design engineer, the human factors analyst, or even the casual user to create, animate, and evaluate human performances. Especially in an engineering rather than artistic environment, users will need an effective motion design and analysis tool without feeling pressed to become overly involved in the mechanism of producing animations.

In actuality we must be careful that reducing the inherent complexity of human animation by simplifying one dimension does not squeeze the difficulty into another. We counter this in two ways: first by providing motion specification tools that move closer to verbal descriptions of tasks and motion characteristics; and second by providing both graphical and textual interfaces to a multiplicity of expressive systems. The consequence of the former is that the more common skill of verbal rather than artistic expression may become a vehicle for task control. The consequence of the latter is that the sheer variety of human movement probably precludes any single simple method or interface. Thus it is rather pointless to argue the *general superiority* of dynamics, kinematics, key parameters, local motor control, etc.: each method has its individual strengths but all are necessary. Instead of seeming clumsy and inelegant, the diversity of methods can in fact be nicely embedded in a formal framework (Badler, 1986, Badler and Dadamo, 1988).

THE UNIVERSITY OF PENNSYLVANIA *TEMPUS* PROJECT

The human figure must become just another object to the design system, albeit one with very special capabilities, requirements, and size variability. We have designed, developed, and built a human figure modeling system which acts as an adjunct to a computer-aided design (CAD) system for human figure modeling, animation, and task performance assessment. Over the last six years, this effort has produced a program, called TEMPUS (Badler et al., 1985), and more recently a high performance workstation version, called JACK (Phillips, 1988), with greatly enhanced features. The principal functions of this system

- Provide a high performance graphics workstation for human figure manipulations.
- Provide a consistent, effective, powerful, and extensible graphics interface to human figure models and human factors tools.
- Create and select individual or statistical human figure models and body sizes.
- Provide interfaces to CAD object information for workplace descriptions.
- Position body segments by direct manipulation, workplace point reach goals, multiple goal positioning, constraint processing, and dynamics control.
- Offer a multiple window environment for easy study of body, camera, light, and scene interaction.
- Provide fast and high quality graphics output for both bodies and objects.

We are currently extending this system into a *task analysis* tool for assessing the actions of one or more individuals in a given environment. For example, the tasks to be performed are enumerated and decomposed into simple, primitive tasks such as reach, view, grasp, transport, etc., each of which has an instantiation as a sequence of movements. Given an environment (3D workplace), agent(s) (human or robotic figures to carry out tasks), and the task description, the system can animate the tasks. In addition, the system provides quantitative and qualitative information about the performance of the agents doing the tasks in that environment. By performance we mean

- Reach assessment. For an individual or a population, specify end effector(s) and fixed ends or restraints. Figure must reach a point in space or a workplace point. Show failure distance, reachable objects, and reachable space. Reaches should respect joint and environment limits and be specifiable for multiple reach goals and arbitrary restraints.
- View assessment. For an individual or a population, specify one or both eyes and the viewed point. Show the corresponding view and show or list visible objects.

- Collision and interference detection. Adjacent body segment collisions are checked by joint limits. Non-adjacent segment collisions depend on the particular geometric representation of the body. A real-time display may be used for simple visual assessment without explicit computation.
- Strength or reaction force assessment. Determine the nominal or maximum force or torque achieved at a body part or end effector. Forces must be resisted, maintained, or reacted through restraints.
- Task load. Determine whether or not a task can be executed in some specific circumstances (e.g., time or strength constrained), whether two or more agents can work in parallel, whether fewer agents can get the jobs done, how much motor or psychomotor workload is imposed on each agent, and so on.

There are many components required to realize this task performance analysis system. The TEMPUS system and its evolving suite of programs is directly addressing large scale questions of effective, general purpose, flexible, and usable human factors analysis tools. The original TEMPUS system runs on a DEC VAX system under VMS. It is essentially a stable, frozen software system. The latest generation of software runs under Unix on a Silicon Graphics Iris 4D-GT (or lower capability) workstation. The computer graphics interface software JACK on the Iris provides the development structure for most of the new features and additions to the design, animation, and evaluation environment.

There are many sources of support for this project, each with its own emphasis and application:

- NASA Johnson Space Center and Lockheed Engineering and Management Services: primarily Space Shuttle and Space Station applications, with major interest in animation, strength models, zero-gravity simulation, and language-based task (command) processing.
- NASA Ames Research Center: the A^3I project to simulation all aspects of a helicopter mission is the application, with primary interest in the pilot model, task load, and task simulation from (separate) mission simulators.
- Army Research Office, the Human Engineering Laboratory at Aberdeen Proving Grounds: application to multi-operator vehicles, with a primary interest in evaluation of reach, strength, workload, and cooperative behavior.
- Pacific Northwest Laboratories, Battelle Memorial Institute: application to control a mobile robot mannequin used to test suit designs for permeability to chemical and biological agents, with a primary interest in animation control, safe path determination, collision avoidance, and motion feasibility.
- State of Pennsylvania Benjamin Franklin Partnership: technology development in Artificial Intelligence methods to aid human factors evaluation.
- National Science Foundation: representations and systems to assist in the interactive and automatic generation of natural, animated human motion.

In addition, this project greatly benefits from its home in a Computer Science Department because we feel that usable computational tools are essential for such a broad spectrum of human performance problems and applications. Rather than solve individual analysis problems, we can focus our efforts on longer-term systems design issues.

SYSTEM COMPONENTS

Figure 1 is a block diagram of the structure of the entire task analysis system. In general, boxes denote processes, ovals denote data storage or knowledge bases, and arrows denote data flow (structures or files) or access. Interaction pervades the whole structure. Below we give a summary of the characteristics of each component.

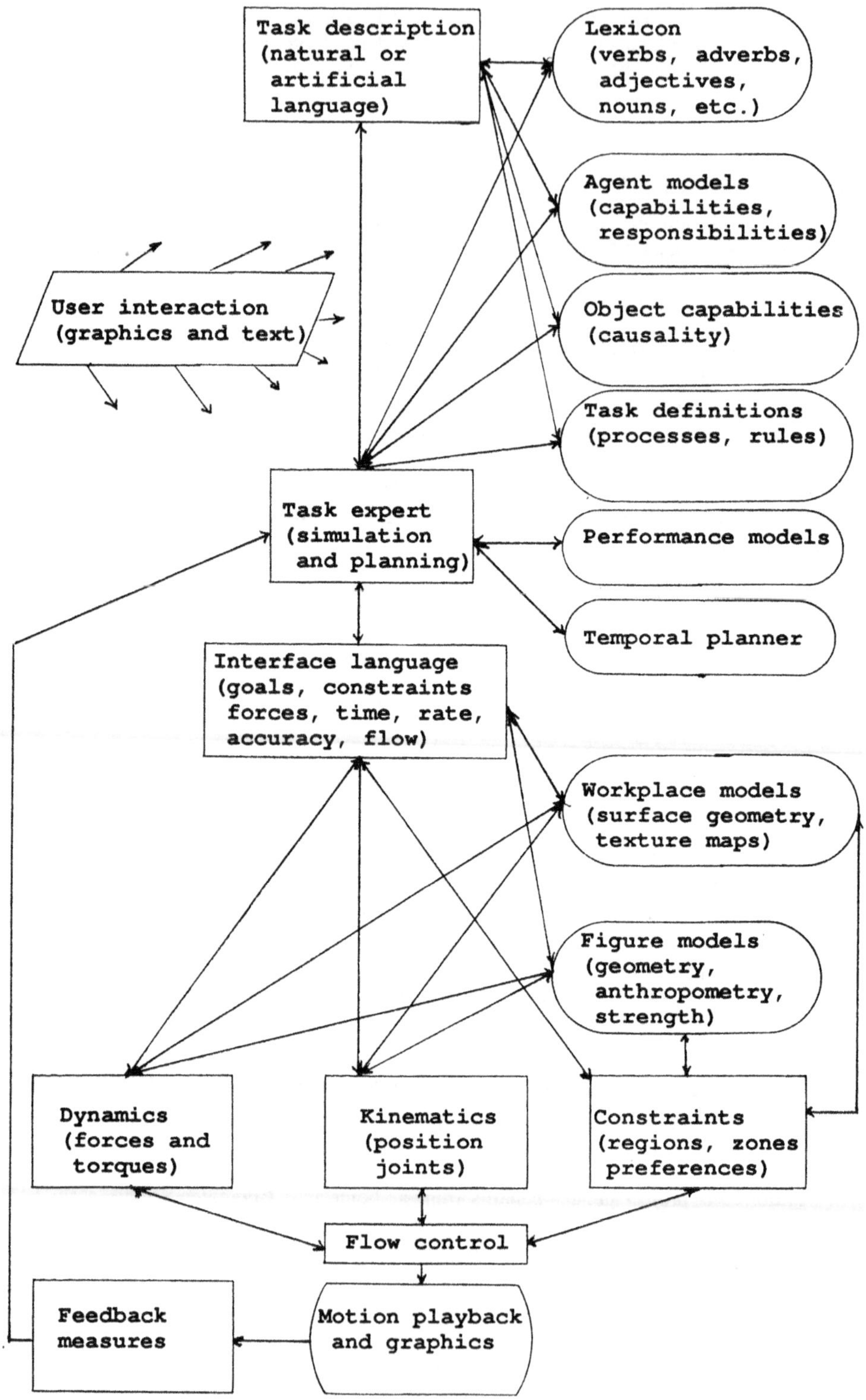

Figure 1. Block diagram of University of Pennsylvania human task animation system.

Workplace Models

Workplace geometry is obtained from an existing internal or external CAD system. By separating object design from human figure modeling, independence from a specific CAD system (and its computer) is assured. Interfaces to CAD systems providing either boundary polygons or constructive solid geometry are available. Internally objects are stored as planar-faced boundary models. Additional surface attributes such as color, transmittance, specular coefficient, and texture may be specified. All workplace models may be displayed in either wire-frame or solid renderings.

The surface models are organized in a database structure, called PEABODY, which represents objects and their relationships in a network of figures, segments, sites, joints, and constraints. Any object may be formed by defining a figure which consists of segments. Segments contain polygon, curved surface, superquadric, etc. geometry models. Joints or constraints at sites (coordinate reference points) are used to connect segments. There is no restriction to hierarchical structures only; arbitrary connections are supported and encouraged giving the designer great freedom in creating the body and environment database. The representation of attached or closed-loop structures is easy: picking up an object or wearing a suit is accomplished by simply attaching the objects through a constraint, while closed loop structures or devices are created with the required joints or constraints. When needed during graphical display, a spanning tree is computed to define a traversal path. The tree is extended through joints before crossing constraints thereby insuring the integrity of the human figure models.

Texture maps are used for a novel function in workplace simulation. Although they can be used simply for visual richness and realism, a more important function is to save geometric storage space for panel-like arrangements of devices. By defining a texture map to be an image of an existing or proposed panel, the tedious and costly modeling of many or all of the contained objects is eliminated. Objects on the texture map are positioned and identified, then become reachable sites on some target polygon in the geometric workplace. During real-time motion display the reachable sites may be indicated by small squares on the polygon (Figure 2); on rendered images the texture map itself appears for accurate visual feedback. We have found that the use of texture maps can reduce the designed model complexity by hundreds of polygons without sacrificing any task animation capability. Moreover, panel texture maps are easily edited on the graphical screen, encouraging panel redesign for improved human performance.

Figure Models

Computer graphics figures with reasonable human-like appearance are provided in TEMPUS. There are at least four different levels of detail that can be used: BUBBLEpeople (Badler et al., 1979), polyhedral figures, and a stick figure. (The stick figure is rather useless.) The most detailed models are BUBBLEpeople: they look surprisingly lifelike and yet are neither expensive nor difficult to move and display. Constructed entirely from overlapping spheres specially rendered to appear smooth and visually continuous across sphere boundaries, the BUBBLEpeople are nonetheless an effective visualization aid in all but the most demanding visual image requirements. There are both detailed and low resolution versions of BUBBLEpeople.

The polyhedral figures come in at least two levels of detail. The lowest resolution polyhedral figure is shown in Figure 2. They are used for fast wireframe positioning, display, and motion playback. The polyhedral figures are used exclusively on the Iris workstation to gain display speed. The models may be customized with additional polygons or spheres to model suits, gear, life-support systems, helmets, etc. All figure models may be solidly rendered to aid visualization of their spatial configuration and workplace fit.

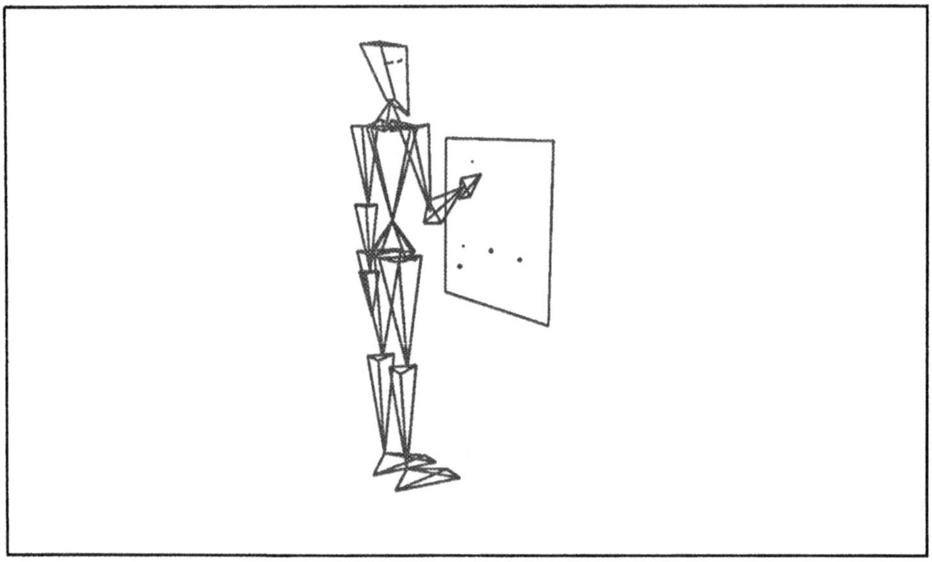

Figure 2. Simple polyhedral figure reaching a site on a texture mapped polygon in the workplace. Other sites on the polygon are indicated by the small dots. The hardcopy output does not show the texture map itself. Other objects and the ground plane have been removed for clarity.

Anthropometry

The models are sized from available anthropometric data. For example, we have been using statistical data from the NASA Manned Systems Integration Standards Handbook. Among the figure data fields are sex, segment lengths, girth values, joint spring and damper values (for dynamics), landmark points, and an indicator telling whether the body represents a real person or a statistical or otherwise specifically constructed generic body. The visualization geometry is not intimately associated with the figure characteristics in the database, but rather is sized when a particular individual is instantiated. Thus body feature locations (sites) are independent of the visualization. If more elaborate and detailed figure models are required, they may be defined in a normalized coordinate system especially designed for body segments and scaled by a set of anthropometric data-defined functions. All bodies may be selected, sized by explicit segment lengths or percentiles, and stored interactively. As many figures as needed may be manipulated concurrently.

Strength Model

A strength model is being constructed which will be used determine reasonable joint torques and forces in a given body position. Based on a degree-of-freedom decomposition of joint torques (whenever possible), this data is used to compute maximum forces at any end-effector. Strength data and forces may be used to determine reaction forces or active forces exertable through the body linkage. In the former, strength data is translated to spring and damper functions for dynamic simulation; in the latter, strength data may be used to assess estimates of task completion times. The strength model will be used in various places in the system; we will return to it later.

Motion Playback

Key postures created by TEMPUS or other animation systems may be interpolated by B-spline curves (Steketee and Badler, 1985). Object file information, key postures, and interpolation parameters are processed to produce an animation. The resulting object, camera, and articulated figure motions are displayed in real-time wireframes on the Silicon Graphics Iris Workstation so that motions may be assessed and tasks validated. The playback software in JACK permits single frame viewing, speed control, camera view control, and single frame rendering. A fully featured graphics display system is included for realistic solid shaded renderings of each frame. This system can shade polygon environments with anti-aliasing, translucency, multiple light sources, and object surface attributes such as texture, glossiness and specular reflection. As part of the JACK interface, image parameters such as light positions, light concentration cones, and the camera position can be interactively set and viewed.

Position Control

An articulated figure is manipulated in several ways. In TEMPUS, positions can be specified as body joint orientations (angles) or by end effector (limb) goals. In either case, joint angles are subject to known joint limits (Korein, 1985). The joint limits are stored in a file and can be adjusted to different situations, such as suits or special capabilities. The limb reach permits positioning the hand tip, grip, or wrist at a point in space while the shoulder is fixed. The remaining degree of freedom permits the elbow to move in an arc while the reach point is held fixed. Similar criteria hold for the legs.

In the JACK interface, any figure segment can be manipulated in translation or rotation independently, including segments representing lights and cameras. The camera view may also be identified with a figure's eye position. There are a variety of user interface tools designed to make this positioning task as straightforward as possible, including on-screen segment picking, real-time feedback, and two-dimensional inputs transformed to three-dimensional rotations around selected axes. Whole figures may be positioned relative to any other object or figure surface, edge, or vertex.

The figure (or object) positioning may also be accomplished by less direct manipulation. Below we discuss some of the alternatives: kinematics, dynamics, constraints, flow, and higher-level task control.

Kinematics

While the TEMPUS reach positioning capabilities are an improvement over joint angle changes alone, single goals and fixed proximal joints are still too limited for general human capabilities. A human or robot figure model must also be kinematically-controlled so that goals and constraints may be used to position and orient the parts and end-effectors (Badler et al., 1987). We developed an algorithm that permits specification of a spatial goal for each body joint. The joint goals are satisfied by a recursive tree balancing algorithm which is iterated until there are essentially no further joint position changes. Goals are described as springs of variable tension connected from selected joints to points in space. The springs move the body joints in such a way as to attempt to minimize the spring energy by simple heuristics.

Though the body is a tree, this algorithm is able to easily handle closed loop situations such as two hands holding the same object. Multiple simultaneous goals are naturally accomodated: for example, a seat belt restraint while the figure is seated and reaching for different objects with each limb, a foot restraint while reaching with the whole body, or a free-floating body reaching with one hand while holding a fixed grip. Figure 3 shows two alternative reaches executed with a figure restrained by a lower torso goal simulating a lap belt. In (a) the figure is given the reach goal for the right hand. In (b) the reach is achieved; notice how the entire torso as well as the arm joints participate in the reach. In (c), the reach is attempted under an additinal constraining goal for the left shoulder (simulating a shoulder belt). The hand reaches toward the goal, but fails; the failure distance would be displayed to the user.

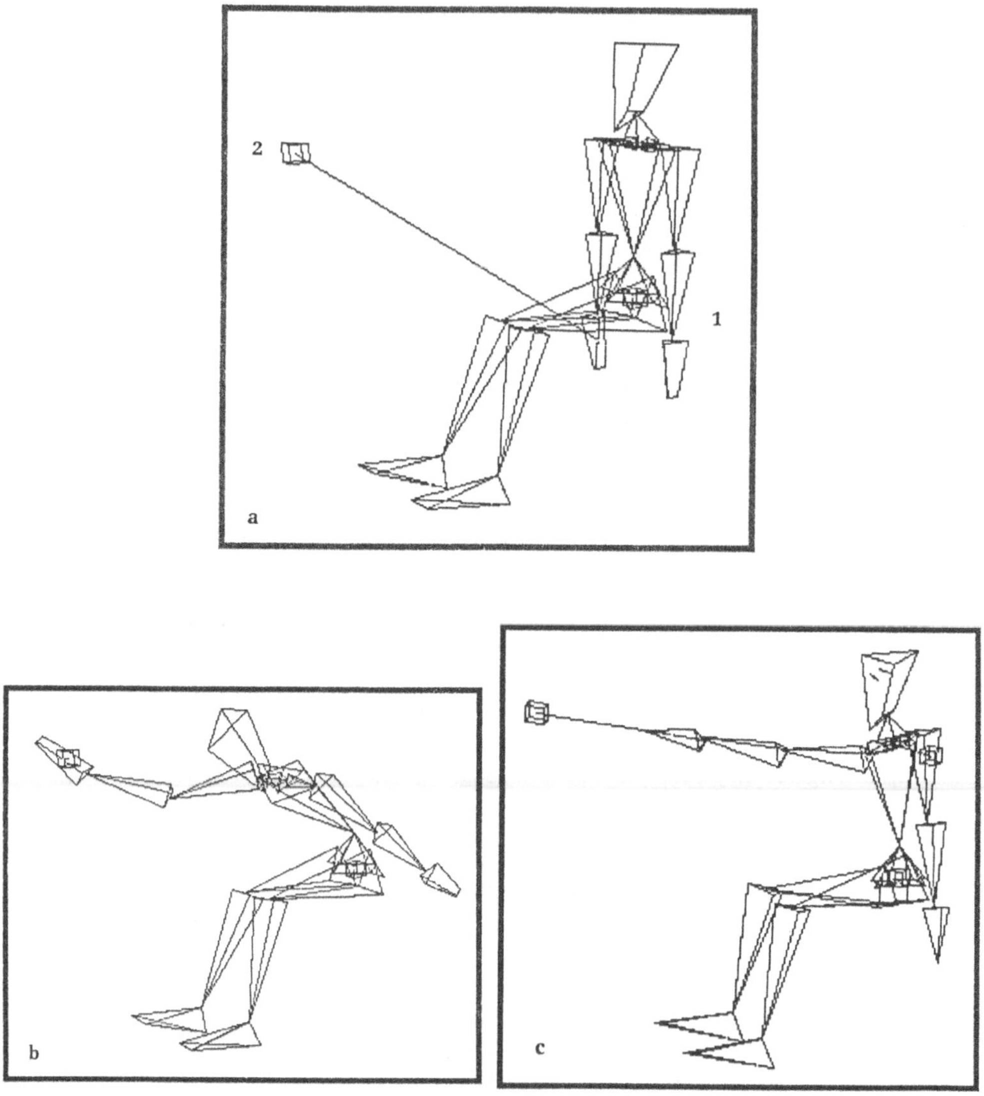

Figure 3. Simple polyhedral figure reaching a goal point in space while restrained by a lower torso goal simulating a lap belt. Other objects have been removed for clarity. In (a), the two goals are shown: (1) is the lower torso goal that tends to keep it in place; (2) is the desired reach position for the right hand. The relative weights of the two spring goals are 100 for (1) and 10 for (2). In (b), the reach is accomplished; all body segments from the right hand through the lower torso are involved in the reach. In (c), the reach is attempted with an additional constraining goal for the left shoulder joint. The reach fails. Notice that the shoulder joint has actually been displaced from its original position, demonstrating the interpretation of the goals as springs.

This feature is being extended to include joint angle limits during the positional goal achievement process. Orientation goals are also being added. The more general algorithm of Witkin *et al.* (Witkin et al., 1987) is being implemented for this and other applications (such as obstacle avoidance).

Dynamics

External or internal forces or torques may be specified through JACK and applied to an articulated figure to produce motion. Dynamic control is most useful for fast motions, for response to external forces (such as gravity), and for incorporating strength models. Our system incorporates a general mechanism simulation system called DYSPAM (Paul and Schaffa, 1985).

As in Wilhelms work (Wilhelms, 1986), we also expect to use kinematics and interpolation to create approximate motions, derive forces and torques, and then adjust the resulting forces and torques to modify the animation. Direct dynamic control (with the exception of restraining forces, environmental obstacles, and joint limits) appears to be much more difficult to specify (Armstrong et al, 1987). We differ though, in the interaction between kinematics and dynamics, preferring to run both in parallel and mix the results according to the requirements of the motion. This animation control method, called Flow, forms the basis of our new animation system TAKE_ONE (Badler and Dadamo, 1988). We expect that the Flow concept will provide a consistent and controllable mechanism for animating complex actions where individual movement styles may vary.

Task Expert

An expert system shell called HIRES (Fishwick, 1986, Fishwick, 1988) transforms task descriptions into kinematics, constraints, and dynamics for execution by the appropriate animation processors. HIRES is a production rule engine with a frame-like (Artificial Intelligence) knowledge base DC-RL. Multiple agents may be utilized. HIRES handles task simulation, agent interaction, and (eventually) motion planning. Its major strength is the general process representation which can be used to animate most any deterministic, stochastic, or rule-based process description. Under revision now, HIRES will be extended to provide more consistent rule syntax, incorporate a recent temporal planner (Kushnier et al., 1988), fully utilize the DC-RL knowledge base for rule storage and application, and provide a task priority, interrupt, and restart facility.

HIRES includes a facility to model the same process at different levels of abstraction. Thus the task does not always require simulation at the most detailed level, but rather at a level which is compatible with user goals. For example, detailed dynamics can be included in one level of a process model, but if that process is being executed "off-stage" then the work need not be actually performed as long as the future state of the system is known or predictable. This is a feature most advantageously exploited in conventional as well as computer animation where complex activities are frequently handled by inference rather than by explicit visualization (Thomas and Johnson, 1981).

Agent Models

Agent capabilities and responsibilities are modeled explicitly. This includes physical attributes such as handedness, strength, and handicaps, and behavioral preferences or characteristics, duties, areas of responsibility (in the workplace), role in a group, etc. Also, general properties of agents may be expressed here, such as the hands being used for most grips, the relationship between the size of the object gripped and the capacity of the gripper, the preferred (normal gravity) support on the feet, the inability to occupy space concurrently with another object, the visual observation of something requiring a gaze or head orientation, etc.

Agent models (other than their anthropometric, strength, and visualization geometry data) are stored in a frame-based knowledge base (DC-RL) accessible to HIRES. Many agent features (hands, view, etc.) are considered as "resources" which may be allocated and freed by HIRES. Conflicts between multiple tasks may therefore by resolved by resource constraints similar to those modeled in computer operating systems.

Task Definitions

Tasks are defined by rules or procedures which are decomposed into simpler acts the system can interpret as goals, constraints, affected objects, paths, directions, etc. Task definitions are built from process models (scripts, Petri nets, data flow diagrams, production rules, or discrete or continuous simulation models) (Fishwick, 1986). The expectation is that a suitable process model will make the specification of a task animation much simpler by capturing the relationships between all the participants (agents as well as objects) and executing the process in a simulation-type (but rule-based) environment.

An important aspect of task description and its simulation by HIRES is the interface language between HIRES and the animation processors. We view this as the "missing link" between Artificial Intelligence knowledge representation systems and the actual animation of the human figure. Additional evidence for this view is also offered by Wilhelms (Wilhelms, 1987) in describing path planning, collision avoidance, and stimulus-response control.

Our YAPS extension of HIRES to better task animation interfaces will include task interrupt control, temporal planning, and task time estimation based on the human strength model and Fitts' law. Task time specification is crucial to the viability and accuracy of a task simulation. Arbitrary time estimates will not do, primarily because the temporal and spatial context of a task is critical to the time duration needed for task completion. For example, a simple button push will be accomplished in rather different durations depending on how close to the button the designated finger is positioned by the previous command. It is unrealistic to expect every action to be accompanied by a departure from and return to some neutral posture.

Task completion times will be specified in one of three ways: by temporal specification, by performance rate, and by target accuracy. In the first case, the time specification (duration or end time) is given and the event can be scheduled to begin immediately and proceed at a rate commensurate with goal achievement at the desired time. In the second case, the performance rate (as a percentage, say) is used as a multiplier of the maximum strength performance of this agent in achieving the goal. The strength model provides as estimate of maximum torques which can be used to compute the duration of the task. The performance rate modifies this duration for the required simulation time. In the third case, the accuracy value is used in a Fitts' Law formula for the generic task type to compute an expected task duration.

Feedback

Critical to the interpretation of the simulation as a task animation is the provision for direct feedback from the figure and the environment models to inform and control the simulation. The information returned includes any desired position, velocity, acceleration, torque, force, or collision. Thus the simulation can take appropriate (rule-based) actions when a collision occurs, when a strength limit would be exceeded, etc. This ability to react to a changing (external) environment outside its high-level knowledge base is not normally associated with Artificial Intelligence systems, though the concept has been developed and is essential for robotics and sensory control applications.

Task Description

Task, action, or process descriptions are provided by programming languages, scripts, or commands in a subset of a natural or artificial language. Certain primitive actions are represented by semantics meaningful to the HIRES simulation, such as move, turn, grasp, look at, etc. More complex actions are expanded to request or determine necessary information such as object referents, to resolve ambiguities such as choosing the proper agent or instrument used by the agent, to supply a default sequence of subtasks, and to establish approximate temporal relationships and timings.

Our first attempt at task description used a subset of natural language or an artifical language (syntactically stylized checklists) to describe tasks in a generic control panel setting (Badler and Gangel, 1986). This system, MVP, uses a parser and a knowledge base of agent and object capabilities to understand the task command and provide a first cut at the subtasks required to execute it. Our initial applications of this task input method focused on panel-type objects: switches, indicator lights, meters, valves, etc. (Gangel, 1985). Recently, the incorporation of more complex tasks and movable objects has been studied (Karlin, 1987). Both systems will produce assertions in the DC-RL representation system which are meant to be interpreted by HIRES.

This natural and artificial language input system is being extended to include additional control constructs with the ultimate intention of processing complete task descriptions with inherent contingencies, repetitions, and alternatives. There is significant human factors material in this form (for example, the NASA Flight Data File cue cards). The ability to use this command data directly to run purely computational human factors and performance data experiments is a realistic goal.

An alternative source of task descriptions is an (external) task simulation. For example, in the A^3I effort, a helicopter mission is simulated by a planner; the tasks required of the helicopter pilot are output in a conventionalized format and transferred to the pilot model in JACK. The tasks are presently a simplified list of reach and view tasks with geometric targets. The timing for each action is determined by the mission simulator's progress. Constraint-based positioning achieves the reach goals as expeditiously as possible in real-time on the Iris. One interesting aspect of this attempt at real-time graphical task simulation is a consequence of driving the graphical simulation too fast. If a task cannot be completed, it is interrupted to begin execution of the next task (since tasks arrive in real-time and in temporal order). The pilot's hands return to a neutral position between tasks only if there is time for that action to occur; otherwise the hands move as fast as the graphical simulation will allow from reach goal to reach goal. Since the tasks are also saved, the task sequence can be replayed after the mission simulation to allow all tasks to complete. At this point various measures of workload could be computed.

Knowledge Bases

Knowledge bases store information shared across system components, such as the geometry data, the anthropometric database, the agent models, the task descriptions, and object capabilities. Object capabilities are used to determine the meaningfulness of a task command and the results of the action on the workplace environment. Sample interaction with control panel objects and their interrelationships have been investigated. For example, turning a dial may change an indicator.

On the Silicon Graphics Iris, all databases are actually in Unix files. Dependence on any specific database system is thereby eliminated. In contrast, our attempts to standardize on a relational database in the TEMPUS VAX system were well intentioned but ultimately failed. In general, our systems are built on the premise that no additional software systems besides the standard language processors and Unix file systems are available. The JACK interface and the accompanying computer graphics is therefore portable to any Silicon Graphics Iris without additional cost or investment in third party software. Likewise, the higher-level functions (HIRES, MVP, and DC-RL) are all written in Commonlisp and run on a VAX, a Symbolics, or even the Silicon Graphics (being tested). We are not dependent on any third-party systems for the AI component. The knowledge base DC-RL, in particular, is quite powerful as knowledge-based systems go. In fact, DC-RL will even allow back-end interfaces through Commonlisp directly to any other existing database, provided that its data schemas and suitable conversion functions are written.

User Interaction

The user may interact with the task animation system at any level. It is expected that different tasks will require utilization of various parts (and maybe all) of the whole system. All interaction is through effective computer graphics interfaces or flexible language understanding processors. We have already reviewed the JACK interface for direct computer graphics manipulation on the Iris; likewise MVP and DC-RL exist for user expression of task commands and world knowledge.

Some programs do not fit so well into these two major interfaces. In particular, the selection of figures and their anthropometry is a separate textually interactive system, and the creation of texture maps uses a different graphical interface. The latter is used to define flat panels of objects as a two-dimensional image with certain named sites identified as panel-type objects (switches, etc.). The panel with its objects is developed interactively through a paint system with generic object icons, or simply read in as a digitized image from a photograph, drawing, or the real thing. Objects may be moved, deleted, or added in either case. Object characteristics are associated with the various image features. When satisfactory, the texture map is stored, the high-level device information is sent to DC-RL, and the geometry of the object locations (as sites) are inserted on a given polygon inside the PEABODY geometric database.

CONCLUSION

All of the system components in Figure 1 are functioning in some form. Though significant efforts remain to broaden the scope of some of the components and build task vocabulary, feasibility has been demonstrated. Moreover, any approach to human performance animation that fails to include all these processes can be shown to have significant weaknesses for certain animation, analysis, and assessment tasks.

There are several ongoing efforts to use our software for actual human performance visualization and assessment tasks. In general, the software is available on a research basis from the University of Pennsylvania Computer Graphics Research Laboratory. While not claiming its universal applicability to all human performance issues, it does offer a substantial, broad, and extensible framework for the investigation and solution of many real problems.

ACKNOWLEDGMENTS

This research is partially supported by Lockheed Engineering and Management Services, Pacific Northwest Laboratories B-U0072-A-N, the Pennsylvania Benjamin Franklin Partnership, NASA Grant NAG-2-4026, NSF CER Grant MCS-82-19196, NSF Grants IST-86-12984 and DMC-85-16114, and ARO Grant DAAG29-84-K-0061 including participation by the U.S. Army Human Engineering Laboratory. This work would not be possible without the assistance of the numerous participants in the Computer Graphics Research Laboratory at the University of Pennsylvania.

REFERENCES

Armstrong, William, Mark Green, and R. Lake. (June 1987). Near-real-time control of human figure models. *IEEE Computer Graphics and Applications*, 7(6), 52-61.

Badler, Norman I. (1986). *A representation for natural human movement* (Tech. Rep.). Philadelphia, PA: Dept. of Computer and Information Science, Univ. of Pennsylvania.

Badler, Norman I. (June 1987). Articulated figure animation. *IEEE Computer Graphics and Applications*, 7(6), 10-11.

Badler, Norman I. and Diana Dadamo. (1988). *The Flow approach to animation control* (Tech. Rep.). Philadelphia, PA: Dept. of Computer and Information Science, Univ. of Pennsylvania. (submitted to The Visual Computer).

Badler, Norman I. and Jeffrey S. Gangel. (June 1986). Natural language input for human task description. *Proc. ROBEXS '86: The Second International Workshop on Robotics and Expert Systems*. Instrument Society of America.

Badler, Norman I., Joseph O'Rourke, and Hasida Toltzis. (Oct. 1979). A spherical representation of a human body for visualizing movement. *IEEE Proceedings, 67*(10), 1397-1403.

Badler, Norman I., Jonathan D. Korein, James U. Korein, Gerald Radack, and Lynne S. Brotman. (1985). Positioning and animating human figures in a task-oriented environment. *The Visual Computer: The International Journal of Computer Graphics, 1*(4), 212-220.

Badler, Norman I., Kamran Manoochehri, and Graham Walters. (June 1987). Articulated figure positioning by multiple constraints. *IEEE Computer Graphics and Applications, 7*(6), 28-38.

Dooley, Marianne. (Nov. 1982). Anthropometric modeling programs -- A survey. *IEEE Computer Graphics and Applications, 2*(9), 17-25.

Fishwick, Paul A. (1986). *Hierarchical Reasoning: Simulating Complex Processes over Multiple Levels of Abstraction*. Philadelphia, PA: Doctoral dissertation, Dept. of Computer and Information Science, Univ. of Pennsylvania.

Fishwick, Paul A. (Jan/Feb. 1988). The role of process abstraction in simulation. *IEEE Trans. Systems, Man, and Cybernetics, 18*(1), 18-39.

Gangel, Jeffrey S. (August 1985). *A motion verb interface to a task animation system*. Philadelphia, PA: Master's thesis, Dept. of Computer and Information Science, Univ. of Pennsylvania.

Karlin, Robin. (December 1987). *SEAFACT: A semantic analysis system for task animation of cooking operations*. Philadelphia, PA: Master's thesis, Dept. of Computer and Information Science, Univ. of Pennsylvania.

Korein, James U. (1985). *A Geometric Investigation of Reach*. Cambridge, MA: MIT Press.

Kushnier, Scott, Jugal Kalita, and Norman I. Badler. (1988). *Constraint-based temporal planning* (Tech. Rep.). Philadelphia, PA: Dept. of Computer and Information Science, Univ. of Pennsylvania. (submitted to AAAI-88 Conference).

Paul, Burton and Ronald Schaffa. (1985). DYSPAM User's Manual. Dept. of Mechanical Engineering and Applied Mechanics, Univ. of Pennsylvania.

Phillips, Cary and Norman I. Badler. (1988). *Jack: A toolkit for manipulating articulated figures* (Tech. Rep.). Philadelphia, PA: Dept. of Computer and Information Science, Univ. of Pennsylvania. (to appear, ACM/SIGGRAPH Symposium on User Interface Software, Banff, Canada).

Steketee, Scott and Norman I. Badler. (1985). Parametric keyframe interpolation incorporating kinetic adjustment and phrasing control. *Computer Graphics, 19*(3), 255-262.

Thomas, Frank and Ollie Johnston. (1981). *Disney Animation: The Illusion of Life*. New York: Abbeville Press.

Wilhelms, Jane. (1986). Virya - A motion editor for kinematic and dynamic animation. *Proceedings*. Vancouver: Graphics Interface '86.

Wilhelms, Jane. (April 1987). Toward automatic motion control. *IEEE Computer Graphics and Applications, 7*(4), 11-22.

Witkin, Andrew, Kurt Fleisher and Alan Barr. (1987). Energy constraints on parameterized models. *Computer Graphics, 21*(3), 225-232.

SAFEWORK: SOFTWARE TO ANALYSE AND DESIGN WORKPLACES

Robert Gilbert, Robert Carrier, Jean Schiettekatte,
Christian Fortin, Bernard Dechamplain, H.N. Cheng,
Alain Savard, Claude Benoit and Marc Lachapelle

Ecole Polytechnique de Montréal
Department of Industrial Engineering
C.P. 6079, Succ. A, Montréal, Canada, H3C 3A7

Les Consultants Génicom Incorporated
279 Sherbrooke West, Suite 207
Montréal, Canada, H2X 1Y2

INTRODUCTION

SAFEWORK is a software package that easily allows the analysis and design of a workplace as well as the various man-machine interfaces. The main objectives of SAFEWORK's development are the following:

a. function in an IBM AT (or similar) environment (possibly using a 80386 and coprocessor),
b. have all the functions required to make it genuinely useful,
c. isolate the user from the complexity of its internal models,
d. be reliable, coherent and robust,
e. be easy to use, and
f. avoid creating long periods of dead time during the execution.

These objectives put together constitute an enormous challenge because, to the best of our knowledge, there are no software programs that meet all those goals. Even the more advanced packages only satisfy two criteria and use large computers.

Evidently to satisfy the above criteria we must bring originality to every dimension of the problem (choice for computer technology, choices for programming languages, choices for 3-D algorithms, modeling, etc).

CHOICES FOR COMPUTER TECHNOLOGY AND PROGRAMMING LANGUAGES

The following set has elements which have been justified by detailed studies:

Material: PC (80286/386)
EGA graphic card
80287/387 coprocessor

Development environment:	Microsoft Windows
Programming languages:	Microsoft C
Graphic Standard:	IRIS
Database format:	DBASE III
Graphic format:	DXF

SYSTEM ARCHITECTURE

The software we call SAFEWORK has been developed at every level with a modular architecture. It consists essentially of three principal modules:

a. a numerical anthropometry module,
b. a CAD module for the workplace, and
c. an analysis and animation module.

DESCRIPTION

The numerical anthropometry module is a software package in itself. It allows the modeling of any human, normal or not. At the moment, the software program uses different correlation matrices for men and women taken from the NASA databank (1978). In the absence of such data for other populations one can use the correlation matrices to approximate the necessary values when the mean values and variance for a given population are provided.

To construct links for the manikins we generate a subset of measures to which are applied two series of transformations (external measures --> length of bones --> length of links) whose equations, in part, come from the results of several authors: Reynolds (1978), Roebuck et al. (1975), Dempster (1955), Dempster et al. (1964) and Trotter et al. (1952, 1958).

Depending on the user's choice, the model links and psuedo-links of SAFEWORK are either represented in orthographic projection or in 3-D. When in 3-D, with the help of a camera moving along the surface of a sphere at different radii, the user can visualize the manikin at its workplace from any angle.

At any time the model links can be completed with the help of body volumes. These volumes are built with the Bezier equations which give a more realistic modeling of the human form than the ellipse, for example. Moreover this module can generate sets of representative models for a given population. This point is crucial in design, and in order to obtain enough precision, a new concept, based on critical anthropometric variables, was developed.

In a workplace it is easy enough to identify certain variables which are critical to the accommodation of a population. Thus, the height between the ground and knees is critical to determining the optimal height of a work table where the work is done sitting down. This height cannot be lower than the highest knees in the population to be accommodated. From such critical variables SAFEWORK can construct a set of manikins that have to be used during the design process.

A specific population can be stored and revised (Figures 1 and 2). Moreover, we have developed a morphological sub-module that can take into account somatotypes, since quite different morphologies may yield identical sets of heights and weights. The choice of somatotype is currently made from seven different profiles (Figure 3). Calculations for volumes of the body segments are done in the animation-analysis

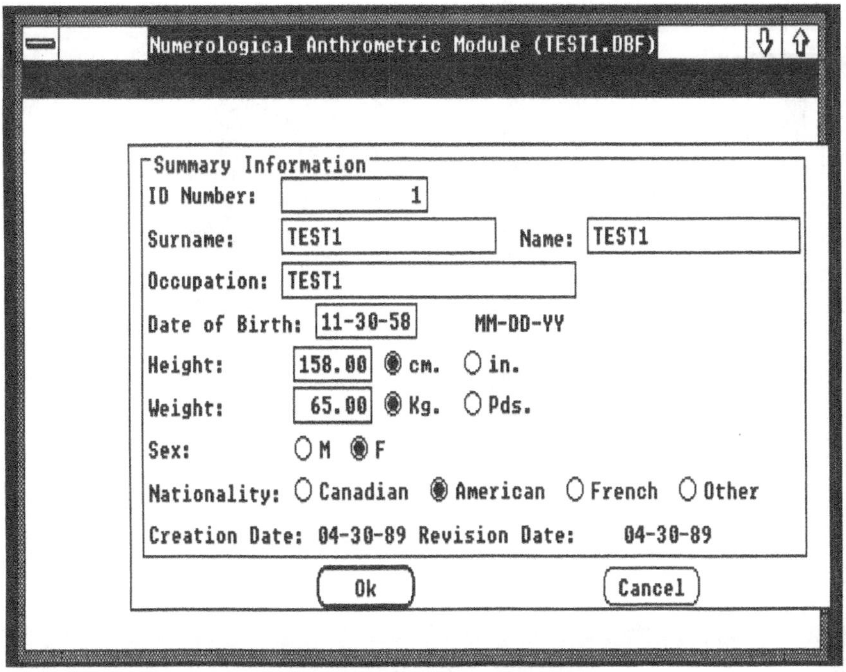

Fig. 1. Individual record (identification).

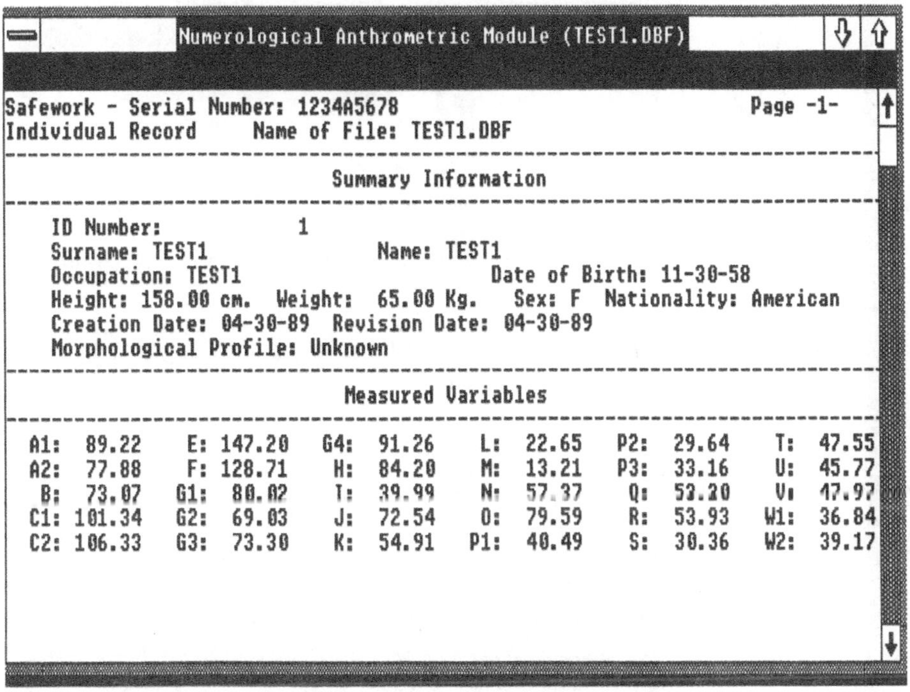

Fig. 2. Individual record (summary information).

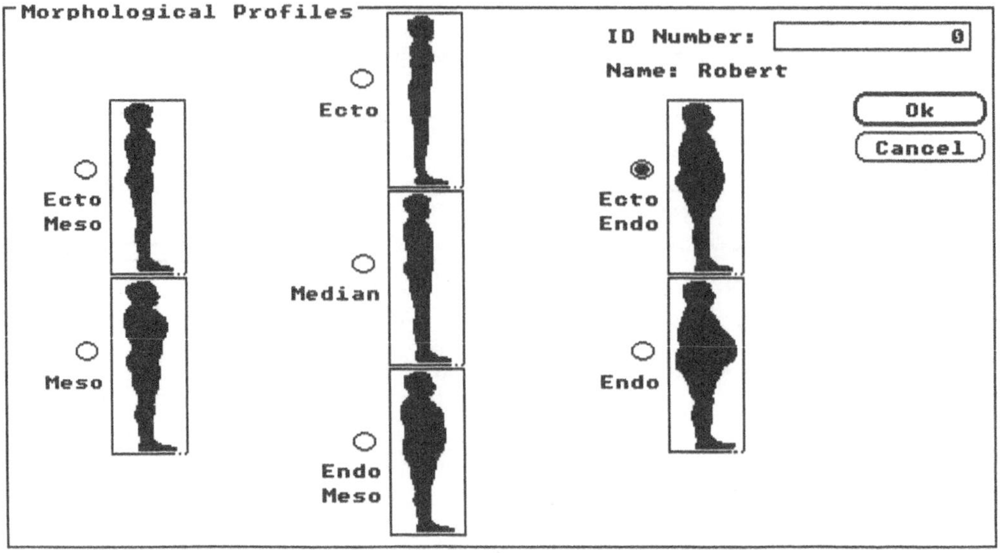

Fig. 3. Choice of profile

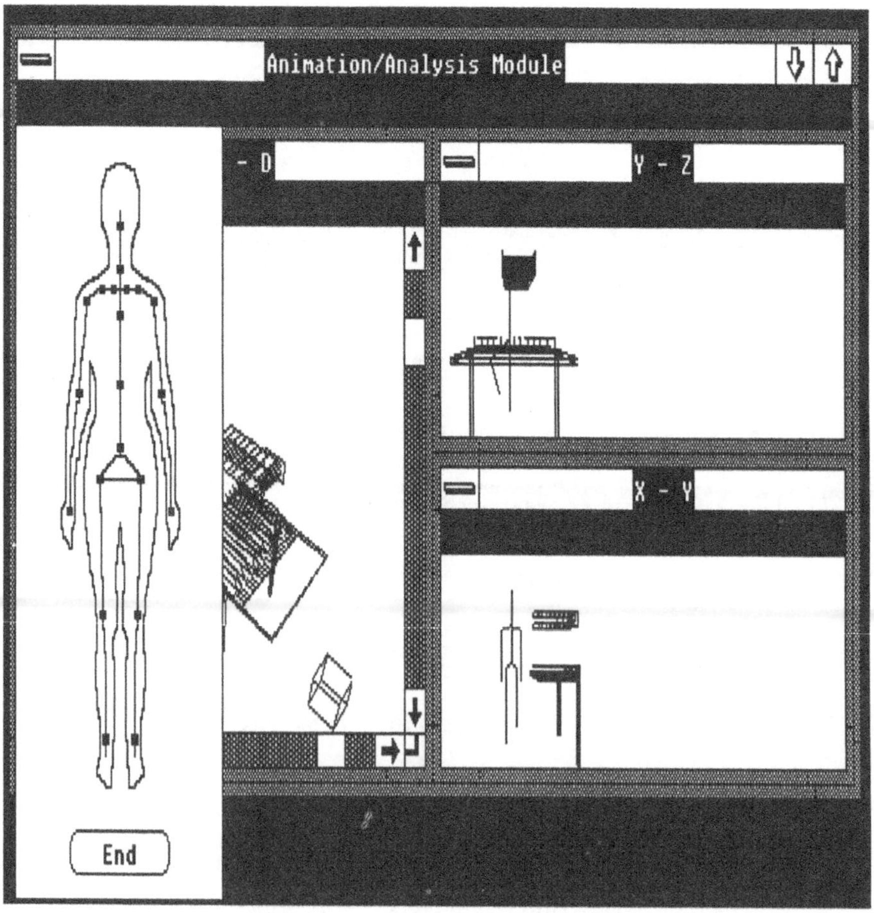

Fig. 4. Selection of body link motion

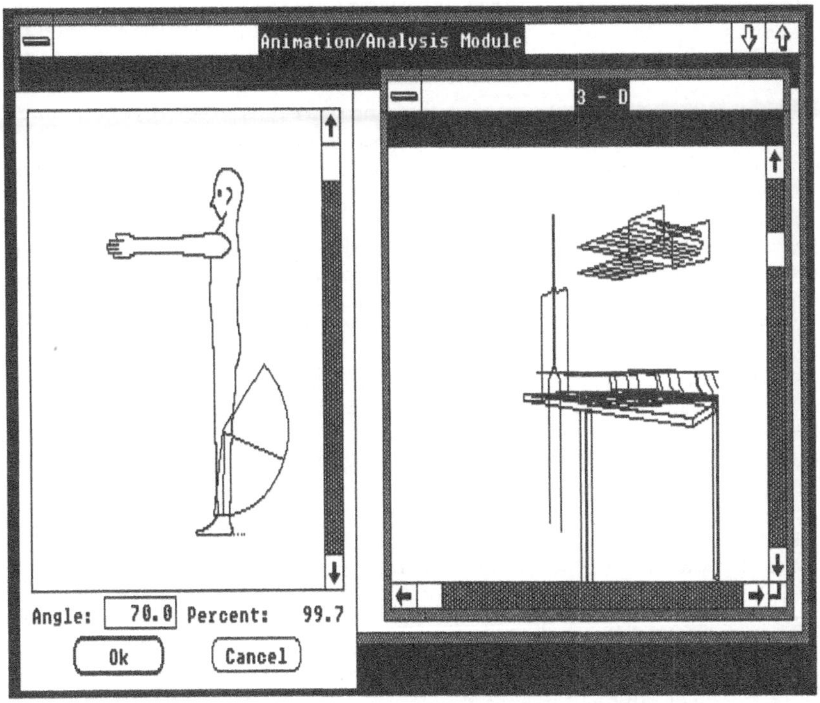

Fig. 5. Parameters of the link motion

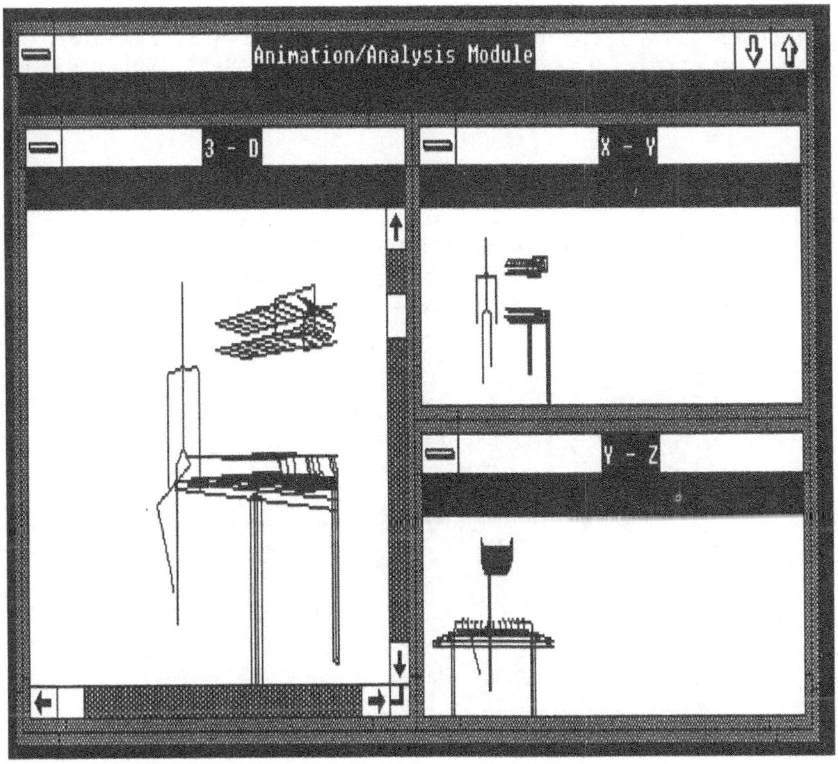

Fig. 6. Result of a leg displacement

module itself. This last point is essentially justified by the fact that movements of the body segments in space can be calculated in a reasonable amount of time if the operations are executed on the links.

The workplace is built with the help of a CAD system such as AUTOCAD and imported in SAFEWORK. The animation-analysis module allows the positioning in 3-D of every link by shifting in the orthogonal planes. It is then necessary to point to the segment with the help of the mouse (Figure 4), choose the desired movement (Figure 5) and choose the angle of movement. SAFEWORK then displays the angle and the population percentile capable of achieving the movement. The angular databank comes from several sources such as NASA (1978), Ryan (1970), Houy (1983), Kapandji (1975), Laubach (1970) and Grieve (1981).

In Figure 6, a movement of the leg is shown. Finally, in Figure 7 the representation of the body volumes with the help of the Bezier equations is illustrated. On this same figure the results of the biomechanical analysis of a specific articulation are graphically displayed.

SAFEWORK also has a biomechanical analysis sub-module which is composed of 14 mathematical models which are managed by a small expert system which leads the user, with simple questions, to the choice of the appropriate model according to the situation. This sub-module calculates the overturning moment (equilibrium), the sliding moment, also the resulting forces and torques in 3-D for every articulation. The user can call upon this module at any time and easily make the desired choice because the program uses the manikin's anthropometric and biomechanical files and the posture in question. The biomechanical models call upon a set of data and mathematical relations (distribution of the body mass, "Body Index", density and volume, body segments, effect of sex, effect of age, somatotype effect, etc, ...) taken from several authors like Drillis et al. (1966), Boyd (1933), Dempster (1955, 1964), Dempster et al. (1964, 1967), Dubois et al. (1964), Dupertuis et al. (1951), Brozek et al. (1963) and Ryan et al. (1970).

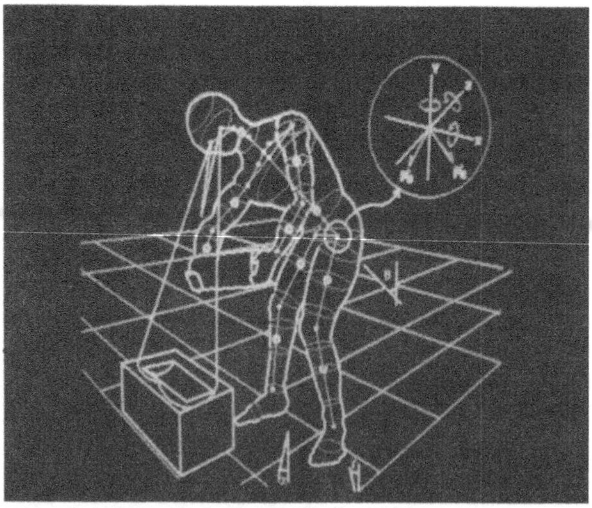

Fig. 7. Full representation of a workplace in SAFEWORK

FUTURE DEVELOPMENTS

1. The development of muscular models, in particular for the back, is currently in progress. These models are already foreseen in the programming of biomechanical models.

2. A vision module will be developed.

3. The module dealing with morphological profiles will be modified to allow the user to make continuous choices between the somatotype extremes.

4. A 3-D pointing animation module will be developed with forced and free reached distances models conditional to the freezing of a given body segment. This sub-module is interesting not only because of its utility but also because the user will no longer need to manipulate every body segment in 3-D separately. Instead the mouse will be used to point to the member and the destination, taking into account the nature of the contact and a fixed body link previously specified by the user.

5. We also intend to represent clothing with selective dilatation techniques.

All these developments and also the representation of body volumes will use the most recent mathematical developments in the theory of Krigeage which constitutes one of the most powerful geometric modeling tools.

ACKNOWLEDGEMENT

Part of this research was supported under CRSNG (CANADA) Grant CRD-3450.

Microsoft C is a trademark of Microsoft Corporation.
Microsoft Windows is a trademark of Microsoft Corporation.
DBase III is a trademark of Asthon-tate.

REFERENCES

1. Boyd W, 1933, The Specific Gravity of the Human Body Human Biology 5: 646-672.

2. Brozek J., Grande F., Anderson J.T. and Keys A., 1963, Densitometric analysis of body composition: Revision of some quantitative assumptions. Annal New York Acad. of Science, 110: 113-114.

3. Dempster W.T., 1955, Space Requirements of the Seated Operator. Geometrical, Kinematic and Mechanical Aspects of the Body with Special Reference to the Limbs. WADC Technical Report 55-159. Wright Air Development Center.

4. Dempster W.T., Sherr L.A., Priest J.T., 1964, Conversion Scales for Estimating Humeral and Femoral Lengths and the Lengths of Functional Segments in the Limbs of American Caucasoid Males. Human Biol., pp. 246-262.

5. Dempster W.T. and Gaughran R.L., 1967, Properties of Body Segments Based on Size and Weight. American Journal of Anatomy, 120: 35-54.

6. Drillis R. and Contini R., 1966, Body Segments Parameters, T.R. No-1166.03 New York University, School of Engineering and Science, NY.

7. Dubois J., Santschi W.R., Walton D.M., Scott C.O. and Hazy F.W., 1964, Moment of Inertia and Centers of Gravity of the Living Human Body Encumbered by a Full Pressure Suit. AMRL-TR-64-110, Wright-Patterson Air Force Base, Ohio.

8. Dupertuis C.W., Pitts G.C., Osserman E.F., Welham W.C. and Behnke A.R., 1951, Relationship of Specific Gravity to Body Build in a Group of Healthy Men. J. Applied Physiology, 3(1): 676-680.

9. Grieve, G.P., 1981, Common Vertebral Joint Problems. New York, Churchill Livingstone.

10. Houy D.R., 1983, Range of Joint Movement in College Males. Proc. Human Factor Soc. Vol. 1 pp. 374-378.

11. Kapandji I.A., 1970, "Physiologie Articulaire", Librairie Maloine, Paris.

12. Laubach L.L., 1970, Characteristics of the Range of Joint Motion and its Relationships to Selected Anthropometric Dimensions and Somatotype Components, Ph.D. dissertation, the Ohio State University, 108 pp.

13. NASA, 1978, "Anthropometric Source Book", Volume I: Anthropometry for Designers. Volume II: A Handbook of Anthropometric Data. Volume III: Annoted Bibliography of Anthropometry. NASA reference publication 1024. Webb Associates (Eds). National Technical Information Service (NTIS). U.S. Department of Commerce.

14. Pheasant S., 1986, "Bodyspace", Taylor and Francis.

15. Reynolds H.M., 1978, In NASA, 1978, "Anthropometric Source Book", Volume I: Anthropometry for Designers, Chap IV of NASA reference publication 1024.

16. Roebuck J.A., Kroemer K.H.E. and Thomson W.G., 1975, "Engineering Anthropometry Methods", John Wiley and Sons. Wiley Interscience Publication.

17. Ryan P.W., Springer W.E. and Hlastala M.P., 1970, Cockpit Geometry Evaluation (JANAIR REPORT 700202), Boeing Military Airplane Systems Division.

18. Stewart, L., 1985, Communique, HFAC, Vol 15.

19. Trotter M. and Gleser G.C., 1952, Estimation of Stature from Long Bones of American Whites and Negroes. American Journal of Physical Anthropometry, Vol. 10, pp. 463-514.

20. Trotter M. and Glesser G.C., 1958, A Re-evaluation of Estimation of Stature Based on Measurements of Stature Taken During Life and of Long Bones After Death. American Journal of Anthropometry, Vol. 16, pp. 79-123.

MODELS OF TRAINING AND SKILL RETENTION

INTRODUCTION: MODELS OF TRAINING AND SKILL RETENTION

Eduardo Salas

Human Factors Division
Naval Training Systems Center
Orlando, FL

OVERVIEW OF THE PAPERS

Over the past few years, researchers and practitioners from many disciplines (e.g., engineering, education, psychology) have increasingly depended on more complex methodologies for the assessment of human behavior. Indeed, research and development (R&D) has intensified since there is an imperative need to make more effective use of human performance data. In response to this need, much of the human performance research has been focused on developing uniform concepts, definitions, categories, and measures to allow better generalization of research findings to an operational environment such as military training (Gagne, 1965; Fleishman, 1982; Levine, Romashko, and Fleishman, 1973; Vreuls and Obermayer, 1985).

The military, which devotes considerable effort and resources to the enhancement of training systems, has benefited from the aforementioned R&D efforts. These efforts have contributed to the military training community by specifying ability requirements for certain tasks, deriving taxonomies and feedback systems, aiding in design decisions for man-machine systems, and developing models of human performance for training management (i.e., what and how to train, which skills are easy or difficult to learn or retain, etc. See Fleishman, 1975; Peterson and Bownans, 1982; McCormick, 1976). However, given the increased sophistication of emerging weapon and training systems, further understanding of human performance models -- especially those that deal with training and skill retention -- is critical for analyzing, designing, and evaluating man-machine systems.

The training models discussed here are somewhat different from the other models presented at the Workshop. That is, training models have a very specific purpose (training design, effectiveness or evaluation) and focus on specific systems (training devices or simulators).

Five papers are presented. The first describes how learning curves are used for modeling and predicting human performance. Towill advocates the use of the most simple model and focuses on the time-constant model and its relation to "industry learning" (e.g., repetitive tasks, continuous flow industries). Three sources of prediction error are discussed, and three ways of incorporating information systems are described.

The acquisition and retention of skills is the focus of the second paper. Here, Rose argues that several fundamental principles and variables affect skill acquisition and retention, and demonstrates that certain task variables (e.g., number of steps, time constraints, etc.) are correlated highly with retention performance. From this, a model was built which predicts retention levels of a task. The model is easy to use, is reliable, and has been validated.

Knerr, Sticha, and Blacksten emphasize the need for improved methods of task analysis, which is the basis of performance models. This is especially true for complex perceptual-motor skills. Quantitative modeling and simulation would benefit from analyses of both procedural and complex cognitive skill, which will in turn enhance training system design. Knerr and colleagues used MicroSAINT to develop and simulate Air Force flight tasks. Learning and attention models were incorporated to examine part-task and whole-task training. Part-task training was recommended for those tasks which had high attention loads. The modeling language, SAINT, was used to evaluate Army procedural tasks. The resulting model predicted performance improvement and performance decay.

Martin and Rose discuss the Automated Simulator Test and Assessment Routine (ASTAR) model. This in an analytic method for use during the design and acquisition of training devices to predict the effectiveness of the system. The model aids system designers and developers in evaluating design alternatives and in identifying possible design problems. ASTAR was used to evaluate several training devices to determine its usefulness. The evaluations demonstrated that the model has distinct and flexible applicability, resulting in good forecasting of training system effectiveness during the acquisition process.

In the final paper, Sticha discusses a model for the Optimization of Simulation-Based Training Systems (OSBATS). The OSBATS model is a decision-making aid for the designers of training systems. The framework requires numerical data because it is a quantitative model. OSBATS contains normative and descriptive models and five modeling tools which aid the designer. Overall, the model is designed to develop an effective, cost-efficient training system.

DISCUSSION

The discussion began with a comment from Dr. Salas who raised the issue of model integration. That is, should the training models be integrated into other model types or should the tools and procedures of the other models be integrated into the training models? Dr. Sticha indicated that the idea of integrating models is reasonable, but as stated earlier, one needs to realize that different models have different goals. For example, the developers of the OSBATS model were concerned with learning and transfer of training. However, the situations in which the model was applied required addressing hundreds of tasks leaving little opportunity for validation. As a result, representations of learning were kept as simple as possible. In other situations in which the interest is in a single task (i.e., part-task training), more opportunity for validation exists. Consequently, a more complex representation of the process can be made. As a result of these differing needs, models that deal with the same phenomenon may use different representations of a single process for different purposes.

A comment was made that the power of integration is not so much in the specific representation or in the integration of models, but in the use of the same data by persons in different fields. In sum, Dr. Salas remarked that if the models are to contribute to system and training design, other concepts (e.g., Wickens' model, workload issues) need to be addressed by the training models. As they become available, those data should be incorporated into the models.

Dr. Rose stated that there are a series of issues relating to the early part of the design process, including how people learn and individual characteristics. These issues should influence how a system is designed. The models that Rose and Martin presented are evaluative rather than prescriptive. He concluded by emphasizing that training objectives should be made explicit at the beginning of the design process.

Sticha wrapped up the discussion by suggesting that what had been presented by the panel encompassed 75% of the pieces that would be required for one to say, "If you design your equipment this way, it will have a certain dollar or hour impact on training costs down the line." The components that remain are large. In addition, there is the problem of integrating them, but that would be a good place to start.

REFERENCES

Fleishman, E.A. (1975). Toward a taxonomy of human performance. American Psychologist, 30, 1127-1149.

Fleishman, E.A. (1982). Systems for describing human tasks. American Psychologist, 37, 821-834.

Gagne, R.M. (1965). The Conditions of Learning. New York: Holt, Rinehart & Winston.

Levine, J.M., Romashko, T., Fleishman, E.A. (1973). Evaluation of an abilities classification system for integrating and generalizing findings about human performance: The vigilance area. Journal of Applied Psychology, 58, 147-149.

McCormick, E.J. (1976). Job and task analysis. In: Dunnette, M.D. (Ed.) Handbook of Industrial and Organizational Psychology. Chicago: Rand-McNally.

Peterson, N.G., Bownans, D.A. (1982). Skill, task structure and performance acquisition. In: M.D. Dunnette and E.A. Fleishman (Eds.) Human Performance and Productivity, Human Capability Assessment, Vol 2, Hillsdale, N.J.: Lawrence Erlbaum.

Vreuls, D. and Obermayer, R.W. (1985). Human-system performance measurement in training simulators. Human Factors, 27, 241-250.

SELECTING LEARNING CURVE MODELS FOR HUMAN OPERATOR PERFORMANCE

D.R. Towill

Dept. of Mechanical and Manufacturing Systems Engineering
University of Wales Institute of Science and Technology
P.O. Box 25, Cardiff, Wales U.K.

INTRODUCTION

In using learning curves for modelling and prediction in the human factors scenario, we seek to identify a number of patterns in the basic data, each of which is an important source of information. These patterns may be classified as follows;

(a) A trend-line, which in some "best" sense, can be used for predicting future performance. This trend-line can be influenced by proper design and planning of the task.

(b) "Normal" scatter about the trend-line, which constitutes a natural and acceptable variation, and which can be used for setting upper and lower bounds.

(c) "Abnormal" scatter about the trend-line, which results in an unacceptable variation. It indicates an avoidable loss in performance which can be traced to an assignable cause and hence eliminated by management control.

(c) "Deterministic" changes in the trend-line. These may be long or short term, and have an assignable cause. An example of a management-induced cause is a planned change in the size or constitution of a team.

To derive a learning curve model which will cope with these four patterns simultaneously is a difficult problem. The author believes there are considerable advantages in selecting the simplest model which is adequate for the purpose of efficient control of a particular activity and will review a procedure for doing so. Understanding and implementing a simple model can often be more profitable than using a complex model, the significance of which is difficult to grasp. The paper concentrates attention on the time constant model, and its variants, as found appropriate to "industry learning" (as distinct from "product learning")[1] .

PRESENTATION OF LEARNING CURVE DATA

There are many papers in the engineering and management literature which record an experience-linked improvement in performance typically observed when plotting a suitable performance index as a function of time. An example is the percentage time-on-target for a tracking task.

These observations have been made for both long and short cycle-time tasks, and appear to apply to both individual operators and to groups of operators, and also in a wide range of industries. It is then common practice for such experimental data to be curve-fitted with an equation which it is hoped will adequately describe the trend line through the inevitable scatter in results. Such curves have variously been called "learning curves" [2], "start-up curves"[3], "progress functions"[4], and "improvement curves" [5]. The axes used to display the data also vary widely according to the particular industry and task being studied. Three typical sets of data are shown in Fig.1 and we note;

(i) In the first case, the axes used are quantity produced per unit time versus cumulative time spent on the task.

(ii) In the second case the axes are cumulative average time per task versus cumulative number of tasks performed (the latter plotted on a log scale).

(iii) In the third case we have task time plotted versus the logarithm of cumulative number of tasks performed.

A number of variants to these graphs also exist in the literature. This paper is concerned with the display corresponding to Fig.1(a) only.

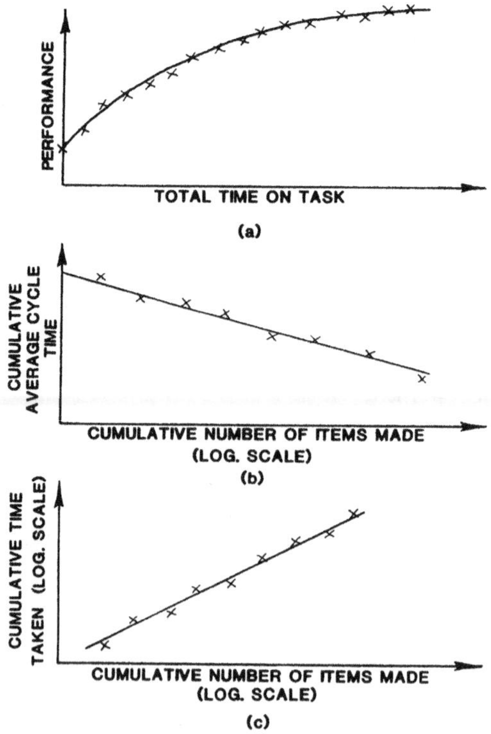

Fig. 1 Three Alternative Learning Curve Displays

FAMILIES OF LEARNING CURVES

Depending on the nature of the experimental data, various "laws" or trend equations describing the data have been proposed, these generally bearing the author's names; for example the de Jong model[6]. As in all empirical curve fitting, the best type of equation is determined by trial and error. The best values of the equation parameters can then be chosen on the basis of a computer algorithm which minimises the least square error of the curve fit. Table I gives a representative set of such models which formed the basis of a rigorous statistical comparison of curve fit adequacy on a wide range of published data [7].

In general, however, when using Table I, there is no logical progression from one type of equation to another, should the initial choice be found wanting, although it is to be hoped that the progression will always start with the simplest plausible solution available. The one exception to this empirical approach is to be found in the family of Industrial Dynamics models which includes as the focal point the time-constant model, and for which the data needs to be in the performance index versus time form. The vertical axis can be any convenient (but agreed and well understood) performance index. Since human factors experimental results are frequently presented in this manner[8 9 10 11], the Industrial Dynamics approach is particularly useful.

Table I Learning Curve and Progress Function Equations as Tested by Hackett[7]

Designation	Corresponding Equation
de Jong model	$y_i = B - Ax_i^{-n}$
Wiltshire model	$y_i = c - ke^{-\alpha x_i^n}$
Time-constant model	$y_i = Y_c + Y_f(1-e^{-t_i/\tau})$
Replacement model	$y_i = a - (a-b)(a-\theta)^{n_i-1}$
Accumulative model	$y_i = \dfrac{b + \theta a(n_i - 1)}{1 + \theta a(n_i - 1)}$
Mathematical model	$y_i = b - \left[\dfrac{1}{c + gx_i}\right]$
Log-mathematical model	$\log y_i = b - \left[\dfrac{1}{c + gx_i}\right]$
Gompertz model	$y_i = ka^{b^{x_i}}$
Second-order model	$y_i = Y_c + Y_f(-(1+\omega_o t_i)e^{-\omega_o t_i})$

THE TIME-CONSTANT MODEL

An exponential learning curve model was first proposed at least as far back as 1950 [12] but was restated in transfer function form with a suitable physical analogue in Ref.8. Since then a number of studies have been published which indicate the utility of the time-constant model. If we denote $Y_M(t)$ = model output at time t; Y_c = model output at time t=0; (Y_c+Y_f) as the model output at time = ∞, and τ as the model time constant, the equation relating model output to time is,

$$Y_M(t) = Y_c + Y_f \cdot (1 - e^{-t/\tau}) \qquad (1)$$

Table II Dynamic Gain and Time Constant For a Number of Industrial Tasks[14]

Industry	Task	Size of worker group	Relative degree of machine pacing	Normalised dynamic gain (Y_f/Y_c)	Time Constant τ (weeks)
Pharmaceuticals	Packaging	Large	Medium	0.6	24
Printing	Startup of two colour press	2	High	0.23	7
Steel	Startup of rolling mill	Medium	Medium	1.26	20
Chemical	Sampling and adjustment of product mix	2	Low	0.71	14
Cigar making	Leaf selection and processing	1	Medium	1.35	3
Electrical	Switch assembly	1	Low	5.50	3
Watchmaking	Watch train assembly	1	Low	1.26	3
Heavy engineering	Anneal plates	3	Medium	0.37	8

which is the equation of the curve shown in Fig.1(a). Hackett in Ref.7 applied the time-constant model and 13 alternative learning curve laws to 88 sets of data recorded on widely different tasks. He found that the model was on average as good a curve fit as any competitor, the only model which gave a better curve fit to some sets of data failing completely in about half the test cases! Such lack of robustness for widespread use is a serious fault in any model,rendering its use suspect in all but the most closely defined circumstances [13]. The time constant model is therefore a preferred choice on the grounds of adequacy of fit, and reliability.

Y_c, Y_f and τ are functions of the task and many other variables. To give some guidance on typical ranges of model parameters, Table II is based on a world-wide collection of industrial case studies [14]. It can be seen that the parameters vary enormously from situation to situation, suggesting that adequate modelling monitoring and forecasting techniques must be available to help avoid the total system becoming unstable due to poor response of a constituent part.

Bohlen [15] has correlated the time-constant model parameters estimated for individual operators with the scores they obtained on the Purdue pegboard and five other tests. He validated his procedure by following the progress of 54 industrial operators working in Illinois light assembly industries. The results are sufficiently good to suggest consideration of the incorporation of his ideas into operator selection and training procedures for a much wider range of tasks.

MODELLING ERRORS

Although the time-constant model has found wide application, it does not work in absolutely every case. With modification of the type suggested later in this paper, it does seem to work in most applications which may be described as "industry learning." The latter term is defined by Harvey [1] as describing performance improvement in repetitive tasks and in the process or continuous flow industries as distinct from situations where the unit of production, e.g. one aircraft, is clearly apparent.

At any time t the observed data, Y(t), will not agree exactly with the model output. The difference between the two is the model residual, N(t), defined by $N(t) = \{Y(t) - Y_M(t)\}$, and this time series is also a source of information on task performance, management, and design.

The residuals are, of course, also the curve-fit errors, which, in some curve-fit procedures, are chosen to minimise the sum of squared error calculated over all data points [7, 16]. The lower the average value of $(N(t))^2$ the better will be the curve fit. Predictions made via the model may also be enhanced via the LSE approach.

In general, there will be three main sources of prediction error,

(i) Errors due to "natural" fluctuations in performance, with the fluctuations random (uncorrelated with each other) or deterministic such as a sinusoidal oscillation. Random errors usually show up as quickly varying scatter which is often Gaussian in pattern [16].

(ii) Deterministic errors usually vary more slowly and include plateaus, for which there may well be physiological, psychological, or environmental causes.

(iii) A complete description of the experimental data is only achieved by taking account of modelling errors; that is, the form of the model selected may not permit adequate description of the trend line.

As an example of modelling errors which may occur, an exponential trend could be curve-fitted by a straight line, as shown in Fig.2. However, there will be considerable modelling error at almost every point on the curve. A statistical analysis, such as the run test [16] would show up this phenomenon as a strong correlation between the model residuals.

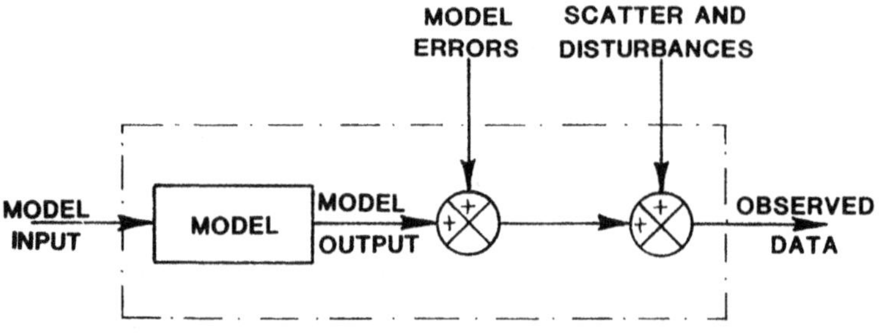

(a) RELATING MODEL AND OBSERVED OUTPUTS

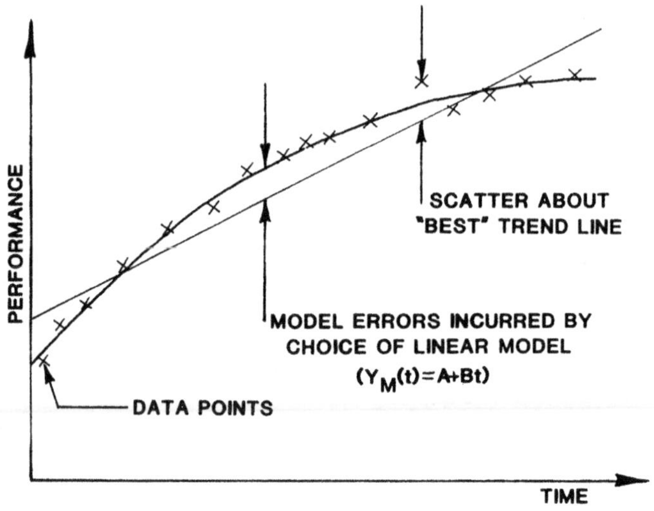

(b) EXAMPLE OF MODEL ERRORS

Fig.2 Fitting the Learning Curve Model to Observed Performance Data

IS OUR LEARNING CURVE MODEL ADEQUATE?

When using learning curves for forecasting purposes, we need to estimate model parameters early on in the improvement process. If the data is well behaved, as shown in Fig.3(a), modelling and forecasting are reasonably straightforward. If the data (from the modelling point of view) is less well behaved, we need to exercise considerable care. For example, in Fig.3(b), the time-constant model, religiously fitted to the early available data, will give a negative τ , and hence useless prediction. Either an 'S'-shaped curve or an oscillatory mode superimposed on the exponential would be more appropriate at this stage.

Unlike the engineer concerned with hardware design and development, who has some opportunity for associating causality via the laws of physics, the learning curve analyst has little opportunity for discerning whether the oscillation is due to an assignable cause or not, or even whether it is likely to continue or die away. We also have the further complication that repeatability of human processes is more widely varied than for machinery, particularly during the improvement phase, so that we can less readily use the conclusions from one set of data to guide the interpretation of another.

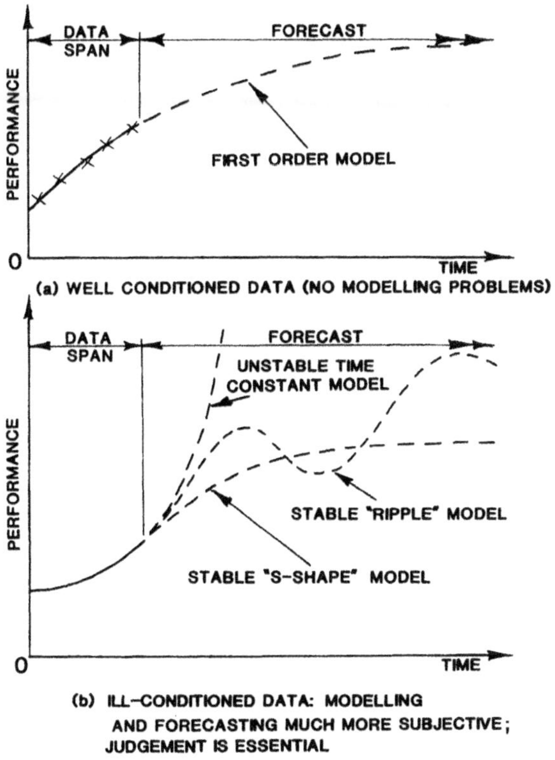

Fig.3 Learning Curve Prediction Problems

In human operator studies, there is added difficulty in ensuring an adequate data collection scheme so that true causalities become apparent, we thus run such risks as concluding that performance is varying throughout the day, whereas the major source of variation may be due to the way in which time is spent by the operators. The true performance (i.e. output/unit of time actually worked) is far less variable [17]. Such factors mean that the time-constant model, adequate though it is for describing many improvement situations, cannot be applied totally blindly during on-line monitoring. Either we need to interact with the modelling process, or build suitable filtering into the data processing to suppress the effects of certain behaviour. When filtering is used, it is recommended that a separate estimate of variability about the trend be made and displayed. Thus the range of observed performance can be used to indicate the state of play and to suggest occasions where further investigation of operating practice is desirable [18].

HIGHER-ORDER LEARNING CURVE MODELS

Suppose the learning curve analyst is ambitious and wishes to use a more complex model, notwithstanding the inherent difficulties likely to be encountered in curve fitting and interpretation. How can transfer functions assist in pointing the way forward? Let us turn to the problem frequently facing the engineer of describing the performance of a physical "black box" in responding to a step stimulus. Here there is a logical path commencing as shown in Fig.4, and which corresponds to a structured search through Laplace Transform tables. In particular we are concerned with the generic family of models described by the equation;

Table III Transfer Function of Dynamic Part of Learning Curve Models

MODEL	TRANSFER FUNCTION
ALL PASS	1
TIME DELAY	e^{-Ds}
TIME CONSTANT	$\frac{1}{(1+\tau s)}$
DELAYED TIME CONSTANT	$\frac{e^{-Ds}}{(1+\tau s)}$
"S SHAPE"	$\frac{1}{(1+\tau_1 s)(1+\tau_2 s)}$
"RIPPLE" MODEL	$\frac{1}{(1+\tau s)(1+2\zeta s/\omega_n + s^2/\omega_n^2)}$

$$Y_M(s) = Y_c + Y_f \left[\frac{e^{-Ds}}{1 + \sum_{i=1}^{i=n} a_i s^i} \right] \tag{2}$$

for which the dynamic part is listed in Table III.

Our first assumption is that the "black box" will faithfully transmit the stimulus undistorted and perfectly timed, thus having a transfer function of unity. The next level of assumption is to assume the "black box" does not distort the stimulus, but delays the stimulus by a time increment D. Then comes the time-constant model, followed by a delayed version of the time constant model. A second-order model with real roots with the characteristic "S" shape then follows. Note that this particular response is frequently advocated in behavioural science descriptions of human performance. Finally, the only third-order system shown is for the only such case which particularly interests us. This is for an oscillation superimposed on the time constant model.

As the transfer function increases in order, the corresponding equation in the time domain for u(t) becomes very much more involved. This increasing complexity is even more noticeable when parameter estimation from scanty data is attempted. It is obvious that for the first-order system, τ must be estimated; for the second-order system τ_1 and τ_2 must be estimated, whilst for the third-order system, τ , ζ , and ω_n must be estimated. This is in addition to the determination of Y_c and Y_f which parameter values are also surrounded by uncertainty.

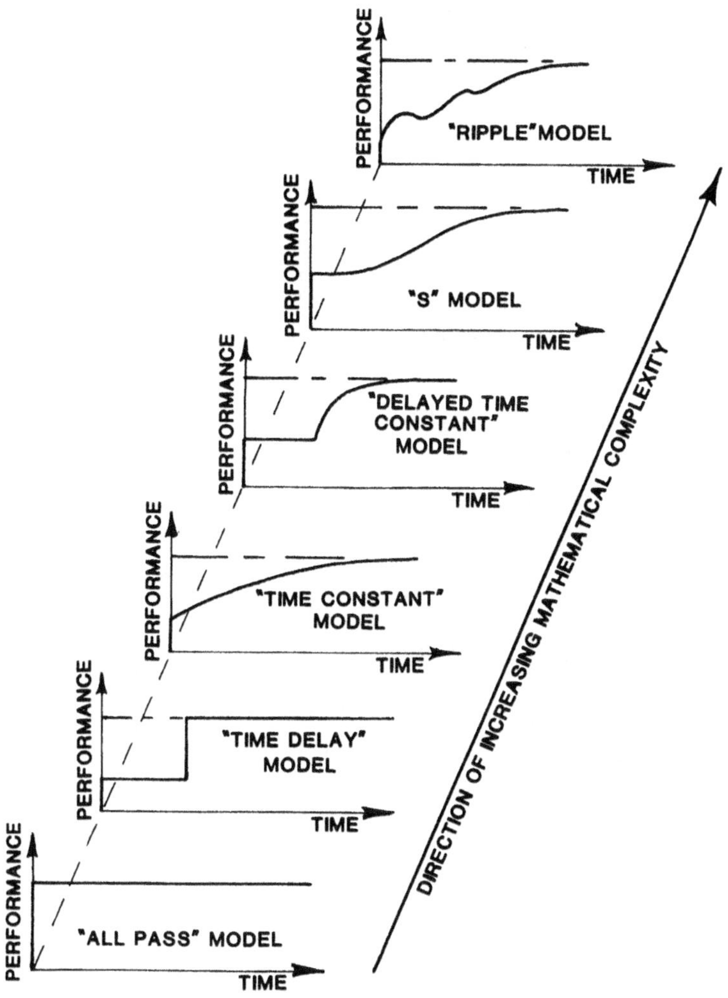

Fig. 4 Family of Industrial Dynamics Learning Curve Models Arranged in Order of Increasing Complexity

The ability of the LSE algorithm to behave as a good predictor depends on the signal-to-noise ratio encountered (i.e. amount of scatter observed); the sampling frequency, and the amount of data available. Table IV lists the accuracy obtained in tests on the Taylor Series based method[21]. This involved prediction of performance of an electrical inspection task at day 70 from data available at earlier points in time. The results show that final performance is estimated to within 13% by day 18, and within 3% by day 38. Better accuracy still can be achieved by building in additional "intelligence", such as might be achieved via the use of activity sampling.

Table IV Prediction of Ultimate Performance for Electrical Inspection Task via LSE Taylor Series Expansion Algorithm[21]

FORECAST MADE AT DAY No:	$\left[\hat{Y}_c / Y_{c\infty}\right]$	$\left[\hat{Y}_t / Y_{t\infty}\right]$	$\left[\hat{\tau} / \tau_{\infty}\right]$	$\left[\hat{Y}_{70} / Y_{70}\right]$
8	0.881	0.754	0.336	0.820
18	0.964	0.962	0.778	0.870
38	0.967	0.978	0.816	0.972
58	1.008	1.001	1.039	1.004
68	1.000	1.000	1.000	1.000

SEQUENTIAL TRANSFER FUNCTION MODELS

Transfer function models may also be used to describe improvement processes where the plateau phenomenon is observed. This is done by using two time-constant models, one of which curve-fits the data up to the start of the plateau. A second model then curve-fits the data subsequent to the initiation of recovery, as shown in Fig.6(a)[20]. Often it is found that the recovery phase model time constant is approximately the same as for the initial phase model. The second curve is then simply the initial curve translated in time by the plateau length. Note that for the recovery phase model, the analyst has the choice of time origin.

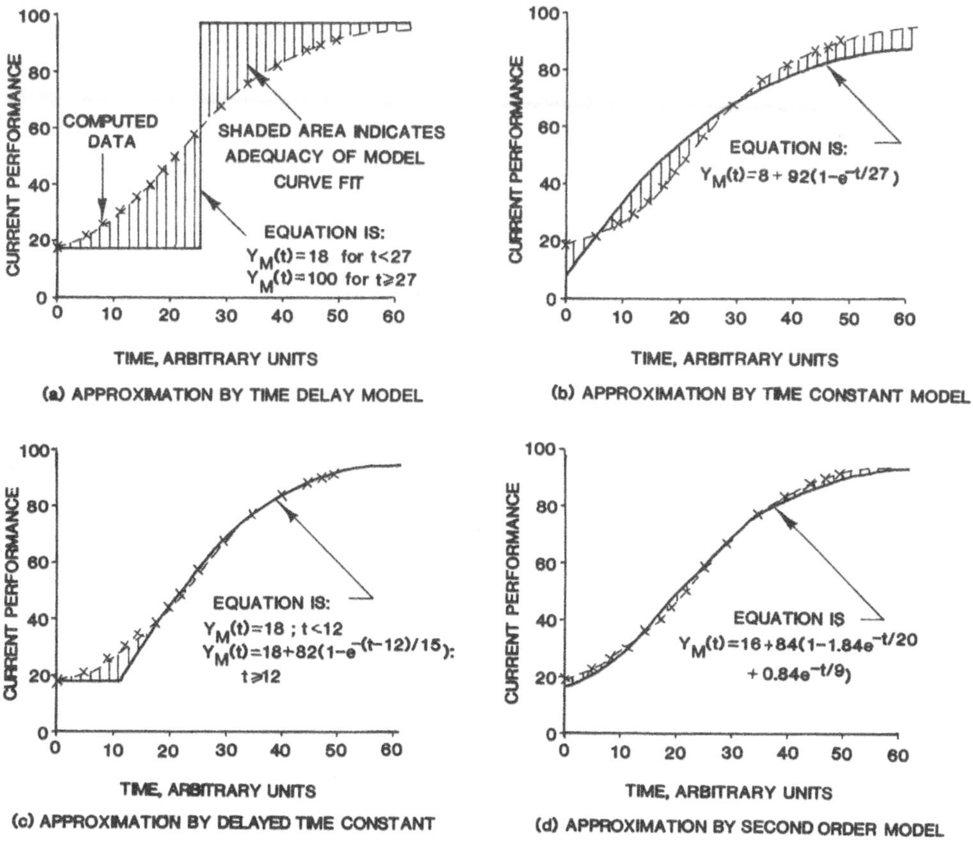

Fig.5 Progressive Increase of Model Complexity Until Good Fit Obtained to Crossman Speed Skill Theory

AN EXAMPLE ON MODEL SELECTION

To emphasise the options open to the modeller, consider the improvement curve resulting from the Crossman theory of the human operator selective process [19]. The data shown in Fig.5 results from the transformation of the original Crossman calculations to the axes necessary for the transfer function approach [20]. We can see a slight "S" curve in evidence (Fig.5(a)) so that a good curve fit attempt is with the time constant model (Fig.5(b)). This results in large errors only near the origin, first negative, then positive in sign.

The second curve fit attempt accepts that there is more to the process than simply the time-constant model, and represents the process by a delayed time-constant model (Fig.5(c)). Finally, Fig.5(a) shows the excellent curve fit achieved using a second order model in which the two time constants are 9 and 20 units respectively.

The majority of the industrial case studies available to this author do not exhibit the "S" curve phenomena, suggesting that the time constant model is adequate for most comparability purposes. A lot does, however, depend on the time origin selected. Inclusion of "pre-training" phases does tend to result in an observable point of inflexion, suggesting that careful thought needs to be given to data interpretation.

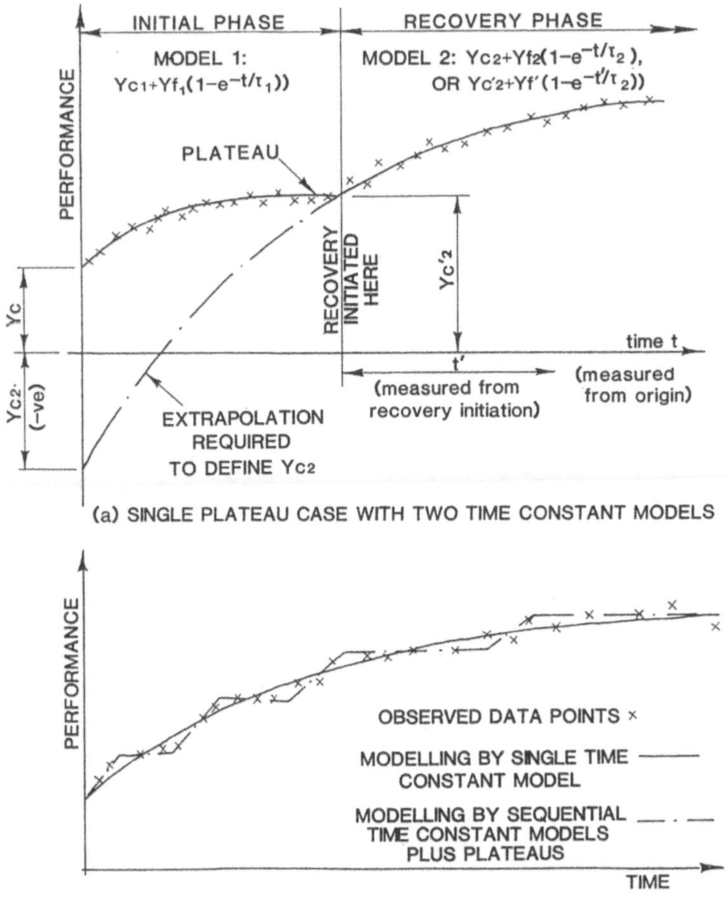

(a) SINGLE PLATEAU CASE WITH TWO TIME CONSTANT MODELS

(b) MULTI-PLATEAU CASE WITH ALTERNATIVE INTERPRETATIONS

Fig. 6 The Effect of Plateaus on Learning Curve Models

The use of sequential transfer functions can be extended to the multiplateau case as shown in Fig.6(b) since computational methods already exist for doing so [20]. However, the learning curve modeller needs to decide whether his purpose is better served by regarding the experimental data as represented by such a sequence of models, or by one model with periodic scatter, since the periodic scatter about the exponential trend may itself be regarded as indicative of a phenomenon worthy of special study [16].

PERFORMANCE MONITORING

It is good supervisory practice to install an information system which monitors learning performance and compares the current improvement and scatter rates with preset targets. These can only be set in one of three ways;

(a) using synthetic learning curve data for individual parts of the task where such an estimate is feasible;

(b) from inter-task comparisons available in the open literature;

(c) from careful use of an in-house data bank, in which previous performance data has been carefully annotated.

The time-constant learning curve model is frequently adequate for methods (b) and (c). Where it is not, the addition of a simple time delay is usually adequate. If periodic variations due to natural, or management induced causes, are suspected, it is suggested that these be allowed for in an additive manner, so that the simplicity of the time constant model is retained. The "event-adaptive" model is a simple way of coping with known contingencies [22].

We thus advocate the setting-up of a learning curve model which represents normal smooth increase in performance, with each significant deviation from the curve being separately accounted for. Once the proper model is in operation, deviations from target can be used to trigger management action. A plateau can suggest onset of bad work methods, a persistent fluctuation about the trend can be the result of a period-based incentive scheme, whilst excessive random scatter can indicate slack standards or poor training procedures. Management needs the simplest time-varying learning curve model which is adequate for this purpose.

CONCLUSIONS

The concept of control engineering analogues for describing learning curves is helpful in the selection of a model appropriate to the data available. It is intuitively satisfying to relate the time varying behaviour of human operator systems to the dynamics of servomechanisms, despite the greater uncertainty that must surround the former. The consequence of this approach is a structured way of choosing the simplest transfer function which will adequately describe the observed operational situation. For many such situations the performance index as a function of time may be reasonably represented by the time constant model. This makes performance comparisons relatively simple to undertake.

REFERENCES

1. R.A. Harvey, "Analysis of Contributory Factors in Aircraft Production Learning", Proc. IERE Int.Conf. on "Industrial Applications of Learning Curves and Progress Functions", London, Dec. 1981 (IERE Conf. Proc. No.52)

2. H. P.Bahrick, P.M. Fitts, and E.G. Briggs, "Learning Curves Facts or Artifacts?", Psychological Bull.54, No.3, 256-268 (1957)

3. N. Baloff, "Startup Management", IEE Trans. on Engineering Management, EM-17, 132-141 (1970)

4. J.H. Glover, "Selection of Trainees and Control of Their Progress", Int.J.Prod.Res., 5, No.1, 43-60 (1966)

5. I. Steedman, "Some Improvement Curve Theory", Int.J.Prod.Res., 8, No.3, 189-204 (1970)

6. J.R. de Jong, "Effects of Increasing Skill and Methods-Time Measurement", Time and Motion Study, 10, 17-24 (1961)

7. E.A. Hackett, CNAA M.Phil. Thesis, Middlesex Polytechnic at Hendon, U.K. (1974)

8. D.R. Towill, "Transfer Functions and Learning Curves", Ergonomics, 19, 623-638 (1976)

9. W.D. Spears, "Measurement of Learning and Transfer Through Curve Fitting", Human Factors, 27(3), 251-266 (1985)

10. Z.A. Sabri and A.A. Husseiny, "Analytical Modelling of Nuclear Power Station Operator Reliability", Annals of Nuclear Energy, 6, 309-325 (1978)

11. R.H. Baran, "A Modified Ainsworth Measure of Learning Efficiency", Proc. IEEE Int.Conf. on Systems, Man. and Cybernetics, 912-916 (1986)

12. A.R. Knowles and L.F. Bell, "Learning Curves Will Tell You Who's Worth Training and Who Isn't", Factory Management and Maintenance, 108, 114-115, 202 (1950)

13. D.R. Towill, "Low Order Modelling Techniques: Tools or Toys?", Proc. IEE Conf. on Computer Aided Control System Design, Cambridge, U.K. 206-212 (1973) (IEE Conf. Pub. No.96, 1973)

14. D.R. Towill, "How Complex a Learning Curve Model Need We Use?", Radio and Electronic Engineer, 52(7), 331-338 (1982)

15. G.A. Bohlen, Ph.D. Thesis, Industrial Engineering Department, Purdue University, Lafayette, Indiana (1973)

16. B. Hitchings and D.R. Towill, "An Error Analysis of the Time Constant Learning Curve Model", Int.Jrnl.Prod.Res., Vol.13, 105-135 (1975)

17. N.A. Dudley, "Work Measurement : Some Research Studies", Macmillan, London (1968)

18. D.R. Towill and M.S. Eler, "Performance Models in a Process Industry", Joint National Meeting of the Operations Research Society of America/The Institute of Management Sciences, Las Vegas, Nevada, USA, TP14.8, 1-1 (1975)

19. E.R.F. Crossman, "A Theory of the Acquisition of Speed-Skill", Ergonomics, 12, 153-166 (1959)

20. H. Sriyananda and D.R. Towill, "Prediction of Human Operator Performance", IEEE Trans. on Reliability, R-22, 148-156 (1973)

21. D.R. Towill, "Forecasting Learning Curves" to be published shortly in the Int.Jrnl. of Forecasting

22. J.E. Cherrington, "Event-Adaptive v. Sequential Modelling of Learning Curves", Proc. V Symp. uber Operations Research, Universitat zu Koln, Aug. (1980)

ACQUISITION AND RETENTION OF SKILLS

Andrew M. Rose

American Institutes for Research
Washington, D.C.

BACKGROUND

A primary responsibility of the military is to train and maintain job skills at levels of proficiency required for successful performance. Rapid, high-quality initial training must be provided. This initial training must produce skills that will endure--the skills must be able to withstand long periods of infrequent use or practice. Likewise, the military must plan for and provide opportunities for periodic retraining; some skills learned during initial training may not be called for on the job for extended periods of time.

Perhaps the key consideration for trainers and training managers is the fact that skills deteriorate if they are not used or practiced. This basic fact affects what is taught, how skills are taught, when skills are taught, and, less directly, who should be taught. This is true for both initial training and refresher training. In both situations, knowing the time course of skill deterioration is critical for selecting which skills to train. Likewise, different methods of training and retraining have different effects on skill retention. Obviously, skill retention functions are vital for determining when to train and retrain skills. And individual differences among soldiers regarding their ability to retain skills affect who should be trained and when they should be retrained.

In this paper, I will touch on four topics. First, I will present a brief conceptualization of the cognitive processes involved in skill acquisition and retention. Next, I will discuss four variables that research has shown to affect retention. Then, I will talk about how all of the above considerations led to a method for predicting skill retention. Finally, I will present further details about the actual method, its utility, and its strengths and weaknesses.

To design effective training, to prevent unnecessary losses in skill, and to predict skill retention, it is important to understand how people acquire and retain skills. Although there are many different theories of learning and retention, practically all cognitive psychologists agree upon some fundamental principles. First is that both acquisition and retention are active processes. When information to be learned is presented, the learner processes it by organizing it, elaborating it, and encoding it to construct a representation of it. These processes do not occur automatically--if learning is to occur, it must be deliberate; if the learner does not take an active part, learning will not occur. The same is true for recall: It is an active process, wherein the person tries different methods to reconstruct or "activate" the representation.

A second fundamental principle is that all information taken in is related to existing knowledge. New learning builds upon old learning; effective training builds upon concepts, principles, facts, and skills that the trainee already knows. This "old" learning includes strategies and techniques for organizing and remembering things.

A third fundamental principle is that information--knowledge and skills--is mentally represented as complex, meaningful structures or networks. For example, the representation of a motor skill includes the actual movements, movement timing, body position, expected feedback, and so on. As learning and practice progress, this representation becomes more elaborate and more organized.

Finally, and perhaps most basic, is that skills deteriorate if they are not used or practiced. There are two basic theories of why this occurs--one, that the memory representation simply "fades out," and two, that other information interferes with recall. From a practical perspective, we may as well assume that both of these theories are true; better learning and retention will occur if the memory representations are better developed (or "stronger") and more elaborate and thus distinctive.

Building upon the concepts just presented, let us consider the variables that affect skill retention. Probably the strongest of these is the degree of original learning: the better the original learning, the better the recall. Or, using the above terminology, the stronger and more distinctive the memory representation the easier it will be for a retrieval cue to activate it. By "stronger and more distinctive," we mean several things. Consider, for example, a procedural skill task such as assembling a rifle. With practice, this task moves from a "list" of individual steps to an organized, coordinated set of movements, where each segment cues the next. In addition, the skill "generalizes" so that the same high level of performance can be attained in different environmental conditions, with similar weapons, and so forth. A corollary to this basic principle is that skill decay can be reduced and delayed by "overlearning" or "mastery training"--that is, continued practice after proficiency has been reached. While this is conceptually true (a memory representation can always be elaborated and strengthened), it is also the case that increased amounts of overlearning facilitate retention at a decreasing rate.

A second variable that obviously affects skill retention is the interval between learning and retention or between one performance and the next. In general, the longer the period of nonuse, the greater will be the decay. There is a classic curve of forgetting: The absolute amount of decay increases with time, while the rate of forgetting decreases. However, this "classic" curve almost never occurs outside of the laboratory. Partly, this is a function of the nature of job tasks: In the lab, we can study memorization of nonsense syllables; military job tasks are far more complex. Most job tasks involve many different components that can decay at different rates. Also, while soldiers may not practice a specific task for a given period of time, they will probably see the task being performed or perform some similar task themselves. There is also good evidence that for several kinds of tasks, "mental rehearsal" is sufficient to maintain proficiency over long periods without actually performing the task.

A third variable affecting skill retention is individual differences among trainees. Of course, we know that some people forget faster than others. However, there has not been much systematic research on the relationship between skill retention and individual difference dimensions. Most conclusions about this variable are derived from its relation to acquisition and original learning. High-ability learners (as measured by ASVAB) tend to reach a higher degree of original learning, and thus retain more, than lower-ability individuals. Also, there is some evidence that lower-ability learners forget a larger proportion of abstract, theoretical material than do higher-ability individuals. I think higher-ability learners have more and better encoding, organizing, and recall strategies and techniques; they make qualitatively better use of equivalent learning time.

A fourth variable that influences skill retention is the task itself. Certain characteristics of a task make it more or less easy to retain. For example, tasks requiring continuous motor control are much better retained than discrete procedural tasks. Tasks with few steps are better retained than those with many steps. Tasks with several "safety checks" are poorly retained. Tasks that involve meaningful material are better retained than those involving less meaningful material (e.g., recalling names of enemy vehicles vs. recalling elements of a SALUTE report).

There are several other task characteristics involved in skill retention. Basically, these characteristics fall into two types. The first type is related to what I call "task organization." Other researchers have termed this "internal organization," "complexity," or "cohesiveness." As we have seen previously, a memory representation is easier to recall if it is organized, distinct, and has internal cues that trigger other parts of the representation. It follows that tasks with these same characteristics will have "good" representations. To illustrate this concept, consider the task of collecting and reporting information. Soldiers are taught to collect and report various pieces of information after observing a suspected enemy position, including the time, location, number of people, what they are wearing and carrying, and what they are doing. To aid the recall of these pieces of information, we can impose some organization by using the SALUTE mnemonic: S=size, L=location, etc. The mnemonic helps to trigger the individual components. This is the type of encoding, elaborating, and organizing process that occurs internally when a "good" representation of a task is constructed.

This concept helps explain some of the effects of the variables we have been discussing. For example, more time to learn during initial acquisition gives trainees more opportunities to construct organized representations. Similarly, "high-ability" learners probably have more and better coding strategies available. Motor skills are better retained because more feedback information from the performance itself can be used to organize the task.

The second type of characteristic affecting skill retention are variables relating to the actual conditions and standards of performance. For example, tasks that must be performed under time pressure or tasks where the steps have to be performed in a specific order place additional demands on memory. Likewise, tasks that have built-in or externally supplied job aids are easier to recall.

Let me briefly summarize some of the main points I have discussed. From the perspective of cognitive psychology, we see the trainee as an active learner who processes information by organizing it, elaborating it, and encoding it into a representation. All information taken in is related to prior knowledge. When instructors or books or whatever institutional medium is used present material to be learned, the learning does not occur automatically. On the contrary, acquisition is a constructive, deliberative process which is significantly affected by such factors as the trainee's existing knowledge, his ability to use strategies and techniques for learning and remembering, the organization and complexity of the task, and the conditions of learning.

- The more the learner can meaningfully integrate new information into his existing knowledge, the better the new information will be remembered.

- The greater the degree of original learning, the slower will be the rate of decay. The degree of original learning can usually be increased if the trainee more "deeply" or "extensively" processes the information (makes more associations, discriminations, uses the information in different contexts, etc.)

- Since memory is largely reconstructive, it is important to teach organizing principles, concepts, and rules. By doing this, we increase the probability that the learner will later be able to generate details that would otherwise be unavailable to memory. Also, this "understanding" serves as an organizational framework which makes a task representation "stronger" and more "distinctive" and easier to recall.

PREDICTING RETENTION

A critical issue for trainers and training managers is whether the course of retention loss can be predicted.

The answer is both a qualified "yes" and a qualified "no." For a particular individual on a particular task, we probably can never know enough about how the person learned the task, the nature of the person's representation, the state of previous knowledge, experiences during the retention interval, and so on, to make an accurate prediction. On the other hand, if we ask the question for group (i.e., unit) performance of a particular task, the answer is that we can predict quite well. Interestingly, many of the variables influencing retention "balance out" when we examine unit performance. For example, the "high-ability" people balance the "low-ability" ones; those who are exposed to the task during the retention interval are balanced by those who are not; and tasks tend to be taught to a group, thereby evening out differences due to different levels of initial learning. What we are left with are task characteristics. As will be presented below, we have been able to successfully predict skill retention by examining a set of task dimensions.

THE SKILL RETENTION MODEL

Under contract to the U.S. Army Research Institute (ARI) and supported by the U.S. Army Training Board (USATB), the American Institutes for Research (AIR) developed, tested, and validated a model to predict how rapidly proficiency on individual tasks deteriorates during intervals of no practice. Details of the project that resulted in the model are published elsewhere (Rose, Czarnolewski, Gragg, Austin, Ford, Doyle, & Hagman, 1984); briefly, we went through the following activities:

1. We conducted a thorough review of the literature to determine the major variables related to skill acquisition and retention (Rose, McLaughlin, & Felker, 1981; Hagman & Rose, 1983).

2. AIR, with support from the Human Resources Research Organization (HumRRO), conducted a series of field studies. Each of these field studies involved training and testing large groups of soldiers on several tasks to determine and equate initial levels of task proficiency, then retesting them periodically (usually after two, four, and six months).

3. Based on the results of these field studies, we attempted to predict retention scores (i.e., unit proficiency levels by task), using a host of task characteristics and individual difference variables (e.g., ASVAB scores, frequency and recency of task performance, etc.) as predictors.

As a result of several iterations, we found that a relatively straightforward set of "task characteristics" variables (e.g., number of steps, time constraints, etc.) correlated quite highly with retention performance. We then built a formal model by determining weights for each task characteristic, developing combination rules, and determining an algorithm for predicting proficiency as a function of the task score and the retention interval. This model was cross-validated: for example, regression weights derived from

Infantryman tasks were used to predict retention results of Artilleryman tasks. The model provided an excellent "fit" for practically all of the tasks for which empirical data had been collected. Correlations in the neighborhood of r=.90 were obtained between actual retention performance and retention levels estimated by the model.

Given the success of the model, several steps were taken to improve its usefulness and acceptability in the military community. First, we wrote an easy-to-use "User's Decision Aid (UDA)", User's Manual, and developed a training course on how to use the model. The UDA, the Manual, and the training course were tested by representatives of all Army Schools in a series of three-day sessions sponsored by USATB. Next, through USATB and the U.S. Training and Doctrine Command (TRADOC), the UDA and User's Manual were published and disseminated as an official document (TRADOC Form 321-R). The model and method were further publicized in other military publications (Hagman, Hayes, & Bierwirth, 1986). Most recently, USATB sponsored a Skill Retention Research and Development Conference, wherein the UDA was further exposed to a wide military audience.

As a result of these efforts, several additional areas of application for the model have surfaced. For example, the Army School of Training Support, Royal Army Educational Corps Center (U.K.), conducted a research effort to determine the generalizability of the model to the British Army. The U.S. Army is currently sponsoring a project to assess the applicability of the model to the Individual Ready Reserves. And the U.S. Army is considering conducting a project that would extend the model to the prediction of collective task performance.

MODEL CONTENT AND PROCEDURES

As mentioned above, the User's Manual contains a detailed description of the model and instructions for its use. This document can be obtained through USATB and AIR; both a paper-and-pencil and a PC-based version of the model are available. A few basic steps are involved in obtaining a retention prediction of unit proficiency on a given task. First, the user answers a series of questions about the task. These questions (with answer options) are:

1. Are job or memory aids used by the soldier in performing this task? (Yes, No)

2. If "Yes," how would you rate the quality of the job or memory aid? (Excellent, Very Good, Marginally Good, Poor)

3. Into how many steps has the task been divided? (One step, Two to five steps, Six to ten steps, More than ten steps)

4. Are the steps in the task required to be performed in a definite sequence? (None are, All are, Some are and some are not)

5. Does the task provide built-in feedback so that you can tell if you are doing each step correctly? (Has

built-in feedback for all steps, Has built-in feedback for most steps, Has built-in feedback for only a few steps, Has no built-in feedback)

6. Does the task or part of the task have a time limit for its completion? (There is no time limit, There is a time limit but it is easy to meet, There is a time limit and it is difficult to meet)

7. How difficult are the mental processing requirements of this task? (Almost no mental processing requirements, Simple mental processing requirements, Complex mental processing reequirements, Very complex mental processing requirements)

8. How many facts, terms, names, rules, or ideas must a soldier memorize in order to do the task? [None, A few (1-3), Some (4-8), Very many (more than 8)]

9. How hard are the facts and terms that must be remembered? (Not hard at all, Somewhat hard, Very hard)

10. What are the motor control demands of the task? (None, Small, Considerable, Very large degree of motor control needed)

Each of the answer options has an assigned numerical value; after answering all questions, the user adds up these values. The total is the "magic number" for the task. Next, this number is used to access a "Performance Prediction Table." This table provides numeric unit proficiency estimates as a function of the "magic number" and the retention interval (i.e., time since last practice).

SUMMARY: UTILITY, STRENGTHS, AND WEAKNESSES

This model can be used to address some important questions faced by training managers and decisionmakers:

- How quickly are specific tasks forgotten? Which tasks will be forgotten most rapidly?

- What proportion of the unit will be able to perform a task correctly after a specific length of time without practice?

- When and how often should refresher or sustainment training be conducted to maintain proficiency at a given level?

In addition, performance predictions, along with criticality, importance, and other pieces of information, are critical inputs for the setting of training priorities. It is not possible to provide sustainment training for all soldiers on all tasks; choices must be made that maximize the utilization of limited time and resources.

The main strengths of this model are its ease of applicability (with a little practice, it should take about five minutes to rate a task), its reliability, and its demonstrated validity. Following our training program, we found interrater reliability for assigning ratings to each task characteristic to be approximately r=.90. High correlations between prediction, and actual performance have been obtained for many different types of jobs and tasks, encompassing a wide range of difficulty.

On the other hand, the model does not generate a prediction for an individual soldier. Also, its applicability to noncombat jobs and tasks and to higher-level "soft" skill tasks has not been explored to any great extent. Finally, we have not explored the extent to which repeated retraining episodes would affect the accuracy of the predictions.

Nevertheless, we believe that this model is a useful and valuable tool for predicting unit proficiency. Its theoretical and empirical support make it a significant advance toward solving an important recurring problem in training management.

REFERENCES

Hagman, J. D., Hayes, J. F., Bierwirth, W. (1986). At the squad through company level a method for estimating task retention: Research pays off, Army Trainer, 54-55.

Hagman, J. D. & Rose, A. M. (1983). Retention of military tasks: A review. Human Factors, 25 (6), 199-214.

Rose, A. M., Czarnolewski, M. Y., Gragg, F. E., Austin, S. H., Ford, P., Doyle, J., & Hagman, J.D. (1984). Acquisition and retention of soldiering skills. Washington, DC: American Institutes for Research.

Rose, A. M., McLaughlin, D. H., & Felker, D. B. (1981). Retention of soldiering skills: Review of recent ARI research. Washington, DC: American Institutes for Research.

HUMAN PERFORMANCE MODELS FOR TRAINING DESIGN

C.M. Knerr, P.J. Sticha, and H.R. Blacksten[1]

Human Resources Research Organization
1100 South Washington Street
Alexandria, Virginia 22314

OVERVIEW AND OBJECTIVES

Mathematical modeling and simulation offer potential methods for analyzing skills and tasks, and for deriving training design guidance. Existing methods provide guidance for academic and technical skills, but not for the complex perceptual-motor skills required for complex military missions. Improvement of task analysis is a critical goal because task analysis is the foundation of training design and the military needs cost-effective training.

This research applied modeling and simulation techniques to enhance task analysis and training design for military skills. It modeled skill acquisition and retention of Army procedural skills, and modeled task performance and training of Air Force flight skills. The Army work integrated theories of two aspects of skilled performance: control of the sequencing of task elements, and learning and retention processes at the task-element level (Knerr, Harris, O'Brien, Sticha, and Goldberg, 1984; Sticha, 1982; Sticha, Edwards, and Patterson, 1984; Sticha and Knerr, 1984).

The Air Force work explored the use of simulation to guide part-task training (PTT) design. Knerr, Morrison, Mumaw, Stein, Sticha, Hoffman, Buede, and Holding (1986) summarized information on task analysis guidance for training design, sequencing of instruction, and PTT strategies. The analyses included all skill types, but focused on the

---

[1]This research was supported by three sources: (a) Contract Number MDA903-81-C-0517 from the Army Research Institute for Behavioral and Social Sciences from the Air Force Human Resources Laboratory (prime contractor: Decisions and Designs, Inc.), (b) Contract No. F33615-84-C-0066 (prime contractor: the University of Dayton Research Institute), and (c) HumRRO funds through Mr. William C. Osborn, Director of the Performance Research Division. The authors are especially grateful to Dr. Randy Mumaw for his work on the model for the dive bomb task. The views in this document are those of the authors and are not to be construed as official government positions unless so designated by other authorized documents.

complex perceptual-motor and tactical skill demands of aircrew tasks. We developed models of two flight tasks and exercised them to expand traditional task analyses, and to recommend performance measures and training strategies (Knerr, Morrison, Mumaw, Sticha, Blacksten, Harris, and Lahey, 1987).

The research developed, validated, and demonstrated the use of models of learning and performance of military tasks. The goal was to create guidelines to specify training characteristics and strategies for military training design, application, and management. The benefits included methods to analyze the complex perceptual-motor, tactical and procedural skills in military tasks, to link that analysis with instructional design guidance, and to specify features for training systems. This paper summarizes the results of that research.

METHODS AND TECHNIQUES

In this research, models and simulations refer to computer programs developed as analogs of human operator and associated equipment performance, or as analogs of the larger human operator training process. Any model may be called a simulation, but the convention we use distinguishes these two concepts. By models we mean representations (mathematical, symbolic, graphic, etc.) that reproduce features of the real system. The model abstracts components and relationships hypothesized as crucial to the phenomena in the system. A model may be static or dynamic (an organizational chart is a static model). The term _model_ is used here to represent any mathematical construct (usually a computerized mathematical algorithm) developed as an analog to some real system or process. Mathematical models express properties of the real system as equations; quantitative measures replace the qualitative verbal or pictorial distinctions and descriptions of relationships among the system components. _Simulation_ refers to a particular type of model that is dynamic and represents a process. Simulation uses models in ways that make them operative or functional. Exercising the model creates representations of the systems by reproducing processes in action.

The first stage in simulation-based training research develops a model of single task performance on the subject operational system, i.e., the operational equipment together with its human operators. The model represents the relevant characteristics of this system, including those components or subsystems that significantly influence the output. The model must accurately portray relationships among the components. This capability to model interactions among components is one of the critical strengths in applying simulation to research on complex missions.

The second stage employs the completed task performance model to aid in the design of empirical research and to formulate guidance and diagnostics for training single tasks. Exercising the model generates synthetic performance data to use in task and training analyses.

Selecting a Modeling Language

Our reviews of human performance modeling described the logic and criteria for selecting a method for a specific application (Knerr, et al., 1987 ; Sticha, 1982). SAINT (Systems Analysis of Integrated Networks of Tasks), a network technique for human-machine systems simulation, appeared to be the best for representing the Army procedural skills and tasks. Selection of SAINT was based on flexibility, generality and the availability of general model-building software (Wortman, Ducket, Seifert, Hann, and Chubb, 1978). SAINT has the general capability for the user to specify psychological process functions, while modeling the human-machine activities and interactions.

The SAINT model represents each procedural step (task-element) and links the steps in a network that represents the constraints on the order in which the steps may be performed. SAINT can represent complex interactions among the steps, modifications of steps by other steps, and various kinds of branching (deterministic, probabilistic, and conditional). SAINT had been successfully used in human factors and training-related modeling (Sticha, Edwards, and Patterson, 1984).

MicroSAINT is a microcomputer version of SAINT (Micro Analysis and Design, Inc., 1985). MicroSAINT incorporates many of the SAINT modeling constructs, and facilitates development and use of task models. It can simulate both flight processes and discrete pilot decisions. MicroSAINT was selected as the simulation tool for the Air Force research, based on these considerations.

Analysis for Model Development

SAINT and MicroSAINT are network modeling languages. They represent activities and their interrelationships by nodes and arcs, respectively. Network modeling decomposes system performance into a series of subactivities or nodes (e.g., a task analysis for the human operator, a functional analysis for the aircraft). Then, the sequencing of activities or functions is defined by constructing a node network.

The level of system decomposition depends on the problem. The system can be defined in as detailed or gross a level as the modeler decides. A network may include several relatively autonomous sub-networks that, while interrelated, are distinct (for example, discrete decision processes are separate from continuous motor control of the aircraft but they influence that control).

Flow charts of the task are an early product of the task analysis. Flow charts are not limited to representing the sequence of task elements; they can represent cognitive processes such as decisions. The flow chart depicts activities and logic flow. It depicts much of the information required to build a task performance model, namely the activities, the temporal relationships between them, and the general flow of information between activities.

Initial stages of modeling usually lack empirical data (e.g., measures of time and performance distributions); however, the conceptual analysis specifies the kinds of relationships expected between skill variables and the performance characteristics of each activity. This specification enables assumptions about the distributions to use during the initial modeling, until collection of empirical data to refine the models. A detailed task analysis provides the basis for estimating much of the information required to build a task performance model and to make the model "run," such as task performance times, performance probability distributions, task component interactions, system and environment status. Actual data on the performance time and accuracy distributions may only be available after empirical research, and need to be assumed to use distributions in the simulation program.

Modeling Levels for Task Analysis and Training Research

Task analysis and training design goals require modeling at multiple levels of detail. Sticha, Knerr, and Goldberg (1984) defined a hierarchy of models for use in training, with (a) task performance models as the lowest level, (b) task training models as the next level, and (c) training system models as the highest level. Task performance models can facilitate communication between the analyst and the military subject matter expert (SME), supplement task analysis, and guide training for single tasks. They portray elements within tasks (e.g., procedural steps, decisions, and continuous-control functions). Further stages of simulation-based training research develop and exercise a task training model. The task training model can detail training system design and help to evaluate training strategies. Training system models can optimize the cost/benefit of integrated, multiple task training systems. This research developed models of the first two types.

Task performance models. A task performance model is an analog of the performance of a human operator in accomplishing a single execution of a task. Values of model parameters control the proficiencies of the simulated human operator in requisite skills. By using different input values for the various proficiency parameters, the model represents operators at different skill levels. Thus the task performance model can be used as a rudimentary laboratory for the training researcher, exploring questions concerning task characteristics and training designs. The task performance model does not predict how much practice is required, but the researcher may be able to address these questions using results from the literature. The task performance model thus complements the researcher's knowledge and allows exploration of training system design issues without incurring the expenses of research with real subjects. Such inexpensive preliminary simulation can focus later experiments with real pilots as subjects.

Task training models. Task training models represent the training history of a task, including skill acquisition, retention, and performance. Skill acquisition considers training strategies such as those of concern to PTT (e.g., task division and reintegration, frequency of practice, information processing loads, and sequencing of instruction). The small amount of research on task training models, conducted in the area of Army initial skill training, showed a positive relationship between Army unit practice and performance. Thus, this level of model has promise for military task-analysis and training design purposes.

Figure 1 presents an example from the Army research. The model starts with initial skill training (Node 1), which sets the performance parameters to values expected after the training. Three timers start when Node 1 is completed. The first timer (Node 2) represents the time the soldier waits between the completion of initial training until his assignment to an operational unit. The second timer (Node 3) represents an external event that interrupts the typical practice schedule. The event is the researcher's decision to measure performance on the task. The third timer (Node 4) represents the time spent between opportunities to practice the task in the unit. Nodes 4 through 6 represent the cycle of periodic practice interspersed with other activities. Node 5 adjusts the performance parameters to reflect the forgetting that occurs during the waiting period, and Node 6 updates the parameters to account for the effects of practice.

The branching logic in the task training model is not entirely represented in the links in Figure 1. Some of this logic is implemented using a feature of SAINT called task clearing. Task clearing is a mechanism by which the completion of one node can interrupt the processing of another node in the network. Thus, completion of Node 2 interrupts the processing of Node 4 and restarts that Node from the beginning. Similarly, completion of Node 3 interrupts the processing of Nodes 2 and 4, and starts Node 5.

Performance of the task is only simulated once in this model. This single performance represents the time that measurements are made by the researcher. Nodes 5 and 6 update performance parameters. If Node 3 has been completed, then the embedded task model is processed, and simulated performance data are generated. The task performance model, embedded in this task training model, is depicted with dashed lines.

This higher level of simulation is much more difficult but potentially of extreme value. At this level the goal is to represent not only how a human operator executes a task, but how he learns from this performance or from associated training.

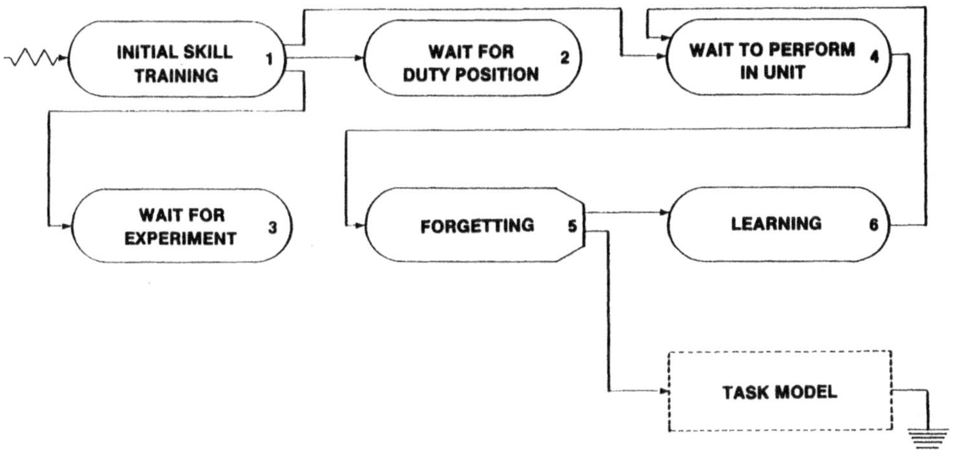

Figure 1. TASK TRAINING MODEL FRAMEWORK

The task training models represent the critical issues in training, including difficulty (e.g., derived from simultaneous loading of information processing on multiple activities), division and reintegration of tasks, comparison of part-task strategies, and effects of amounts of training. The task training models operate at a level appropriate to incorporate mathematical formulas to express human skill acquisition.

For example, the Air Force research produced a prototype task training model for the 30-degree dive bomb task. It assumed that operator performance on continuous processes such as tracking improves according to the power rule of practice. It assumed that operator rule-learning behavior improves according to a Markov process. Two kinds of skill acquisition formulas, therefore, had to be incorporated into the model to represent the dive bomb performance.

Model Verification, Validation, and Refinement

The model is exercised to verify its accuracy, and to provide the basis for drawing conclusions regarding task training and performance measurement. The first runs are for initial credibility assessment, and to identify missed steps or interactions that make the model behave inappropriately. One of the benefits of simulation is that it identifies errors in the task analysis, without the expense of empirical data collection.

Model refinement continues based on a review of the results of model applications. Military SMEs work with an analyst to create or edit the model and understand what occurs, in what sequence, and whether the time estimates are appropriate. Model checkout sessions continue until the reviewers are confident that the model accurately depicts the task or skills. Sensitivity analysis can establish the limits of performance, prioritize skills to train, and identify critical performance measures. Empirical validation is possible if data are available. We performed different levels of validation for the models, depending on the availability of data.

MILITARY TASK MODELS AND SIMULATION RESEARCH

We developed models and performed simulation research for an array of different military tasks. The tasks included eight Army procedures and an Air Force flight task. This section discusses the Air Force task first, then the Army procedural tasks.

Research on Air Force Aircrew Skills

The goal of the Air Force research was to create guidance for PTT strategies and devices. This research provided examples and heuristics for the extension of task analysis and training guidance for aircrew skills. It focused on error analysis and factors that determine the efficacy of PTT compared to whole-task training. The products were (a) a task performance model to use as a diagnostic tool, and (b) a task training model that simulated the influence of training strategies on performance. The task training model represented the discrete and continuous processes in the dive bomb task. It also incorporated an attention mechanism, which may account for some PTT phenomena.

The aircrew task selected for this research was the 30-degree dive bomb for the A-10 aircraft. Knerr, et al. (1986) described the task and conducted the traditional and structural task analyses of the 30-degree dive bomb task. The results described the structure of the task segments, difficulty, and skill types.

Pilots divide this maneuver into the three functional segments (base leg, turn, and final leg). If the pilot determines that he is not on the correct path, he must compensate the flight path errors by adjusting his release altitude. The three major factors that the pilot must consider are dive angle, airspeed, and release altitude (e.g., adjust release altitude to compensate for a shallow dive angle).

Task performance models and simulations. We developed MicroSAINT performance models that contrast between expert and novice performance to provide a diagnostic tool. Figure 2 shows the network for the base leg of the model of expert performance. The right side of the chain tests for the critical event to end the base leg network (reaching the roll-in point). In an analogous network for the final approach, the critical event is reaching the bomb release altitude. These events stop the local network and pass control to the next network or end the task. Control of processing is passed back and forth between the two chains probabilistically; only one event is activated at any time.

The base leg begins with a dummy node (1.1) that requires no time. It allows processing to begin on either of the dual chains; however, the probability is low that processing will begin on the right chain. Nodes 1.2, 1.3, and 1.4 record the difference between current airspeed, altitude, and heading values and the values the pilot believes are optimal. Differences are created by the introduction of normalized variation in the continuous flight function. The monitor nodes are processed serially. As time elapses in the base leg, the probability increases that control is passed to the right chain instead of the next

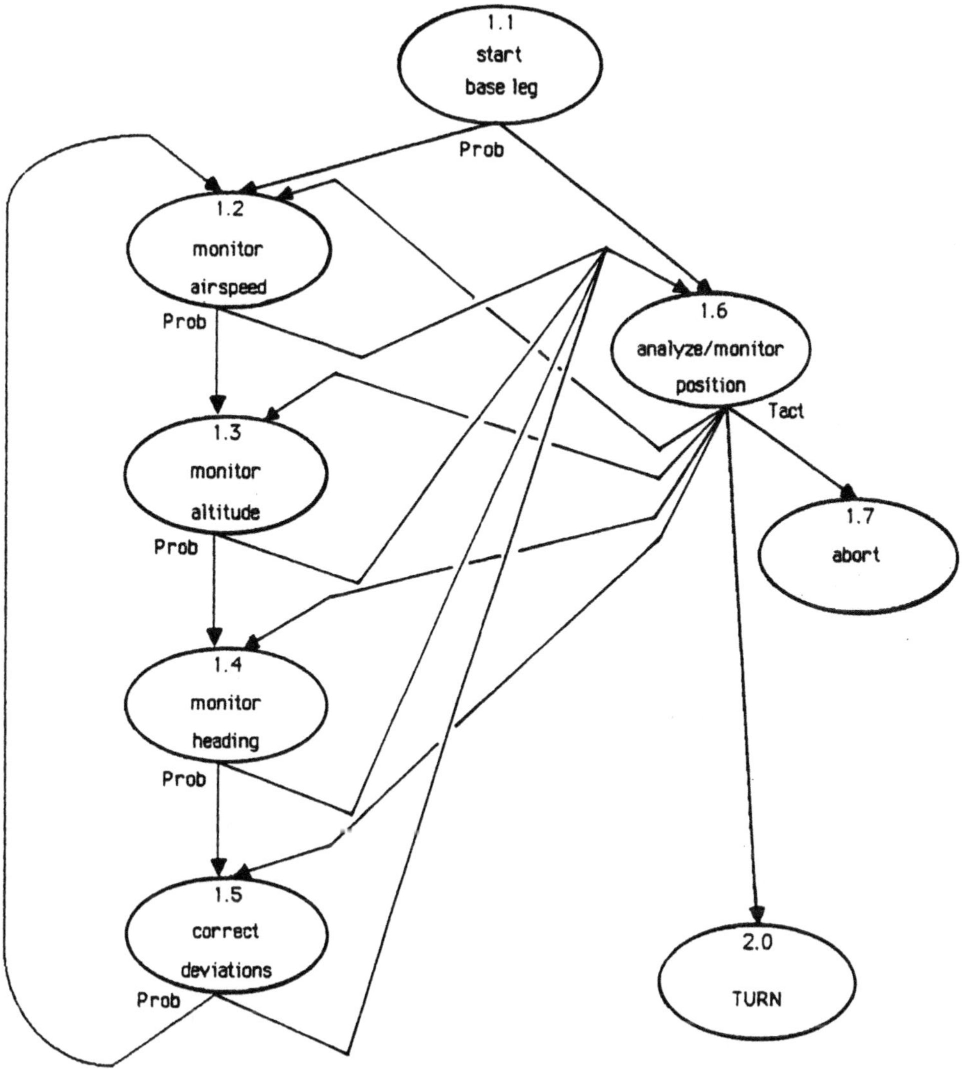

Figure 2. NETWORK FOR BASE LEG OF THE DIVE BOMB TASK

node in the left chain. Execution of Node 1.6 passes control to the next node in the left chain. When Node 1.5 is executed, all flight deviations are corrected to the perceived optimal values by cutting them in half. Control is then passed back to Node 1.2 to continue monitoring the flight path.

The final leg is the most complex of the three task segments. The checks are more likely to lead to a significant correction. Determination of release altitude takes both dive angle and airspeed into account. Knerr, et al. (1987) depict and describe the models for all three maneuver segments.

In the simulation runs, the difference between the expert and novice models was the extent of performance error. We categorized errors into three types:

o Decisional: errors in use of the rules used to control flight path and bomb release (e.g., corrections for dive angle)

o Perceptual: perceptual judgment errors, inability to perceive the correct perceptual cues (e.g., select an aim-off point 1000 feet beyond the target)

o Motor-control: variability errors in steering ("driving" skill).

We simulated degraded performance by systematically introducing the three error types into the three segments of the model. Decisional errors, for example, were represented as rules with two components (direction and amount of adjustment). Decisional errors were introduced by reducing the probability that the pilot knew the rules. Twenty simulated trials were run for each error configuration. The major dependent variable was the average circle error of bomb placement (in meters), with other factors of the simulation held constant.

The results of the simulations showed that the accuracy scores were only affected by a small subset of the possible errors. Perceptual errors had no affect on bomb accuracy, even when they were introduced into all task segments. The other error types affected performance only when they were introduced into the final approach (possibly because errors made in early phases are corrected in the final approach). This result has both performance measurement and training implications.

The bomb scores are final measures of effectiveness for the dive bomb task: getting the bomb on the target is the reason for performing this task. Elements of the task that are closely related to bomb accuracy should be given priority in training. These results identify three aspects of performance that are critical for accurate bomb placement: (a) the ability to control the aircraft during the final approach so that it is consistently flying toward the aim-off point, (b) the ability to perform checks on aircraft attitude and speed rapidly, and (c) the ability to make the proper adjustments in release altitude as a result of deviations in speed and dive angle from their nominal values. These three performance components correspond to the three types of error that produce substantial decrements in performance.

The dive bomb research included empirical performance data collection in the A-10 flight simulator at Williams Air Force Base, Arizona. The objective was to test the effectiveness of part-task training (PTT) strategies. The performance data provided initial validation of the MicroSAINT model of the task, and the basis for building a task training model and exercising it to simulate part-task strategies. The subjects were 68 U.S. Air Force student pilots in their last two months of undergraduate-pilot flight training. The experiment had three training conditions; the background for these conditions, including prior research that the experiment attempted to replicate, are described by Knerr, et al. (1987). The three training conditions produced the same level of proficiency after 35 trials. Average bomb miss distances were very good for the students and were not influenced substantially by training conditions. In addition, students' performance measures were similar to those of the expert pilot. The strongest conclusion was that the dive bomb task presented to these students was not difficult enough for them to allow exploration of PTT phenomena.

Nominal values for model parameters were obtained from the empirical data. The empirical data confirmed the assessments of model parameters that were obtained through discussions with subject-matter experts. Thus, the simulation model gave reasonable results, verified by the data. In situations where no data were available, the model would serve task analysis and training design purposes without requiring empirical data to estimate model parameters.

Task training model. The task training model connected training strategies to learning outcomes, and described performance at different stages of learning, from novice to expert. The model described the learning process as a function of several learning parameters. At the core of the model were two learning functions and a limited-capacity attention mechanism. Learning of decisional knowledge (discrete rules) was characterized by a Markov process describing the transition of individual rule elements from an unlearned to a learned state. Learning of perceptual-motor and control information (continuous scale) was characterized as a power function relating performance variables to the number of training trials. An attention mechanism determined the learning rate as a function of which elements of the task were being trained and how well each element was known.

The learning model allowed us to determine learning parameters analytically from a description of the learning history. The performance simulation, in turn, determined the effects of these learning variables on performance. To implement the learning models, we first developed them analytically on an electronic spreadsheet; it was not necessary to incorporate the calculation of learning parameters within the MicroSAINT simulation. Then, the outputs of the spreadsheet analysis were used as input variables to the MicroSAINT performance model. Knerr, et al. (1987) presented the Markov transition matrices and formulas, and the power functions used to model these skills.

The attention mechanism reflected reasons that justify PTT strategies. Only a limited amount of material can be learned at a single time, depending on task difficulty (difficult tasks require greater attention for mastery than simple ones). The assumption, logic, and formulas for the attention mechanism were given by Knerr, et al. (1987). The attention mechanism can lead to an advantage for PTT strategies over whole-task training in situations in which the attentional limitations are severe. This feature of the learning model was illustrated with the 30-degree dive bomb task using different training strategies and assumptions about attentional limitations.

The parameters of the learning model included initial and final values of control standard deviations, perceptual biases, and speed of performing checks. Also included were initial probabilities of knowing the proper adjustment rules, learning rates, and the maximum attention capacity. Thus, the elements of the learning model corresponded to the classes of errors in the performance simulation. Each task element was characterized by a difficulty measure that indicates the extent to which it requires learning resources. The difficulty values loosely corresponded to the judged difficulty of task elements reported by Knerr, et al. (1986).

Training options were defined according to the variables trained in them (e.g., train only those variables in the final approach). A training strategy is a combination of training options. The complete specification of a training strategy allows one to predict analytically the values of the model parameters at the end of any training trial. Because of the complex relationship between the model parameters and performance variables, one must rely on the simulation model to predict the overall performance resulting from the application of any training strategy.

The procedure to implement the learning models consisted of two steps. In the first step, the training strategies and other learning parameters were defined and the resulting values at each training trial were calculated. These values were used as input to the MicroSAINT simulation. In the second step, the MicroSAINT simulation was used to determine the performance associated with the training strategies. This performance was measured by the accuracy of bomb placement.

We used sensitivity analyses to illustrate ways in which task or trainee variables influence the effectiveness of different training strategies. We ran four simulations that compared whole-task training with PTT. The whole-task training strategy consisted of 40 trials training all task elements. The PTT strategy had 10 trials on the final approach, followed by 15 trials on the turn and final approach, followed by 15 trials training all task elements.

The calculated parameters of the learning model at the conclusion of the 40 training trials were used as input to the performance simulation. The simulation was run 20 times under each condition. The same seed to the random number generator was used for each condition, so that the random number strings are partially correlated between conditions.

The average circular error of bomb placement under the four conditions was the criterion. The results indicated a strong advantage of PTT in the low-capacity condition, and approximately equivalent performance for the two training strategies in the high-capacity condition. The low-capacity condition leads to substantial reductions in learning for the whole-task training strategy. PTT reduces the impact of attentional limitations and consequently produces superior training. In the high-capacity condition, the effect of attentional limitation is minimal. The two strategies produce roughly equivalent training on the final approach, while whole-task training produces substantially greater training on other segments of the task. However, the first two segments have almost no effect on overall bomb placement accuracy. Because the training on the final approach is approximately equivalent, the overall error scores are also approximately equivalent. When this result is coupled with the fact that trials using PTT are shorter than those using whole-task training, an unqualified recommendation for the part-task training strategy is produced.

Summary of dive bomb research. The analyses of the dive bomb task used task performance simulation (for error analysis and performance measurement recommendations), an empirical experiment, and task training simulation to examine PTT issues. Exercising the model with various combinations of errors simulated degraded performance. The error analysis differentiated categories of errors that are likely to occur in many aircrew tasks and missions. The errors encompass decisions, perceptual judgments, and motor control skills. The simulation produced qualitatively different forms of degraded performance given different sets of errors. The error results relate to training guidelines, since such guidelines use task categories and error analysis to draw on human learning literature and research for training design.

The analysis of attention mechanisms is central to PTT. PTT is advocated for training tasks so difficult that the student cannot master them as whole tasks. The information processing demands are prime determinants of difficulty for the cognitive skills in many aircrew tasks. The attention mechanism research can link the PTT results to other domains, such as dual task learning (the parts of the task do, in fact, act like dual tasks in attention demands). Attention mechanisms also have implications for the task loading of skill types, and therefore for the utility of pre-training enabling skills, to reduce the cognitive/attention load. A major contribution of the attention model is that it quantifies the value of sequencing of instruction. Further research should improve the attention analysis methods, perhaps by drawing on information processing literature.

Research on Army Armor Procedures

The Army tasks were selected from those performed by the driver, loader, and gunner of the M60A1 tank (e.g., Stop the tank engine). They represented a range of length, complexity, and extent of practice in the unit after initial training (values on these dimensions, and other details of the data collection are reported by Knerr, et al., 1984). Analysis of the tasks determined the task elements, standards, conditions of performance and characteristics shown in the literature to influence skill acquisition and retention. We used the task information to develop SAINT models of eight tasks.

Validation and cross-validation. The models were validated by comparing their predictions to data from two samples of soldiers. The first sample comprised 471 soldiers from four training companies at Fort Knox, Kentucky, in their fifth to tenth weeks of training. Each soldier performed two of the eight tasks for a total of six trials (five acquisition trials and a retention trial a month later). The second sample comprised 116 soldiers from operational units at Fort Knox who had completed their initial training within 31 months. Soldiers in the unit performed all eight tasks, one time each.

The learning and retention models were separate from the task-element sequencing models provided by the SAINT simulation; therefore, a rigorous statistical evaluation of the models was possible, rather than relying on sensitivity analyses alone. The validation strategy used maximum-likelihood estimates of the model parameters in a general, iterative, unconstrained optimization routine. The necessity and the sufficiency of each model were assessed by comparing the goodness-of-fit of that model to both more and less general models using log-likelihood chi-square tests. Data from each task were divided into groups. Model parameters were estimated from the first half of the data, termed the model-development group. These parameters were then applied to the second half of the data, termed the cross-validation group.

We used performance data from the soldiers in training to test the learning component of the models. Parameters were estimated for learning and retention models of several levels of generality. The basic learning model postulates geometric increase in strength over trials to an asymptote, followed by a strength decrease between the final training trial and the retention trial. The basic model has six parameters (three response thresholds, a learning rate, an asymptote, and a retention proportion).

The results indicated that the basic learning and retention model provided a significantly better characterization of the data than a simpler (three-parameter) model for all eight tasks. This result demonstrated the necessity of the basic model. For four of the tasks, the shape of the learning curve was adequately characterized by the exponential function postulated in the basic model. The remaining four tasks showed significant improvement in fit between the basic model and the model encompassing a more general learning curve. For two tasks, differences in the shape of the learning curve were confounded with differences between task elements.

We applied the basic learning and retention model to predict the results of the training experiment. We compared the learning and retention data to 100 simulated subjects generated by the SAINT model. The fit was impressive, even though it explicitly ignored some of the significant differences identified above (e.g., task element differences). Sticha, Edwards and Patterson (1984) present the results for all eight tasks.

Task-element differences. The validated models were analyzed to examine the relationship between task-element performance differences and task characteristic ratings. The SAINT models of the eight tasks were run using the parameters of the task-characteristic model to simulate 100 subjects for each task. The percentage of task-elements performed correctly was compared to the data from the soldiers, as well as to the predictions of the basic six-parameter model (without the task-element characteristics).

A model predicting task-element differences as a function of five task characteristics provided a significant improvement over a model that assumed all task elements had the same values for the learning and retention parameters. The improvement brought about by the task-characteristic model was especially evident on the first trial, and in some cases, the retention trial. The fit was impressive, but some differences were not predicted by the task-characteristic model. Additional factors (e.g., individual differences) must be considered to account for task-element differences in learning and retention.

The task-element analyses were largely exploratory. The basic models were validated by applying the parameters estimated from one set of subjects to the data from another set of subjects. Since the models developed in this analysis contained considerably more parameters than the simpler models that did not consider task-element or individual differences, the data were insufficient to divide into model development and cross-validation groups. The results should be interpreted with the same care that is required for all correlational analyses. The results of this analysis need to be confirmed with replication studies, or analyses of other acquisition and retention data. Sticha and Knerr (1984) presented details of these analyses.

Discussion of Army results. This research focused on the development, validation, and application of mathematical models of procedural skill learning and retention. The major accomplishments of this research were the development of integrated models of procedural learning and retention, and the incorporation of these models in a complex performance simulation. The model predicted accurately the improvements in performance that occur during training, and the decay in performance shortly after training is completed.

A major purpose of the Army research was to illustrate the application of mathematical models to investigate acquisition and retention of complex military skills. The application to task-element analysis illustrated some of the aspects that characterize the method and distinguish it from traditional methods.

SUMMARY AND CONCLUSIONS

This research augmented, through task modeling and simulation, the traditional methods of analyzing tasks for training design. The military tasks contain a range of skills. The armor tasks were procedural. The dive bomb task had predominantly cognitive and continuous perceptual-motor skills.

Traditional task analysis serves well for academic tasks. Alternative methods for cognitive tasks, reviewed by Knerr, et al. (1986), include a variety of hierarchical, structural, and skill cluster techniques and production systems. Cognitive methods identify the memory organization of the task to segment the task into meaningful parts. The analyst can divide the task according to the structure meaningful to job incumbents, and reintegrate the parts in training.

Other task analytic procedures, without the benefit of simulation, could potentially provide the same level of detail and information needed to generate instructional guidelines. The state-of-the-art, however, indicates that other methods have failed to do so, especially for continuous perceptual-motor tasks. No current method appears to provide thorough analysis for complex psychomotor and decisional tasks. Flow charts and other methods applied to continuous motor tasks do not incorporate quantitative models of the aircraft parameters or environment and they do not generate simulated data. Creating and programming mathematical models of aircraft parameters, and linking them with decisions and perceptual-motor skills, is also the main strength of quantitative modeling. Therefore, we recommend continuing this line of simulation-based research to meet training design needs.

Task simulation should be selective. Rather than trying to build models for all, or even a large number of tasks, a few should be selected (perhaps those problematic for training). Otherwise the effort is too large and labor intensive. Modeling and simulation, given the current state of the techniques, are research tools and not routine task analysis tools.

The two major risks of this approach are the capability of the behavioral simulation to reflect the relevant skills, and the ease of obtaining the data for the simulation. The development of an analytic model requires estimation of performance times and success probabilities in addition to data that are obtained from traditional task analyses. These data may at times be difficult to obtain (e.g., for systems still under development). While certain data requirements must be filled precisely, for other variables, meaningful results may be obtained with only nominal data values. Thus, the possibility that the data requirements of the model are excessive does not necessarily invalidate the approach. In the event that all required data cannot be estimated accurately, the modeler uses the best available data, and determines those conclusions that are relatively invariant with respect to the values of unspecified variables. The variables that cannot be estimated accurately but have a great impact on model recommendations are those variables that must be studied in future research.

Modeling and simulation for task analysis have strengths and weaknesses. This quantitative modeling provides rigorous, detailed task information that users can verify and validate. In situations without data, the models can be examined to ensure that the model plausibly represents the task and skills, then can be used in sensitivity analyses to explore task characteristics and task component interactions.

Task training models explore training options and strategies. This phase of the research focused on skill acquisition, and thus incorporated mathematical models of human learning. Key issues also include skill retention and transfer. Sequencing of instruction and PTT effectiveness rely on transfer concepts; therefore, the analysis needs to incorporate these in future research.

A concept that is central to the predicted benefits of PTT strategies is the concept of a limited learning capacity that is allocated to specific skills to be learned. PTT has its maximum benefits when the learning demands of the task are great relative to the attentional capability of the learner. The extent of the effect is mediated by the complex relationships among task elements that determine their relative criticality in producing the output of the task. Thus, in the 30-degree dive bomb task, PTT and whole-task training produced equivalent performance as measured by bomb accuracy when attentional capacity was high, even though WTT produced greater learning on the majority of the performance parameters. The overall performance of the two methods was equivalent because both produced roughly the same learning on the critical performance parameters.

The above example illustrates the importance of both the learning model and the performance simulation in evaluating training strategies for complex tasks. The learning model allows us to predict the effect of training strategies on the individual skills that compose a task. The performance simulation is required to translate information about individual skills to an overall prediction of task performance.

REFERENCES

Knerr, C.M., Morrison, J.E., Mumaw, R.J., Sticha, P.J., Blacksten, H.R., Harris, C., & Lahey, G. (1987). Analysis and design of part-task training for aircrew skills (Final Report 87-2). Alexandria, VA: Human Resources Research Organization.

Knerr, C.M., Morrison, J.E., Mumaw, R.J., Stein, D.J., Sticha, P.J., Hoffman, R.G., Buede, D.M., & Holding, D.H. (1986). Simulation-based research in part-task training (AFHRL-TR-86-12, AD-B167 293). Williams AFB, AZ: Operations Training Division, Air Force Human Resources Laboratory.

Knerr, C.S., Harris, J.M., O'Brien, B.K., Sticha, P.J., & Goldberg, S.L. (1984). Armor procedural skills: Learning and retention (Technical Report 621). Alexandria, VA: U.S. Army Research Institute. (ADA153227)

Knerr, C.M., & Sticha, P.J. (1984). Application of SAINT simulation to investigate task-element differences. Proceedings of the National Aerospace and Electronics Conference, NAECON (pp. 849-852). New York: Institute of Electrical and Electronic Engineers.

Knerr, C.M., & Sticha, P.J. (1985, Jan). Models of learning and performance of armor skills. Proceedings of the symposium on the military value and cost-effectiveness of training (pp. 491-512). Brussels: NATO Defense Research Group Panel on the Defense Applications of Operational Research.

Micro Analysis and Design (1985). MicroSAINT User's Guide. Boulder, CO: Author.

Sticha, P.J. (1982). Review of analytical models of procedural learning and performance (Report No. DDI/PR 82-14-334). McLean, VA: Decisions and Designs, Inc.

Sticha, P.J., Edwards, T.D., & Patterson, J.F. (1984). An analytic model of learning and performance of armor procedures (ARI Research Note 84-12). Alexandria, VA: U.S. Army Research Institute for Behavioral and Social Sciences.

Sticha, P.J., & Knerr, C.M. (1984). Task-element and individual differences in procedural learning and retention: A model-based analysis (Research Note 84-1). Alexandria, VA: U.S. Army Research Institute. (ADA136789)

Sticha, P.J., Knerr, C.M., & Goldberg, S.L. (1984). Application of simulation and modeling to Army training management. Proceedings of the 28th Annual Meeting of the Human Factors Society (pp. 1023-1027). Santa Monica, CA: Human Factors Society.

Wortman, D.B., Duket, S.D., Seifert, D.J., Hann, R.L., & Chubb, G.P. (1978). Simulation using SAINT: A user-oriented instruction manual (Report No. AMRL-TR-77-61). Wright-Patterson Air Force Base: U.S. Aerospace Medical Research Laboratory.

FORECASTING TRAINING DEVICE EFFECTIVENESS USING ASTAR

Mary Frances Martin and Andrew M. Rose

American Institutes for Research
Washington, D.C.

INTRODUCTION

The purpose of this paper is to discuss applications for the Automated Simulator Test and Assessment Routine (ASTAR) in the design and acquisition of training systems. ASTAR is a computer-based, analytic method for forecasting the effectiveness of training devices; it is designed to be used throughout the training device design and acquisition process.

This paper contains four main sections. The Background section briefly describes the process of acquiring training devices and why a need exists for analytic techniques such as ASTAR. The second section describes the content of the ASTAR analyses and how the technique is applied. The third section presents potential applications for ASTAR and three recent applications of the technique. The Summary section discusses the advantages and limitations of the technique.

BACKGROUND

The military and private industry have come to rely heavily on the use of device-based training systems to prepare people to perform their jobs. There are several reasons for this. The rapid growth in technology in recent decades has resulted in the development of increasingly complex systems that people are required to operate or interact with. Training devices are used to prepare people to operate these systems because training on the actual system is often prohibitively expensive or dangerous. Also, training on a system with features designed to enhance learning increases training efficiency and decreases overall training time.

However, training devices are costly to design, build, and produce. Thus, it is critical to assess or systematically consider the actual or potential effectiveness and cost implications of alternative device design approaches throughout the device development cycle. The earlier in the cycle these assessments can be begun, the more likely will be the development of cost-effective training devices.

Ideally, training systems should be designed based on detailed information about the operational system, the operational tasks, the capabilities of the trainees, and clear and precise training and operational performance objectives. For training systems that are tied to a major weapon system, the military has a formal system in place to accomplish this through empirical training system evaluations occurring during each phase of the acquisition process. The military acquisition process and its effect on training system design has been examined by a number of authors, including Kane and Holman (1982), Nutter and Terrell (1982), and Goldberg and Khatri (1985); we briefly summarize this process in the following paragraphs to serve as a background for the development and application of ASTAR.

The four main phases of the military acquisition process are preliminary evaluation of alternative design concepts, validation of the selected concept, full-scale development of the design, and production/deployment of the system. The first phase, evaluation of alternative design concepts, involves developing and testing preliminary system designs and choosing a set of preliminary concepts. In the second phase, the preliminary concepts are tested and a design is chosen. In the full-scale development phase, the design is further refined and models of the system are built and tested. In the production/deployment phase the system is manufactured and fielded.

Theoretically, the design of a training system follows the design of the operational system; models of both systems are constructed and tested for effectiveness at each stage of the acquisition process. Unfortunately, the logistics of the acquisition process frequently interfere with this ideal process for training system development. For example, the training device developer usually is required to produce and field a training system before the operational system is deployed. Changes to the operational system design in the final stage of the development process must be incorporated into the training device design. Time constraints in this phase of the process often preclude empirical evaluations of changes to the training device design. Also, empirical training system evaluations are costly and often are not conducted due to resource constraints.

The result of these constraints in the acquisition system is that many training system design decisions are based on factors other than whether the design will produce an effective trainer. For example, a designer has a requirement to produce a high-fidelity pilot trainer. If some feature in the operational equipment design is changed in the latter phase of acquisition, the change may be incorporated into the training device design regardless of whether the change increases or decreases the effectiveness of the trainer; the developer usually does not have the time to ascertain the effects of including or not including the change. Another factor is the influence of "state-of-the-art" training system technology. Suppose that a device developer must choose between including a new, state-of-the-art visual system and an older model visual system in a training device design and that no data on the training effectiveness of the new visual system are available. If the developer does not have the time or resources to test the effectiveness of the system, the decision will be made on the basis of information supplied by the manufacturer of the new system and cost factors.

The main goal in training system design is to produce systems of maximum training effectiveness within the constraints of the acquisition process. Because of the factors discussed above, many of the decisions regarding system design are based on analytic information rather than on empirical training system studies. However, there are few formalized techniques for analytic training system analyses; design decisions are often based on the developer's best judgment. The ASTAR technique was developed to address the need for a systematic analytic process to aid in training device design and acquisition.

THE ASTAR TECHNIQUE

In this section, we present an overview of the theoretical content of ASTAR and a description of how ASTAR evaluations are conducted. ASTAR is a direct extension of the Device Effectiveness Forecasting Technique (DEFT). The development and operation of DEFT/ASTAR and assessments of the procedure have been previously reported (Rose, Evans, & Wheaton, 1987; Rose & Martin, 1984; Rose & Wheaton, 1984a, 1984b). ASTAR is a series of interactive, menu-driven, computer programs that guide an analyst through the evaluation of a training device-based system. There are several computer programs that support the building and maintenance of data files and the conduct of analyses. Each is written in COBOL and is designed for use on an IBM (or compatible) Personal Computer.

ASTAR converts information and judgments about various facets of a training system into forecasts of device effectiveness. An analyst provides a number of judgments or estimates in response to a variety of rating scales. These scales facilitate consideration of different kinds of information about the training system and its parent equipment.

Theoretical Basis of ASTAR

ASTAR is based on a multidimensional view of training system effectiveness. The model examines the global training effort: Given a trainee population with specific capabilities and limitations, how well the entire training system promotes the acquisition of the skills and knowledge required for proficiency both on the training device and in the operational situation. This perspective is in contrast to other training effectiveness models, which focus exclusively on transfer of training as the sole criterion of effectiveness. This narrower focus ignores the acquisition component of training effectiveness--the comparison of what is required and trained by the device and what is required for operational proficiency. Further, ASTAR examines not only what is trained, but also how well the device-based system is designed to promote effective and efficient training and transfer.

ASTAR is intended to provide training system designers and developers with various kinds of information about the potential effectiveness of a training device-based system. ASTAR is not designed to produce a single "Figure of Merit." Our approach to effectiveness analysis is to provide a framework in which device developers can compare devices for effectiveness and diagnose potential problems in a system design.

Four major analyses are conducted during an ASTAR training system evaluation. The first is an analysis of the Training Problem to define the skill and knowledge deficiencies that trainees have relative to criterion performance on the training device. As part of the same analysis, the difficulty trainees will have in overcoming these identified deficits is estimated. The second analysis examines Acquisition Efficiency. The quality of training provided by the training device is studied by analyzing the instructional features and training principles that have been incorporated in the device to help trainees overcome their deficits. These two analyses are the "acquisition" component of ASTAR.

The third and fourth analyses are analogues of the first two. In the third, an assessment of the Transfer Problem is undertaken to determine the deficiencies trainees will have with respect to operational criterion performance after they have achieved criterion performance on the training device. The difficulty in overcoming these residual deficits is determined and any effects of physical and functional dissimilarities between the training and operational equipment are assessed. The fourth analysis is Transfer Efficiency; it indicates how well use of the training device will promote transfer of the learning that has occurred to the parent or actual equipment. The Transfer Problem and Transfer Efficiency analyses are the "transfer" component of ASTAR.

Conducting ASTAR Evaluations

An ASTAR evaluation can be conducted at three different levels of analysis, ranging from micro to macro in detail. The level that is chosen depends upon the kind and amount of information available and upon the degree of diagnosis desired. If analysts have very detailed information about a training system (descriptions of subtasks, displays, controls, instructional features, data on the trainee population) and want an in-depth evaluation, they would choose the micro level or ASTAR Level 3. ASTAR Level 3 analyses are performed at the "subtask" level, and involves analyses of individual displays and controls contained in both the training and parent equipment.

If analysts have less detailed information consisting of general task and hardware descriptions and/or they want a less diagnostic evaluation, ASTAR Level 2 would be chosen. ASTAR Level 2 analyses are performed at the "task" level, and require somewhat more global judgments than do ASTAR Level 3.

If analysts have only general information about the components of a training system and are interested only in a global evaluation, they might choose the least detailed version--ASTAR Level 1. ASTAR Level 1 analyses are performed at the "training system" level, and do not involve the specification of tasks, subtasks, or controls in either the training or parent equipment.

```
        ASTAR Main Menu
(1)  Performance Deficit
(2)  Learning Difficulty
(3)  Quality of Training - Acquisition
(4)  Residual Deficit
(5)  Residual Learning Difficulty
(6)  Physical Similarity
(7)  Functional Similarity
(8)  Quality of Training - Transfer
(9)  Evaluation Summary
Enter Option Number ____
```

Figure 1. ASTAR main menu.

The ASTAR main menu is shown in Figure 1; it is identical for all three levels of evaluation. Once a level of evaluation is selected, it is carried out for all eight ASTAR analyses. The first two analyses, Performance Deficit and Learning Difficulty, define the Training Problem. The third analysis, Quality of Training-Acquisition, is the Acquisition Efficiency portion of ASTAR. The fourth through seventh analyses shown in Figure 1 make up the Transfer Problem component. These analyses include Residual Deficit, Residual Learning Difficulty, Physical Similarity, and Functional Similarity. The eighth and final analysis, Quality of Training-Transfer, is the Transfer Efficiency portion of ASTAR.

Within each major analysis, the number and kind of ratings that are required vary as a function of the ASTAR level. ASTAR Level 1 analysis requires eight ratings based on general information about the device and parent equipment. To conduct an ASTAR Level 2 or 3 analysis, the evaluator builds a database that includes lists of tasks, subtasks, controls, and displays in the training device and the operational equipment. ASTAR Level 2 analysis entails 13 ratings for each training or operational task under consideration. ASTAR Level 3 analysis requires 35 different ratings, most of which are keyed to each training or operational subtask. Ratings are entered on a computer keyboard. The sequence in which ratings are completed is determined by the ASTAR menus.

At the end of the evaluation exercise, an analyst receives numerical estimates of device effectiveness and diagnostic information on potential strengths and weaknesses of the device in an Evaluation Summary. The program calculates values for the four major analyses previously described--Training Problem, Acquisition Efficiency, Transfer Problem, and Transfer Efficiency. These values are further combined into Acquisition and Transfer scores; these two scores are added to produce a Summary Score of effectiveness. These latter three scores are intended to be correlates of the time necessary to reach criterion performance on the training and operational systems. In situations where training time is constrained (e.g., first-trial transfer), these scores are intended to reflect estimated proficiency levels of trainees.

APPLICATIONS FOR ASTAR

In this section, we describe several applications for ASTAR in training system design and acquisition. We present future uses that we envision in all stages of the device acquisition process and then present three recent applications of the technique to serve as examples.

Potential Applications for ASTAR

ASTAR is intended for use throughout the acquisition process. The successively detailed levels of analysis naturally lend themselves to applications at key points in the training device design and acquisition process. ASTAR Level 1 analyses typically would be conducted very early in the acquisition process to compare broad training concept alternatives. ASTAR Level 2 would be used early in the acquisition process to compare alternate device designs or to ask "what if" questions about a design. ASTAR Level 3 would be used when a training system is in the final design phases of the acquisition process or to evaluate existing, off-the-shelf devices, especially if empirical (e.g., transfer of training) evaluations are infeasible.

There are other applications for the ASTAR technique that are less obvious. Although ASTAR was not designed to be used in a prescriptive manner, it could be used during the formulation of design concepts. For example, a designer might need to choose between two or more training methods for a set of tasks; an ASTAR comparison of concepts employing each method could aid in making this decision. Similarly, ASTAR could be used to rapidly evaluate the potential impact of proposed design changes.

Recent Applications of ASTAR

We recently completed a project for the Naval Training Systems Center; the purpose of this project was to assess the feasibility of implementing ASTAR in the training system acquisition processes of each branch of the U.S. Armed Forces. The main goal of this effort was to apply ASTAR to a variety of training systems. We present a summary of the methods and results of these evaluations in this section to serve as examples of successful applications of ASTAR.

We evaluated three training devices during this effort: the Portable Aircrew Trainer (Martin & Rose, 1988), the Combat Talon II Avionics Subsystems Maintenance Trainer (Rose & Martin, 1988), and the Precision Gunnery Training System (Rose, Martin, & Wheaton, 1988). As illustrated in Table 1, these systems differed on a number of key dimensions.

The Portable Aircrew Trainer (PAT) was developed for the U.S. Navy to train a P-3 aircraft crewmember, the Tactical Coordinator (TACCO), in airborne antisubmarine warfare tactics. The mission of a P-3 crew is to track submarines from the aircraft using sonar buoys and other detection devices. TACCOs have the complex job of deciding how to deploy the buoys, estimating the course of the submarine, determining the course of the aircraft, deciding when to deploy torpedoes and other weapons, and coordinating the activities of other crewmembers. They perform these tasks in the aircraft with the aid of a

Table 1. Characteristics of Three Trainers Evaluated using ASTAR

System Dimension	Portable Aircrew Trainer	Combat Talon Maintenance Trainer	Precision Gunnery Training System
Task Type	System Operation	System Maintenance	System Operation
Task Complexity	Very Complex	Fairly Simple	Moderately Complex
Amount of Task Trained	Part	All	Part
Stage of Acquisition	Fielded	Initial Design	Final Design
Typical Trainees	Experienced	Experienced	Inexperienced
Device Size	Small	Medium	Medium

sophisticated computer system; they interact with the system through a workstation containing numerous displays, switches, buttons, and a keyboard.

The PAT is a desk-top, microcomputer-based system; it replicates most of the control and display functions available to the TACCO in the actual aircraft. Two versions of the system are available; one is equipped with a low-fidelity keyboard and the other is equipped with a higher-fidelity version of the TACCO controls. The two versions use the same CRT display and provide the same set of control functions.

TACCO training, at the time of our evaluation, consisted of practice and qualification exercises in a full-crew Weapons System Trainer. One PAT system with the low-fidelity keyboard was being used informally to supplement TACCO training. The Navy wanted to determine whether the PAT should be formally included in the TACCO training curriculum. Specifically, they were concerned with whether an instructor was necessary during PAT training and whether the high-fidelity keyboard provided better training than the low-fidelity keyboard.

In order to conduct the ASTAR evaluation of the PAT, we made an extensive study of the TACCO's task, the operational equipment, and both versions of the PAT. We developed a detailed list of tasks and subtasks and four PAT "utilization scenarios." The four scenarios were the low-fidelity PAT used with an instructor, the low-fidelity PAT used without an instructor, the high-fidelity PAT used with an instructor, and the high-fidelity PAT used without an instructor. In all scenarios, the TACCO trainee completed a series of PAT training exercises and then transferred to the Weapons System Trainer.

Three AIR researchers evaluated these utilization scenarios using all three levels of ASTAR.

The conclusions and recommendations that emerged from these analyses were that: 1) training TACCOs under any of the four PAT utilization scenarios before they trained on the full-crew Weapons System Trainer would be more effective than training them only on the Weapons System Trainer; 2) the higher fidelity PAT with an instructor would be the most effective utilization of the trainer; and 3) training with the high-fidelity PAT without an instructor and training with the low-fidelity PAT with an instructor would be approximately equal in effectiveness.

<u>The Combat Talon II Avionics Subsystems Maintenance Trainer (CT-II)</u> is being developed at Wright-Patterson Air Force Base to train technicians to perform system-level maintenance on the avionics subsystems of the Combat Talon II aircraft. The trainer and the aircraft were in the early stages of the acquisition process at the time of our evaluation; a preliminary specification for the trainer had been published. The Combat Talon II avionics subsystems design includes an integrated diagnostics capability that will achieve 100% fault detection at the Line-Replaceable Unit level. This will be accomplished through built-in test equipment and a computerized maintenance program.

The maintenance technician's job will be to perform routine maintenance and to detect and repair reported faults. The CT-II trainer will teach students to use the computerized maintenance program and Technical Orders to perform fault detection and isolation. The preliminary design for the trainer was a medium-fidelity concept. Some pieces of equipment were to be exact replications of their counterparts in the aircraft while others were designed to be much lower in physical fidelity. The trainer functionally replicated the aircraft computer maintenance program and all the possible faults in the avionics systems.

The developers of the CT-II wanted to investigate the relative effectiveness of higher and lower fidelity versions of the trainer. In order to accomplish this, we worked with the CT-II developers to formulate two alternative design concepts. The high-fidelity concept was essentially a full-scale representation of the operational aircraft system. The low-fidelity concept was a microcomputer-based, interactive video approach. It included representations of all components on the video display and the trainee would interact with the system through a standard keyboard. Both of these alternate concepts replicated the aircraft computer maintenance program and faults in the avionics systems to the same extent as the original design concept. The training and operational tasks, performance objectives, utilization of the trainer, and characteristics of the trainees were the same for the three design concepts.

Three AIR researchers and one of the CT-II developers compared the three CT-II design concepts using ASTAR Levels 1 and 2. (Because of the early stage of design, sufficient information was not available for an ASTAR Level 3 analysis.) The main finding was that the predicted effectiveness of the three CT-II design concepts did not differ significantly; all of the concepts were judged to adequately address the training problem.

The Precision Gunnery Training System (PGTS) is being developed by the U.S. Marine Corps to improve the performance of TOW and Dragon gunners. The TOW and Dragon are optically-tracked, wire-guided antitank missiles. These missiles are fired by individual gunners; the TOW is normally mounted on the back of a jeep and the Dragon is shoulder-mounted. The two weapons operate on the same principle--the gunner tracks a selected target through an optical sight, places the sight crosshairs on the desired point of missile impact, fires the missile, and continues to track the target until missile impact. As the missile travels downrange, it sends information about its position to the optical sight via an infrared signal and receives position correction commands along a wire-link that guide it to the point at which the gunner is aiming.

The trainers currently used to train TOW and Dragon gunners are the M70 and the LET. These trainers are actual expended missile launchers specially equipped for training. They utilize an infrared target source and a monitoring set; the target is mounted on a vehicle which moves along a path in front of the trainee's firing position. As the trainee tracks the target, the monitoring set records the deviation of the gunner's line-of-sight from the ideal line-of-sight for a hit. The M70 and LET are also equipped with simulations of the noise and weight shift of the missile firing. The PGTS TOW and Dragon trainers also physically replicate the operational systems but training is based on a very sophisticated visual simulation. These trainers will be able to present numerous, very realistic targets under a variety of engagement conditions in the trainee's sight. They will realistically represent the weight shift, noise, and blast effects of the operational systems.

When we conducted our evaluation, the PGTS trainers were nearing the final stages of the design process. A very detailed specification had been published and the design was nearing completion. The Marine Corps developers of the PGTS were interested in comparing the new trainers with the existing TOW and Dragon trainers. To conduct this evaluation, we developed a detailed task list and made an extensive study of the new trainer designs and the existing trainers. Three AIR researchers evaluated the PGTS TOW trainer vs. the M70 and the PGTS Dragon Trainer vs. the LET using all three levels of ASTAR.

On the basis of our analyses, we concluded that the PGTS trainers would address substantially more of the operational performance requirements than the existing trainers. Also, the PGTS trainers included instructional features that would lead to more efficient training and transfer.

Summary of ASTAR Applications

The PAT, CT-II, and PGTS evaluations illustrate three very different applications of ASTAR:

1. Investigating the most effective utilization of an existing device,

2. Evaluating the effectiveness of alternative design concepts in the early stage of system acquisition, and

3. Comparing the potential effectiveness of new training system designs with the effectiveness of existing systems.

For all three applications, findings were consistent across all levels of ASTAR analysis. Interrater reliability improved as we gained experience with actual devices: Although high to begin with (e.g., interrater correlations for the PAT evaluation ranged from r=0.64 to r=0.88, depending upon the particular analysis and ASTAR level), by the third application (the PGTS) we were in almost complete agreement on all ratings. For example, interrater correlations for all ASTAR Level 1 ratings were above r=0.97. We believe that this reliability is generalizable to other ASTAR users: We have formalized rating procedures and clarified definitions so that ASTAR would be more useable by non-psychologists.

These applications demonstrate that ASTAR is a flexible and potentially valuable tool that can be employed in a wide variety of situations. It allows for the systematic consideration of important issues in the selection, design, or purchase of training devices. And despite obvious limitations--lack of sufficient validity data, lack of a cost component, and the fact that it has not been applied by non-AIR staff--we believe that ASTAR would fill a vital need for effectiveness forecasting in the training device acquisition process.

References

Goldberg, I. & Khatri, N. (1985). A review of models of cost and training effectiveness analysis (CTEA), Volume I: Training effectiveness analysis (Contract No. MDA903-82-C-083, Task 8). Washington, DC: Consortium of Washington, D.C. Universities.

Kane, J. J., & Holman, G. L. (1982). Training device development: Training effectiveness in the Army system acquisition process (SAI Report No. 82-02-178). Minneapolis, MN: Honeywell, Incorporated, Systems and Research Center.

Martin, M. F. & Rose, A. M. (1988). Implementation of ASTAR: Evaluation of the Portable Aircrew Trainer (Purchase Order No. C-44045). Washington, DC: American Institutes for Research.

Nutter, R. V. & Terrell, W. R. (1982). A management system for RDT&E funded training device acquisitions in the Naval Education and Training Command (Technical Report No. 118). Orlando, FL: Department of the Navy, Training Analysis and Evaluation Group.

Rose, A. M., Evans, R., & Wheaton, G. R. (1987). Methodological approaches for simulator evaluations. In S.M. Cormier & J.D. Hagman (Eds.) Transfer of learning: Contemporary research and applications. San Diego, CA: Academic Press, Inc.

Rose, A. M. & Martin, A. W. (1984). Forecasting device effectiveness: III. Analytic assessment of DEFT (Contract No. MDA 903-82-0414). Washington, DC: American Institutes for Research.

Rose, A. M. & Martin, M. F. (1988). Implementation of ASTAR: Evaluation of the Combat Talon II Maintenance Trainer (Purchase Order No. C-44045). Washington, DC: American Institutes for Research.

Rose, A. M., Martin, M. F., & Wheaton, G. R. (1988). Implementation of ASTAR: Evaluation of the Precision Gunnery Training System (Purchase Order No. C-44045). Washington, DC: American Institutes for Research.

Rose, A. M. & Wheaton, G. R. (1984a). Forecasting device effectiveness: I. Issues (Contract No. MDA 903-82-0414). Washington, DC: American Institutes for Research.

Rose, A. M. & Wheaton, G. R. (1984b). Forecasting device effectiveness: II. Procedures (Contract No. MDA 903-82-0414). Washington, DC: American Institutes for Research.

NORMATIVE AND DESCRIPTIVE MODELS FOR TRAINING-SYSTEM DESIGN

Paul J. Sticha[1]

Human Resources Research Organization
1100 S. Washington St.
Alexandria, VA 22314

INTRODUCTION

The increasing cost of training and limitations in the military training budget have led to increased emphasis on training cost-effectiveness. In addition, advances in instructional technology have greatly increased the options that are available to the training-system designer. Current training system design processes do not address the cost-effectiveness of the wide range of training-device and simulator options available to the training designer. This paper describes a system of models for the optimization of simulation-based training systems (OSBATS). The OSBATS system contains both normative and descriptive modeling components. The normative modeling components provide a structure for the decision-making processes involved in training-system design. The descriptive modeling components support the decision process, and characterize the effectiveness, efficiency, and costs involved in training-device development and use. The OSBATS system provides a coherent set of procedures for decision making and a set of tools to aid the designer in following these procedures.

The Cost of Military Training

The U.S. military invests a considerable amount of resources for training, both by training institutions and in operational units. This training provides soldiers the skills required to operate and maintain complex modern weapon systems. According to the Military Manpower Training Report for Fiscal Year 1988 (Office of the Assistant Secretary of Defense, Force Management and Personnel, 1987), the cost of military

---

[1]The author would like to acknowledge the individuals who developed the OSBATS model, particularly H. Ric Blacksten and Dennis M. Buede, who had primary responsibility for the definition of three of the OSBATS modules. The author would also like to thank Michael J. Singer for his comments on an earlier draft of this paper. This research was supported by contract MDA903-85-C-0169 from the U.S. Army Research Institute for the Behavioral and Social Sciences. The opinions expressed are those of the author, and should not be interpreted as representing official policies, expressed or implied, of the U.S. Army Research Institute or the Department of Defense.

training conducted by training institutions for fiscal year 1988 is estimated to be more than $18 billion. This figure includes $7.1 billion for training areas related to weapon-system operation and maintenance. Analyses of the total military budget indicate that the magnitude of unit training is at least as great as that of institutional training (Training Data Analysis Center, 1985). Thus, the total annual cost for institutional and unit training probably exceeds $34 billion, with perhaps $14 billion of this training directly related to the operation and maintenance of weapon systems. Given the magnitude of military training, the importance of cost-effective training is clear. An improvement in training efficiency as small as 1% could save $340 million annually.

Many of the reasons for the high cost of military training are obvious. Weapon systems required for hands-on training are expensive to procure and operate. Other required equipment, such as ammunition, is also expensive. In addition, training of many tasks requires special conditions that replicate the battle environment, equipment malfunctions, opposing force activities, and special environmental situations that provide critical cues for weapon system operation and maintenance. Associated with the cost of producing these special training conditions are limitations on the availability of training ranges, ammunition, and so forth, as well as safety considerations.

Advances in Training Technology

Advances in instructional technology, such as computer-generated imagery, computer-assisted instruction, interactive videodisc, and simulation technology have made simulation-based training possible for a wider range of skills. The result of these advancements has increased the number of options available to the training designer. The overall effect of the increased number of options has been to make the design task more difficult. The designer must consider different training strategies (that is, a part-task training strategy, a full-mission simulator, or actual equipment training possibly enhanced with embedded training), more or less sophisticated training-device designs, and specific allocations of training times to training devices. The training-system designer needs to have a formal training-system design process and tools to aid in the performance of this process.

The OSBATS Model

We have developed a system for the optimization of simulation-based training systems (OSBATS) that incorporates several modeling techniques to aid the training-system designer. The planned user of this model is the system engineer who is responsible for the formulation of a training concept for the design and use of a training device.

The models encompassed in OSBATS consist of normative and descriptive models. The normative models provide the structure to the training-system design problem, specify a decision process, and specify the requirements for data content and format. The descriptive models predict the performance and provide the input to the normative models. They define methods for aggregating available data to obtain values for the parameters of the normative model. In doing this, they replace some of the data requirements of the normative model with their own data requirements. Ultimately, the descriptive models provide a simple description of the complex processes that occur as a result of skill training.

The following two sections of this paper describe the normative and descriptive modeling components, respectively. The section following those describes the tools that have been developed to guide the decision process. The final section discusses our results, and comments on what has been accomplished and what remains to be done.

NORMATIVE MODELING COMPONENT

The normative model defines the training-design problem, provides a structure for decomposing the overall problem into manageable units and specifies a procedure for aggregating the solutions of the individual units to obtain an overall solution. The components of the training-design problem determine an overall strategy for the employment of classes of training devices, develop candidate designs for individual training devices, and evaluate combinations of training device designs to determine which combination can be employed to meet training objectives at the minimum cost. The resulting decision process is comprehensive, logical, and defensible.

A second product of the normative models is a statement of requirements for data content and format for effective decision making. Thus, the normative models specify what the training-system designer must know to produce optimal designs. Sometimes the system designer will already have this information; sometimes the system designer will be able to obtain the information from subject-matter experts. Still other times the system designer will have to rely on other, descriptive models to provide the desired information from more basic subjective or empirical data.

The Overall Modeling Framework

The overall modeling framework is based on methods that attempt to define the training strategy that meets the training requirements at the minimum cost. This framework was originally described by Roscoe (1971) and has been extended by Povenmire and Roscoe (1973), Carter and Trollip (1980), Bickley (1980), Cronholm (1985), and our own work (Sticha, Blacksten, Buede, & Cross, 1986; Sticha, Singer, Blacksten, Mumaw, & Buede, 1987). In its simplest form, the method compares the ratio of effectiveness of two training alternatives to the ratio of cost of the options. For example, if a training program that employs one hour of training on a simulator saves 30 minutes of training on actual equipment, and the hour of simulator training costs as much as 20 minutes of training on actual equipment, then the simulator will meet the training requirement at a lower cost than actual equipment. Thus the approach addresses the tradeoff between the increased training time that is usually required by the use of a simulator and the decreased cost of that time.

This simple formulation of the cost-effectiveness of training may be used to provide recommendations regarding the optimal mix of simulator and actual-equipment training when the effectiveness of training on the simulator is expressed as a function of the amount of simulator training. In general, the effectiveness of simulator training is a decreasing function of the amount of training on the simulator. That is, the first hours of simulator training replace more training time on actual equipment than subsequent hours do. The optimal mix of simulator and actual-equipment training involves training on the simulator until the marginal cost savings from reduced actual-equipment usage equals the marginal cost of the simulator training. Cronholm (1985) generalized this finding by splitting the transfer function into three components representing learning on the simulator, transfer to the actual equipment, and learning on the actual equipment, respectively.

We have extended the basic modeling framework in two ways in the OSBATS model. The two extensions are implemented in two modules of the OSBATS model that address the selection of training devices from multiple candidates for multiple tasks, and the allocation of training time to the selected devices. Both extensions make the same assumptions about learning and transfer processes. The first extension makes simplifying assumptions about training cost so that it can provide an interactive environment for addressing training-device selection alternatives. The second extension relaxes some of the assumptions to allocate training resources to training devices considering both discrete purchase costs and device use constraints.

This basic framework, as we have extended it, provides the overview to the decision process. However, it is limited in that it does not aid either the determination of an initial training strategy or the design of training devices. Straightforward incorporation of these processes with the basic framework opens the process up to the potential for combinatorial explosion of options, which would almost certainly preclude the success of the modeling process. Our response to this concern has been to develop submodels that reduce the number of options that need to be addressed by the overall framework. An iterative procedure allows us to revise the submodel solutions in light of the results of other submodel solutions. Although this approach does not guarantee optimality of the overall solution, it provides a method to obtain a good solution with reasonable effort. Since we expect that the objective function will have a flat maximum, we should be able to obtain a good result even if it is not optimal. The two other methods that are incorporated into the OSBATS model include a preliminary clustering method and a training-device design method.

Preliminary Clustering Method

The goal of the preliminary clustering method is to review task requirements, simulation needs, and cost of simulation capability in order to define clusters of tasks that have similar simulation requirements. The method currently defines the following three classes of training devices: (a) a full-mission simulator (FMS) that simulates many or all of the subsystems of the actual equipment, (b) one or more part-mission simulators (PMSs) that simulate selected equipment subsystems, or (c) actual equipment.

This evaluation examines device-unique capabilities, such as training in unsafe situations, and cost savings to establish the value of training with some sort of training device. In addition, the task requirements for fidelity are used to estimate the development cost that would be required to achieve the required fidelity for each task. Using the assessed costs and benefits, the model sorts the tasks into three clusters: (a) those tasks that should be trained on actual equipment because the benefits of simulation do not justify the expense required to develop an effective training device, (b) those tasks for which training in a simulated environment is cost-effective and which have limited cue and response requirements so that they require only a PMS, and (c) those tasks for which training in a simulated environment is cost-effective, and which require an FMS because they require a high-fidelity representation of the environment on several dimensions.

The benefits of simulation are assessed directly from ratings of safety concerns, special performance conditions, and training effectiveness factors. Operating-cost savings is determined by comparing the cost of training on actual equipment to the projected cost for an optimal mix of training on both actual equipment and a representative training device. The task fidelity requirements along several cue and response dimensions are used to estimate the cost to develop a training device with the capability required by the task.

The method evaluates simulation options for each task by comparing the benefits of simulation due to improved safety or operating-cost savings to the cost required to develop task-specific effective simulation capability. For tasks in which the benefit of simulation is high, or for which simulation is required for safety concerns, a high-cost simulator may be justified if it is required to train the task. For tasks in which the benefit from simulation is moderate, a simulator would be justified only if it is less expensive to develop.

The model makes its major recommendation regarding whether a task should be trained on actual equipment or some kind of training device by comparing the required development cost of the training device to the potential operating-cost savings brought about by its use. If the operating-cost savings is sufficient to recover the development cost over the life cycle of the weapon system, a training device will be recommended. Otherwise, actual equipment will be recommended. The recommendations of the economic analysis are overridden, however, if a training device is required for safety considerations.

Training-device Design Methodology

The task clusters defined by the above procedure provide the requirements for individual training devices. The task at this point is to develop training device designs that have the fidelity and instructional features required to meet the training requirements for the tasks in a single cluster while avoiding extraneous or inefficient features. We have applied a general design methodology to the analysis for training-device design. This methodology addresses problems in which there are a large number of alternatives formed by the factorial combination of several dimensions. We have developed two applications of this methodology. The first application addresses the instructional features that should be included in the training device; the second application addresses the fidelity features that should be included.

Instructional feature selection model. Instructional features are viewed as elements of training devices that can improve training efficiency on individual tasks. That is, instructional features reduce the time or cost required to achieve a given performance level on a training device. They do not affect the ultimate level of actual-equipment performance that can be reached by using a training device. The number of tasks aided by each instructional feature forms the basis of an index of benefit for the feature. The analysis proceeds by comparing the benefit to the cost of incorporating each instructional feature into the training device. The analysis then orders the features according to the ratio of benefit to cost. This order specifies a collection of optimal features as a function of the total budget for instructional features. The appropriate budget for instructional features, given a total training-device budget, is determined in the Fidelity Optimization Model.

Fidelity optimization model. The same modeling framework is then used to address how much should be invested in the fidelity of the training device being designed. The model considers several dimensions of fidelity that describe task cue and response requirements. The task requirements on the fidelity dimensions are compared to the cost of meeting these requirements to determine the dimensions for which increased fidelity is justified by increased training effectiveness. The output of this model is a set of possible training-device configurations applicable to the task set, each of which offers the greatest effectiveness for its cost.

The model makes its selection based on the incremental benefit/cost ratio of the fidelity dimension levels. The costs are calculated from the fidelity levels, and represent development costs. The benefits are calculated from the number of tasks for which each level of the fidelity dimensions would be adequate, based on the technical performance associated with each option and the cue and response requirements of the tasks from the fidelity dimensions.

DESCRIPTIVE MODELING COMPONENT

The descriptive models in the OSBATS model provide a simple description of complex processes involved in skill acquisition and transfer. The output of these models provides the critical information that is used by the normative models. These models, in turn, provide logical methods for aggregating more basic task-analytic and empirical data, and thus affect the data requirements.

In developing the descriptive modeling component, we must be concerned with both validity and parsimony. Validity is a concern because the models are predicting the critical values upon which the optimization is based. Inaccuracy of the descriptive models would produce a concomitant inaccuracy in their recommendations of the normative models. Parsimony is also a critical concern, because of the scope of the problem being addressed by the OSBATS model. For example, addition of a single task parameter would increase the total number of parameters in the model by the total number of tasks being addressed, which may well number in the hundreds.

The OSBATS system contains models that describe human performance variables and provide training cost estimates. This paper concentrates on human performance description, which has the following three components: (a) overall characterization of acquisition-transfer processes, (b) prediction of transfer of training, and (c) predictions of training-efficiency.

The Acquisition-Transfer Function

The acquisition-transfer function predicts performance on actual equipment as a function of training time on a training device (which may be actual equipment). We describe this process using the following function:

$$P_{Ti}(t) = a_{Ti}\{1 - [1 + m_{Ti}c_T(s_{Ti} + t)]^{-k}\}$$

where

$P_{Ti}(t)$ = the predicted performance on task T after training on device i for time t,

a_{Ti} = the asymptote of function; that is, the maximum performance level on task T that can be obtained by training on device i,

m_{Ti} = the time multiplier representing the efficiency with which device i can train task T, relative to actual equipment,

c_T = a time scaling constant for task T,

s_{Ti} = the head start that sets performance at the entry level for $t = 0$, and

k = the exponent of the function.

The function represents acquisition and transfer processes by a power function. The power function is characterized by an initially high learning rate that decreases with increasing training (see Figure 1). This representation is consistent with a long line of research that has been recently summarized by Newell and Rosenbloom (1981).

The acquisition-transfer function has five parameters. Two of these parameters, the asymptote and the time multiplier, are critical to the analysis. The asymptote (a_{Ti}) provides a measure of transfer of training from device i to the actual equipment. It represents the maximum level of actual-equipment performance that can be attained with unlimited training on device i. This critical parameter is estimated from the cue and response requirements of the task and the fidelity of the training device using a function described later. The time multiplier (m_{Ti}) is a measure of training efficiency that takes into account reduction in training time due to reduced setup time as well as improvements in training efficiency due to instructional support features.

The other parameters are not critical in distinguishing among training-device options, although they are important in defining the shape of the function. The time scaling factor can be viewed either as a measure of task difficulty (along with the exponent) or as a factor that expresses time on a constant scale. This parameter is constant across training device options. The head start describes the entry performance level in terms of equivalent training time. This parameter is determined from the entry performance level and the values of the other parameters. Finally, the exponent gives another measure of task difficulty. The

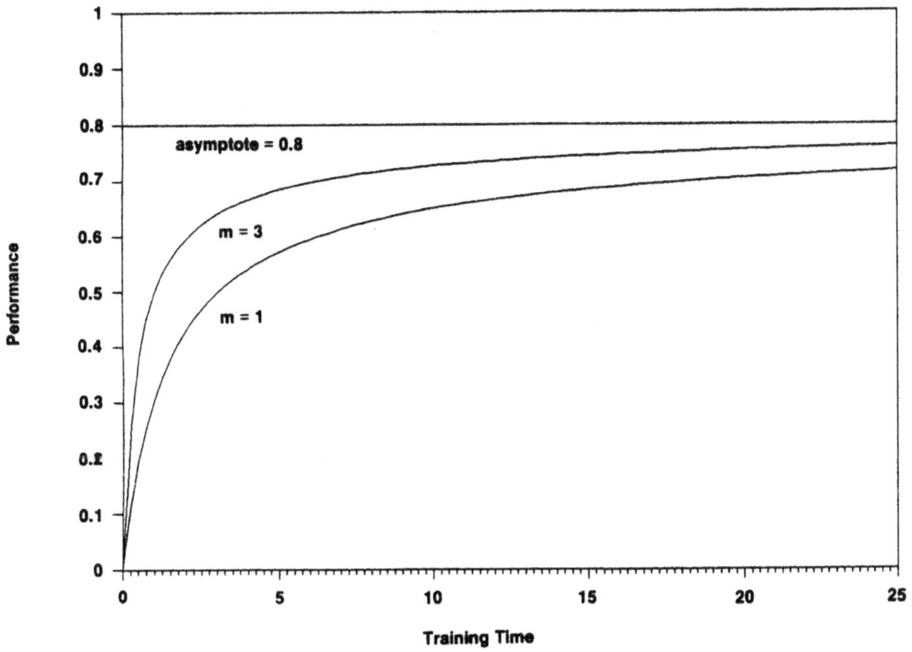

Fig. 1. An illustration of the acquisition-transfer function with $c_T = 1$, $s_{Ti} = 0$, $k = 0.7$.

OSBATS model assumes a constant exponent ($k = 0.7$) so that we may reduce data requirements, and because there is literature suggesting relatively little variation in this parameter (Card, Moran, & Newell, 1983).

The OSBATS model estimates the values of the parameters directly from expert judgment or empirical data for one baseline situation--a situation in which all training occurs on actual equipment. The model then adjusts the critical parameter values depending on the characteristics of the training device using the functions described in the following sections.

Estimation of the Asymptote

The asymptote of the acquisition-transfer function is estimated by comparing the fidelity of the training device to the requirements of the task along several fidelity dimensions. The estimation is made according to the following equation.

$$a_{Ti} = q_T[\prod_j (C_{ij} + b_{Tj})^{R_{Tj}}]^{Q_T}$$

where

a_{Ti} = the asymptote of the acquisition-transfer function,
q_T = the multiplier of the asymptote equation for task T,
C_{ij} = the capability of training device i measured on fidelity dimension j,
b_{Tj} = the cue and response reference point for task T and fidelity dimension j,
R_{Tj} = the cue and response requirement for task T and fidelity dimension j, and
Q_T = the exponent of the asymptote equation for task T.

The function is a product of the capabilities of the training device on the fidelity dimensions. The task cue and response requirements are used as an exponent so that low requirements are not as important as high requirements. The other three parameters adjust the function so that the following conditions are met: (a) perfect fidelity leads to perfect transfer, (b) meeting all requirements exactly allows the training device to be used to train to the performance standard, and (c) reducing the fidelity on one dimension to 0.0 while keeping others at perfection reduces transfer to a dimension-specific value that expresses the criticality of that fidelity dimension.

Estimation of the Time Multiplier

Two factors are considered in estimating the time multiplier of the acquisition-transfer function. The first is the overall capability of a training device to reduce training time that would be otherwise spent performing tasks that are already known to criterion, such as taking off, flying to the exercise area, and so forth. The second is the capability of the instructor support functions to make training more efficient. The time multiplier is the product of these two efficiency factors.

We assume a fixed percentage of setup time will be saved by a training device, independent of its characteristics. Analysis of instructional features is more complicated, and considers task characteristics and the number of appropriate instructional features a device possesses for each task. The method for determining the adjustment for instructional features is briefly described below.

We assume that each instructional feature is appropriate for the subset of the tasks that have characteristics matching the capabilities of the instructional feature. Consequently the process for evaluating the effectiveness of instructional features begins by examining the types of skills involved, trainee abilities, and other factors for each task. This information is used to specify the subset of tasks for which each instructional feature is appropriate. By comparing the instructional features possessed by a training device with the list of appropriate features for a task, the model determines how many appropriate instructional features the device possesses. The estimated improvement in training efficiency from the use of the device is calculated from the number of appropriate features. The calculation assumes that (a) there is a maximum improvement in training efficiency that can be obtained using instructional features, (b) the first feature has the greatest benefit on training efficiency for a particular task, while additional features have smaller effects, and (c) after some number of instructional features have been added, further addition of features will not lead to further improvement in training efficiency.

THE DECISION PROCESS AND DECISION SUPPORT TOOLS

The OSBATS model currently consists of the following five modeling components:

1. Simulation Configuration Module. A tool that defines clusters of tasks according to the categories of part-mission training devices, simulators, and actual equipment.

2. Instructional Feature Selection Module. A tool that analyzes the instructional features needed for a task cluster and specifies the optimal order for selection of instructional features.

3. Fidelity Optimization Module. A tool that analyzes the set of fidelity dimensions and levels for a task cluster and specifies the optimal order for incorporation of advanced levels of these dimensions.

4. Training Device Selection Module. A tool that aids in determining the most efficient family of training devices for the entire task group, given the training device fidelity and instructional feature specifications developed in the previous modules.

5. Resource Allocation Module. A tool that aids in determining the optimal allocation of training time and number of training devices needed in the recommended family of training devices.

The concept of operation for the OSBATS model is based on the iterative use of the five model tools to make recommendations regarding the definition of task clusters, the design of training devices, and the allocation of training resources among selected training devices. Both the subset of tools that are used and the order in which they are used may vary depending on the requirements of the problem and the preferences of the user. Although the tools may be used in a variety of orders, the most natural order is the order in which the tools were listed above. An application of the tools in that order is described in the following text.

The Simulation Configuration Module would be used first to examine the tasks to be trained and to provide a preliminary recommendation for the use of either actual equipment or one or more training devices. The result of this analysis would be three clusters of tasks. Two of these clusters define tasks for which a full-mission simulator or part-mission training device should be designed.

The analyst would then use the task clusters defined by the Simulation Configuration Module as the basis for the application of the Instructional Feature Selection and Fidelity Optimization Modules. These two modules would be used to define candidate training system designs for each task cluster. The output of the two modules is a range of options that vary in cost. Thus, the overall results of the application of these modules would be a collection of training device designs specifying for each design the level of fidelity on each fidelity dimension and the collection of instructional features included in the design. The analyst would select several of these designs for further examination.

The Training Device Selection Module evaluates the training device designs produced in the previous process. The analyst would exercise this module several times using different combinations of training devices. For each combination, the module would determine the number of tasks assigned to each training device, the number of hours each task would be assigned to each device to meet the training requirements at the lowest cost, and the optimal training cost given the selected combination of training devices. This model makes the simplifying assumptions that the hourly cost of a training device is fixed and that all devices are fully utilized. These assumptions allow the Training Device Selection Module to determine a solution in less than one minute.

When the analyst was relatively confident of the solution of the Training Device Selection Module, he or she would then investigate the solution using the Resource Allocation Module. It could be that the recommendations of the Training Device Selection Module would require the procurement of more training devices than would be feasible, or would provide some training on actual equipment for tasks in which such training violated safety regulations. The Resource Allocation Module allows the analyst to impose constraints such as these on the training system and examine the resulting optimal solution. The Resource Allocation Module also relaxes the simplifying assumptions that were used by the Training Device Selection Module to estimate training device cost, leading to a more accurate cost function. As a result of its increased generality, the Resource Allocation Module takes several minutes to reach a solution, several times longer than the Training Device Selection Module.

At many points in the analysis process the analyst has the option of returning to modules that were used previously to refine the analysis, change assumptions, or choose different solutions. For example, the analyst might change the definition of the task clusters based on the results of Training Device Selection Module, or may use those results to select different candidate device designs for evaluation.

DISCUSSION

The problem of optimizing the design of training systems is filled with complexity. The work described in this paper defines a decision process for training-system design that decomposes the overall problem into more manageable subproblems, and then solves the subproblems. Using this top-down system approach, we have defined five procedures that help a training device designer cluster the tasks to be trained, define training device designs, and allocate training to training devices. The procedures that were developed involve both normative models that provide the problem structure, and descriptive models that estimate the critical values used by the normative models.

We have implemented the decision process in software that provides an interactive environment to guide the engineer or education specialist who must formulate a training concept incorporating training devices and operational equipment. The software allows the training system designer to conduct the analyses included in the OSBATS model. Our concept of operations for the model supports the iterative nature of the concept-formulation process by allowing the user to pass the result of one module to be used as input by other modules.

Data Requirements

All methods of training system design require a good front-end analysis. The OSBATS model is no exception to this rule, and requires information about training requirements, task characteristics, trainee population skills, candidate training-device instructional features, and fidelity dimensions. In addition, because the model is quantitative rather than qualitative, it requires numerical estimates for many of its parameters. Ultimately, the source of most of the data will be the judgments of subject-matter experts (SMEs), that is, instructors, training-device designers, and training developers. Consequently, the design of the OSBATS model has been made to obtain the required data as easily as possible.

A major factor affecting the quantity of judgments required by the model is the level of aggregation of the basic data. In the prototype version of OSBATS, we obtained SME judgments at intermediate level of aggregation. For example, a team consisting of a psychologist and two instructor pilots made direct assessments of the cue and response requirements for each task along eleven fidelity dimensions (Sticha et al., 1986). While this procedure minimized the number of assessments required, it forced the SMEs to make judgments that were outside of their area of expertise. We are currently investigating procedures to infer the cue and response requirements from a limited set of more basic questions. We are implementing the assessment procedures as an expert system to minimize the number of assessments that would be required.

The use of numerical models as the basis of OSBATS adds to the robustness of the model in light of the likelihood of errors in the parameter assessments. That is, small errors in the inputs to the OSBATS model will likely have small effects on the recommendations of the model. However, use of quantitative models requires numerical assessments of parameters by various SMEs. Scaling procedures will need to be developed so that numerical assessments may be made as easily as possible. When these procedures have been developed, the requirement for numerical assessments should not increase the load on the SME.

Future Needs

The OSBATS model has been developed and represented by prototype software. There are many future needs for model expansion, data base development, development of scaling procedures, development of production software, validation, and technology transfer. This section concentrates on the last two of these issues.

Validation. Because of the complexity of the OSBATS model, validation of the model as a whole is probably impossible. Other aspects of the model preclude validation of major sections of the model with empirical data, such as from a transfer-of-training study. The OSBATS model is a model of what is possible; it addresses ideal training systems. When we evaluate real training systems, the result we obtain depends on the effectiveness of the design and on how well the design was implemented. If negative results are obtained, we are open to the criticism that the features we were evaluating are useful but that they were implemented poorly.

Probably a better strategy is validation of submodels and attempts to determine key model parameters. Such a validation should concentrate on the key assumptions of the model, such as the extent to which instructional features can improve training efficiency or the asymptote calculation function. We are currently performing analyses on the model to determine the model parameters that are the major determiners of the overall results, and thus would deserve the bulk of the validation effort.

Technology transfer. The OSBATS model represents a new process for making decisions in the design of simulation-based training systems. Consequently, transfer of this technology to the user community depends on the integration of the new methods with existing requirements. We must ensure that the OSBATS model captures the richness of the decisions and the variety of possible interactions between individual decisions. To maximize the likelihood of successful technology transfer, we must work with decision makers. If there is one thing that we know for sure, it is that successful development of a decision support system depends on the involvement by the users in its development (Adelman, 1982; Sprague & Carlson, 1982).

REFERENCES

Adelman, L. (1982). Involving users in the design of decision-analytic aids: The principal factor in successful implementation. Journal of the Operational Research Society, 33, 333-342.

Bickley, W. R. (1980). Training device effectiveness: Formulation and evaluation of a methodology (Research Rep. 1291). Alexandria, VA: U.S. Army Research Institute for the Behavioral and Social Sciences.

Card, S. K., Moran, T. P., & Newell, A. (1983). The psychology of human-computer interaction. Hillsdale, NJ: Erlbaum.

Carter, G., & Trollip, S. (1980). A constrained maximization extension to incremental transfer effectiveness, or, how to mix your training technologies. Human Factors, 22, 141-152.

Cronholm, J. N. (1985). The optimization of training systems. Proceedings of the Symposium on the Transfer of Training to Military Operational Systems (pp. 237-259). Brussels, Belgium: NATO.

Department of Defense Training Data and Analysis Center (1985). Training resources: Macro level training funds prototype data base preliminary results. Orlando, FL: Author.

Newell, A., & Rosenbloom, P. (1981). Mechanisms of skill acquisition and the law of practice. In J. R. Anderson (Ed.), Cognitive skills and their acquisition. Hillsdale, NJ: Erlbaum.

Office of the Assistant Secretary of Defense, Force Management and Personnel (1987). Department of Defense military manpower training report for FY 1988. Washington, DC: Author.

Povenmire, H. K., & Roscoe, S. N. (1973). Incremental transfer effectiveness of a ground-based general aviation trainer. Human Factors, 15, 534-542.

Roscoe, S. N. (1971). Incremental Transfer effectiveness. Human Factors, 13, 561-567.

Sprague, R. H., Jr., & Carlson, E. D. (1982). Building effective decision support systems. Englewood Cliffs, NJ: Prentice-Hall.

Sticha, P. J., Blacksten, H. R., Buede, D. M., & Cross, K. D. (1986). Optimization of simulation-based training systems. Volume III: Model description (Final Report 86-13). Alexandria, VA: Human Resources Research Organization.

Sticha, P. J., Singer, M. J., Blacksten, H. R., Mumaw, R. J., & Buede, D. M. (1987). Optimization of simulation-based training systems: Year II report (Interim Report 87-30). Alexandria, VA: Human Resources Research Organization.

REVIEW AND CRITIQUE

INTRODUCTION: REVIEW, CRITIQUE AND KEYNOTE ADDRESS

Grant R. McMillan

Armstrong Aerospace Medical Research Laboratory
Wright-Patterson Air Force Base, OH
USA

OVERVIEW OF THE PAPERS

Any review of human performance models would be incomplete without a discussion of the strengths and weaknesses of this technology. Clearly, this volume demonstrates that there have been advances in recent years. In most of the modelling areas reviewed, important enhancements have been made in model availability and usability. The development of powerful personal computers, and the rehosting of modelling software on these machines, has made many of the tools accessible to a wide range of users. There has been a parallel effort to make PC-based models easier to learn and use, with notable success in areas such as network modelling.

Those familiar with the history of human performance modelling may argue that there have been few fundamental theoretical advances in recent years. This appears to be largely correct. In the modelling areas reviewed in this Workshop, there appear to be two notable exceptions to this statement. First, the multiple-resource/attentional-demand approach to workload prediction represents a genuine advance over the serial time-line techniques. Although further development and validation are required, this is a promising approach. Second, the tools for predicting training effectiveness and skill retention are largely new developments. While they may not represent a theoretical breakthrough, they do provide users with a new class of tools.

As the papers in this session point out, human performance modelling is still plagued by most of the same weaknesses it has had for many years. The range of human behaviors which models adequately address is still very limited. Not nearly enough effort is being spent to determine the validity and generality of available models. A great deal of technical knowledge is still required to use most models. Most models are normative and do not adequately represent individual differences or the sources of error in operator performance.

The first two papers in this session discuss these and other modelling limitations in more detail. Sanders argues that the fundamental aim of modelling is to establish human performance limits for system design. He analyzes some existing models and shows how the model assumptions may lead to over- or underestimates of these limits.

Hulme and Hamilton also address model limitations, but from the perspective of model users. They discuss their experience with anthropometric, network, visual interface, and cognitive assessment techniques, and enumerate the host of theoretical and practical problems they have faced in trying to use the tools. On a positive note, they continue to use human performance models in their design work, despite these difficulties.

Card outlines the approach his firm has taken in trying to exploit human performance theory and models in the development of new systems. He shows how simple models such as Fitts' Law played a key role in the design of the computer mouse, and how more complex theories have provided leverage points when cut-and-try design got stuck. He points out that while intuitive design is an important part of interface development, it is not the only way to make progress. Theory-driven design can help the designer reconceptualize a problem, suggest leverage points, and assist the intuitive process to take higher payoff steps.

Cody and Rouse summarize the results of the model evaluation process conducted by the Workshop audience. The paper provides insights into potential user opinions about available models. It also suggests differences between the criteria users say are important and the criteria they actually use to make their judgements. These findings may be of use to model developers who are trying to improve their products.

The final paper in this volume is the Keynote Address delivered at the Workshop. In their thought-provoking paper, Alluisi and Moses discuss another application area that has great need of human performance data: the simulation of combat for force structuring and training purposes. They argue that the skills of model developers are badly needed in these areas, as well as in system design. The potential impact of breakthroughs in combat modelling is significant indeed.

DISCUSSION

The discussion period addressed four primary issues:

(1) Point design versus the screening of design options.
(2) Problems in the transition of models from developer to user.
(3) Cost-benefit of model use.
(4) The lack of representation of organizational and social factors.

There seemed to be a consensus that users should not expect models to provide detailed design solutions or final selection of the best design option. Rather, potential users should think of models as tools to help reduce a large number of design options to a manageable subset for evaluation in mockups or experiments. Mr. Hamilton stated that they have used models such as SAMMIE in this fashion, and were able to quickly discard a number of "nonstarter" options and reduce mockup costs. Dr. Card emphasized that, in his experience, the qualitative payoff from models is more valuable than the quantitative result. As an example, he reiterated the use of Fitts' Law in the design of a computer mouse. While the model does make specific quantitative predictions, the primary benefit is the identification of design parameters that matter versus those that do not. This is a form of design screening, but at a different level. It appears, however, that quantitative results are important when a designer is defending human factors requirements to the design team. Unless the concepts and benefits can be made quite specific, e.g. defended with numbers, they tend to be ignored.

Dr. Rouse initiated the discussion of the model transition topic with the assertion that model developers typically are not particularly interested in, or skilled at, putting an end product in the hands of users. Dr. Baron disagreed and responded that lack of funding from sponsoring agencies is the primary reason that many models are not completed as finished products. In his experience, research organizations which fund model-based analysis have specific research aims to satisfy, and are not interested in product development. Several members of the panel felt that model developers need to: (1) provide better information on the limits of the tools that they provide, and (2) initiate and maintain mechanisms by which users can contact them with problems and suggestions for improvements.

The third issue was opened by an audience member who felt that while the cost of implementing and using models can usually be determined, it is difficult for potential users to estimate the benefit. He voiced the need for a "lessons learned" data base that users can access to get estimates of model benefits. Dr. Card argued that cost should not be an overriding issue. If it can be shown that a theory is "correct", the cost of not using it can be tremendous. The ensuing discussion pointed out that the validity of the theory underlying most models is not known, and as a result the cost-benefit of using the theory in the form of a tool is a real issue.

The final topic was initiated with the question, "What is the relevance of organizational and social psychology issues to design". Dr. Sanders stated that these issues are major factors in the introduction and acceptance of new technology, but agreed with Dr. Baron that they are seldom accounted for in human performance models. Dr. Card noted that a common reason for the failure of office automation is that informal procedures, the way things really work, are not taken into account. Dr. Card's point may be related to the distinction between rule-based and knowledge-based behavior. When attempting to model a man-machine system, it is difficult not to reduce everything to rule-based constructs. As suggested by Dr. Card, this may disregard a significant component of the operation of real systems.

HUMAN PERFORMANCE MODELS AND SYSTEM DESIGN

A.F. Sanders

Department of Psychology
RWTH Aachen
Aachen, F.R.G

INTRODUCTION

The paper discusses aspects of modelling human performance with respect to establishing human limits to be taken into account in system design. It is noted that the term "model" is used in several ways, ranging from an imprecise reference to an explanatory concept, to well defined and computer-simulated general purpose process models. The simulations have the advantage that they allow a full simulation of the complete system; validation is their bottleneck, in particular since behavioural science is not characterised by a variety of well established laws. The weaker models suggest many heuristics for system design which lead to combinations of task analysis and checklist techniques. In addition, dynamic task simulation provides fruitful checks of the appropriateness of a system. The human factor is more dependent on weak models as the system is more concerned with higher cognitive activities, such as in C-3 systems. It is noted that energetical and motivational human limits are often ignored although systems are not exclusively meant for optimal circumstances.

ON COMMUNICATION GAPS

The task of writing on a topic as general as that mentioned in the title of this paper is extremely hard. Within the limitations of the allocated space, one could merely superficially touch on or enumerate a number of models and systems and, hence, run the risk of being either trivial or incomprehensible. As an illustration of my problem, I may refer to the recently published two volumes of the Handbook of Perception and Human Performance (Boff, Kaufman, and Thomas, 1986), to the Handbook of Human Factors (Salvendy, 1987), and to the three volumes of the Engineering Data Compendium: Human Perception and Performance (Boff and Lincoln, 1988). The first mentioned volumes provide 45 excellent reviews of selected topics from the basic literature on perception and performance. They cannot be recommended, however, to a non-expert because they require considerable preknowledge. Even within the various areas of perception and performance different types of expertise are required, making the literature hard to read for someone who is not directly an insider. Some of the basic issues return in the Human Factors volume, which then, of course, concentrates on techniques and applications in 1880 pages of highly condensed writing. I think that, in fact, the huge number of "models" contained in these handbooks should in principle satisfy the various needs of designers, as summarised in Rouse and Cody's (1989) introductory paper to this conference, yet they are ineffective in actual practice. Why is this so?

Despite the presence of the materials, one usually hears of a lack of accessibility of human factors data to designers. Indeed, the texts do not provide simple rules of thumb, but reflect a science in development with many open ended questions. Perhaps, then, the inaccessibility is due to the fact that human factors is a different discipline. Although one should obviously try to improve on accessibility, it might be basically wrong to think that the accumulated knowledge of applied experimental psychology can be simply transferred to a designer. In the same way, nobody would claim that the technical know-how of a designer can be easily transferred to an engineering psychologist!

It might be better, therefore, to improve the communication between designers and various breeds of human factors specialists, in order to make the assumptions and computations of the models understandable and acceptable. One of the obvious requirements is that the various groups involved have some notion of each other's way of thinking, know-how, and interests. There exist serious gaps, distrusts and misunderstandings, not only between engineering and behavioural scientists (see Meister, 1987, for a discussion), but even among various breeds of human factors specialists.

MODELS, MODELS . . .

A major misunderstanding arises with respect to the usage of the word "model", and on this issue I will concentrate in this paper. Depending on the background of the investigator, the word "model" has quite different meanings, which may be said to vary from weak to strong (e.g., Pew and Baron, 1983). At the extreme weak side, it is used in the loose sense of "a set of concepts", qualitative in nature, and not properly specified, which are thought to underlie certain empirical phenomena. A first step toward a predictive and specified model is made when processes are formulated which are thought to make up the functioning of the concepts and their mutual relations. There is wide variation with respect to the precision with which the relations are described. Again, they may be mere qualitative statements, but the ultimate goal is to arrive at a quantitative and mathematical description and at detailed rules enabling computer simulation of the phenomena concerned. Most models in psychology are qualitative in nature with, at best, initial attempts toward quantitative descriptions of limited scope.

Indeed, once a quantitative statement is available the question arises which range of phenomena the model can cover. If the range is sufficiently wide and if there is sufficient precision, the model may be used to "generate" behaviour in a larger scale simulation of a man-machine system without actually measuring behaviour. I would like to make some remarks about this strong type of model - of course from a psychologist's perspective - and then devote the rest of the paper to the relevance of the "weaker" models for system design.

A PRELIMINARY REMARK

As a preliminary remark valid to all human performance modelling, I think it is evident that the research has the aim of establishing estimates of human limits: limits of perception, of memory, of motor behaviour, of selective attention, of sustained performance, of decision, of choice and, more generally, of strategical flexibility, and finally, limits of motivation and coping with stress. Elsewhere, (Sanders, 1984) I have outlined a taxonomy of types of performance limits and suggested a number of their distinguishing features. Thus, limits can be structural, such as sensory limits and certain motor limits; they are functional when concerned with cognition and reasoning in any form; they are energetical when concerned with basic aspects of energetical supply to ongoing activities, and finally, they are motivational when concerned with attitudes and morale. The major problem in establishing these limits is that they are often quite adaptive, variable, and contingent on small environmental changes. Errors are bound to occur when performance limits are exceeded and their prevention is the main aim of any performance modelling.

GENERAL SIMULATION MODELS

General simulation models which include the human are the obvious dream of the designer, but there is the problem that validated quantitative descriptions of a sufficiently wide range of human behaviour are usually not yet feasible, even if one limits oneself to operator control (e.g., Chubb, Laughery and Pritsker, 1987). An often proposed alternative could be to start from an analytic point of view. Thus, there are attempts toward larger scale modelling of human behaviour using some logic. Some of these logics have the implicit claim of complete models of at least some portions of human behaviour. Models of manual and supervisory control which have their root in engineering rather than psychology are a case in point. Other logics are largely interested in a description of activities and less in the underlying dynamics of behaviour. Task-network analysis constitutes an example. It attempts to subdivide the human task in subunits and sequences, operational attributes of which are traced and programmed. One may say that this approach is a modern combination of time line and task analysis. In this sense, it lays only a network foundation for a more general performance analysis which needs to be completed and weighed by actual task elements, which as such are not part of the model. Although the network approach attempts to refrain from assumptions about behaviour, it is not completely successful. For example, a classical assumption in time line analysis is strict single channel processing which excludes the possibility of doing two things at once (Sanders, 1979). Although Pert Charts are more flexible and allow for parallel activities, serial and parallel relations must be fully specified prior to an application - and that means the introduction of behavioural assumptions.

I think that at least at present, strong performance models are more useful as they rely less on behavioural assumptions in view of the validation problem. As an illustration I will discuss in more detail the HOS model developed by Lane and coworkers (Lane, Strieb, Glenn, and Wherry, 1981). This strong simulation model contains a significant amount of explicit "psychology" including micromodels and the rules that pertain to them, in addition to strategical principles of the dynamics of human information processing.

VALIDITY OF SIMULATIONS: THE EXAMPLE OF HOS

The major problem for most engineering models is validation. To what extent is human performance really simulated and does one obtain a realistic and unbiased picture of the performance limits under investigation? Even a first approximation might do to satisfy the needs of a designer. If a model does not commit grave errors, the designer would still be satisfied. After all, psychology is not their main concern.

I am not convinced, however, that the absence of grave errors is guaranteed in present-day performance modelling. Let me consider the primary assumptions of the HOS model in some more detail. First, it contains a single channel notion of human information processing. This was once thought to constitute a general limit to attention (Welford, 1967), in particular with respect to time sharing abilities. However, more recent research has shown that single channel processing is certainly not a general characteristic of human performance, but is limited to time sharing of highly similar tasks (e.g., Wickens, 1984). Even then, performance appears to be highly adaptive in that human limits decrease as processing becomes more automatic (e.g., Shiffrin and Schneider, 1977). Simple application of a single channel principle will therefore usually underestimate human attentional capabilities and draw too rigid a picture of human performance. Yet, a single channel principle may still suffice as a first approximation and not lead to gross errors in the sense of overestimating human capabilities.

This cannot be said of the assumption of HOS that trained operators rarely forget procedures or otherwise commit procedural errors. There are numerous observations in accident analysis which demonstrate the occurrence of procedural errors as a major source of human error. One of the most famous examples is that one of the causes leading to the Three-Mile Island near-accident was that the reserve feedwater supply system pipe had been blocked off by maintenance personnel who forgot to reinstall the system when going off duty. Human memory for automatic and self-evident actions is quite unreliable. In car accidents, procedural errors are a dominant factor; not because procedures are forgotten, but because the driver has the opinion that the situation permits violation. There are indications that procedural errors in driving increase as a driver is more practiced. It seems paradoxical, but our greater strategical flexibility has the side effect of introducing limits in following standard procedures.

A further assumption of HOS is that human behaviour can be described as evolving from a sequence of micro-events, such as information absorption, recall of information, etc. It should be admitted that the model allows for some flexibility and "top-down" processing in that, prior to an action, a combination of functions is selected. The question arises - which are the limits of strategical freedom? Again, it remains to be seen whether the detailed models of the micro-events correspond to human information processing in more than a highly limited range of actions, but this is open to tests in traditional performance studies.

A final comment concerns the assumption of HOS that human behaviour is predictable and goal-directed. As a general statement nobody will probably deny this. Yet it does not mean that the human operator is noise-free; in contrast, the variability of human performance is considerable. In comparison to any physical system, the human is much more error-prone and, in addition, he is quite sensitive to stress, emotions, and fatigue which renders his behaviour relatively unstable. I referred to these aspects as energetical and motivational limits. In HOS, they are only represented by "relaxation", i.e., moments that the system is fully blocked. The alternative, that there is a larger probability of ill-considered actions, seems not to be in the model. It is noteworthy that most systems are designed for normal conditions. Little attention is usually paid to problems of anxiety and panic in emergency situations.

In conclusion, it can be said that models developed on the basis of the HOS behaviour theory are liable to gross errors, particularly in less clearly defined conditions.

SMALL SCALE SIMULATION MODELS

A major shortcoming of general simulation models could be that they are too general, i.e., that they cannot provide information about human functioning at a somewhat more detailed level. Yet, this is not a basic issue since more specialised satellite models may be added for serving special aims, at least as long as the basic axioms are not mutually exclusive. Such more limited models are also separately applied. Examples are instrument scanning (e.g., Senders, 1983, for a review), optimal control of the human operator (Baron and Levison, 1980), and failure detection (Rasmussen and Rouse, 1981), which were all developed with special purposes in mind. Again, these models stem mainly from engineering, but, due to the very fact that they address more specific situations, they are usually easier to test than the large scale simulation models.

So far, the models have withstood experimental tests with a varying degree of success. For instance, manual control models do quite well in describing manual tracking behaviour (Poulton, 1981). In contrast, the predictions of scanning models correspond poorly to what a human observer does. This is true for Senders' classical reconstruction models of scanning - which were never intuitively plausible - but still a fortiori for the more recent conditional sampling models (Hannen, 1987). People not only scan in a different way than prescribed by the model, they also do it much less consistently and their strategy may depend on minor changes in condition. The general result of research is that, as top-down elements increase, the variability in performance increases.

The question is, of course, what is the value of the model if it fails to pass critical tests? This is merely one step from the question of what is the value of a model which, as yet, has not been validated or which is too complex to be validated. I realise here that I have hit upon an area of disagreement. On the one hand are engineers who feel that, despite a lack of behavioral validity, a model can still be "quite useful" (Rouse, 1981). On the other side are psychologists who feel that an untested model is at best speculation, and who are not impressed by the observation that "it works" under certain conditions. On the one side is the engineer who feels that in fact an engineering model is beyond discussion, since it represents a tool that always works when the parameters have been properly set. On the other side is the behavioural scientist who tends to be more inductive and data oriented. And further, you have the engineer who wants to design a system in which the human is just an element, and the psychologist who is interested in mental functioning more than in the system as such.

Not all simulation models stem from engineering. In fact, in basic behavioural research, models have the advantage that they usually address a considerable body of experimental data and, hence, are in close contact with how people behave. Small scale simulations of mental processes are rapidly expanding. Particularly in the area of memory, there are a number of such models, the results of which are very promising (e.g., Anderson, 1981; Raaymakers and Shiffrin, 1981). In order to be properly testable, they are usually characterised by a limited range, such as retention following paired-associate learning and free recall. Again, not all small scale behavioural simulations have been successful. Witness, for instance, the normative Bayesian decision models which have hampered rather than promoted progress in the field of how people arrive at decisions.

In summary, this section is not meant as a criticism of simulation models of human performance. If properly validated, they are very useful both as a theoretical tool in the study of performance and as a means of predicting human behaviour in a system. If model descriptions deviate from actual behaviour, they may still be acceptable as long as they do not lead to gross errors, or when they are somewhat conservative. The situation becomes dangerous when the model promises more than can be expected from the human, and indeed, if there are important factors about which the model has nothing to say. Given the extreme complexity and time-varying character of human performance, it is recommended to aim at small scale simulations for fairly specific tasks rather than attempting to generate the general purpose simulation. In addition, much attention should be devoted to testing the validity of the axioms of a simulation. This is more than parameter estimation and, indeed, concerns the applicability of a model's basic concepts to the simulation of human behaviour. It is my contention that application of simulations to system design without attempts toward validation may lead to serious design errors.

WEAK PERFORMANCE MODELS AND DATA

As mentioned in an earlier section, the term "performance model" is used in various ways, ranging from a simple qualitative reference to some concept to a more precise mathematical formulation. There is a tendency in circles of engineering performance modelling to underestimate the relevance of the "weaker" models. Indeed, one might wonder whether the word "model" is perhaps misused when used in too wide a sense. Yet, ignoring the empirical data underlying the weak models would mean that bodies of relevant evidence are disregarded for the simple reason that they are not in a suitable format to be included in simulation models. The weak and strong models complete rather than compete.

In fact, the weak performance models provide many answers - or at least suggestions - to important questions in the prevention of suboptimal design. For example, one should have avoided the 50 dBA ventilation noise on my new Apple Macintosh SE computer, since it is known that continuous noise of that intensity evokes aversive reactions in some proportion of people in conditions where they are used to working in quiet (McCormick and Sanders, 1982). Again, one should have avoided in my 1987 Ford Escort a left turn of the key in order to lock the doors, since many people have a bias to the reverse. This concerns a violation of the well known principle of S-R Compatibility in performance theory. Some standard programmes for text processing require sheer paired-associate rote learning of relations between codes and effects, and in addition, codes may have different meanings under different conditions (e.g., Sanders, 1987). In the design of such systems there is full neglect of the principles of S-R Compatibility and furthermore, of different resource capabilities which the human is thought to have at his disposal (see Wickens, 1984). In other examples complex 200 page handbooks must be studied for text processing, which fully neglect principles of human knowledge acquisition, and probably are written by an expert!

It would take little effort to continue this blacklist with further examples; some will be provided at appropriate places in the rest of this paper. The models underlying the effects of noise annoyance and of S-R Compatibility are actually still very much a matter of debate in basic research (e.g., van Duren and Sanders, in press), but this does not affect the empirical validity of the phenomena. With respect to Rouse and Cody's (1989) criteria, the above examples should not pose any problems, since the costs of not using the underlying principles may actually be quite high. The consequences of human errors in nuclear power stations or in traffic are a case in point, at least to the extent they could have been prevented by better designs.

In addition to formulating heuristics, there is a trend toward developing new techniques of measurement. The last decades have seen rapid developments in areas such as scaling human judgement and choice, so that subjective estimates have gradually obtained the status of serious data. More recent are attempts to revive the classical techniques of introspection in the analysis of human knowledge with direct application to expert systems.

Let me devote some comments to introspective techniques in knowledge extraction. It should be realised that protocol analysis and thinking aloud traditionally belong to the weakest methods in the analysis of behaviour. One problem is that many, usually more overlearned and automatised, types of human activity are little open to introspection and hence such activities are liable to be absent in the protocol (e.g., Broadbent, 1977). The confounding of introspection and retrospection is also a classical problem for protocol analysis. It is at least doubtful, therefore, whether the present procedures used in extracting knowledge and rules of operation from experts deliver a fair and sufficiently faithful picture of their cognitive schemata. This is not to say that protocol techniques are useless, but only that it is insufficiently known how and to what extent the results are biased. It is, of course, not excluded that the techniques succeed in obtaining the primary factors that account for most of the variance. Much more research should be devoted to this question, which obviously includes the extent to which one really wants to simulate human behaviour in expert systems or the extent to which one wants to improve upon it.

One of the main problems of the weak performance models is that their status is less clear and that their realm of validity may be undefined. In fact, the major problem of behavioural science is that small and seemingly trivial changes in work environment often exert considerable and unexpected effects on actual behaviour, thus preventing formulation of many general and widely valid behavioural principles. If anything, behaviour is more characterised by higher order interactions than by simple main effects.

This poses a problem for strong as well as for weak models. The consequence for system design is that detailed forms of micro-task analysis are required in order to detect possible sources of system failures due to deficient man-machine communication. One of these techniques is described in detail by Drury, Paramore, van Cott, Grey and Corlett (1987). A description of a detailed example - aligning a lamp in the lampholder used in a photocopying machine - is presented by Drury (1983). It consists basically of a highly detailed description of all elementary actions, and a listing of the requirements in the interaction between man and machine elements involved in solving the task. In the case of the lampholder, this is mainly a problem of how to design displays and controls to arrive at an optimally efficient solution. It is probably fair to say that many important principles of standard "knobs and dials" design are known and ready to be applied.

Much of this knowledge is contained in the handbooks cited in the introduction. Rather than attempting to summarise this knowledge in large scale performance models, I think that at least for standard cases, it is better to work toward computerised checklists of performance limits - and in particular, of higher order interactions between them - accompanied by recommendations on how to proceed in specific circumstances. Yet, in my own practice, I have seldom met a case in which the final outcome was ideal, principally due to conflicting interests leading to the inevitable compromise. I do not think, therefore, that checklists can be simply handed over to the designer as a cookbook for application.

It is quite feasible, though, to use the data on heuristics as input to models of the network type in an attempt to provide the network contents to act upon. Such developments are found in combination with larger scale modelling, including supervisory control (e.g., Pew and Baron, 1983).

RAPID CHANGES IN APPLICATIONS

An important problem to all performance modelling is the rapid development in the nature of the applications. Small scale models are usually too rigid to be simply transferred to quite different conditions. The main tools of the human factors analyst in addressing new applications may be task analysis and a set of heuristics about the type of problem and, perhaps, a network type framework. Let me briefly describe some examples.

Despite the abundant research on perceptual-motor limits of performance, there are also continuously new types of developments mainly due to rapidly developing computer technology. Thus, computer technology has created ample possibilities for integrated and diagnostic displays which are gradually replacing the classical arrays of dials in, for instance, the cockpit of an aircraft. Similar developments in driving appear to open a fully new research area known as "car electronics", with the idea of providing the driver various types of new useful information, ranging from route information to instructions about ongoing maneuvers. In all kinds of supervisory systems, such as ship traffic control, new data processing systems are being installed which include more sophisticated and automatic display of, say, ship movements and ship berths. For the cockpit, the advantage of integrated displays could be that more rapid answers are provided in diagnosing certain problems, given of course that there is a representation of the problem at hand in the display system. On the other hand, not "everything is available at the same time" so that totally relying on integrated displays means that the computer does at least part of the "thinking".

The same can be said about problems of data presentation in the car. Thus, electronic route guidance systems may eventually be helpful in preventing uncertainties in maneuvering, yet they are also likely to create problems when operated during driving due to possible failures of divided attention. Research on dual task performance provides ample suggestions for optimal solutions. Similarly, new displays of the occupancy of ship berths face the problem that simple listings of ships are not satisfactory in comparison with classical solutions with written strips, positioned on a wide and very large sized display which mimics the harbour systems. The problem is that the spatial code is no longer represented in the listing and that, accordingly, the operators complain that they have lost the "overview". Again, there is ample evidence about the relevance of complementary spatial and verbal codes in optimizing man-machine interaction. I do not anticipate that the underlying principles can be either properly summarised in simulated human performance models, or in cookbook types of checklists ready for application in whatever condition. Very detailed task analysis may guide discussions to proper evaluation of the performance limits.

DYNAMIC SIMULATION

In addition, I would like to stress the possibilities of dynamic simulation of complete tasks in situations where actual designs do not yet exist. I am referring here to large scale simulators. Examples concern maneuvering ships in narrow fairways, car driving, and flying; the simulation of which has reached high levels of perfection and is well validated. Thus, the possibility to actually maneuver a six element push barge in a simulator through a fairly narrow and curvy part of the Waal River on its way from Rotterdam to the Ruhr area is an excellent way of finding an answer to the problem of whether such traffic can be allowed or not. I do not see how either analytical simulation, task analysis or a combination of both, could at present replace such studies, at least not in more complex task settings. At best, they might provide hypotheses and, hence, suggest the most interesting conditions for testing in a dynamic simulation.

COMMAND, CONTROL AND COMMUNICATION

Dynamic simulation is also of interest for tasks, the requirements of which go far beyond perceptual-motor functions and, which pose special problems to modelling as well as measuring human performance. Higher order information processing, reasoning, decision making, and problem solving are the main elements in Command, Control and Communication (C-3) systems. Moreover, such systems are usually characterised by cooperation between several people. Their main input comes from display units, which usually produce large amounts of information of varying reliability and diagnosticity, constituting the data base for a C-3 system.

It is clear that perceptual-motor aspects are not absent in such systems - for instance, in problems of radar and sonar interpretation - but the demands on complex decision making are far beyond the perceptual and motor aspects of performance. In addition, a C-3 system is not one task, but constitutes a multi-man-machine interaction. Thus, a major set of problems concerns, on the one hand, how to optimally divide the work among machines, and on the other hand, among various persons. For instance, in air-traffic control one may have a "planner" of the various operations and another operator who controls and executes the actual commands. In military command systems, one may have various operators who evaluate different types of information and decide whether a bit of material is sufficiently relevant to be forwarded to higher command. In that case, only a fraction of the incoming data plays a role in the ultimate decision of how to act. The criteria whether or not to forward and how much to forward constitute a major problem.

Alternatively, the operator may weigh all the various bits of incoming information with regard to a limited number of predefined dimensions and merely forward status estimates of the dimensions to the higher level. It is obvious that, in weighing, both utility and probability factors play a role. It has the possible disadvantage that the higher levels do not "see" actual data any more. As yet, the human factors involved in such systems are little specified, and modelling is hampered by a lack of knowledge about how to proceed.

However, the interest in C-3 systems has evoked much research on the dynamics of human decision making with the result that a considerable number of heuristics and biases have been described (e.g., Wright, 1985), suggesting much stronger limits than previously anticipated. It is astonishing to note the many deviations from rationality in human decision making and the sensitivity of choice behaviour to seemingly small changes in environment. I do not think that in this area the time is ripe for the overall engineering type of performance models, including those concerned with supervisory control. On the other hand, C-3 systems do certainly benefit from computerized decision aiding techniques.

It is doubtful whether large scale dynamic simulation studies are useful in complex C-3 environments, given the experiences of the past (Parsons, 1972). The attempts of the fifties have been extremely costly while delivering mainly trivial results. It is probable that computer technology has somewhat eased this situation, but in particular when cooperation between people is involved, dynamic simulations become very complex. In my view, experimentation and modelling should be limited to manageable bits of systems in combination with small scale basic research on elementary aspects related, first, to optimal channeling, formatting and presentation of data, so as to avoid biased treatment of the various data; second, on operational criteria with respect to the relevance of a datum; and, third, on the side of the executive decision maker, the question of what constitutes a "good" decision should receive more attention.

THE ENERGETICS OF PERFORMANCE

As usual in many discussions of performance modelling, I have not yet devoted any attention to the question of energetical and motivational limits. Their neglect could be due to the fact that such limits seem not to be relevant to a design, yet nobody will deny that factors such as noise, heat or cold, sleep-loss, boredom, and anxiety affect performance. It is striking that most systems are designed for everyday circumstances only. For instance, sonar equipment usually does much less well in operational conditions; and energetical factors related to boredom and sleep-loss, are a major cause of accidents and human error in traffic. A suboptimal design may normally remain unnoticed, but become critical in stressful conditions.

There is relatively little research on issues of energy and motivation (Hockey, Gaillard and Coles, 1986). Perhaps there is the dated notion that the effects of energetical and motivational factors are aspecific and, hence, occur irrespective of the design. There is now ample evidence that this is not the case (Sanders, 1983), and the operational circumstances should be heavily weighed. I was reminded of this when I once stayed in a skyscraper type of hotel which had emergency cords in each room to allow escape in case of fire. Attached was a user's manual that was so complex that one may doubt its effectiveness when actually needed.

Under conditions of stress, people are over-aroused or activated; they tend to resort to well practiced and available responses and to bypass well-considered reasoning. The result is that many more errors are committed than one may expect under normal circumstances. In contrast, stress arising from boring situations leads usually to slow responses or even to omissions. Yet, this is a rough generalisation, a kind of first approximation, which is strongly affected by a whole score of structural and functional variables.

IN SUMMARY

In this paper, I have briefly discussed how I conceive of models in establishing human performance limits in system design. As a psychologist, my first interest is the study of human performance, not of system design. To an engineer, probably the reverse applies. Hence, the discussion about performance simulators is tuned by interest. In addition, the engineer feels less at home with the weaker qualitative models of the psychologist and with his research techniques and problems of measurement. Yet, the engineer is interested in validation to the extent that false premises may render his design useless. In turn, psychologists have discovered that simulation is a highly powerful tool in building explicit theory. I assume that engineers will admit limits of general purpose performance models, and accept preliminary heuristics about performance limits, as contained in the weak models. I would like to end by expressing the hope that gradually the time will pass when the human factor is considered as "easy" and "trivial", something the designer knows from intuition with a "little thinking", although I am afraid that I still regularly meet this attitude in my own contacts.

REFERENCES

Anderson, J.R. (1981) "Cognitive Skills and Their Acquisition." Erlbaum, Hillsdale.

Baron, S. and Levison, W.H. (1980) The optimal control model: Status and future directions, In: "Proceedings of the IEEE", Boston.

Boff, K.R., Kaufman, L., and Thomas, J.P. (1986) "Handbook of Perception and Human Performance," Wiley, New York.

Boff, K.R. and Lincoln, J.E (1988) "Engineering Data Compendium: Human Perception and Performance," Harry G. Armstrong Aerospace Medical Research Laboratory, Wright-Patterson Air Force Base.

Broadbent, D.E. (1977) Levels, hierarchies and the locus of control, The Quarterly Journal of Experimental Psychology, 29, 181-201.

Chubb, G.P., Laughery, K.R., and Pritsker, A.A.B. (1987) Simulating manned systems. In: G. Salvendy (Ed.) "Handbook of Human Factors", Wiley, New York.

Drury, C.G., Paramore, B., van Cott, H.P., Grey, S.M., and Corlett, E.N. (1987) Task analysis, In: G. Salvendy (Ed.) "Handbook of Human Factors", Wiley, New York.

Drury, C.G. (1983) Task analysis methods in industry, Applied Ergonomics, 14, 29-38.

Hannen, P. (1987) Systematik in der Reihenfolge der Beobachtungen Mehrerer Unabhangiger Binarer Stochasten, Diplomarbiet, RWTH Aachen.

Hockey, G.J., Gaillard, A.W.K., and Coles, M.G.H. (1986) "Energetics and Human Information Processing," Nijhoff, Dordrecht.

Lane, N.E., Strieb, M.I., Glenn, F.A., and Wherry, R.A. (1981) The human operator simulator: An overview. In: J. Moraal and K.F. Kraiss (Eds.) "Manned System Design: Methods, Equipment and Applications," Plenum, New York.

McCormick, E.J. and Sanders, M.S. (1982) "Human Factors in Engineering and Design," McGraw-Hill, New York.

Meister, D. (1987) A cognitive theory of design and requirements for a behavioral design aid, In: W.B. Rouse and K.R. Boff (Eds.) "System Design: Behavioral Perspectives on Designers, Tools, and Organizations," North Holland, Amsterdam.

Parsons, H.M. (1972) "Man-Machine System Experiments," The John Hopkins Press, London.

Pew, R.W. and Baron, S. (1983) Perspectives on human performance modelling, Automatica, 19, 663-676.

Poulton, E.C. (1981) Human manual control, In: V.B. Brooks (Ed) "Handbook of Physiology: The Nervous System," 1337-1387, American Physiological Socioty, Bethesda.

Raaymakers, J.G.W. and Shiffrin, R.M. (1981) Search of associative memory, Psychological Review, 88, 93-134.

Rasmussen, J. and Rouse, W.B. (Eds) (1981) "Human Detection and Diagnosis of System Failures," Plenum, New York.

Rouse, W.B. (1981) Human computer interaction in the control of dynamic systems, Computer Surveys, 13, 71-99.

Rouse, W.B. and Cody, W.J. (1989) This Volume.

Salvendy, G. (1987) "Handbook of Human Factors," Wiley, New York.

Sanders, A.F. (1979) Some remarks on mental load, In: N. Moray (Ed.) "Mental Work Load," Plenum Press, New York.

Sanders, A.F. (1983) Towards a model of stress and human performance, Acta Psychologica, 53 61-97.

Sanders, A.F. (1984) Structure and interconnections between human performance limits, In: "Performance Limits," NATO meeting, Toronto.

Sanders, A.F. (1987) Problems of a novice: some introspective observations, In: G. Salvendy (Ed.), "Second Conference on Human-Computer Interaction," Hawaii.

Senders, J.W. (1983) "Visual Scanning Processes," Tilburg University Press, Tilburg.

Shiffrin, R.M. and Schneider, W. (1977) Controlled and automatic human information processing, Psychological Review, 84, 127-190.

van Duren, L. and Sanders, A.F. (in press) On the robustness of the additive factors stage structure in blocked and mixed choice reaction designs, Acta Psychologica.

Welford, A.T. (1967) Single channel operation in the brain, Acta Psychologica, 27, 5-21.

Wickens, C.D. (1984) "Engineering Psychology and Human Performance," Charles E. Merrill, Columbus.

Wright, G., (1985) "Behavioral Decision Making," Plenum Press, New York.

HUMAN ENGINEERING MODELS: A USER'S PERSPECTIVE

A.J. Hulme and W.I. Hamilton*

British Aerospace plc
Richmond Road
Kingston-upon-Thames
Surrey, KT2 5QS
United Kingdom

*Sowerby Research Centre (FPC 267)
British Aerospace plc
PO Box 5, Filton
Bristol, BS12 7QW
United Kingdom

INTRODUCTION

British Aerospace is the United Kingdom's major producer of military and civil aircraft, of guided weapons, and of electronic and satellite systems. Many of these are man/men-machine systems and consequently the Company is very concerned with the optimisation of the human-system relationship.

The process of designing any man-machine system raises a variety of human factors issues. The human factors specialist on a system design team will be asked to answer questions on such wide ranging topics as the anthropometric suitability of the design, the quality of the visual interface, and the functionality and semantic complexity of the cognitive interface.

Human factors specialists have often argued strongly that they should be involved in the design and evaluation of the man-machine system right from its earliest conception. In order to realise this laudable aspiration, he must be equipped with the technology to enable him to predict the performance consequences of the interface design in advance of any prototype being available for empirical evaluation. Over the past decade or so we have witnessed the emergence of a variety of engineering models/modelling techniques for use in the evaluation of one or other aspect of the man-machine interface. Some of these tools are on view at this workshop.

Most of these systems allow the designer to simulate in some way the physical and functional nature of the proposed design, and then to evaluate it through the application of some appropriate database or empirical model of performance. In this way it should be possible to identify possible design problems before any irreversible engineering commitment is made.

This paper reports on experiences gained in using, or in some cases attempting to use, certain human engineering models. The techniques have been classified according to the area of ergonomic assessment to which they are intended to contribute. In each case the usability of the technique is considered along with the quality of the assessment it provides. Where appropriate, recommendations have been made for their improvement.

ANTHROPOMETRIC ASSESSMENT

SAMMIE

Experience and benefits. SAMMIE (System for Aiding Man-Machine Interaction Evaluation) is a three-dimensional CAD workspace modelling system which uses a 3-D model to evaluate the physical or anthropometric aspects of a workspace design (Compeda Ltd.). The system was evaluated by British Aerospace in 1981 to assess its potential usefulness as a tool for evaluating jig designs (a jig is the structure which supports an airframe while it is being built) - see Howson (1981).

The system's utility was demonstrated and it was shown to offer potential benefits for reducing development costs. In jig design it is not normally the policy to produce a mock-up but one was developed for the purposes of comparing the design produced by this method with that produced using SAMMIE. It was found that the cost of the evaluation using SAMMIE was approximately 25% of that for the mock-up. The similarity of the final designs suggests that, in this case, nothing is lost in terms of validity when using the model-based approach.

Problem and recommendation. The firm who market the SAMMIE system was taken over by another company who had an interest in selling dedicated hardware with the system. Negotiations eventually failed when British Aerospace was unable to purchase a system which could be run on its existing computers.

This experience shows that a potentially useful tool can be rendered impractical (not to mention made more expensive to acquire) simply because it can only be run on a specific type of machine. It is therefore recommended that, whenever possible, system variants for use on different machine types should also be made available. The extra cost of the development of the tool would soon be recovered due to the greater sales potential.

The Crewstation Assessment of Reach Model (CAR)

Application experience. The Crewstation Assessment of Reach Model (developed by Analytics Inc.) is a design tool for use in determining whether or not a candidate population of operators can see the displays and reach the controls within the constraints of a workstation design.

In 1981 British Aerospace and McDonnell Douglas Aircraft Co. were involved in the development of the Hawk VTXTS (a carrier-based navy trainer aeroplane). This was an adaptation of the RAF Hawk which has been in service since 1976. Although the cockpit geometry of both aircraft types was essentially the same, a number of changes were planned in order to make the aircraft more suitable for use by U.S. Navy personnel. These added up to a redefinition of the layout of the cockpit instrumentation and controls.

It was decided that the CAR II program should be applied to the formal evaluation to decide the final configuration for the VTXTS cockpit. However, the results from running CAR II on the Hawk cockpit geometry gave only low aircrew accommodation levels for what was in fact a proven layout. Subsequently, the program was given the measurements of a sample of Hawk pilots in the 90th plus percentile range. The reach envelopes for these same individuals were already known to be acceptable. In spite of this the program calculated that they should be unable to reach the throttle and stick when these are in their forward positions.

Problem and recommendation. There followed a great deal of correspondence between British Aerospace and Analytics as an explanation for these discrepancies were sought. The ensuing investigation included an examination of the data used by the model, such as the mapping of the three-dimensional co-ordinate systems used in the aircraft design and that used by CAR II, the validity of the Monte Carlo sample generation and the original statistics from which it was produced, the measurement statistics, and seat geometry. For a full report on these tests see Johnson (1982).

Some of the results from this work suggested that there may have been a problem with the way in which the program handled the links in the operator's limbs (especially the shoulders) when the model is "moved". At that time a recommendation was made to Analytics to look at this further.

The main problem, however, was thought to be in the way CAR II had defined the Hawk seat geometry and, specifically, the way in which it related the operator to the seat reference point (SRP) and the design eye point (DEP) before calculating accommodation levels. The program appeared to define the SRP, not as the actual point of contact between the operator and the seat, but as the point of intersection of the planes formed by the seat pan and seat back. Clearly, the distance from the operator's buttocks and the SRP, defined in this way, changes as a function of the seat pan/seat back angle. If the program does not compensate for this it would have the effect of placing the operator "into" rather than onto the seat. This effectively puts him lower down that he would actually be. This in turn means that the program has to "adjust" the seat up further than would actually be necessary for the operator to acquire the cockpit DEP. This would place the controls out of reach for many operators which might explain the low accommodation levels.

British Aerospace also noted that accommodation levels were being affected by unrealistic assumptions about the operational use of some of the controls. For example, "stick full forward and to the left" obtained very low accommodation levels, but this position is, according to the aircrew, never likely to be used.

British Aerospace can also cite a number of cases where an operator has used a switch or control in a most unexpected fashion. For example, because of the need to stabilise hand movements in a high vibration environment such as the cockpit, many switches designed for use with the index finger are in fact operated with the thumb while the palm is braced against the adjacent panel.

In our view any further developments of CAR or similar systems should be based on information from dynamic and more lifelike assessments. Without this their utility may be limited only to identifying gross incompatibilities.

Following this experience British Aerospace, in a subsequent cockpit development programme, resorted to the more traditional method of evaluation with an actual physical mock-up. It may be argued that having identified some of the problems inherent in the model it would have been better to have rectified them and persevered. However, by this stage faith in the system was at a low ebb and it was felt that the effort which would be required to improve the program could not be justified. This is an example of how a potentially powerful tool can be left unused because of certain flaws which severely limit its validity.

VISUAL INTERFACE ASSESSMENT

Predictive Modelling of the Visual Interface

Models to predict visual performance at the man-display interface have been under continuous development in British Aerospace for around 15 years.

This work began during a surface-to-air missile system development programme as an attempt to predict target detection probabilities as a function of the design characteristics of optical sights (field of view and magnification), and the nature of the image of the target (size, range, contrast, etc.). The result was a mathematical model of threshold detection performance called ORACLE - see Cooke (1984). Other sub-programs have since been developed to calculate the image from various electro-optical devices and this has enabled ORACLE to be applied to the assessment of detection performance with these.

More recently, ORACLE has been developed to predict the visibility of suprathreshold and coloured objects and this has enabled it to be applied to predict visual performance at the man-display interface in the cockpit. The full procedure for this has been described by Johnson (1987). In brief, however, a physical model is applied to calculate how the spectral output of a display is modified in both luminance and chromaticity before reaching the eye. The model considers spectral attenuation through filters, from around the cockpit, and the inclusion of ambient illumination. The main outputs of this model are the spectra for the display image and its background as seen at the eye. These may be input to the ORACLE model (along with other parameters) which then calculates suprathreshold visibility which is similar to the ratio of suprathreshold contrast to threshold contrast. The output of this is a set of graphs and tables of predicted visibility against peripheral angle.

This technique has been used to good effect by the British Aerospace military aircraft division where there are a variety of problems affecting the optimisation of displays in the cockpit. The designer must satisfy the conflicting requirements for display legends to appear sharp and legible and of the appropriate colour under conditions of very high ambient illumination, and yet to be comfortably visible in low light levels without adversely affecting the performance of night vision goggles. Johnson (1987) reports that the models can be applied repeatedly to evaluate the effects of different combinations of phosphors and contrast enhancement filters. Because this assessment can be carried out in the absence of any physical prototype, it has been possible to examine a wide range of options very cheaply and quickly. Obviously, flight trials are still necessary but Johnson argues that model-based evaluation should increase the probability of getting the design right "first time". Also she points out that with such a system it is possible to evaluate more innovative ideas with the minimum of risk.

Problems

The main drawback to the use of ORACLE is the fact that the program demands a certain amount of expertise on the part of the user. He must be knowledgeable about vision and display design in order to simulate and test only practicable engineering solutions. In addition to this the program to run ORACLE is very "user unfriendly". The physical model must be tailored to every application and at present this can only be achieved by rewriting that aspect of the software. Consequently, ORACLE can only be successfully operated by a limited group of experts who have been involved in its development.

A further limitation to its utility is the lack of a database which gives ranges of acceptable ORACLE visibilities for various tasks such as symbol and character recognition and reading. This needs to be tackled through the development of appropriate cognitive user models, such as a model of recognition as a feature extraction process, and to introduce workload and expectation factors into the assessment.

Future Developments

Presently, the program which allows ORACLE to be applied in this way has been configured in a modular design and has been programmed in Pascal to run on an IBM PC. It is a state-of-the-art model and is undergoing continual development and validation. Work is currently under way to validate the colour and suprathreshold predictions of the ORACLE model. Preliminary results from this work are very encouraging. Other work is being conducted to develop an extension to ORACLE to consider the effect of night vision goggles on observer performance. Work is also needed to extend the physical model to cover a wider range of display types as currently it is limited to emissive displays. Also, the model is specific to one aircraft type, the Tornado, although in principle, it could be adapted to any.

COGNITIVE INTERFACE ASSESSMENT

The optimisation of the cognitive (or man-machine information) interface involves the consideration of the following:

i) the effect of the coding and organisation of the information displayed on information assimilation by the operator;

ii) the effect of the design of the controls and displays on task performance time and operator workload; and,

iii) the compatibility of the semantic qualities of the interface design with the operator's understanding of the task and the system.

These three areas will now be addressed in turn.

Early Assessment of Assimilation Performance

Display design has traditionally been approached by developing a prototype of the system's display formats and then showing these to potential operators. The assessment is usually based on subjective data gathered through asking "Can you understand this?" Alternatively, although rarely, the assessment has been performed in the manner of a behavioural experiment designed to observe assimilation behaviours such as eye-movements and fixations, search, and reading. Either way, there are two important restrictions on this. Firstly, the evaluation has to be delayed until a prototype is available. The major disadvantage of this is that, after prototyping, it may then be difficult to effect any substantial changes to the overall design concept - this is particularly true in the development of aircraft systems. What is needed are rapid prototyping tools which enable the designer to simulate the information displays without the need for the system software to drive them. Secondly, empirical assessment can be somewhat problematic with the results often being too subjective and altogether inconclusive. The designer needs a more objective means of evaluating assimilation performance, preferably one which can be applied in advance of the prototype stage of design. Both rapid prototyping techniques and assessment based on human performance models will be discussed in this Section.

REGENCY. REGENCY (Regency Systems Inc.) is a microcomputer based system derived from the PLATO computer-aided classroom training aids programme. The current REGENCY R2-C system is a twin floppy disc machine with a two-colour touch sensitive CRT display. It is programmed using its own high-level language called USE, and claims to combine a simple programming technique with a powerful graphics capability.

The system was designed to enable instructors, or any other non-programmer, quickly to input simple simulations for use in testing the level of a student's understanding of the task and system. The potential advantage of such a system for early evaluation of cockpit displays soon became apparent.

In use in this way the system has proved itself to be an invaluable aid. In advance of the construction of any mock-ups, and before there is any commitment to specific hardware or software, control panels and multi-function displays can be simulated. Consequently, the dialogue between operators and the designers can begin at the earliest possible date.

REGENCY has been used successfully by British Aerospace. In less than 10 weeks it was used to mimic a set of cockpit display formats. This enabled the designers to work out the switching philosophy for mode changing with sufficient precision to be able to place detailed requirements on the equipment suppliers.

We believe that the reason for its successful use is the fact that it requires little expertise in programming to create interactive control panels and display formats.

The potential benefits of systems like REGENCY are most clearly in evidence from the fact that a large number of format generation packages are now marketed, or have been developed in-house by those concerned with the design of information displays. One limitation, however, is the fact that we found it difficult to extract a programme specification which could be applied directly in programming the displays. A development of the system to facilitate the translation of the USE programme into the implementation language for the final system would be a valuable improvement.

Model-based evaluation. Using a sample of operators to assess the adequacy of display format designs can be problematic. In the first instance it may be difficult to gain sufficient access to a representative group of aircrew, either because they are otherwise occupied or because it is a novel application for which there are no operators with relevant experience. Secondly, subjective assessment by any individual will be strongly biased by his personal experience.

A possible solution to this problem is to devise some way of predicting the operator's assimilation performance from a specification of the features of the format design. This necessitates having available a set of empirical models to predict performance as a function of display design. Ideally this would be configured as an evaluative software package which is driven directly from the format prototyping tool.

Thomas Tullis (1985) has developed a PC compatible display analysis package which is not unlike this. His package accepts as input an ASCII file containing a literal example of the screen to be analysed. The program then assesses the format with respect to six characteristics:

i) overall character density;

ii) local density of characters around each individual character;

iii) the number of distinct character groups;

iv) the average visual angle subtended by those groups;

v) the number of distinct labels or data items; and,

vi) the average uncertainty of the positions of items on the screen.

At the time of writing, British Aerospace was in the process of acquiring this package.

Performance Time and Workload

British Aerospace has come into contact with three techniques for the prediction of operator performance time and workload. These are SAINT, MicroSAINT, and GOMS. Experiences with each of these are described below.

Systems Analysis of Integrated Networks of Tasks (SAINT). SAINT is not itself a model, it is a task simulation package which enables an analyst to model the operator's task as a network of elements (Wortman, Duket, Seifert, Hann and Chubb, 1978). The elements are individual steps or actions which have to be performed to accomplish the task goals. The elements form nodes which are connected so as to represent the sequential nature of the task.

In 1983, British Aerospace acquired SAINT for the purpose of exploring the impact on pilot performance of a variety of interface designs for advanced cockpit systems. A programme of work was proposed to validate the utility of SAINT in this respect. This involved the development of a SAINT simulation of a bombing mission, and the comparison of performance predictions from this with real data from simulator runs of an identical mission profile.

A number of difficulties were encountered in using SAINT. To start with the human factors specialists called in to input the task analysis and performance data found that significant software support was required just to construct the network. Secondly, the manuals, and in particular the worked example, fell a long way short of providing the sort of back-up which one might expect. For example, eleven statistical distributions were offered without a really adequate description of what they were or what aspects of behaviour they were appropriate for. Furthermore, it seemed that the creation of the network models was very complex requiring a great amount of detail to be provided from the outset.

The result of all these difficulties was that the assessment was never carried out. This is an example of a potentially useful simulation tool being rendered completely useless by poor user interface and instruction manual design.

MicroSAINT. MicroSAINT (Micro Analysis and Design Inc.) is a revamp of the SAINT software to run on any IBM PC compatible machine. British Aerospace recently acquired this package for use in a study of performance modelling techniques in the design of cockpit systems.

We have found MicroSAINT to be a very usable system. The model development is quite straightforward, and the user support documentation is almost a step by step guide. Overall, the system appears to provide an excellent tool to support the performance of the sort of task analysis which we consider to be essential in the early phases of system development.

The MicroSAINT system also claims to provide a means of assessing the mental workload experienced on the task (Laughery, Drews and Archer, 1986). Workload assessment is based on subjective estimates (given by the modeller) of demand on the various subcomponents of the human information processing system. This is not supported by any reliable theory or guidelines by which it is possible to gauge the severity of these loadings. Consequently, it is, in our view, too crude to be of much practical value in optimising the design in order to balance the loads imposed on the respective information processing subsystems.

The last point represents just one aspect of the principal weakness of both SAINT and MicroSAINT. That is, they are not associated with an appropriate model of human information processing and performance capabilities. This means that the user of this software has to rely on actual empirical observation in order to derive estimates of performance time for the individual elements of the task. Consequently, the applicability of SAINT/MicroSAINT will be limited to tasks and behaviours for which the user already has a performance time database.

This severely restricts its application to new systems, since the performance time estimates for all-new task elements would hardly be reliable enough. Thus, the greatest potential benefit of the SAINT/MicroSAINT models, performance time prediction, is impossible to attain.

A possible solution to this limitation would be to construct a model of human performance capabilities which could be applied to the task model in order to derive estimated task performance times. The task description would have to specify fundamental behaviours which could be referred directly to the performance model. In turn, this performance model would have to specify precisely how the execution times for these fundamental behaviours would be affected by the design of the interface. While this would obviously be a large undertaking it is by no means beyond the capacity of modern applied behavioural science, and the rewards in terms of a marketable product would be enormous. The Human Operator Simulator (HOS - Strieb, Glenn and Wherry, 1978) is an example of a task simulation environment which is supported by a model (or a series of models) of human performance capabilities. At the time of writing British Aerospace had been trying for some time, and without success, to acquire a copy of HOS. Because of this we turned our attention to an alternative, and more readily available, performance modelling technique - GOMS. Our experience with this technique is described in the next Section.

GOMS. Card, Moran and Newell (1983) developed the GOMS (Goals,

Operators, Methods and Selection rules) task analysis technique out of the established theories of human problem solving, and skilled cognitive behaviour. It was intended for use in the analysis of tasks performed at the human-computer interface and its application is supported by a model of the human information processing system which can be used to supply data for the calculation of performance times for simple keyboard tasks. We believe that it is also possible to augment this model for use in determining the information processing demand which the task design imposes on the various subcomponents of the operator's cognitive system.

In a research project, British Aerospace set out to establish the validity of the GOMS analysis framework for the description of routine cockpit tasks. The results of this work are anticipated to be available after March 1988, however, initial impressions are that the analysis technique can be adapted successfully to this domain of tasks.

At the same time we realised that the GOMS information processing model was inadequate in certain respects. For example, it would be incapable of predicting the variance in performance times for certain simple information processing behaviours as a consequence of the presentation of information. Also, it does not represent the mainstream thinking on parallel processing and limitations in attention. Because of this we established a programme of experimental work to extend the predictive power of the science base.

From the work completed so far it would seem that GOMS is a relatively simple performance modelling technique which is capable of being generalised to tasks other than word processing. It also has an associated cognitive science base which, although basic, could be extended to meet an analyst's particular requirements. Its major drawback is that it is, as yet, uncomputerised. Both the development of the task model and the calculation of performance times have to be done "by hand". This is very laborious and time-consuming.

If GOMS does prove to be useful for the description of cockpit tasks it is anticipated that the information processing model could be computerised. This would be programmed in such a way as to accept as input a GOMS type task analysis for a given interface (which ideally would be generated automatically), and to provide as output calculated performance times and demands on cognitive subsystems for task components such as: assimilation, central processing, and keying behaviours. A 3 to 5 year programme of work is planned for the development of this evaluation tool.

Summary and recommendation. British Aerospace found MicroSAINT to be a very usable tool. Unfortunately, like SAINT before it its applicability is limited to tasks for which performance data is already available. HOS represents an important addition to this class of tools as the performance predictions are derived from a set of human performance models. British Aerospace has had difficulty in obtaining this package and so we have resorted to trying to develop our own similar system based on the GOMS analysis technique. So far the GOMS analysis technique itself appears to be valid for the tasks with which British Aerospace is concerned but the human information processing model associated with it is rather underdeveloped.

Due to the importance of this form of assessment in design, there is a genuine market for tools like HOS and for computerised versions of GOMS. Tools of this kind now need to be developed and actively marketed.

Optimising Semantic Compatibility

Theory and techniques. The problem here can be thought of as attempting to predict what knowledge the operators will need in order to use the system, and whether or not they will be likely to comprehend the information which is presented to them and the way in which the system will work.

It can be argued that people operate a system on the basis of a mental model of the task they want to perform and of how the system works. These internal models determine how operators will interact with the system and how they will attempt to solve problems and accomplish their task goals. The internal models also determine what expectations the operators will have about the system's behaviour under all the circumstances within the scope of their experience.

Many techniques have emerged which are intended to externalise and formally represent these mental or user models. Some of these techniques are intended to represent the operator's conceptions of the task domain for which the system is being developed (e.g. CLG - Moran, 1981; TAKD - Johnson, Diaper and Long, 1985; PAD - Keane and Johnson, 1987). If this can be achieved it should be possible for the system programmer to design the task directed aspects of the system interface around the features of the operator's concept of the task domain. For example, the operator's understanding of the objects in the task space (e.g. target, label, weapon, etc.), and of the functions which may be performed with them (e.g. initial, move, arm, etc.), could be applied to the development of an appropriate command language or key set. Models of this type have already been used successfully in the development of a menu of word processing functions (McDonald, Stone, Liebelt and Karat, 1982), and in the development of an interface for a transport timetabling system (Sharratt, 1987).

A further application of user models is in the formal representation of the man-machine interaction. Models of this type have been used to explore the mapping of the internal (mental) tasks performed by the operator, to the external task which is defined and constrained by the design of the interface - see Moran (1983). By externalising the operator's concept of the system (obviously this could only be attempted after initial experience with at least a prototype) the designer could see his misconceptions or confusions about the system's functioning, and on the basis of this he could modify the dialogue or documentation so as to rectify this. At this stage user models of the system and its behaviour can be used to measure the apparent complexity of the system to the operator (Kieras and Polson, 1985). Such measures may also be used to anticipate training requirements, and to predict learning time.

Problems in applying user models to interface design. British Aerospace is interested in applying user models to the design of the cognitive interface of computer systems which are intended to support complex problem solving and planning tasks, such as in the command and control of military systems. Unlike task procedural knowledge, however, the type of knowledge represented by user models is thought to be held by the operator in long-term-memory in a declarative format (i.e. knowledge is held as rules which probably have an IF<condition>THEN<action> structure) - see Anderson (1983). According to Sticha (1987), network type models (such as SAINT/MicroSAINT) are of limited utility for representing declarative type knowledge, and he claims that it can be encoded more conveniently using a production system representation.

The problems for the application of user models to the design process are threefold:

i) knowledge elicitation;

ii) complexity of the computing languages; and,

iii) mapping of the features of the models into system design requirements/specifications.

The first of these problems concerns the process of obtaining the knowledge to be modelled. The particular techniques used for this depend on whether or not the knowledge has to be acquired from a human or from an alternative source such as the interface specification. This can be regarded as a separate area of study which it will not be possible to discuss here.

The second is a problem of relevant expertise. Few human factors specialists have the necessary knowledge of production system languages to be able to programme their user models. This means that, although models are often developed they are not formalised in a way that can be communicated and reasoned with.

The third problem is closely related to the second. The ergonomist has to be able to highlight the features of the model which are relevant to design. The programmer then has to translate these into elements of a design plan. All too often the ergonomist and the programmer speak two different technical languages and the mapping of requirements into a design specification is never achieved (Pew, Sidner and Vittal, 1980). A potential solution to both the second and third problems, and one which is a logical development of a suggestion made by Pew et al., would be the development of modelling tools or shells. The shell would support the development of user models by the ergonomist, much in the way that an expert system shell supports the development of a knowledge base by a non-programmer. Once developed the model could be communicated to the programmer as a formal description of the task. This description could then be interpreted by the designer (or an appropriate interpreter system) into an initial specification of the interface program.

It is the absence of these tools, more than anything else, which is restricting the application of user models to the design of the semantic aspects of the man-machine interface.

CONCLUSIONS

This review has examined a range of modelling tools and techniques which are intended to help the designer contribute as early as possible to the design of the man-machine system. Three principal areas of design optimisation have been addressed and in each case the availability, validity and usability of relevant engineering models has been considered. At each stage specific views and recommendations have been forwarded based on British Aerospace's experiences with particular commercial tools or theoretical techniques. To conclude we will attempt to draw these points together in order to identify more general recommendations. These will be presented as development goals for tool builders.

GOAL 1: Maintain an Awareness of the Requirement for Tools

Tool builders should consider what problems interface designers have to address and whether or not the tools to support them in respect of these are available. For example, in the area of performance time and workload assessment the behavioural theory necessary for this kind of assessment is already emerging (see Section on GOMS), but the tools to support its application to design are not yet available.

Similarly, the optimisation of the semantic compatibility of an interface can, in principle, be addressed through the application of production rule models of the user's knowledge. These can be employed to define the requirements for the interface dialogue, to identify areas of operator difficulty, and to anticipate training requirements and training time. Unfortunately, their application in practice suffers from a lack of tools to support the development of the models by non-programmers, which, in turn, impedes the translation of the features of the models into requirements for the design.

GOAL 2: Ensure that the Tool Satisfies the Analyst's Objectives in Applying it

Tool builders should consider the role which the tool will play. For example, is it a rapid prototyping tool or an evaluation tool? The products required from each of these, although not entirely independent, are quite different.

Rapid prototyping tools should ideally provide the designer with a product which can be a direct input into the design process. For example, in the area of display software design it should be possible to use the software for the prototype to develop the implementation program. Since most tools will use a higher-level language than would be employed for system design this will necessitate the creation of some kind of software translation package. Systems of this type are already emerging and will undoubtedly be an important aid to the interdisciplinary communication process.

Evaluation tools on the other hand, operate in parallel to the mainstream design activity and provide an indirect input to it. The primary concerns here are for completeness and validity. Firstly, the tool should have everything needed to fulfil its role. SAINT and MicroSAINT are examples of incomplete tools which rely on the analyst to supply the most crucial part of the task model, the execution times for the individual actions. Secondly, if the tool already has some kind of science base this must contain reliable and valid data. The problems which British Aerospace experienced with CAR II appeared to have been due to validity problems in the way the program handled certain reach calculations and in the way it related the operator to his workstation.

As an alternative to human simulation, rapid prototyping techniques offer the possibility of interface simulation which is simple and cost-effective. The use of such interface simulations for empirical evaluation is considered by some to represent a sufficient approach to interface evaluation; especially in the area of semantic compatibility where user modelling techniques are not yet fully proven. While this is true to some extent it cannot be considered to be a complete approach. For example, it does not overcome the problems of aircrew availability and subjectivity which were mentioned previously. The ideal solution would appear to be a rapid prototyping tool which also has built-in evaluation functions.

GOAL 3: Ensure Usability

It cannot be overstressed that it is essential that tool builders consider the usability of their product by the targeted user group. Clearly, usability will be affected by both person and machine constraints. A tool's usability may be severely restricted by the fact that it is machine specific; this was the case with SAMMIE. Developers should anticipate the machines and languages available to their targeted users and tailor the systems accordingly.

With respect to usability by the end user, developers should avoid forcing the analyst to relive the agonies suffered during the development of the model itself. ORACLE is a case in point here: the physical model has to be tailored to every application and this has to be performed at the level of rewriting the aspects of the program which relate to it. Consequently, this is a task which can only be performed by a few experts. The program also demands that the user understands the ORACLE equations and is familiar with the engineering constraints relevant to display design. These factors severely limit the usability of the model.

SAINT also demands that the user has an appropriate level of programming expertise. Developers are encouraged to remember that for many potential users this expectation is often unrealistic and could result in the tool remaining unused. Developers should design tools so as to be less demanding of their users. Extensive use should be made of on-line help facilities and, where appropriate, developers should capitalise on new interactive techniques such as adaptive interfaces.

GOAL 4: Provide After Sales Service

Even when the previous three goals have been achieved there is no guarantee that the tool will function free of problems. This is especially true with these kinds of scientific applications where the user may need the program to be customised to his particular requirements. At the very least users need to have a hot-line to the developers with a sympathetic ear at the other end. In addition to this, however, the tool builders should make every effort to produce a complete and fully tested user guide.

REFERENCES

Anderson, J. R., 1983, "The Architecture of Cognition," Havard, Cambridge M.A.

Card, S. K., Moran, T. P., and Newell, A., 1983, "The Psychology of Human-Computer Interaction," Lawrence Erlbaum Ass., New Jersey.

Cooke, K. J., 1984, "The ORACLE Acquisition Model Equations," British Aerospace plc, JS 10090, Bristol.

Howson, J., 1981, "Evaluation of 'SAMMIE' Computer Man Modelling System," British Aerospace plc, M.D.D. 015, London.

Johnson, D. F., 1987, "A Model to Predict Visual Interface Performance at the Man-Display Interface in the Cockpit," British Aerospace plc, Bristol.

Johnson, P., Diaper, D., and Long, J. B., 1985, Task analysis in interactive system design and evaluation, *in*: "Analysis, Design and Evaluation of Man-Machine Systems," G. Johannsen, C. Mancini and L. Martensson, eds., Pergamon Press, Oxford.

Johnson, S., 1982, "Hawk VTXTS - CAR II Assessment of Crewstation Accommodation," British Aerospace plc, BAe-KSE-N-HAR-2450, London.
Keane, M., and Johnson, P., 1987, Preliminary analysis for design, in: "People and Computers III," D. Diaper and R. Winder, eds., Proceedings of the 3rd conference of the British Computer Society Human-Computer Interaction Specialist Group, Cambridge University Press, Cambridge.
Kieras, D., and Polson, P. G., 1985, An approach to the analysis of user complexity, International Journal of Man-Machine Studies, Vol. 22, pp. 365-394.
Laughery, K. R., Drews, C., and Archer, R., 1986, A MicroSAINT simulation analysing operator workload in the LHX helicopter, in: "Proceedings of NAECON Meeting," Dayton, Ohio.
Moran, T. P., 1983, Getting into a system: external-internal task mapping analysis, in: "CHI'83 Proceedings," ACM Publishers, New York.
Moran, T. P., 1981, The command language grammar: a representation for the user interface of interactive computer systems, International Journal of Man-Machine Studies, Vol. 15, pp. 3-50.
McDonald, J. E., Stone, J. D., Liebelt, L. S., and Karat, J., 1982, Evaluating a method for structuring the user-system interface, in: "Proceedings of the Human Factors Society, 26th Annual Meeting."
Pew, R. W., Sidner, C. L., and Vittal, J. J., 1980, MMI design documen-ation: representing the user's model of a system, in: "Proceedings of the Human Factors Society, 24th Annual Meeting."
Sharratt, B., 1987, The incorporation of early interface evaluation into command language grammar specifications, in: "People and Computers III," D. Diaper and R. Winder, eds., Proceedings of the 3rd conference of the British Computer Society Human-Computer Interaction Specialist Group, Cambridge University Press, Cambridge.
Sticha, P. J., 1987, Models of procedural control for human performance simulation, Human Factors, 29:4.
Strieb, M. I., Glenn, F. A., and Wherry, R. J. Jr., 1978, "The Human Operator Simulator: HOS Study Guide," Technical Report TR-1320, Vol. 9, Analytics, Willow Grove, P.A.
Tullis, T. S., 1985, A computer-based tool for evaluating alphanumeric displays, "Proceedings of INTERACT'84," IFIP, Amsterdam.
Wortman, D. G., Duket, S. D., Seifert, D.J., Hann, R.L., and Chubb, G. P., 1978, "Simulation Using SAINT: A User-Oriented Instruction Manual," (AMRL-TR-77-61), USAF Aerospace Medical Research Laboratory, Dayton, Ohio.

THEORY-DRIVEN DESIGN RESEARCH

Stuart K. Card

Xerox Palo Alto Research Center
Palo Alto, California

INTRODUCTION

In the computer industry, it is commonly assumed that breakthroughs in interface design are, and will be, largely a matter of intuition, possibly aided by some loose empirical observations in the manner of rapid prototyping. ("Many of the great breakthroughs of science and invention were the result of intuitive leaps," Tagnazzini in Herot, 1987) . In this view, there is not much useful that cognitive scientists can do for advancing the development of the technology--the most that can be hoped for is some evaluation of other peoples' designs.

Yet, if it can be said that many of the breakthroughs of technology were the results of intuitive leaps, it can equally be said that many were not. The question is not whether intuitive design will play a role in the progress of interface design--of course, it will. The question is, rather, can we ever do better than just pure intuitive design for human-machine interaction? Is that all that is possible? For the moment, let us group with intuitive design not only "seat-of-the-pants" design, but also pure empirical testing as in rapid prototyping.

This question is really a version of the broader and older issue concerning the relative effectiveness for technological progress of cut-and-try engineering vs. science. It might be thought that by this time the history of technology itself would provide strong patterns that would settle the matter. The problem is that the history of technology provides *many* patterns. Almost all interactions are found in one history or another (Ziman, 1976).

Take, for example, the bicycle. There is some rather subtle physics in the bicycle. If the front fork is bent slightly up, as in almost all modern bicycles, then the bicycle is stable. If, however, it is bent the other way, then the bicycle is unridable. But the bicycle evolved to its mature form almost exclusively by cut-and-try engineering. Science, even though demonstrably relevant, was not an important driver of technological progress. On the other hand, one would never have gotten the transister with only cut-and-try engineering.

So, as a sort of intellectual first-aid until the historian arrives, consider the following: Science is likely to pace technological progress when:

1. Technology gets stuck, or
2. Science can identify leverage points or key constraints--places where modest effort suddenly unlocks progress, or
3. Science provides tools for thought, either conceptually or mechanically. That is, science can be effective when it helps us see a new way for conceptualizing the design space or it allows us to make inventions that themselves aid intuitive design.

THE THEORY-DRIVEN DESIGN RESEARCH PARADIGM

To take the next step in this argument, it is important to distinguish research from product development. Research is about advancing the state-of-the art. Development is about making things economically that satisfy some goal. The important thing to understand is that techniques too expensive to be used directly in product development can nevertheless be successfully used in research--and can lead to commercially successful products.

An example is the Xerox "TypeRight (TM)" system that goes into a typewriter and squeezes 100,000 English words, more or less, plus all the code for the algorithms to check spelling into 64K bytes--more than twice as compressed as any other system. The system was done by Ronald Kaplan, a psychologist, and is based on both theoretical linguistics and theoretical computer science. One would not want to be the hapless engineer stuck with the task of trying to reverse engineer this device without knowing the theory.

TypeRight is an example of how science, cognitive science even, can contribute to the march of technological progress. In this case, there is a cognitive science theory that provides a leverage point. It would normally be impractical to develop this theory and technology directly as part of the critical path to a product. *But once in hand*, the theory makes it possible to zip out products faster than cut-and-try engineering because one can skip a lot of the cuts and the tries. There are now versions of the TypeRight technology that work with popular word processing programs, a version that connects directly to a PC keyboard, and a pocket version with a sleek case that can be purchased out of airline in-flight catalogues. Furthermore, TypeRight was developed in 6 months, from idea to product shipment, demonstrating that far from being a leisurely luxury, theory can be even quicker than "quick and dirty" engineering methods.

The example of TypeRight is an example of a sort of paradigm for using science to advance systems technology. This paradigm was first suggested by John Seely Brown and has been used with some success. In this paradigm, we try to get a combination of four things:

1. a well-defined *problem*,
2. a *theory* that gives insight into the problem,
3. an *artifact* that embeds the theory in the service of the problem, and
4. a *reuse* of the theory or technology for solving other problems.

This last part, reuse, is especially important for focusing on the leverage points of science for technology (and, incidently, helping to validate that the theory really is a substantial insight into nature). For shorthand, let us refer to this paradigm as the *theory-driven design research paradigm.* It is primarily a way of organizing human-machine systems *research.* But a frequent side benefit is a prototype for product *development.* Examples of the use of this paradigm show how research and theory can augment intuitive design for advancing interface technology.

Example 1: The Mouse

A first example is the mouse. The *problem* is to make a device that allows reference to parts of the interface by pointing. The *theory* of pointing in this case is just Fitts's Law, which tells us the time to point with the hand is a simple log function of the ratio of the distance to the target diameter:

$$Movement\text{-}time = constant + I_D \log_2 (D/S + 0.5)\,.$$

The *artifact* is the mouse. The discovery that pointing with the mouse is also governed by Fitts's Law and with approximately the same slope parameter I_D, tells us that the limitations on pointing speed are in the human information processing, not in the mouse device (Card, Moran, & Newell, 1983). Hence there are not dramatic gains to be made in pointing time by other devices. This argument was, in fact, used directly in deciding on the commercial introduction of the mouse. The theory can be *reused* to tell us how to do rapid testing of mice and other pointing devices as well as how to make devices superior to the mouse.

Here science has been used to reconceptualize the design space: a winning device should have a lower curve on the Fitts's Law slope than the mouse, or be the same but be cheaper, lower error rate, etc., or be able to beat the mouse on some other task, such as fine drawing.

Example 2: The Rooms Window Manager System

Another example is the design of the Rooms system (Henderson & Card, 1987). Here the *problem* is overhead in using windows. The *theory* is a version of the Denning virtual memory working set theory and successors. In terms of the theory, the reason there is so much overhead, especially with overlapped windows, is an intense resource contention brought on by the small size of the screen. This resource contention causes "window faulting"--scrolling, popping overlapped windows to the top, moving, or resizing windows in order to see wanted information that is not currently showing. But windows exhibit "locality of reference" like memory locations in a computer program (that is, most user references are to just a few recently referenced windows). Furthermore, references tend to come in clusters, organized around some task like reading mail. Switching to another task changes the set of windows likely to be referenced. This analysis suggests an *artifact*, the "Rooms" system, that exploits locality of reference by analogy with techniques used in virtual memory operating systems to reduce average access time.

Rooms is a window manager system that extends the desktop metaphor. Each Room is a cluster of windows on the screen at the same time. Each Room tends to be identified with some task the user does, such as read his mail or write a particular paper or work on a particular program. Actually, Rooms is a window analogue of a virtual memory preloading policy, accomplishing the inevitable window faults automatically more rapidly than a user could manually by expanding or shrinking icons on demand. The theory can be *reused* to analyze user task switching. Studies of users show that they shift frequently among different tasks (Bannon et al, 1983), behavior not well supported in most workstation environments. The same basic theory can be reused to show that the Rooms system provides an excellent basis for supporting task-switching (Card & Henderson, 1987). Here theory has provided the conceptual tools of thought that make the invention of a novel artifact and its extension relatively straightforward.

Example 3: Mackinlay's Thesis--APT

A last example is Jock Mackinlay's PhD thesis (Mackinlay, 1986). The *problem* is to construct an interface that automatically presents data. The *theory* is a formal mathematical model of the form and semantics of a graphical data presentation: Consider a set of data, such as car fuel mileage as a function of car price. There are a large number of ways these same data may be presented: scatter plots, bar graphs, pie charts, etc. Each of these can be thought of as a sentence in a graphical language. Mackinlay's theory formalizes two important notions: (1) what it means for a graphical presentation to express some meaning, and (2) what it means for a graphical presentation to be effective doing it. For example, the meaning of the points and axes of a scatter graph of car price against mileage would be formally described in terms of relations between dataset and graphical elements in the expression like the following:

Encodes (*s*, {*Price* (*Accord*, 5799), *Mileage* (*Accord*, 25)}) iff
Point(*s*, *Pnt*) AND
Axis(*s*, *HorizAxis*) AND
Axis(*s VertAxis*) AND
Encodes(*Pnt*, *Domain*(*Accord*)) AND
Encodes(*HorizAxis*, *Domain*(5799)) AND
Encodes(*VertAxis*, *Domain*(25)) AND
Encodes(*Position*(*Pnt*, *HorizAxis*), *Price*(*Accord*, 5799)) AND
Encodes(*Position* (*pnt*, *VertAxis*), *Mileage*(*Accord*, 25)) .

A presentation is said to expresses an intention if and only if the graph encodes all the relations intended and none that are not. One of Mackinlay's insights was to note that graphical design is hard because while it may not be difficult to encode an intended meaning in a graphical representation, it often is very difficult to prevent the graphics from conveying additional meanings not intended.

Even if the intended meaning and only the intended meaning is conveyed, graphical presentations will still differ, according to Mackinlay, depending on how "effective" they are. Using psychophysical data, he derived tables of relative effectiveness for different encodings of the data. For example, length is relatively effective for showing quantitative data, but relatively ineffective for ordinal data. The reverse is true for color hue.

Mackinlay's formalism allows him to develop a composition algebra for combining the elements of a graph based on his formal semantics and his tables of relative effectiveness. That is, this theory structures the design space for graphical presentations and even makes it possible to generate figures of merit for the points in the design space.

The *artifact* that the theory enables is a system called APT (for A Presentation Tool) that automatically generates reasonable presentations of data for users. *Reuse* of this theory looks promising for other aspects of interface design. Here again, we have a case where theory can be used to structure the design space, this time giving enough structure actually to permit automatic design.

In the examples above, it was possible to improve upon intuitive design by using science to get insight into the structure of the problem or design space. In each of the examples, it was possible to point to an artifact that if not already a product is readily made into one. To make the argument that it is possible to do more than just intuitive design, it is not necessary to argue that this is possible every time, only that it is some times and that these get more frequent as the science matures.

COMPUTATIONAL ERGONOMICS

In the cases above, theory was used during *research* to advance the state of the art. It was then used analytically at design time for product development. Mackinlay's thesis suggests the possibility of using theory more directly, computationally, in workstations for computer-aided engineering. Making this a reality is very demanding of theory to the point of interacting with the development of theory itself: Theory-driven design often benefits from design-driven theory.

In many disciplines, computers are making a fundamental difference in the way science is done and in the relationship between science and application. In fact, it has been suggested that a whole new paradigm of computational science is evolving, enabling discoveries through computation in contrast to theoretical and experimental science (Bell, 1988). This new paradigm is reflected in new, computationally-oriented branches of old disciplines. Thus we have computational chemistry and computational linguistics. We have as well at least the possibility of computational ergonomics.

Of course, there have been attempts to use computers to simulate human performance in the service of designing human-machine systems for many years now. SAINT and other network models, the various dialects of time-line analysis, and HOS are examples. But, the rapid development of computing technology and increased scientific understanding in the human sciences mean that it may become feasible to build much better, and much more widely accessible, tools as part of engineering workstations. The most recent attempt is the A^3I project pursued jointly by NASA and the Army for designing helicopter cockpits. An estimated 70% of the lifecycle cost of a helicopter is determined in the conceptual design phase, well before it is possible to build prototypes or do much in the way of human testing. Most of the important part of the design is therefore unreachable by the classical prototype and test paradigm. NASA is trying to build a CAD/CAM workstation in which embedded human performance models would allow information about human performance to be predicted at conceptual design time, leading to better designed systems and filtering the things that have to be tested with simulation. The idea is to try to see if it is possible to do more of the human engineering up front by doing computational ergonomics.

To help choose the human performance models for this workstation, NASA sponsored a study (ongoing at the time of writing) by the Committee on Human Factors of the National Research Council, the operating arm of the National Academy of Sciences. This is one step towards a computational ergonomics. The task is not an easy one. Classes of models considered, for example, include the following:

Motion detection	Surface quality from shading
Temporal detection	Shape from shading
Spatial detection	Object recognition
Spatial coding	Work load
Optical flow	Learning
Occlusion contours	Decision making
Reflectance contours	Knowledge representation
Edge detection	Errors

Working on such studies gives one a chance to assess the promises and difficulties of computational ergonomics as a method of theory-driven design. From this vantage point, three difficulties suggest themselves: (1) we need a better understanding of design, (2) many theories will never be useful for design, and (3) transfer of theoretical knowledge to practice is hard for most engineering disciplines.

1. We Need a Better Understanding of Design

The tacit assumption in the modeling world has always been that science and modeling is hard, but that if one can get some results the design, while a little scruffy, is relatively easy. In fact, one of the greatest difficulties in this National Research Council study has been in figuring out what designers could use models for and what is a reasonable design methodology that would incorporate models. Theorists have been observed to undergo severe culture shock when required to extract from their model or theory some piece of information necessary to answer some design question. Systems builders have been observed to ask for models without a clear idea for what or how they would use them, not realizing that the details of use are what determine which idealizations in the models are viable or impractical. But the needs of design determine tradeoffs in models and somewhat vice versa. Furthermore, the process of design and even the structure of a design in terms of its rationales are not well understood. Even if human performance models were demonstrably useful in design, CAD/CAM workstations containing computational ergonomics might be beyond reach because of lack of knowledge for how to make it a practical part of the design process. There is no possibility of a viable computational ergonomics without a better understanding of design.

A confusion is that there are at least two related understandings of design that are ultimately needed: (1) understanding of *design in the small* and (2) understanding of *design in the large*. Design in the small refers to design techniques at the engineering analysis level. Design in the large refers to the organizational and social issues of organizing the design activities of large groups of people. These two are easily confused because an engineering artifact may serve roles in both areas simultaneously. For example, an engineering diagram serves as both the external memory to keep track of details for an analysis and also as the means of communicating the design and its constraints around an organization. Both are important, but the notion of theory-driven design in this paper is concerned principally with the former, where the direct impact of models and theories would be expected to be felt.

2. Many Theories Will Never Be Useful for Design

One view of the relationship between science and application is that science is like a fruit tree. The fruit on the various branches eventually will ripen. If a human performance model is not very useful yet, just water the tree, give it sunshine and be patient; eventually the fruit will fall into one's hands. Unfortunately, a look at what is available in various models against what is needed to do engineering suggest this view is not accurate, and for five reasons:

(1). *Content validity.* The problem with many psychological theories is, as Andries Sanders has pointed out (Sanders, 1983), content validity. Small-scale laboratory experiments do not necessarily produce true theories that brick by brick build the house of knowledge. And unfortunately, ecologically valid observations and simulations are often so hopelessly confounded that it is impossible to determine what led to what. Sanders prescription of back-to-back experiments--a controlled experiment, an ecological experiment, and a formal argument for their relations--is still an excellent idea. Otherwise, the chances are that some small laboratory effect will be swamped by other effects in the real environment or be artifactual. Of course, lack of content validity can have highly visible consequences during design. Theory-driven design is one tactic for going after content validity of psychological theory.

(2). *Conflicting idealizations.* Engineering use and scientific theory-building sometimes have different aims resulting in conflicting idealizations. Engineering uses often require broad coverage of phenomena, whereas scientific models are often aimed at uncovering the mechanisms behind narrow phenomena. Woods et al (1986) have noted the

> *. . . tradeoff between the formal, applicable, and scope dimensions of models. In general for the behavioral sciences, the more formal a model, the narrower the coverage of and applicability to real world tasks. (Woods, Roth, and Hanes, 1986, p. 29).*

The use of idealizations means simplifying a phenomena so that inference about it is tractable. The simplification is achieved by dropping out details that will have little effect on the outcome. But which details will have an effect may depend on whether one is interested in broad coverage or in subtle mechanisms. The theory-driven design research paradigm is a heuristic for keeping in the idealizations of the theory those details that will matter for some class of design.

(3). *Model integration.* Isolated human performance models cannot necessarily be integrated together to give a larger model. Trying to match up the miscellaneous hodge-podge of inputs and outputs into an integrated whole is one of the chief problems in trying to build a computational ergonomics out of isolated models from the literature. This is the real advantages of global models that attempt to explain multiple phenomena.

(4). *Lack of formalization.* Verbally stated models have their use, but that use is not in computational ergonomics.

(5). *Lack of use pressure.* It is rather unsettling that even the authors of many psychological models seem to assume no one will ever actually *use* their models for anything. This would seem to be a departure from most sciences. Since the models are not used, there have not been the iterative refinements to make them usable easily. In fact, there are extreme circumstances where one suspects the real knowledge may be less in the model and more in the unexplicated skill of the modeler.

The above are really, as much as anything else, matters of attitudes on the part of the theorist/modeler. One implication is that the way to get human performance theories that are usable for design is specifically to fund theoretical activities that are designed for engineering use: *design-driven theory.* This is not an argument that all theory should be of this sort, only that some should be. Actually, the history of cognitive psychology has shown that research on applied problems has been one of the most effective ways of advancing the basic science.

3. Transfer of Theoretical Knowledge to Practice is Hard for Most Engineering Disciplines

Rouse and Cody (1989) have shown that the route from basic research to engineering use of human performance research in the aerospace industry is lengthy and uncertain. Among other things, most design practitioners are not technically able to read the academic journals. Most design questions are answered not from the literature, but from colleagues with nearby offices. Notions of computational ergonomics or theory-driven design need to take into account the realities of actual design practice. While Rouse and Cody's study was of aerospace human engineering, in fact it would be surprising if the situation were not somewhat similar in other engineering disciplines as well. This fact suggests particular attention to the method in which theoretical knowledge for a supporting science is packaged for the practicing engineer.

Incremental, cut-and-try intuitive design is an important part of human interface design development, but it is not the only way to make progress. Theory-driven design research is also part of the engine of technological progress, and we now have instances of it as a successful method. In fact, both approaches can be used together. What theory-driven design research can do is to help the designer reconceptualize the design space, it can change the very way he or she thinks about the problem, it can show the designer where the points of leverage are or how to overcome difficulties insurmountable by cut-and-try techniques alone, it can show the designer how things work. In this way each cut-and-try can be a bigger step with higher likelihood of hitting the sweet spot in the design space. Theories also have potential for being embedded in design tools directly as part of a sort of computational ergonomics. Three difficulties currently arise in attempting to do this: (1) we need a better understanding of of design, (2) many theories will never be useful for design, and (3) transfer of theoretical knowledge to practice is hard for most engineering disciplines. All are matters that would reward targeted research.

REFERENCES

Bannon, L., Cypher, A., Greenspaan, S., a& Monty, M. L., 1983, Evaluation and analysis of users' activity organization, transcript of talk delivered at

the ACM CHI '83 Human Factors in Software Conference, SanFrancisco, December, 198).
Bell, G., 1988, Another member of the supercomputer faimily--the graphics supercomputer, lecture at Stanford University, September 8, 1988.
Card, S. K., Moran, T. P., & Newell, A., 1983, "The Psychology of Human-Computer Interaction," Hillsdale, Lawrence Erlbaum Associates, Hillsdale, New Jersey.
Card, S. K., & Henderson, A., Jr., 1987, A multiple, virutal-workspace interface to support user task switching, in: 'Proceeding of CHI '87 Human Factors in Computing Systems ", 53-59, ACM, New York.
Henderson, A., Jr & Card, S. K., 1987, The use of multiple virtual workspaces to reduce space contention in a window-based graphical user interface, ACM Transactions on Graphics.
Herot, C. F., 1987, Where do user interfaces come from, SIGGRAPH Conference Proceedings, reprinted in Computer Graphics, 21(4):321-324.
Mackinlay, J, 1986, Automating the design of graphical presentations of relational information, ACM Transactions on Graphics , 5(2):110-141.
Rouse, W. and Cody, W., This Volume.
Sanders, A., 1983, in: "Attention and Performance X"., Lawrence Erlbaum Associates, Hillsdale, New Jersey.
Woods, D, Roth, and Hanes, L.,, 1986. Technical Report. Pittsburgh, Pennsylvania: Westinghouse Corp.
Ziman, J., 1976, "The Force of Knowledge," Cambridge University Press, Cambridge, U.K.

A TEST OF CRITERIA USED TO SELECT HUMAN PERFORMANCE MODELS

William J. Cody and William B. Rouse

Search Technology, Inc.
4725 Peachtree Corners Circle, #200
Norcross, Georgia 30092

INTRODUCTION

Earlier in this volume, Rouse & Cody (1989) discussed the nature of complex system design and designers' use of human performance models. We noted that in our experiences designers show little interest in using the types of analytical model that were the subject of the present workshop. These impressions are based on interviews that we conducted with over 60 crew system designers in field studies of the aerospace system design process (Cody, 1988; Rouse & Cody, 1988). We hypothesized that designers apply seven specific criteria when evaluating information sources, and that models are generally perceived to be weak along these criteria relative to their alternatives.

The present paper is a companion report that has three purposes. First, we sought to test Rouse and Cody's hypotheses about the criteria that designers apply when choosing sources of human performance information. Second, we wanted to explore the characteristics of present and potential model users. Third, we hoped to provide feedback to model developers on how they might increase the attractiveness of their products to designers.

Background

To reiterate briefly the basis for the earlier paper, our interview studies demonstrated to us the enormous variety of technical information, including information about human performance, that aerospace designers access during the course of producing a crew system. Designers said they acquire human performance information from three principal sources: human judgment, the archives and models. Human judgment refers to the designer's own experience and recollections as well as the opinions of colleagues, domain experts and system users. The archives include past designs, standards, practices, regulations, and the scientific, technical and trade literatures. Models include both empirical studies and analytical tools. In empirical studies, the system is simulated with more or less fidelity but the system operator or maintainer is a human subject who is "in-the-loop." In contrast, analytical models refer to computer-based simulations in which both the system and human components are represented computationally or symbolically.

Designers also claimed that they do not access these three sources equally often for human performance information. Rather, human judgment is by far the most frequently accessed source followed by the archives with models a distant third. Within models, empirical studies are used more often than analytical models to produce human performance information.

Table 1. Criteria that Designers Apply to Select Information Sources (Rouse & Cody, 1989).

Applicability:	Does the source produce information that is directly relevant to the problem?
Credibility:	Do co-workers and customers perceive information from the source to be valid?
Availability:	Is the source commercially accessible (versus in an experimental state or proprietary)?
Cost:	Is the financial cost of obtaining and using the source acceptable?
Interpretability:	Are outputs from the source directly usable or easily transformed for the purpose at hand?
Learnability:	Can one become proficient in producing information with the source in an acceptable period of time?
Usability:	Once mastered, is the source easy to configure and use?

Why do analytical models not compete well with these other sources? In Rouse & Cody (1989), we suggested that designers' information seeking behavior might be understood from the perspective of consumer psychology (cf. Bettman, 1979, 1986). Consumers have been shown to trade off the expected benefits and costs of seeking information to support purchase decisions (e.g., Meyer, 1982). Furthermore, the likelihood that consumers will access a particular source has been shown to vary with source attributes (Meyer, 1981). As information consumers, designers' reluctance to use models, therefore, suggests that these sources are perceived to compare poorly with other sources in terms of a cost-benefit relationship.

In Rouse & Cody (1989), we hypothesized that this relationship is composed of the seven criteria defined in Table 1. Moreover, we suggested that system developers apply these criteria as a series of gates in roughly the sequence given. That is, if an information source fails to exceed some threshold value along a criterion, further consideration along the remaining criteria is suspended and an alternative information source is then sought.

The present workshop appeared to offer an excellent opportunity to test these claims. We assumed that workshop speakers were primarily model developers who were interested in having their tools and techniques applied in system development efforts. We also assumed that the audience for the most part would consist of system developers who were seeking models and methods to apply in their design work. As the data will show, our assumptions were not fully met; about half of the workshop attendees described themselves as researchers, not system developers. While not conclusive for this reason, the study did suggest factors that govern the use of models in system applications.

To take advantage of this opportunity, we developed a questionnaire for attendees to complete as they listened to the papers presented during the workshop. Due to time constraints, neither the speakers nor the audience had seen the questionnaire in advance. In addition to requesting background data, the questionnaire contained an evaluation form constructed around the seven criteria. Respondents were asked to evaluate each model and method along the seven criteria, e.g, "how applicable is this model to your needs?" We used the data from this exercise to explore five primary issues.

Study Objectives

First, we were interested in whether designers agree with the assertion that they select information sources based on the seven criteria. We established this by asking them to rate the importance of each criterion to their selection of information sources as part of their work. As discussed further below, their agreement does not guarantee that designers in fact apply these criteria. The psychological literature in judgment, decision making (e.g., Dawes, 1979; Nisbett, Krantz, Jepson, & Kunda, 1983) and self-report (e.g., Nisbett & Wilson, 1977) is replete with findings that show individuals' behavior and their explanations for their behavior often conflict. However, if respondents roundly disagreed with our notions of what motivates their information seeking behavior, less confidence would attach to any advice we could offer model developers.

Second, based on attendees' ratings, how did the models fair along the seven criteria? Arguably, the papers and demonstrations presented at this workshop represented the state of the art in human performance modeling. Moreover, having invested the resources to attend this workshop, the members of the audience should represent the population most likely to apply this technology if they perceive it to be worthwhile. Hence, ratings assigned to the models by this group merit close attention by model developers.

Third, we were interested in how respondents' backgrounds affected ratings. Therefore, we explored the relationships between ratings and educational discipline, current job role, level of model use, and pre-workshop familiarity with models. Model developers should find such information useful for tailoring presentations of a model to audience characteristics. For example, empirical results from consumer research show information seeking and product selection are related to familiarity with a product category (Kiel & Layton, 1981; Punj & Staelin, 1983; Reilly & Conover, 1983) and uncertainty about product attribute values (Meyer, 1981).

Fourth, of special interest was whether workshop attendees were likely to act on the basis of information provided by the speakers. More specifically, for each paper we asked respondents to judge how likely it was that they would: a) request more information about the model that was discussed; b) advocate its use to others; and c) personally use the model. These three types of behavior represented increasing levels of commitment on the part of the respondent. Our interest in these behavioral expectations was twofold.

The behavioral ratings would indicate the extent to which the workshop stimulated interest in human performance modeling. Without a follow-up study to determine whether attendees did, in fact, act as they said they would, we cannot verify their claims. However, the behavioral ratings should provide some insight into the workshop's impact.

We also sought to determine how well the ratings people assign to the models predict how they expect to behave. For example, if an individual gives a model high marks for applicability and credibility, but poor marks on the remaining criteria, will he request more information about the model or not? This analysis tested our notion that designers' apply the seven criteria sequentially. More importantly, however, the analysis detected discrepancies between what the respondents said were important criteria and which criteria they would actually use to act. Hence, this particular issue has psychological significance beyond the results of the workshop.

Under the fifth issue, we were concerned with how to stimulate wider adoption and use of human performance modeling technology. On which criteria should model developers concentrate? Which criteria are most highly related to behavioral expectations? Which criteria tend to be positively and which negatively associated with a model or technique?

METHOD

Questionnaire Design

The 5-part questionnaire solicited the following information. Numbers of questions and ratings per part are shown in parentheses:

1. Background data (20 questions).

2. Extent to which the Rouse & Cody criteria were perceived to be important in selecting information sources (7 ratings). These ratings were subsequently used in statistical models to weight respondents' ratings of workshop papers.

3. Ratings of individual models along the seven criteria and ratings of the likelihoods of three types of behavior (29 papers X 10 ratings = 290 ratings). Figure 1 shows a section of the evaluation form.

4. Ratings of model groups that were discussed in each session (6 sessions X 7 ratings = 42 ratings).

5. Workshop review and critique (6 ratings).

For present purposes, analyses are reported for the first four parts. Hence, a completed questionnaire contained 359 items of interest. The form also contained a "service card" that attendees could fill out to request more information about particular models and demonstrations.

Where ratings were requested, a 5-point scale was used throughout the questionnaire. End-point values differed as a function of the question (e.g., never-often; unimportant-very important; poor-excellent).

SESSION I
WORKSPACE DESIGN -- ANTHROPOMETRIC AND BIOMECHANICAL APPROACHES

(1) (2) (3) (4) (5)
Unimportant-------> Very Important

1 2 3 4 5 **12. How important are the human system <u>design issues</u> that have motivated the development of this group of models and techniques?**

13. How does the model or technique discussed by each presenter in this session rate on each of the criteria proposed by Bill Rouse? (Write your ratings in the appropriate cells of the matrix.)

(1) (2) (3) (4) (5)
Poor--------------------------> Excellent

PRESENTERS

CRITERION	Bonney	Kroemer	Badler	Rothwell	Evans
Applicability					
Credibility					
Availability					
Cost to Access					
Interpretability					
Learnability					
Usability					

Based on what you have heard in the presentation, how likely do you consider each of the following behaviors?

(1) (2) (3) (4) (5)
Unlikely----------------------> Very Likely

Likelihood you will request more information on this model or technique					
Likelihood you will advocate the use of this model or technique to your colleagues					
Likelihood you will use the model or technique yourself					

Figure 1. Example of the evaluation form

Data Collection Procedures

Attendees learned of the evaluation task at the introductory session of the workshop. Moreover, although workshop speakers were informed that an evaluation of current modeling technology would be conducted, they did not know beforehand the criteria definitions nor that the evaluation would be conducted in the manner that it was. Papers and talks were not prepared specifically to address the criteria and, therefore, may not have contained all of the information that the audience members felt they required to formulate a judgment.

The Rouse & Cody (1989) presentation discussed the criteria as well as the rationale for the evaluation task. The questionnaire was then distributed and specific instructions provided for its completion. We emphasized to the audience that they should respond to the questions and ratings from the perspective of their personal interests and needs, and not as they thought designers in general might respond. Also, if a speaker presented insufficient information to make a judgment along a criterion, respondents were instructed to leave the rating blank. Respondent anonymity was protected through a coding scheme.

After the instructions, the audience completed the background data and rated the importance of the seven criteria to their source selection behavior. The remainder of the evaluation form was then filled in over the course of the workshop. The procedure was to record judgments for each of the 29 papers/models immediately following its presentation. At the end of each of the six sessions, respondents rated the models as a group. Review and critique items were answered at the close of the workshop. Reminders to complete the evaluation form were provided at regular intervals.

Data Reduction and Analysis

Our primary thrust was to provide the model development community with insights on how to effect more model use by designers. Therefore, we examined the data from two different perspectives. On the one hand, increasing the use of models might require enhancements to existing techniques or entirely new types of model. Overall ratings of the models and ratings of behavioral expectations were examined to investigate this possibility.

On the other hand, model use might also be increased by "packaging" existing technology in different ways to enhance the perceived value to people who are not already using models. We explored this possibility in two ways. First, we dichotomized respondents along four dimensions to explore the relationship between opinions about modeling and subject factors. These dimensions were: 1) disciplinary background (behavioral sciences vs. engineering); 2) job role (researchers vs. system developers); 3) expertise with modeling (model users vs. non-users); and 4) pre-workshop familiarity with particular models (familiar vs. unfamiliar). For these comparisons, we computed t tests on the group mean rankings. This practice acknowledges that an interval scale is not a prerequisite to making a statistical inference based on a parametric test (cf. Anderson, 1961; Boneau, 1960). Second, we developed statistical models of the respondents' expected behavior as a function of their ratings along the evaluation criteria. Our interest was in the relative importance of the seven criteria to respondents' expected behavior. Differential importance would suggest on which dimensions the modeling community should concentrate.

RESULTS

Questionnaire Return Rate and Missing Data

Ninety three questionnaires were returned from the 139 workshop attendees, for a 66% return rate. Four were discarded for providing background information but no ratings. Across the 89 usable questionnaires, only 14 contained responses to all 359 items of interest. Response rates for each group of questions were as follows:

1.	Background information	97%
2.	Ratings of Rouse & Cody criteria	90%
3.	Ratings of individual models	68%
4.	Ratings of model groups	76%

Table 2. Percentages of Ratings Provided For Each Criterion and Behavior on Partially and Wholly Answered Questions.

Criterion/Behavior	Individual Papers	Sessions
Applicability	98	100
Credibility	97	99
Availability	90	96
Cost	80	88
Interpretability	93	99
Learnability	91	97
Usability	92	97
Request More Information	96	--
Advocate Use	93	--
Personally Use	94	--

As these data suggest, response rates for ratings of individual models and model groups were comparatively low. Respondents may have failed to rate a paper or group for either of two types of reason. On the one hand, they may have skipped a paper or session, forgotten to bring the form, become tired of the rating task, and so on. On the other hand, respondents may not have received enough information from a speaker or session to form a judgment. For our purposes, if the respondent filled out none of the ratings for a given paper or session, we assumed the first type of reason. If he or she filled out some but not all of the ratings for a particular paper or session, we assumed the second reason. This distinction was important because, for some analyses, we included partial responses in overall average ratings. Thus, average ratings on the criteria and expected behaviors were sometimes based on unequal sample sizes within a paper or session.

The pattern of missing data was informative in its own right. Table 2 shows the percentages of ratings given per criterion and behavior for both individual papers and groups for partial and wholly answered questions. All questions on which individuals recorded none of the ratings were removed from the denominators of these percentages.

For individual papers, three groups of response rates emerged. The top group consisted of judgments of applicability and credibility drew rates in excess of 97%. Hence, respondents received sufficient information from the presentations to judge models along these dimensions. In contrast, cost judgments drew an 80% response rate, suggesting that presentations offered less information to estimate this dimension. The remaining criteria clustered around a 92% rate. Table 2 shows that the cost information deficit appeared in the session ratings as well.

This finding about cost judgments is not surprising in light of the fact that presenters had not been instructed to provide specific information along each of the seven dimensions. It may also suggest that cost to potential users was not a concern to presenters. Perhaps more interesting, however, the data suggest that the audience <u>did</u> judge the cost characteristics of models in the majority (80%) of cases even though specific cost numbers were not given. Hence, it appears that the audience often "filled in" missing information about models, a finding that corroborates results from consumer behavior research (cf. Bettman, 1986; Meyer, 1981).

<u>Background Information on Respondents</u>

Figure 2 shows the distribution of educational backgrounds of the workshop attendees. Approximately 82% of the respondents were trained in engineering, behavioral and cognitive science, or human factors engineering. For subsequent analyses, respondents were divided into two groups according to educational background. The Behavioral

Science group was composed of behavioral/cognitive scientists, human factors engineers and those with degrees in education (n=47). The Engineering group combined engineering, computer science, physical science, mathematics, and biology (n=39).

Figure 3 presents respondents' current employment. The majority (76%) were employed in business and industry or in the Department or Ministry of Defense. Academicians accounted for about 15% of the respondents. Non-defense governmental positions (e.g., NASA, FAA) accounted for 5%. For our purposes, this distribution was encouraging under the assumption that the business and military communities are motivated to use modeling technology toward improved products and not solely to develop models for scientific and theoretical ends. Combined with the disciplinary profile of attendees, these data indicated that workshop attendees would represent a promising mix of potential model users interested in applications to system design.

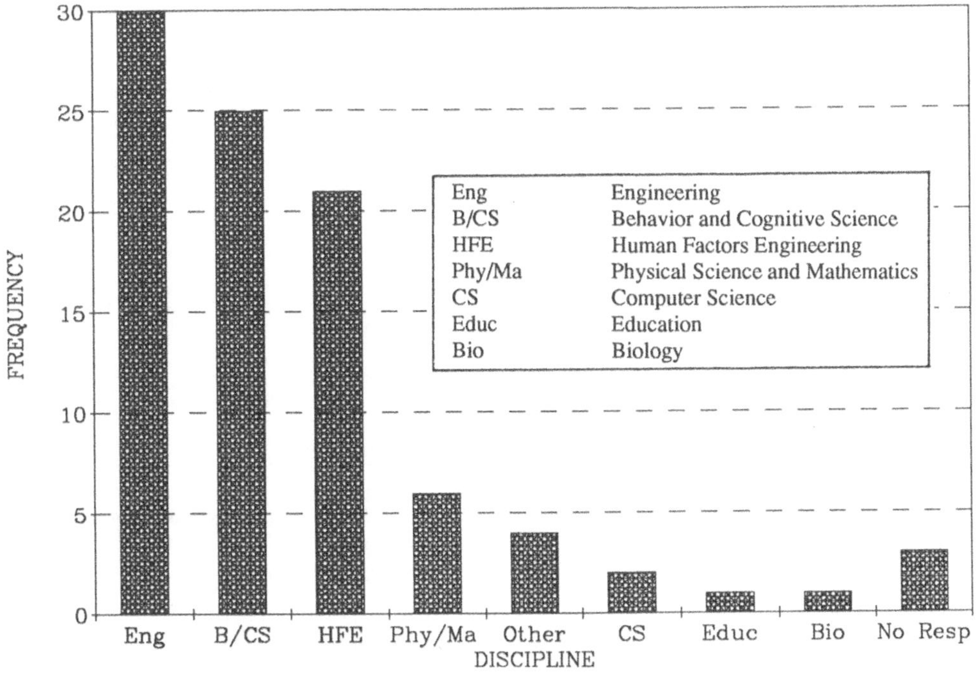

Figure 2. Disciplinary specialties of workshop attendees

The data shown in Figure 4, however, initially reduced our expectations. Over half of the respondents identified research and technology development (R&D) as their primary job role whereas only 15% identified system design (DSG) as primary. These data introduced two types of uncertainty. First, the information selection criteria that we developed from interviews with crew system designers may not be relevant to this group. Second, workshop attendees may not represent the design practitioners that model developers would be most interested in reaching to increase model use in system design. Thus, the selection criteria and model preferences shown by workshop attendees may contain little advice for model developers.

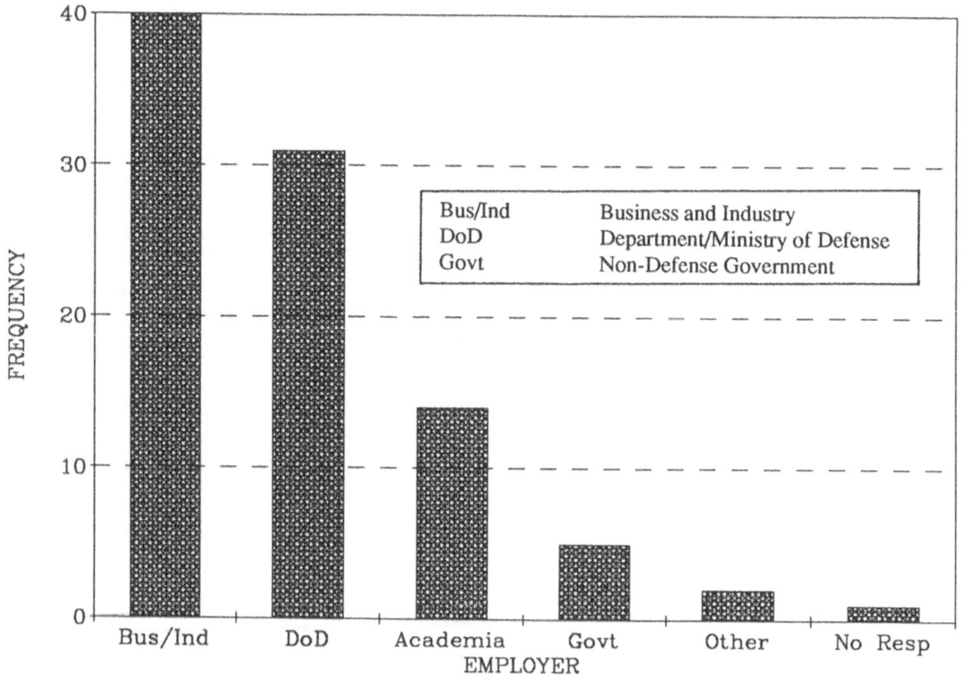

Figure 3. Current employment of attendees

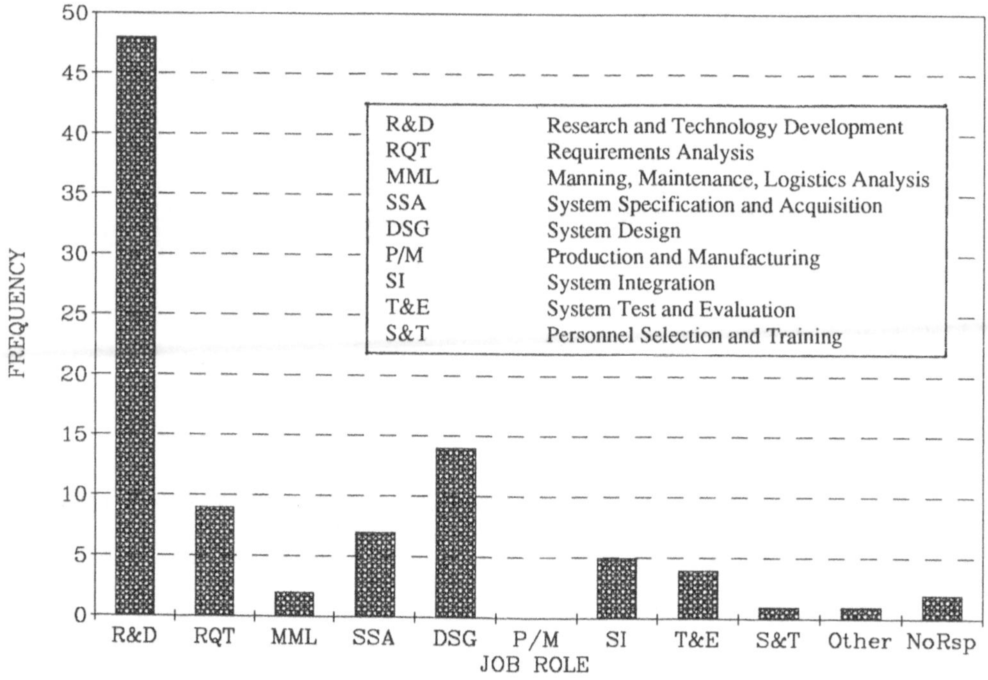

Figure 4. Primary job roles of attendees

Table 3. For Each of Three Questions, Mean Ratings and Mean Rating Differences for Three Pairs of Subject Classifications: Behavioral Scientists vs. Engineers; Researchers vs. System Developers; Model Users vs. Non-users.

			Rating Differences		
Question	N	Mean Rating	Behav -Engr	Resrch -Syst	User -NonU
1. Information Sources?	(1=Never 5=Often)				
Own Experience	88	4.23	.25	.49**	.09
Technical Literature	88	4.18	.21	.43*	.42*
Colleagues	88	3.69	.14	.26	.36
System Users	87	3.41	.46*	-.24	.10
Analytical Tools	88	3.09	.46*	.61**	(2.30)
Empirical Methods	88	2.99	.34	.64**	.23
2. Reason for Attending?	(1=Unimportant 5=Very Important)				
Receive Overview	88	4.55	.09	-.11	-.17
Learn Specific Tools	88	4.28	-.37	-.25	-.12
Determine Cost	88	2.83	.04	-.14	-.51*
Give Feedback	86	2.56	-.10	-.11	-.23
3. Pre-Workshop Familiarity?	(1=Unfamiliar 5=Very Familiar)				
Anthro & Biomech	88	3.07	.34	-.15	.53*
Workload & Task Alloc.	87	3.37	.44	.62**	.64***
Individual Tasks	86	2.98	.42	.74**	.85***
Multi-Task Situations	85	2.79	.34	.88**	.82***
Crew Performance	86	2.92	.45	.73**	1.12***
Training	86	2.47	.87**	.18	.35

* $p < .05$
** $p < .01$
*** $p < .001$

From a different and more positive perspective, however, Figure 4 also shows that 46% of the respondents identified their primary job roles as something other than R&D. In addition to design in particular, the non-R&D jobs that were listed constitute the standard elements of systems engineering (cf. Blanchard & Fabrycky, 1981). These include requirements analysis, manpower and logistics analysis, maintenance analysis, system specification, system integration, production, test and evaluation, personnel selection and training. If one expands the potential user population from system designers to system developers, then Figure 4 shows that nearly half of the respondents could be considered potential model users for system applications. To explore the effect of this distinction in subsequent analyses, we split the respondents into two groups: Researchers (n=48) and Systems Developers (n=40). The latter group included all job roles other than R&D.

Table 3 summarizes respondents' answers to questions about where they typically obtain human performance information, why they attended the workshop, and how familiar they were with models prior to the workshop. Values under the "Mean Rating" column are averages across all respondents. Values in the remainder of the table show the differences in mean ratings between members of each of three dichotomies. For example, behavioral scientists rated the frequency with which they rely on their own experience for human performance information at 4.36 on a 5-point scale; engineers rated this source at 4.11. The rating difference (4.36-4.11 = .25) is shown under the Behav-Engr column associated with "Own Experience." Statistical significance of differences was established with t tests.

The Behavioral Science vs. Engineering and Researchers vs. System Developers dichotomies were described above. A third dichotomy shown in Table 3 distinguishes between Model Users and Non-Users. This classification was based on respondents' ratings of how frequently they use analytical tools including human performance models in the course of their work. Model Users (n=35) gave a frequency rating of 4 or 5 to analytical tools; Non-Users (n=53) gave a 1, 2 or 3 rating to this question. As can be seen in Table 3, the rating difference between Users and Non-Users on analytical tools was 2.30.

The top portion of Table 3 shows judgments of how frequently respondents access human performance information from each of six sources. Mean ratings indicate that personal experience and the technical literature are used most often and with equal frequency. This finding contrasts with other reports (cf. Allen, 1977; Rouse & Cody, 1988) that suggest designers rarely seek information from the technical literature. Colleagues and system users were judged the next most frequently accessed sources. Analytical tools including human performance models and empirical methods (e.g., manned simulation) were fifth and sixth overall, respectively.

Regarding the three dichotomies, only the Researcher vs. System Developer dimension seemed to have a fairly regular effect on information source selection. Researchers claimed more frequent use of personal experience, technical literature, analytical and empirical sources for human performance information than did System Developers. Although not significant, System Developers access system users more frequently than Researchers access this source, a trend that is sensible given the contrasting goals of the two communities.

The middle section of Table 3 deals with reasons for attending the workshop. Receiving an overview of modeling technology and learning about specific tools that might be applied in their work were most important to attendees. Determining the cost associated with using models and providing feedback to model developers about specific application needs were less important overall. Regarding the dichotomies, the only significant difference emerged between Model Users and Non-Users. Determining the cost of using models was a more important reason for attendance to Non-Users than to Users. Finally, seven respondents added that presenting a paper or poster and receiving feedback from potential users were important reasons for attending the workshop.

The bottom portion of Table 3 summarizes ratings of pre-workshop familiarity with each of the six classes of model that were discussed. Overall familiarity ratings were mid-scale, ranging from 3.37 for Workload and Task Allocation methods to 2.47 for Training and Skill Retention methods. Rating differences show that Researchers and Model Users rated themselves more familiar with most model classes than their dichotomy counterparts.

Answers to the third question in Table 3 also defined a fourth respondent dichotomy. We grouped respondents into Familiar (ratings of 4 or 5) and Unfamiliar (ratings of 1, 2 or 3) categories for each class of model. The numbers of respondents in the Familiar and Unfamiliar groups for each model class were as follows:

Model Class	Familiar	Unfamiliar
Anthropometric/Biomechanical	32	56
Workload & Task Allocation	41	46
Individual Tasks	28	58
Multi-Task Situations	24	61
Crew Performance	26	60
Training & Skill Retention	13	73

Ratings of Rouse & Cody Criteria

Figure 5 shows how respondents judged the importance of the seven criteria for selecting an information source. In general, they gave strong endorsements to these criteria. Except for cost, importance ratings averaged above 4 on the 5-point scale for all criteria. Applicability of an information source was a slightly more important dimension than the others. In subsequent analyses, we found no differences in importance ratings as a function of the four respondent dichotomies.

Informal comments from workshop attendees suggested that our definition of cost may not have captured the most meaningful aspect of this criterion, and hence, the somewhat depressed importance rating on this criterion. They suggested that the cost/benefit relationship, not the absolute dollar cost, of a source is what is important.

Twenty-two respondents added criteria that were not among the seven we offered. Eight people listed validity in one form or another. (However, note in Table 1 that our definition of credibility is perceived validity.) Responses that were classified as validity included these variations: "proof of correctness;" "extent to which model outputs match human behavior;" and "realness of the data with respect to human behavior." Four respondents mentioned compatibility with other tools as an important information selection criterion. Other suggestions included adaptability, expandability, flexibility, maintainability and sensitivity. Respondents who offered additional criteria did not provide definitions of them.

Ratings of Model Groups

Figure 6 shows how each group of models was rated along each criterion. Several findings of interest to model developers emerged from these data.

Applicability: All model classes were rated above mid-scale in terms of applicability to respondents' work. The average applicability rating across model groups

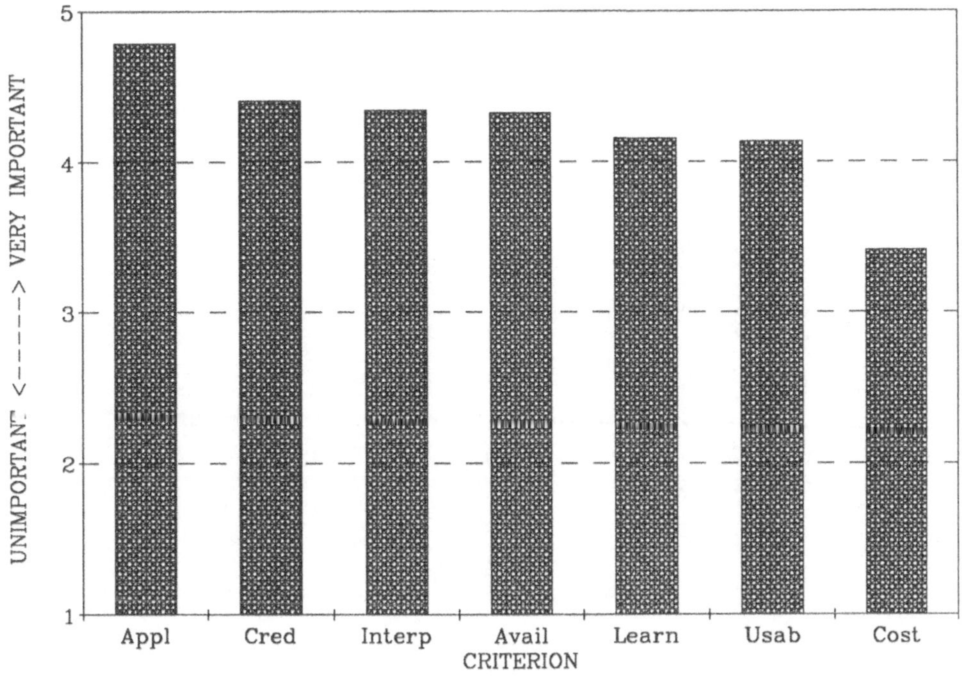

Figure 5. Mean importance ratings of the seven criteria for selecting information sources

was 3.81 on the 5-point scale. Models of Multi-task Situations and Workload methods were judged most applicable; their mean ratings exceeded 4.0. This is of particular interest because the tools discussed under these classes included modeling languages and frameworks (e.g., MicroSAINT, timeline analysis, W/INDEX), and not just specific models. This suggests that respondents may find tools for creating their own context-specific models more attractive than off-the-shelf models that require modification to produce information of direct relevance.

Credibility: Respondents generally found the outputs of all model classes to be credible. The mean rating on this dimension was above mid-scale at 3.63. No significant differences emerged across models classes.

Interpretability: The mean rating on this criterion was 3.30. Models of Individual Tasks were rated significantly lower than the other five classes on this dimension. There are at least two plausible explanations. First, understanding the outputs from tools in this class requires greater mathematical sophistication (e.g., optimal control theory) than for tools in other classes. For example, methods for Multi-task Situations and Workload Analysis, which were rated most easily interpreted, require little more than knowledge of arithmetic and rudimentary probability theory. Alternatively, people may understand the outputs from Individual Task methods but not know what to do with them in system design which is an inherently multi-task domain. Unfortunately, the present data do not permit our distinguishing between these explanations.

Availability: Models averaged 3.24 in terms of their availability. Multi-task methods, which included commercially accessible software products, were rated more available than the mean. Anthropometric tools were rated less available. This may reflect a perception that the latter class of tools requires more sophisticated computing resources than most other model classes.

Learnability: The mean learnability rating was 3.00. Two groups of model classes emerged along this criterion. Workload Analysis, Multi-task Situations, and Training models

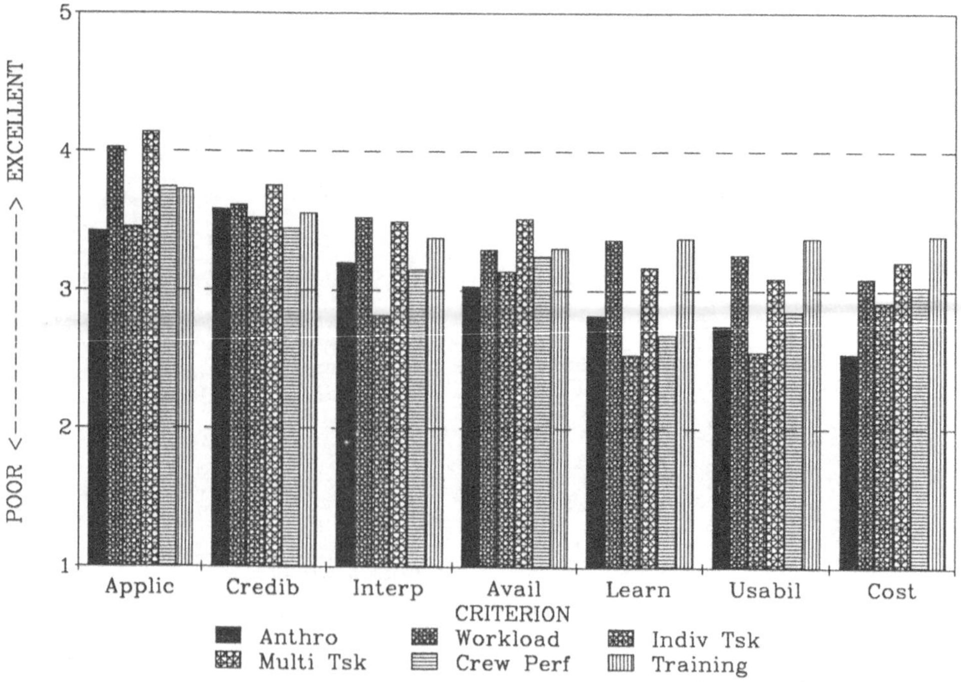

Figure 6. Average ratings of model classes along each criterion

were rated easier to learn than Anthropometric, Individual Task and Crew Performance methods. As suggested under interpretability, this difference may reflect the different levels of mathematical ability required to understand outputs from the two groups of models. The split may also reflect a perceived difference in domain knowledge that is required to become proficient in tool use. For example, workshop speakers (e.g., Rothwell) noted this as a specific requirement for present Anthropometric tools.

Usability: The mean usability rating was 3.00, and the pattern of ratings across model classes was similar to that for learnability. Hence, if tools were judged to be relatively difficult to learn, they were also perceived to be difficult to use even after mastery.

Cost: The average cost rating across all model classes was 2.98. Models for conducting Anthropometric Analyses were judged more costly than the remaining model classes. Indeed, this group included techniques and models that currently rely on advanced graphics workstations and mini-computer levels of computer power. In contrast, most tools in the other model classes were available to PC users.

We also examined the model group ratings as a function of each of the four dichotomies. Out of 168 comparisons (7 criteria X 6 model classes X 4 dichotomies), only 15 comparisons exceeded a .05 probability level of significance. This number falls within an expected margin for spurious significant comparisons and, therefore, suggests that the subject variables captured in the dichotomies had little or no effect on model class ratings.

Turning to respondents' judgments of their own behavior, Figure 7 shows the mean likelihood ratings associated with each of three behaviors per model class. The behaviors we asked respondents to estimate were 1) requesting more information about a model, 2) advocating the use of a model or technique to colleagues, and 3) making the investment to personally use the model or technique. Our assumption was that these behaviors fall on a rough scale of increasing commitment to a model. Recall that we asked respondents to make these behavioral judgments for each individual paper/model, not according to groups of models. Thus, the data shown in Figure 7 are averages computed from the mean ratings across the papers in a session for each respondent.

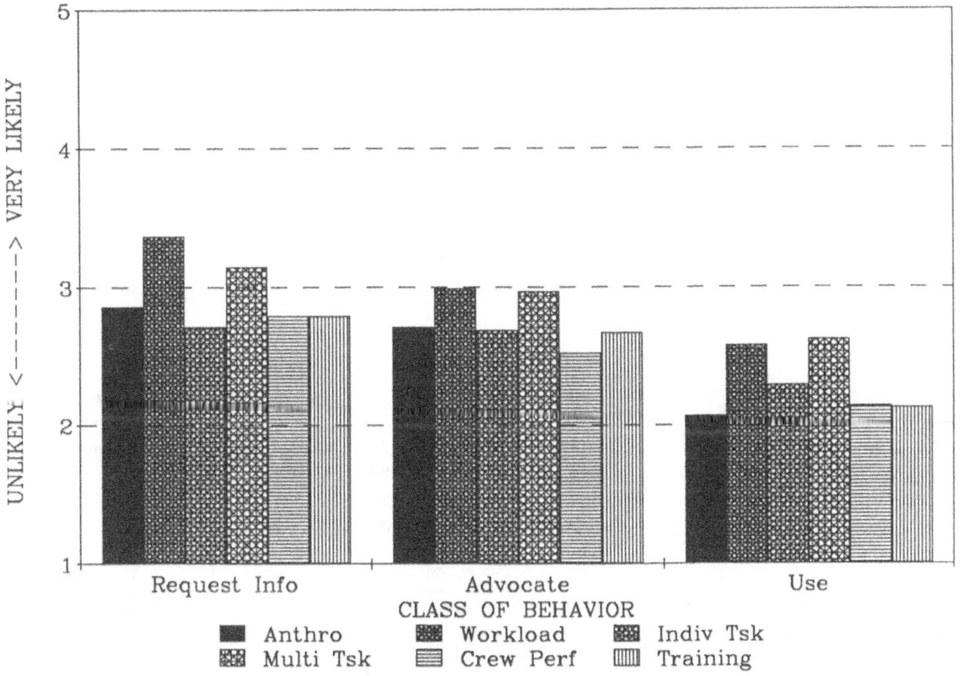

Figure 7. Judged likelihood of three types of behavior with respect to each class of human performance model

Table 4. Mean Likelihood Ratings and Mean Rating Differences for Four Pairs of Subject Classifications: Behavioral Scientists vs. Engineers; Researchers vs. System Developers; Model Users vs. Non-users; People Familiar vs. Unfamiliar with Specific Classes of Model.

		Rating Differences			
Expected Behavior	Mean Rating	Behav -Engr	Resrch -Syst	User -NonU	Famil. -Unfam
Request More Information					
Anthropometry	2.81	0.25	0.28	0.03	0.74**
Workload	3.37	0.08	0.39*	0.21	0.12
Individual Tasks	2.60	0.17	0.16	0.38*	0.56*
Multi-Task	3.15	0.01	0.30	0.21	0.03
Crew Performance	2.77	0.08	0.18	0.20	0.18
Training	2.80	0.34	0.25	0.25	0.61**
Advocate Model Use					
Anthropometry	2.70	0.23	0.31	0.28	0.60**
Workload	3.00	0.23	0.16	0.39*	0.14
Individual Tasks	2.63	0.05	0.07	0.39*	0.62**
Multi-Task	2.96	0.21	0.34	0.24	0.23
Crew Performance	2.52	0.18	0.36	0.08	0.01
Training	2.66	0.34	0.40*	0.13	0.50*
Personally Use Model					
Anthropometry	2.03	0.35	0.18	0.13	0.78**
Workload	2.58	0.37	0.24	0.13	0.12
Individual Tasks	2.21	0.03	0.04	0.48*	0.15
Multi-Task	2.61	0.29	0.38	0.38*	0.09
Crew Performance	2.15	0.22	0.16	0.13	0.10
Training	2.12	0.37*	0.35	0.35	0.62*

* $p < .05$
** $p < .01$
*** $p < .001$

Figure 7 suggests that the likelihood of respondents accessing and using models based on the workshop presentations is moderate at best. Mean likelihood values were 2.82, 2.68, and 2.21 for requesting more information, advocating model use, and personally using a method, respectively. The only likelihood ratings that exceeded mid-scale were requesting additional information about Workload Analysis methods and models of Multi-task Situations. These same model classes instigated slightly higher ratings on the other behaviors as well.

Table 4 lists comparisons of the three behavioral ratings for each of the four dichotomies as well as the overall mean ratings per model class. As can be seen, relatively few significant effects emerged. Familiarity with models prior to the workshop seemed to have the only consistent effect. Individuals who were already familiar with Anthropometric methods and Training models appeared more likely to act than people who were less familiar with these tools. However, in general, one is more struck by the absence of effects from these dichotomies than by the few differences that did emerge.

On the one hand, these data seem at odds with the relatively higher ratings of model classes along the seven criteria. Although we cannot assume that the different scales are directly comparable, respondents appeared to show more enthusiasm in their criteria ratings than their behavioral ratings would lead one to suspect. On the one hand, this might be because respondents already knew about, would advocate, or are personally using models. Hence, their behavioral ratings might merely reflect their desire to avoid redundant behaviors. However, the demographic information discussed earlier shows this is not the case. Respondents said that they make little use of models and knew relatively little about them before the workshop (see Table 3). Thus, there seems to be a disconnect between people's opinions about the models and whether these opinions reflect how they will subsequently act. The lesson for the model developer is that expressions of enthusiasm by a potential user do not indicate a definite "sale."

In summary, three important points emerged from ratings of the model classes. First, if a mid-scale rating can be taken to mean "good," then respondents were generally positive about models presented at the workshop. In particular, applicability and credibility ratings tended to be above mid-scale. Ratings that fell below mid-scale might prompt model developers to examine their products along specific dimensions. For example, perceptions of models of Individuals Tasks might be more favorable if ways to increase the interpretability of their outputs and learnability of model configuration could be developed. Second, subject variables that one might assume play an important role in model use (e.g., disciplinary background, job role, etc.), had minor effects on judgments. Third, there appears to be a disconnect between respondents' evaluations of models and the actions that they will take.

Statistical Analyses of Behavioral Expectations

In light of these results, we took an alternative tack to exploring which dimensions are likely to govern model use in system design. The thrust was to determine how people's ratings of individual papers/models along the seven criteria predicted judgments of their own behavior. We accomplished this with two types of analyses.

First, we conducted three multiple regression analyses for each of the 29 papers/models, one analysis per type of behavior. These analyses took the following form:

$$\text{Behavioral Rating}_i = \text{constant} + \sum (a_j \times \text{Rating}_j),$$

where:

i	=	the class of behavior (i.e., requesting more information, advocating model use, personally using).
j	=	criterion (i.e., applicability, etc.).
a_j	=	the respondent's weighting of the criterion in general (see Figure 5).
Rating_j	=	rating respondent assigned to the paper/model on criterion j

Figure 8 shows the key finding from this analysis. It plots the minimum, average and maximum values of the multiple correlation coefficient (R^2) across all 29 statistical analyses per behavior. As can be seen, the simple linear analysis was fairly robust in accounting for the variance in respondents' behavioral ratings.

The second approach examined the relationship between criteria and behavioral ratings more directly. We performed a discriminant analysis to determine the coefficients of a linear combination of the criteria ratings that maximally distinguished high from low behavioral ratings. For each respondent, we separated the papers/models into those the respondent assigned his highest behavioral rating (4 or 5) and those he assigned his lowest behavioral ratings (1 or 2). This was done independently for each of the three behaviors. We then averaged criteria ratings across models for each criterion to yield 14 means per respondent per behavior, 7 associated with high likelihood behavior and 7 associated with low likelihood behavior. Canonical coefficients for the discriminant functions were then computed using the data from half of the respondents drawn at random. Finally, we tested the reliability of the discriminant functions using them to predict behaviors both for respondents' who were used to compute the discriminant functions and for those whose data were not used in the computation.

Table 5 lists the canonical coefficients associated with each criterion for each type of behavior and also the multiple correlation coefficients of the discriminant functions. Looking first at the R^2 values, the functions were robust, accounting for approximately 85% of the variance in the behavioral judgments. Put simply, this means that given a respondent's ratings along the seven criteria, these functions should accurately predict what the respondent will say when asked if he or she is likely to request more information, advocate the use or personally use the model.

The results presented in Table 6 verify this. These data are the frequencies with which respondents' behavioral judgments were accurately classified by the discriminant functions. The "Training" column shows the frequencies for respondents used to compute the functions. The "New" column shows the data from remaining subjects, and the "Combined" column shows their cumulation. For all behaviors, the discriminant functions led to very accurate behavioral predictions.

For example, consider the matrix of four cells under the Training column in the first row. These frequencies show that for the 33 respondents whose data were used to compute the discriminant function in Table 5, only 4 cases were classified incorrectly. Using the function, one would predict that four respondents were very likely to request more information based on their criteria ratings when in fact the respondents claimed the opposite. Similar levels of predictive accuracy can be seen down the remaining rows under the Training column.

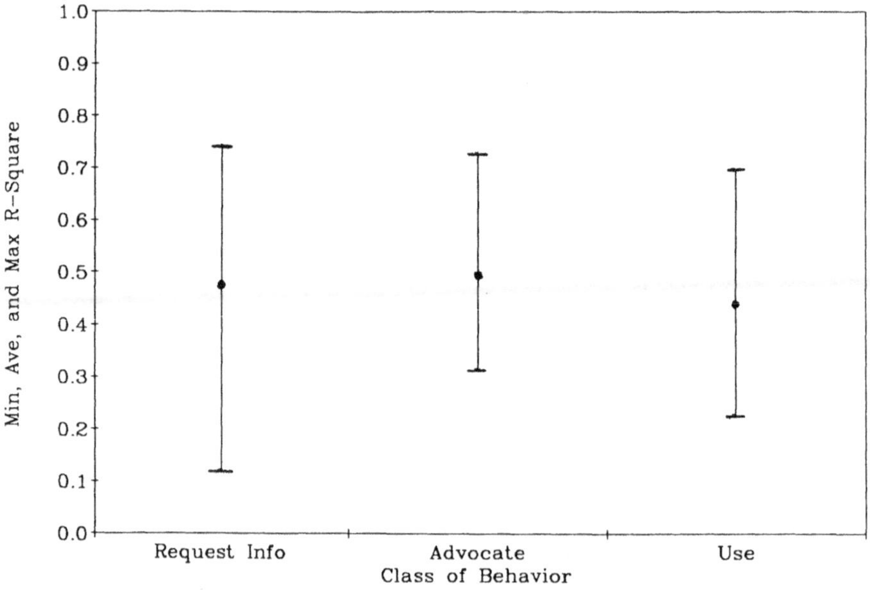

Figure 8. Means and ranges of variance accounted for in behavioral judgment by linear regression.

Table 5. Canonical Coefficients of Criteria and R-Square Values for Discriminant Analyses, One Per Type of Expected Behavior. (Each Analysis Based on Half of the Respondents Selected at Random.)

Criterion	Request More Information (n=33)	Advocate Model Use (n=32)	Personally Use (n=29)
Applicability	0.887	0.777	0.746
Credibility	0.090	- 0.071	- 0.205
Interpretability	- 0.161	0.339	0.281
Availability	0.615	0.326	0.467
Learnability	0.199	- 0.336	- 0.314
Usability	- 0.016	0.158	0.381
Cost	- 0.432	0.005	- 0.012
R-Square	0.845	0.845	0.865

Table 6. Split-Half Reliability of Discriminant Functions: Frequencies of Actual and Predicted Responses to Three Types of Behavior for Half of Respondents Used to Compute Discriminant Functions ("Training"), Remaining Half (New), and All Respondents (Combined).

		"Training"		New		Combined	
Behavioral Rating	Predicted: / Actual	Low	High	Low	High	Low	High
Likelihood of Requesting More Information	Low	29	4	28	5	57	9
	High	0	33	2	31	2	64
Likelihood of Advocating Model Use	Low	30	2	29	3	59	5
	High	2	30	2	30	4	60
Likelihood of Personally Using Model	Low	27	2	27	1	54	3
	High	0	29	3	25	3	54

As the New column in Table 6 shows, behavioral predictions were nearly as accurate for the randomly selected respondents whose data were not included in the discriminant analysis. Very few misclassifications occurred. This split-half reliability confirms the predictive validity of the discriminant functions and, therefore, lends confidence to the values of the canonical coefficients. We now return to Table 5 to consider these values.

Canonical coefficients are standardized scores. Hence, their magnitudes are directly comparable. Moreover, the magnitude and sign indicate, respectively, the strength and direction of the relationship between criteria and behavioral ratings. Hence, scores near zero indicate no relationship. For instance, coefficients associated with requesting more information suggest that applicability (.887) was positively related to this behavior and was the most important dimension for discriminating high from low behavioral ratings. Furthermore, applicability was approximately 1.4 times "more important" than availability (.615), the second most important dimension. Note, since the three discriminant functions were computed on different samples of respondents, comparisons of coefficients across behavioral categories are not meaningful. However, within each behavior, these data help to answer two questions of interest to this report.

First, in contrast with what respondents said about the importance of the seven criteria for selecting information sources (see Figure 5), the discriminant coefficients show that only a few dimensions controlled behavioral ratings. If one selects coefficients closer to zero than $\pm.30$ as reflecting dimensions that were relatively unimportant to respondents, then each behavior can be predicted with only three or four values.

Requesting more information was positively related to applicability and availability, and negatively related to cost (-.432). Thus, respondents felt that they would be very likely to probe further into a model if they perceived it was relevant to their work, easy to obtain, and relatively expensive. The cost relationship is an unusual one and difficult to explain. Interestingly, the credibility, interpretability, learnability and usability of models appear less important to respondents who wanted more information.

To advocate its use to others, respondents had to perceive a model as applicable (.777), yield outputs that could be easily interpreted (.339), be somewhat difficult to learn (-.336), and be easy to obtain (.326). Except for the learnability dimension, the other relationship seem reasonable. Perhaps respondents felt that anything worthwhile should require some investment to learn and understand.

Finally, to say they would actually use a model or technique, respondents had to perceive it as applicable to their work (.746), readily available (.467), easy to use day to day (.381), but relatively difficult to learn (-.314).

Values of these coefficients lend themselves to a variety of interpretations, some more plausible than others. For our purposes, the first conservative conclusion is that all criteria were not equally important to governing respondents' stated intentions. Moreover, different behaviors were related to different subsets of the seven criteria. Hence, model developers who hope to influence the behavior of potential users can use these data to tailor presentations of their material.

The second question concerns the hypothesis that information seekers in general apply the seven criteria sequentially in the order presented in Table 1, i.e., applicability followed by credibility followed by availability, etc. The patterns of canonical coefficients partially supported this hypothesis. Values suggest that respondents did judge applicability first, but thereafter applied the criteria in no uniform order. Complete support for the hypothesis would have required the absolute values of the coefficients to scale with the order of the criteria.

DISCUSSION

The discussion focuses on three aspects of these findings: 1) potential criticisms and caveats, 2) implications of the data for tool needs, and 3) implications of the results for "marketing" modeling technology to prospective users.

Criticisms and Caveats

Any interpretation of these findings must acknowledge two potential criticisms with the data. First, there were many missing data for individual papers and sessions. We were gratified that the rating task, an unusual request to levy on attendees of a workshop, could be accomplished and may have enhanced the workshop for some. Nonetheless, the many missing responses per questionnaire require that conclusions about audience opinions, both positive and negative, toward modeling technology be treated as tentative and subject to further verification.

Second, based on the demographic data, one could argue that audience members did not represent the community of human-system designers that Rouse and Cody (1989) described and for whom the questionnaire was designed. Attendees claimed to be researchers and, thus, could not be characterized as design decision makers in actual systems. This may be a legitimate criticism but one that need not diminish the value of the present results to model developers. Respondents also said that a very important reason for attending the workshop was to find specific tools that they could apply in their work (Table 3). Although they may not perfectly match our original image of the system designer, attendees were bonafide "information buyers" whose opinions merit the attention of model developers who wish to have their products used. Moreover, one could argue that if the present audience reacted negatively to modeling technology, then designers who did not attend would be even less disposed to accept and use this technology. Given these caveats, the data provide several insights.

Implications of Results for New Tools

The results suggest that potential users believe that there is room for improving present-day modeling technology. Reactions to models presented at the workshop were, for the most part, lukewarm. Average ratings along the seven dimensions were mid- scale (Figure 6) and behavioral ratings (Figure 7) were even lower. Given that attendees represent the best prospects for using models, the data offer insights for model developers on what to provide. Two findings are most pertinent.

First, the results demonstrate unequivocally that potential users must be convinced of the applicability of a tool even to seek more information about it, let alone use it. The importance ratings of criteria (Figure 5) and discriminant functions (Table 5) support this claim. Second, although we did not report the ratings of individual papers in detail, there were some exceptions to the lukewarm reaction. Wickens' multiple resource theory, North's W/INDEX, and Laughery's MicroSAINT generated very positive reactions. Each of these tools is a conceptual framework or a means for building models rather than a specific model.

These results suggest a mismatch between potential users' perceived needs and available modeling products. Workshop attendees may have been seeking context-specific tools, a reasonable conclusion given the high weighting assigned to applicability. Two general types of tool were offered: context-specific tools and frameworks. Our hypothesis is that the context-specific tools that were presented either did not match users' contexts or else were not perceived to generalize across contexts. Hence, users turned to frameworks, perhaps not as their first choice, but because no directly applicable tools exist.

For example, suppose that the user wishes to model supervisory control in a specific process control plant under particular conditions. In this case, he will judge as irrelevant any model that does not contain the details peculiar to the process control system, disturbance conditions and human characteristics specific to his system application. Our hypothesis is that most workshop attendees perceived their needs from just such a perspective. This perception places rather stringent demands on the model developer that are difficult to meet without intimate knowledge of the user's applied problem. Very specific models had little chance to match more than a few people's circumstances. Frameworks and model-building tools were perceived to involve some work, but ultimately offered the flexibility to create context-specific models.

If this explanation is correct, it suggests two strategies for model developers who seek to have their products accessed and used. First, developers can improve existing frameworks and model-building environments. Their flexibility gives these tools broad appeal. Second, developers can increase use of existing models by demonstrating their relevance in the user's specific context. This latter strategy is easy to state but may not be adopted for at least two reasons.

First, when a model has some generality across domains, example applications that developers concoct still may require many domain specific features to convince users via face validity of the tool's relevance. Unfortunately, acquiring sufficient domain knowledge to succeed may involve more effort than the tool developer is willing to invest.

The second reason stems from a deeper problem. Tool developers, especially in the human factors community, have been criticized for producing data and instruments that have little practical value to system developers (cf. Boff, 1988; Rouse, 1987). This may be because tool developers simply do not know or care to know what system developers need. Indeed, some model developers hold that the goal of modeling is foremost to produce better theoretical explanations of human behavior. These explanations may have utility for other endeavors, including system design. However, for these individuals, this utility is a by-product and not the primary motivation for their efforts. We excluded these individuals from the above strategies by emphasizing the motive to have products used by practitioners. If the model developer shares this goal, then he or she must better understand what system developers perceive to be their needs.

Implications of Results for Tool "Marketing"

Several findings reveal aspects of the consumer psychology of potential model users. Considering these issues should help developers package their products in ways that promote greater user interest and recognition of value.

Respondents rated the Rouse & Cody (1989) criteria as important determinants of their information selection habits, but their own behavioral expectations contradict this solidarity. Comparing the criteria importance ratings (Figure 5) with the discriminant functions (Table 5) reveals large and regular discrepancies between what respondents said mattered to them and perceived characteristics that would actually govern their behavior. If we trust people to act as they say they will (which itself is not clear), then the mismatch between opinion and action in these data corroborates findings from the decision making literature in general (Dawes, 1979) and consumer behavior research in particular (Bettman, 1986). The lesson for model developers is to focus on dimensions that control (expected) behavior rather than on what people say are important product characteristics.

Results also show that potential users fill in information about models that presenters do not overtly provide. The best example of this phenomenon emerged along the cost dimension. Very few presenters made any explicit statements about cost, yet 80% of opportunities to judge cost were completed by respondents. Similar findings have been reported in the consumer psychology literature (e.g., Huber & McCann, 1982; Meyer, 1981).

The present results do not indicate what specific values respondents assigned to attributes that presenters failed to fill in for them. Hence, model developers should not conclude that the best strategy is always to provide exhaustive information along all attributes without first establishing how potential users fill in missing values. Indeed, the consumer literature shows that people fill in missing values according to multidimensional internal rules. These rules include past experiences with similar products as well as the values that are presented on accompanying attributes (cf. Bettman, 1979; Meyer, 1981). The main point is that model developers who are interested in having their products applied should make an effort to understand this phenomenon about consumers.

As a final observation, this exercise showed that the rating task could be performed and may have affected how people usually attend to presentations at workshops. Based on the informal comments from workshop attendees, the task appears to have increased their vigilance for the information needed to judge models along the seven criteria. From most comments, attendees found this organizing structure to be an aid despite the inconvenience of the task itself.

REFERENCES

Anderson, N.H., 1961, Scales and statistics: Parametric and nonparametric, Psychological Bulletin, 58: 305-316.

Bettman, J.R., 1986, Consumer psychology, Annual Review of Psychology, 37: 257-290.

Bettman, J.R., 1979, An Information Processing Theory of Consumer Choice, Addison-Wesley, Reading, MA.

Bettman, J.R., and Park, C.W., 1980, Effects of prior knowledge and experience and phase of the choice process on consumer decision processes: A protocol analysis, Journal of Consumer Research, 7: 234-248.

Blanchard, B.S., and Fabrycky, W.J., 1981, Systems Engineering and Analysis, Prentice-Hall, Englewood Cliffs, N.J.

Boff, K.R., 1988, The value of research is in the eye of the beholder, Human Factors Society Bulletin, 31 (6): 1-4.

Boneau, C.A., 1960, The effects of violations of assumptions underlying the *t* test, Psychological Bulletin, 57: 49-64.

Cody, W.J., 1988, "Recommendations for Supporting Helicopter Crew System Design," Letter Report No. 351, U.S. Army Human Engineering Laboratory, Aberdeen Proving Ground, MD.

Dawes, R.M., 1979, The robust beauty of improper linear models in decision making, American Psychologist, 34: 571-582.

Huber, J., and McCann, J., 1982, The impact of inferential beliefs on product evaluations, Journal of Marketing Research, 19: 324-333.

Kiel, G.C., and Layton, R.A., 1981, Dimensions of consumer information seeking behavior, Journal of Marketing Research, 18: 233-239.

Meyer, R.J., 1981, A model of multiattribute judgments under attribute uncertainty and information constraint, Journal of Marketing Research, 18:428-441.

Meyer, R.J., 1982, A descriptive model of consumer information search behavior, Marketing Science, 1: 93-121.

Nisbett, R.E., Krantz, D.H., Jepson, C., and Kunda, Z., 1983, The use of statistical heuristics in everyday reasoning, Psychological Review, 90: 339-363.

Nisbett, R.E., and Wilson, T.D., 1977, Telling more than we can know: Verbal reports on mental processes, Psychological Review, 84: 231-259.

Punj, G.N., and Staelin, R., 1983, A model of consumer information search behavior for new automobiles, Journal of Consumer Research, 9: 366-380.

Reilly, M.D., and Conover, J.N., 1983, Meta-analysis: Integrating results from consumer research studies, Advances in Consumer Research, 10: 510-513.

Rouse, W.B., 1986, On the value of information in system design: A framework for understanding and aiding designers, Information Processing and Management, 22: 217-228.

Rouse, W.B., 1987, Much ado about data, Human Factors Society Bulletin, 30(9): 1-3.

Rouse, W.B., and Cody, W.J., 1989, Designers' criteria for choosing human performance models, this volume.

Rouse, W.B., and Cody, W.J., 1988, On the design of man-machine systems: Principles, practices and prospects, Automatica, 24: 227-238.

KEYNOTE ADDRESS

WARGAMING: APPLICATIONS OF HUMAN PERFORMANCE MODELS TO SYSTEM DESIGN AND MILITARY TRAINING

Earl A. Alluisi and Franklin L. Moses

Office of the Secretary of Defense, U.S.A.
and U.S. Army Research Institute for the
Behavioral and Social Sciences

There is increasing recognition of a need -- a need that can be met with the help of the community represented by the participants at this workshop. The need is for valid, data-based, computer-aided, adaptive, human-performance models that can be inserted into wargaming simulations. The need is to increase the validity and utility of such simulations, especially if the current and future technological capabilities are to be exploited to provide realistic battle-management aiding and training for military leaders, their staffs, and those they command.

WHAT IS WARGAMING?

A <u>wargame</u> is a simulation of a military operation, involving opposing forces, using information designed to depict actual or real-life situations.

Some wargames are fully automated; i.e., they are computer simulations that do not involve human intervention while running. Other wargames are interactive; they involve <u>man-in-the-loop</u> simulations. Interactive wargames differ in the degree of human intervention permitted or required.

Wargaming simulations have been constructed, demonstrated, and used to represent a broad range of situations. Some represent very large global and theater-level conflicts. Others represent smaller sectors of larger conflicts, such as a naval battle group or an army corps air-land battle zone. Still others, especially designed for training, represent smaller collective battle elements such as an army tank battalion, company, or platoon, or even individual battle elements such as a single aircraft, ship, or army tank.

There is increasing interest in the networking of wargaming and simulations to provide for the interactive, and potentially quite realistic, training of collectives (crews, groups, teams, and units, or <u>CGTUs</u>) as well as individuals.

USES OF WARGAMING

There are four primary uses to which wargames have been put:

(1) They have been used for the analysis and evaluation of proposed new weapon systems. The typical application is to employ a fully automated simulation with changed parameters representing the new or changed capabilities of the proposed system. The outcome (and the way the outcome is achieved) is analyzed and evaluated further, and inferences are drawn regarding the effectiveness or likely benefits of the new system either as an addition to, or a replacement for, an existing capability. A weakness in the typical application is the relatively static representation of the doctrine, or actions of the military commander, which admittedly is likely to change in light of the new capability that the proposed system would provide.

(2) Wargaming has been used in the development of doctrine and tactics, and in the planning of operations. The typical application for these purposes is to use an interactive game with the "players" being the commanders, or decision-makers, and at least parts of their senior staff. The weakness here is the "ironhorse" or non-interactive, doctrinaire representation of the enemy commander and his actions and forces.

(3) A third use of wargaming is for military education and training. The former is meant to provide a broadening experience that replaces ignorance with knowledge of doctrine and procedures. It is rightly the province of the services' senior schools and colleges. The purpose of <u>training</u>, however, is to provide for the acquisition, refinement, and maintenance of battle-winning skills. Combat training, in all services, is regarded as the province of the operational commander. Educational experiences may be as infrequent as a once or twice-in-a-lifetime event, but training experiences have to be sufficiently frequent to support not only the initial acquisition, but also the further refinement and high-level maintenance of the skill. Until recent years, wargaming has been used relatively effectively for education, but nearly not at all for training. High cost and limited availability have been limiting factors that technological advances are beginning to overcome.

(4) Finally, with these same advances in technology, primarily in computer technology, the uses of wargaming in battle-management aiding and training is increasing. The prediction is that this will be the fastest growth area in future wargaming development, with an increasingly "fuzzy" distinction between "aiding" and "training," and a growing realization that it provides potentially a most cost-effective contribution to "readiness" and should be regarded as providing a real force-multiplier effect.

POTENTIALS OF WARGAMING AND TRAINING

Is there evidence that the provision of wargaming simulations for training would produce beneficial results? There is evidence in the open literature. Some of it is direct, but much is indirect. For example, during 1987, U.S. Army tank platoons stationed in Europe were trained with a network of tank simulators. The skills they gained in the simulator training transferred positively to the NATO-exercise range, and the U.S. Army tankers took first prize in the annual competition. This was the first U.S.A. win in ten years.

That there are vast differences in the battle-effectiveness of commanders and their units is a widely recognized fact among both warriors and military historians. Naval commanders and historians know of "good ships" and "bad ships," often identical hardware in ships of the same class. Army commanders and historians know of "good divisions" and "bad divisions," identically equipped and often fighting side-by-side. Data are not impossible to find. For example, one NATO-sponsored study of WW-II German U-boat commanders found that 32.9% failed to engage any targets, and 13.1% failed to hit any targets they found. Thus, 46% were ineffective. On the other hand, the best 10% of the commanders sank 45% of the allied ships sunk. Such differences are usually explained with reference to some combination of the "human factors," e.g., selection and training, both of which can be impacted by wargaming and training.

A third and final example is drawn from the WW-II air-battle experiences of the German and American forces in the European theater. This is shown by the air combat loss rates represented in a non-classified form in Figure 1. The German pilots experienced near 40% first-mission losses, whereas the American first-mission loss rate was half that, or 20%. Essentially, the figure shows a five-mission advantage for the Americans relative to the Germans. This advantage has been attributed fully to the differences in training. The Germans were short of fuel, and although they trained their pilots well to fly the aircraft, they did not have fuel "to waste" on the training of air-combat maneuvering skills. The Americans did and the results are obvious. Evidence obtained with the Advanced Simulator for Pilot Training (ASPT) at the Operations Training Division of Air Force Human Resources Laboratory has more than suggested that similar advantages can be gained by augmenting flying training with air-combat simulator training.

There is sufficient evidence, in theory as well as fact, to indicate that combat experience is the best trainer of combat skills. Combat veterans have out-performed "green" troops in every recorded engagement. The technological capability is now present to be combined in training systems that can produce combat veterans and aces in peacetime! Can that capability be prudently ignored? If not, how can the technology be applied to the training of combat-relevant skills?

TRAINING APPLICATIONS OF WARGAMING

A few of the better-known wargaming simulations used by the U.S. military services, and the Joint Chiefs of Staff (JCS), are listed in Table 1. They cover a range of applications in military education or training. Typically, these simulations address the owning service's requirements, sometimes with a second service at least represented, but seldom with a third or fourth. On the other hand, the JSC-sponsored simulations are moving more and more towards representations of combined land, sea, and air forces as they would be represented under the actual conditions of conflict.

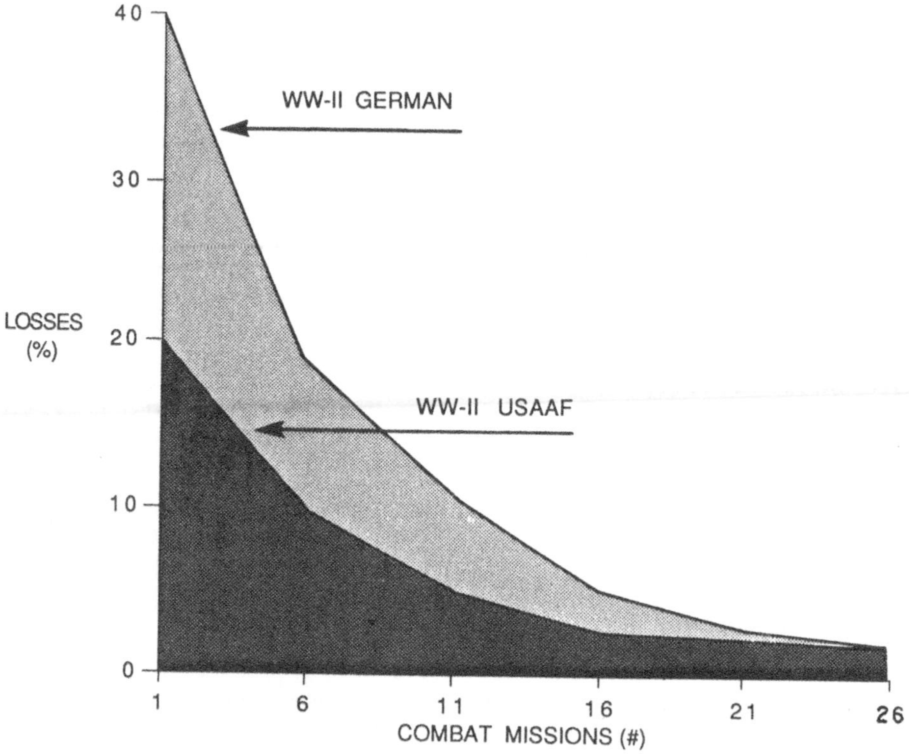

Figure 1. Representation of air combat loss rates experienced by German and American pilots during World War II in the European theater of operations. The German pilots were given very little pre-combat training, relative to the Americans, in air combat maneuvering.

Table 1. Some better-known wargaming simulations used by the U.S. Services and the Organization of the Joint Chiefs of Staff.

ARMY	NAVY
• CORPS BATTLE SIMULATIONS • CONEWAGO WARGAME • Army Training Battle Sim. System (ARTBASS) . . .	• ENWGS • GLOBAL WARGAME • Joint Land, Aerospace & Sea Simulation (JLASS) . . .
AIR FORCE	**JCS**
• BLUE FLAG • BIG STICK/FAST STICK • COMMAND READINESS EXERCISE SYSTEM (CRES) . . .	• JESS • JTLS • STATE-OF-THE-ART CONTINGENCY ANALYSIS (SOTACA) . . .

The three columns of Table 2 list, generally, who is trained in the first column, the techniques used in wargaming and training in the middle column, and in the third column, what technologies are available for use in the construction of training systems based upon these capabilities. Some interesting observations can be made from the listing.

Table 2. Representation of what technologies are applied (right-most column), in what ways (center column), and who is trained (left-most column) in computer applications to training and wargaming.

WHO TRAINED	HOW APPLIED	WHAT TECHNOLOGIES
CINCS & STAFFs	---------	PROCESSORS/MEMORIES
COMPONENT COMMANDERS	SIMULATIONS	DISKS
	---------	GRAPHICS/DISPLAYS
LOWER-ECHELON COMMANDERS	WARGAMES	COMMUNICATIONS
UNITS	---------	DATABASES
CREWS/TEAMS	CPXS	OPERATING SYSTEMS
OTHER GROUPS	---------	INTERFACES
INDIVIDUALS	EXERCISES	PROGRAM LANGUAGES
	---------	EXPERT SYSTEMS
		SIMULATIONS

Who Trained

Those to be trained range from the global or theater-level commanders-in-chief (CINCs) and their staffs, through their component and lower-level commanders (and staffs), CGTUs, and individuals. Nearly all, and certainly the greater majority of all, training and training technology has been devoted to the lower end of the column (individual training), with very little devoted to the upper levels.

The situation is changing, however, and a recent report of the Defense Science Board Task Force on Computer Applications to Training and Wargaming has recommended to the Chairman of the Joint Chiefs of Staff that major efforts be taken now to employ current capabilities and, by judicious networking and rapid prototyping, to develop additional capabilities to address the training of joint commanders and their staffs and component commanders.

What Technologies

The right-most column of Table 2 lists the enabling technologies for wargaming and training. Those at and near the top are generally the hardware-based technologies about which most is known, risk is lowest, and progress is greatest (in terms of increasing capability and reducing costs). These are truly the enabling technologies.

On the other hand, those at and near the bottom are generally the "cognitive-based" technologies about which least is known, risk is highest, and progress is questionable, at best. These are truly the limiting technologies. And these are the very technologies to which the conference participants can contribute most!

How Applied

The computer technologies are applied to training through simulations, wargames, command post exercises (CPXS), and field (or sea) exercises, with increasing costs from the first to the last named. Effectiveness is another matter! There are few, if any, public and well-documented studies of the effectiveness of military training. That is still another area in which the conference participants could contribute!

THE OPPORTUNITY

Military "readiness" depends on the training of the commanders, staffs, and troops -- both as collectives (CGTUs) and individuals. Simulation-based training is effective, and provides cost-effective augmentation to field and sea training. The opportunities for applications of relevant research and development of training and wargaming are uniquely high at the present time: (1) the services are ready to accept and use it, and will be even more so in an era of reducing budgets, (2) the technology is here and coming, driven by a vast civilian market, (3) affordable wargaming/simulation networking is the key, and (4) credible data on effectiveness and cost are crucial.

CHALLENGES AND ISSUES

What can research and development contribute to establish how best to use the technologies -- computers, models, wargames, simulations, networks, etc., (1) to analyze and evaluate proposed new systems, (2) to develop tactics and doctrine, and plan operations, (3) to educate wartime leaders and train warriors, and (4) to aid and train commanders to win battles? What can the future research and development of the conference participants contribute?

What can research and development contribute to establish a testbed, a rapid-prototyping program, (1) to develop timely prototypes, (2) to monitor and harness technological advances, (3) to provide new simulation and training techniques, (4) to ensure user involvement, utility, and use, and (5) to coordinate the emerging networks (SIMNET, AIRNET, JETNET, CATTS, WARNET, and "JOINTNET")?

APPENDIX A - PARTICIPANTS

Ainsworth, James
US Army Res Inst
HQ TEXCOM
Ft Hood TX 76544 USA

Aldrich, Theodore
Anacapa Sciences, Inc.
P.O. Box 489
Ft Rucker AL 36362 USA

Allen, Wade
Systems Technology, Inc.
13766 S. Hawthorne Blvd
Hawthorne CA 90250-7083 USA

Alluisi, Earl
OUSD (A) R&AT (E&LS)
The Pentagon, Room 3D129
Washington DC 20301-3080 USA

Anderson, Mary
NAVSSES
Bldg 619, Code 03ZE
Philadelphia PA 19112-5083 USA

Armstrong, Nancy
US Army FSTC
220 Seventh Street N.E.
Charlottesville VA 22901 USA

Astley, James
Computing Devices Co.
P.O. Box 8508
Napean Ontario K1G 3M9 CAN

Badler, Norman
Univ of Pennsylvania
Computer & Information Science
Philadelphia PA 19104-6389 USA

Baer, Clarence
Martin-Marietta
P.O. Box 179, MS T5742
Denver CO 80201 USA

Banks, Dennis
Grumman
South Oysterbay Road
Bethpage NY 11746 USA

Baron, Sheldon
BBN Laboratories, Inc.
70 Fawcett Street
Cambridge MA 02238 USA

Bateman, Bat
Boeing Military ARP
P.O. Box 7730, M/S K78-08
Wichita KS 67277 USA

Beevis, David
DCIEM
P.O. Box 2000
Downview Ontario M3M 3B9 CAN

Bernotat, Rainer
FGAN/FAT
Neuenahrer Str 20
D5307 Wachtberg-Werthhoven GER

Biferno, Michael
Douglas Aircraft Co.
3855 Lakewood Blvd
Mail Code CIE24 78-73
Long Beach CA 90846 USA

Binder, Laura
McDonnell Aircraft Co.
P.O. Box 516
St Louis MO 63166 USA

Bonney, Maurice
University of Nottingham
University Park
Nottingham, NG7 2RD UK

Broughton, Blythe
Royal Military College
EE Department
Kingston Ontario K7K 5L0 CAN

Bubb, Heiner
Kath Uni Eichstaett
Ostenstrasse 26
D8078 Eichstatt GER

Bubb, Peter
BMW-AG
Petuelring 130
D8000 Munchen 40 GER

Card, Stuart
Xerox PARC
3333 Coyote Hill Road
Palo Alto CA 94304 USA

Carrier, Robert
Genicom Consultants
279 Sherbrooke St W, Suite 207
Montreal Quebec H2X 1Y2 CAN

Cherry, Peter
Vector Research, Inc.
P.O. Box 1506
Ann Arbor MI 48106 USA

Chubb, Gerald
SofTech
460 Trotten Pond Rd
Waltham MA 02154 USA

Cody, William
Search Technology, Inc.
4725 Peachtree Corners Cir
Norcross GA 30092 USA

Collins, Dennis
HQDA, ODCSPER
HQDA, DAPE-MR, 2E733
Washington DC 20310 USA

DeSwart, Herman
Royal Netherlands Army
Behavioral Sciences Dept
P.O. Box 90701
2509 LS The Hague NL

Dick, A.O.
Analytics, Inc.
2500 Maryland Road
Willow Grove PA 19090 USA

Dintruff, Diane
IBM
54Y/458 IBM
Kingston NY 12401 USA

Doring, Bernhard
FGAN/FAT
Neuenahrer Str 20
D5307 Wachtberg-Worthhoven GER

Enos, Wayne
Human Factors
MS 138-210, GE-ESD
Moorestown NJ 08057 USA

Evans, Susan
Vector Research, Inc.
P.O. Box 1506
Ann Arbor MI 48106 USA

Fegan, Ben
Hay Systems, Inc.
2000 M Street, S.W., Suite 650
Washington DC 20036 USA

Gilbert, Robert
Genicom Consultants
279 Sherbrooke St W, Suite 207
Montreal Quebec H2X 1Y2 CAN

Goddard, Daniel
AFWAL/FIGDA
Wright-Patterson AFB OH
45433-6573 USA

Greene, Janettarose
Naval Air Development Center
Code 6022
Warminster PA 18974

Gribbon, Steven
Oerlikom Aerospace
882 Champlain
St Jean, Quebec, J3A 7BY CAN

Guadagna, Lisa
HSD/YA OL-AC (CAT)
Wright-Patterson AFB OH
45433-6573 USA

Hamilton, Ian
British Aerospace
FPC 267, P.O. Box 5, Filton
Bristol, BS 127QW UK

Harris, Steven
Naval Air Test Center
Patuxent River MD 20670 USA

Hart, Sandra
NASA Ames Research Center
MS 239-3
Moffett Field CA 94035 USA

Hartzell, Earl
US Army
Mail Stop 239-9
Moffett Field CA 94035 USA

Harvey, Roger
Admiralty Research Est.
Queens Road, Middlesex
Teddington UK

Hazeltine, Alice
Martin Marietta
P.O. Box 179, MS D6013
Denver CO 80201 USA

Hefley, William
Software Engrg Institute
Carnegie-Mellon University
Pittsburgh PA 15213 USA

Hendy, Keith
DCIEM
P.O. Box 2000
Downview Ontario M3M 3B9 CAN

Henning, Klaus
HDZ/KDI - RWTH Aachen
Rolandstrasse 7-9
D5100 Aachen GER

Hennessy, Robert
Monterey Tech, Inc.
P.O. Box 223699
Carmel CA 93922 USA

Holley, Charles
Sequitur Systems, Inc.
7525 Nine Mile Bridge Road
Fort Worth TX 76135 USA

Howell, Ned
Pritsker & Associates
1305 Cumberland Avenue
West Lafayette IN 47906 USA

Huntoon, Richard
Rockwell International
400 Collins Rd, N.E.
Cedar Rapids IA 52498 USA

Jagacinski, Richard
Ohio State University
Dept of Psychology
404 D.W. 17th Avenue
Columbus OH 43210 USA

Jarboe, Ralph
USAARENBD-ATRD
P.O. Box 336
Ekron KY 40117 USA

Jonsen, Gordon
Boeing Aerospace Company
P.O. Box 3707
Seattle WA 98124 USA

Judge, Carolyn
B1B FOT&E USAF
Det 1 49th Test Squadron
Dyess AFB TX 79607 USA

Kaiura, Richard
US Army Research Institute
P.O. Box 2086
Ft Benning GA 31905 USA

Kleinman, David
University of Connecticut
Dept of ESE-U157
Storrs CT 06268 USA

Kloke, Bernhard
Inst Arbeitsphysiologie
Ardeystrasse 67
D4600 Dortmund GER

Knerr, Mazie
HumRRO
1100 S Washington Street
Alexandria VA 22314 USA

Kolodrubetz, Steve
Auto Research System
4401 Ford Avenue, Suite 400
Alexandria VA 22310 USA

Kopp, Ullrich
Tech Universitat Berlin
Institut fur Luft und Raumfahrt
Marchstrasse 12-14
D1000 Berlin 10 GER

Kroemer, Karl
IEOR Dept, VPI & SU
551 Whittemore Hall
Blacksburg VA 24061 USA

Kunz, Bernhard
Dornier GmbH
Postfach 1420
D7990 Friedrichshaven GER

Laughery, Ronald
Micro Analysis & Design
9132 Thunderhead Drive
Boulder CO 80302 USA

Laurig, Wolfgang
University of Dortmund
Ardeystrasse 67
D4600 Dortmund GER

Lauzon, Armand
CAE-Canadair
8585 Cote De Liesse
Montreal Quebec H4L 4X4 CAN

Levison, William
BBN Laboratories, Inc.
70 Fawcett Street
Cambridge MA 02177 USA

Liebert, Philippe
DRET SDR G9
26 Boulevard Victor
Paris 7505 FR

Lindsey, Dana
Naval Training Center
Code 712, NTSC
Orlando FL 32813-7100 USA

Linton, Paul
Sikorsky Aircraft
6900 Main Street, MS 326A1
Stratford CT 06601 USA

Luczak, Holger
Tech Universitat Berlin
Ernst-Reuter-Platz 7
D1000 Berlin 10 GER

Mackey, William
Spar Aerospace Ltd.
1235 Ormont Drive
Weston Ontario M9L 2W6 CAN

Madden, Ned
Boeing Helicopter
1123 Crestover Road
Wilmington DE 19803 USA

Maher, Frank
Unisys Corporation
4140 Linden Avenue, Suite 200
Dayton OH 45432 USA

Makadi, John
Canadian Marconi Co.
415 Leggett Drive
Kanata Ontario K2K 2B2 CAN

Malcolm, Scott
Litton Systems, Ltd.
1290 Martin Grove Road
Etobiocoke Ontario M9W 4X3 CAN

Martin, Ed
ASD/ENETA
Wright-Patterson AFB OH
45433-6503 USA

Martin, Mary F.
American Inst for Research
1055 Thomas Jefferson, NW
Washington DC 20007 USA

Mather, Mike
US Army Research Institute
5001 Eisenhower Avenue
Alexandria VA 22333 USA

Matheson, Edward
ASD/ENECH
Wright-Patterson AFB OH
45433-6503 USA

Maxwell, Kenneth
General Dynamics
P.O. Box 748, MZ 1555
Fort Worth TX 76101 USA

McMillan, Grant
AAMRL/HEF
Wright-Patterson AFB OH
45433-6573 USA

Milgram, Paul
University of Toronto
Dept of Industrial Eng
Toronto Ontario M5S 1A4 CAN

Mohney, Jack
ASD/ALH
Wright-Patterson AFB OH
45433-6573 USA

Morgan, Ben
Univ of Central Florida
Psychology Department
Orlando FL 32816 USA

Myers, Kent
Auto Research Systems
4401 Ford Avenue, Suite 400
Alexandria VA 22310 USA

Nelson, Bill
Idaho National Engr Lab
P.O. Box 1625
Idaho Falls ID 83415 USA

Nicholson, Nigel
Army Research Institute
HQ TEXCOM, Attn: PERI-SH
Ft Hood TX 76544 USA

Nontasak, Tatree
Naval Aero Med Rsch Lab
Naval Air Station
Pensacola FL 32508 USA

North, Robert
Honeywell, Inc.
3660 Technology Drive
Minneapolis MN 55418 USA

O'Brien, Larry
Dynamics Research Co.
60 Concord Street
Wilmington MA 02184

Osga, Glenn
NOSC
Code 441
San Diego CA 92152 USA

Paetzold, Hans
BMVg
Postfach 1328, RuFo3
D5300 Bonn GER

Papin, Jean Paul
DAT
Etas BP4107
49041 Angers Cedex FRA

Parks, Donald
Boeing Aerospace Company
P.O. Box 3707, MS 33-22
Seattle WA 98124 USA

Patterson, Debbie
Computing Devices Co
P.O. Box 8505, MS420
Napean Ontario K1G 3M9 CAN

Peckham, Donald
Central Florida Rsch Park
12350 Research Highway
Orlando FL 32826 USA

Perkins, Mike
Litton Computer Service
P.O. Box 2498
Columbus GA 31905 USA

Plocher, Thomas
Honeywell S&RC
3660 Technology Drive
Minneapolis MN 55418 USA

Potts, Fred
Computing Devices Company
P.O. Box 8508
Napean Ontario K1G 3M9 CAN

Quill, Rose
NAVSSES
Bldg 619, Code 032E
Philidelphia PA 19112-5083 USA

Ramirez, Tammy
Battelle Columbus Division
5100 Springfield St, Suite 210
Dayton OH 45431 USA

Reichelt, Werner
Department E6W
Daimler Benz AG
Postfach 60202
D7000 Stuttgart GER

Repa, Brian
General Motors Res Labs
1151 Crooks Road
Troy MI 48084 USA

Rose, Andrew
American Inst for Research
1055 Thomas Jefferson St NW
Washington DC 20007 USA

Rose, David
Naval Air Development Ctr
Code 6022
Warminster PA 18974 USA

Rothwell, Patti
DCIEM
P.O. Box 2000
Downview Ontario M3M 3B9 CAN

Rouse, William
Search Technology, Inc.
4725 Peachtree Corners Cir
Norcross GA 30092 USA

Ruessmann, Till
Inst fur Fahrzeugtechnik
Tech Univ Berlin
Strasse 17. Juni No. 135
1 Berlin 12 GER

Salas, Eduardo
Naval Training Systems Center
Code 712
Orlando FL 32813 USA

Sanders, Andries
RWTH Aachen
Jaeggerstrasse 17/19
D5100 Aachen GER

Schaub, Karlheinz
Inst fur Arbeitswissenschoft
Petersenstrasse 30
D6100 Darmstadt GER

Schendel, Peter
Erno Raumfahrttechnik GmbH
Hunefeldstrasse 1-5
D2800 Bremen GER

Schick, Fred
DFVLR, Flughaven
Inst fur Flugfuhrung
D3300 Braunschweig GER

Schiettekatte, Jean
Genicom Consultants
279 Sherbrooke St W, Suite 207
Montreal Quebec H2X 1Y2 CAN

Schumacher, Wilfried
Fraunhofer-Inst IITB
Sebastian-Kneipp-Str 12/14
D7500 Karlsruhe 1 GER

Sepehr, Michelle
Technical Univ of Berlin
Strasse de 17.Juni No. 135
1000 Berlin 12 GER

Smith, Terry
Biokinetics
1481 Cyrville Road
Ottawa Ontario K1B 3L7 CAN

Smootz, Edwin
Army Research Inst
Texcom, Attn: PERI-SH
Ft Hood TX 76544-5065 USA

Stager, Paul
Dept of Psychology, York Univ
4700 Keele Street
Toronto Ontario M3J 1P3 CAN

Stark, Edward
Singer-Link
Binghamton NY 13902 USA

Stassen, Henk
Park Berkenoord 40
2641 CZ Pynacker NL

Stein, Willi
FGAN/FAT
Neuenahrer Str 20
D5307 Wachtberg-Werthhoven GER

Stica, Paul
HumRRO
1100 S. Washington St
Alexandria VA 22314 USA

Strub, Michael
US Army Research Institute
P.O. Box 6057
Ft Bliss TX 79906 USA

Sutton, Robert
Royal Naval Eng College
Manadon Devon PL 53AQ UK

Sweezy, Robert
Science Applications
1710 Goodridge Dr
McLean VA 22102 USA

Thome, Franz
Ministry of Defense
Konrad Adenauer Ufer 2-6
D5400 Koblenz GER

Tijerina, Louis
Battelle
505 King Avenue
Columbus OH 43201 USA

Tousignant, Jacques
Oerlikon Aerospace, Inc.
225 Boul du Seminaire Sud
St Jean sur Richelieu,
Quebec, J3B 8E9 CAN

Towill, Dennis
UWIST
P.O. Box 25
Cardiff CF1 3XE UK

Van Breda, Leo
Inst for Perception -TNO
Kampweg 5, Postbus 23
3768 ZG Soesterberg NL

Van de Graaff, Rene
National Aerospace Lab-NRL
A Fokkerweg 2
1059 CM Amsterdam NL

Waldron, Les
Dept of Natl Defence
Ottawa Ontario K1A 0K2 CAN

Wallersteiner, Ulrika
ERGO Systems Canada
535 Robin Hood Road
West Vancouver BC V7S 1T4 CAN

Welch, Michael
Pacific Missile Test Center
Code 400T
Point Mugu CA 93042-5000 USA

White, Lou
Canadian Marconi Co
415 Leggett Drive
Kanata Ontario K2K 2B2 CAN

Wickens, Chris
Institute of Aviation
University of Illinois
Savoy IL 61874 USA

Willumeit, H.P.
Tech Univ of Berlin
Strasse 17. Juni No. 135
1 Berlin 12 GER

APPENDIX B - POSTER PRESENTATIONS

POSTER SESSION I

A First Investigation of Human Body Posture and Movement
Heiner Bubb

WERNER - A Small Computer Based Tool for Ergonomic Design
Wilhelm Bernhard Kloke

ERGON EXPERT: Occupational Health Oriented Tool for Ergonomic Modelling
Wolfgang Laurig

Workspace Evaluation and Design: Using the 3D Man Model "HEINER"
Karlheinz G. Schaub

Crew Assessment of Reach (CAR)
A.O. Dick

"SAMMIE" Crew Station Design
Laura Binder

Workload Analysis for the Canadian Forces
John Makadi and Lou White

Workload Prediction
Bat Bateman

POSTER SESSION II

Driver-Vehicle Manual Control Modelling and Simulation
Werner Reichelt

Generic System Analyst Workstation (GENSAW)
Frank Maher and Robert Cathcart

GrafSAINT: A Grafical User Interface to SAINT
K.F. Kraiss and H. Küttelwesch

Speech System Performance
Diane Dintruff

Cockpit Automation Technology
Lisa Guadagna

Using MicroSAINT and SOFTWARE X to Model UCI Performance
Glenn Osga

A Method for Modeling Human Response to Abrupt Changes in System Dynamics
Mike Welch

POSTER SESSION III

SLAM II and Its Use in Modelling Air Defense Crew Performance
James Ainsworth

Human Engineering Performance Requirements and Operational Model
Dennis Banks

Functional Models of Complex Human Performance
William Nelson

Computational Human Factors in Support of the Designer
Earl Hartzell

INDEX